Progress in Optical Science and Photonics

Volume 28

Indexed by Scopus

The purpose of the series Progress in Optical Science and Photonics is to provide a forum to disseminate the latest research findings in various areas of Optics and its applications. The intended audience are physicists, electrical and electronic engineers, applied mathematicians, biomedical engineers, and advanced graduate students.

Mingjiang Zhang · Jianzhong Zhang · Lijun Qiao · Tao Wang

Novel Optical Fiber Sensing Technology and Systems

清華大学出版社
TSINGHUA UNIVERSITY PRESS

Springer

Mingjiang Zhang
Taiyuan University of Technology
Taiyuan, China

Jianzhong Zhang
Taiyuan University of Technology
Taiyuan, China

Lijun Qiao
Taiyuan University of Technology
Taiyuan, China

Tao Wang
Taiyuan University of Technology
Taiyuan, China

ISSN 2363-5096 ISSN 2363-510X (electronic)
Progress in Optical Science and Photonics
ISBN 978-981-99-7148-0 ISBN 978-981-99-7149-7 (eBook)
https://doi.org/10.1007/978-981-99-7149-7

Jointly published with Tsinghua University Press
The print edition is not for sale in China (Mainland). Customers from China (Mainland) please order the print book from: Tsinghua University Press.

This Springer imprint is published by the registered company Springer Nature Singapore Pte Ltd.
The registered company address is: 152 Beach Road, #21-01/04 Gateway East, Singapore 189721, Singapore

Preface

Since the implementation of the 13th Five-Year Plan for Science and Technology Innovation of the People's Republic of China, sensing and monitoring technologies such as national major infrastructure security guarantee, intelligent marine environment security, rapid identification and risk prevention, and control of major geological disasters, deep earth exploration and environmental protection engineering have ushered in unprecedented development opportunities and challenges. Besides, the "Smart Cities" with artificial intelligence monitoring network as the core has become the development trend of the times. As an important part of the monitoring network, optical fiber sensing technology has been widely used in petrochemical industry, civil engineering, electrical transmission, aerospace, transportation, geotechnical security, and other major fields.

In recent years, the author has presided over major projects such as National Major Scientific Research Equipment, National Natural Science Foundation of China Project, Shanxi Province Key Research and Development Program, and Shanxi Province Key Technologies Research and Development Program. In addition, aiming at the bottleneck problem of the contradiction between the sensing distance and the spatial resolution in the existing optical fiber sensing system, the novel optical fiber sensing technology and application research have been carried out and obtained some achievements.

This book is based on the research results achieved by the authors in recent years, including:

Chapter 1, Generation and Control of Chaotic Laser. The results of the study have revealed the mechanism of chaotic laser bandwidth enhancement and time delay suppression, and the chaotic laser with the widest bandwidth without time delay characteristics has been generated in the experiment, which solves the problem of high-quality chaotic signal source necessary for long-range and high-spatial-resolution chaotic laser sensing.

Chapter 2, Photonic Integrated Chaotic Lasers. The photonic integrated broadband chaotic signal generator and broadband chaotic laser source have been developed to promote the engineering application of chaotic laser in the field of sensing.

Chapter 3, Chaos Brillouin Distributed Optical Fiber Sensing. The chaotic Brillouin scattering characteristics were analyzed, and the chaotic Brillouin optical coherence domain reflection sensing technology and optical coherence domain analysis sensing technology were proposed to realize temperature and strain monitoring with long sensing distance and high-spatial resolution.

Chapter 4, Brillouin Distributed Optical Fiber Sensing Based on Disordered Signals. The Brillouin optical coherent domain sensing technology based on noise signal, pseudo-random sequence and physical random code modulation is proposed to realize distributed fiber optic sensing with high spatial resolution.

Chapter 5, Chaotic microwave photon sensing. A new technology of chaotic microwave photon long-range ranging and water-level monitoring is proposed, which realizes remote water-level monitoring and multi-object distance sensing.

Chapter 6, Distributed Fiber Optic Raman Thermometer and Applications. A number of sensing solutions have been proposed for major engineering applications. Two new distributed fiber optic Raman thermometers have been developed and applied to local major projects for health monitoring, realizing real-time monitoring and accurate early warning of disasters such as fires in long tunnels and leaks in gas pipeline networks.

Chapter 7, Narrow Linewidth Fiber Laser for Fiber Sensing. A new structure of narrow linewidth fiber laser has been proposed, which can achieve single wavelength and dual wavelength output, and has good single frequency characteristics and stability.

In the process of writing the English version of this book, I would express my gratitude to two assistant professors, Jian Li and Yahui Wang, and the author's graduate students, Kangyi Cao, Jing Chen, Yang Chen, Zijia Chen, Bowen Fan, Xiaopeng Ge, Zhiyong Guo, Yijia Liu, Hui Liu, Jinglian Ma, Lintao Niu, Jiaxin Peng, Xiaona Wang, Chenyang Zhang, Fan Zhang, Weiyi Zhang, Yongqi Zhang, Yuting Zhang, Yaoxin Zhu, Yaqi Zhu, and so on, assisted in translating and proofreading the English chapters, which made the book available in English!

Thanks to Prof. Yuncai Wang and Prof. Tiegen Liu for their guidance and advice on the research of this book!

Special thanks to Prof. Yanbiao Liao of Tsinghua University for recognizing our research work and recommending this book to be included in the "Transformative Light Science and Technology Series" planned by Tsinghua University Press!

Due to the limitations of the author's knowledge and the short time to compile this book, there are inappropriate and even wrong points in the writing. Please do not hesitate to enlighten us.

Taiyuan, China
June 2023

Mingjiang Zhang
Jianzhong Zhang
Lijun Qiao
Tao Wang

Contents

Chapter 1
Generation and Control of Chaotic Laser

Chaos is the phenomenon of an unpredictable, seemingly random irregular motion of a deterministic system due to its unusual sensitivity to initial value conditions [1, 2]. Initial value sensitivity indicates that the system for a weak change in initial conditions will generate a large cascade of chain reactions. The "butterfly effect" graphically illustrates the chaotic phenomenon of long-term unpredictable motion caused by a change in initial conditions [3]. However, the random-like motion of a chaotic dynamical system is essentially different from the random volatility in a stochastic process. Although chaotic dynamics exhibit irregular random motion, it is determined by deterministic systems. The study of chaotic properties has provided scientists in various fields with new ideas and methods, including information science, space science, life science and other related fields, promoting the rapid development of modern society.

1.1 Characteristics of Chaotic Laser

The semiconductor laser is a non-linear system. Its output can be divided into three forms: steady state, non-steady state and chaotic state. When the semiconductor laser is disturbed, the output is chaotic under certain conditions. At this time, the output (light intensity, wavelength, phase) is no longer steady state in the time domain, but a noise-like random variation. In this case, the dynamic characteristics of the laser can also be described by a defined rate equation. However, the extreme sensitivity to the initial conditions makes the output vary randomly [4, 5].

In 1980, Japanese researchers have found that external optical feedback can cause instability and chaos in semiconductor lasers [6]. However, in the earlier study, researchers focused on how to maintain the steady-state operation of lasers and suppress laser noise and unstable output [7, 8]. In 1980–1990, the dynamic properties of optical feedback (and optical injection) lasers were gradually clarified and the

M. Zhang et al., *Novel Optical Fiber Sensing Technology and Systems*, Progress in Optical Science and Photonics 28, https://doi.org/10.1007/978-981-99-7149-7_1

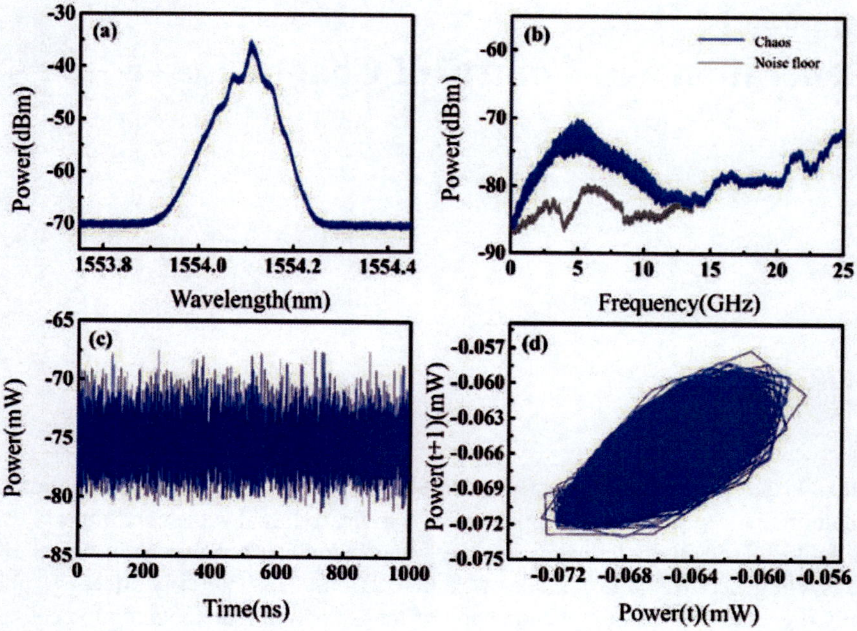

Fig. 1.1 Characteristic of chaotic lasers **a** optical spectrum; **b** RF spectrum; **c** time series; **d** phase diagram

evolution of the laser from low-frequency undulation to the chaotic state and from multiplicative period to the chaos was discovered [9–13]. Until 1990, researchers at the University of Maryland and the U. S. Naval Laboratory have developed the concepts of chaos control and chaos synchronization [14, 15]. This has led to the recognition that chaotic lasers may have important applications.

The characteristics of a chaotic laser are shown in Fig. 1.1. Figure 1.1a shows the spectrum of a chaotic laser. The chaotic laser exhibits wider spectral characteristics and lower coherence than the general continuous laser. And the corresponding power spectrum in the frequency domain has flat and broadband characteristics, as shown in Fig. 1.1b, it behaves as a noise-like random variation in the time sequence, as shown in Fig. 1.1c; the phase diagram is a chaotic attractor, indicating that the chaotic laser generation process is a random process, as shown in Fig. 1.1d.

1.2 Typical Ways of Generating Chaotic Lasers

Currently, typical generation methods of chaotic lasers include external cavity optical feedback, external optical injection, photoelectric feedback [6, 16–29], etc.

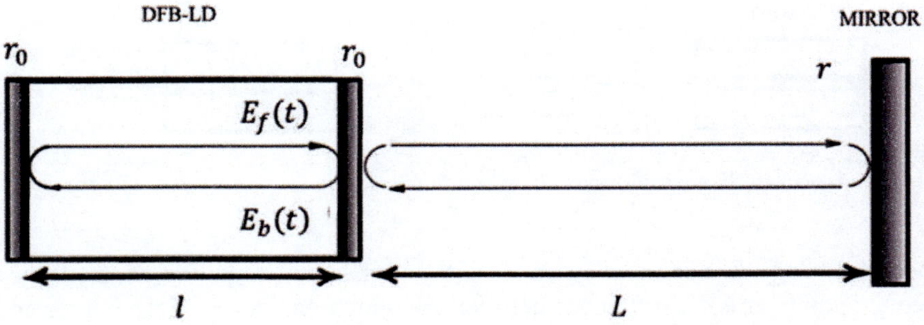

Fig. 1.2 Schematic diagram of external cavity optical feedback structure

1.2.1 External Cavity Optical Feedback

Figure 1.2 shows the schematic diagram of the external cavity optical feedback method. The left side shows the internal schematic of the semiconductor laser. The active optical resonant cavity is called the inner cavity. It is used to control the frequency and phase of photons in the cavity, etc. The right side is an external light reflector. It is used to provide external cavity optical feedback. The passive part between the laser's light output surface and the external light reflector is called the outer cavity. After the single-mode laser output from the laser is transmitted to the external light reflector, part of the output light is reflected into the laser cavity in the original way disturbing to the laser. Lasers exhibit non-stationary or chaotic states within a certain range of optical feedback. In the process, the chaotic state can be optimized by adjusting the optical feedback intensity and the length of the outer cavity. The optical feedback intensity is the ratio of the optical power fed back to the laser by the light reflector to the output power of the laser itself. The outer cavity length is the distance between the laser and the light reflector, which is characterized by time delay.

In addition, feedback light can be provided by the external light reflector, but also by the fiber ring self-feedback, Fabry–Perot (F–P) standard with filtered feedback [18], etc. The laser exhibits rich chaotic dynamics by controlling different variables through different external cavity optical feedback methods.

1.2.2 External Light Injection

The schematic diagram of the external light injection method is shown in Fig. 1.3. It uses two semiconductor lasers to inject the output light from the main laser into the slave laser in one direction. Chaotic laser generation is achieved by introducing degrees of freedom to perturb the slave laser through external light injection and controlling the frequency detuning and injection intensity of the two semiconductor

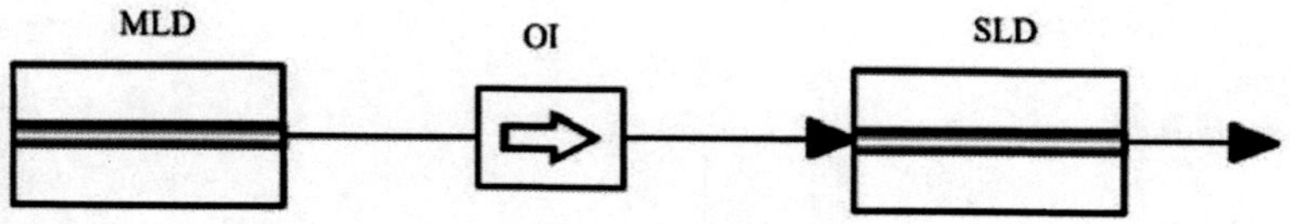

Fig. 1.3 Schematic diagram of external light injection structure

lasers to change the nonlinear dynamics of the slave laser. In addition, the frequency detuning amount is the difference between the optical frequency of the master and slave semiconductor lasers, and the injection intensity is the ratio of the optical power injected into the master laser from the laser to the output power of the slave laser itself.

At present, researchers have improved the output state of the chaotic laser based on the external light injection method in order to further optimize it. For example, the output light from a laser is injected into a chaotic laser with external cavity optical feedback, thus achieving bandwidth enhancement. Li Nianqiang et al. proposed a cascaded coupled semiconductor laser, consisting of an external cavity master laser, an intermediate laser and a slave laser, to achieve the suppression of time delay characteristics [23].

1.2.3 Optical Feedback

The schematic diagram of the optoelectronic feedback method is shown in Fig. 1.4. The light signal from the semiconductor laser (laser diode, LD) first passes through the photoelectric detector (photoelectric detector, PD) to realize the conversion of light signal to the electric signal. After amplification by the microwave amplifier, the electrical signal modulates the current drive of the semiconductor laser and perturbs it. After the laser is perturbed, the carrier density and photon number density and the phase of the optical field are dynamically changed, showing nonlinear dynamics, which results in a chaotic laser.

Among the three chaotic laser generation methods of optical feedback, optical injection and photoelectric feedback, the optical feedback method and optical injection method use an all-optical structure. The feedback laser and the injection laser act directly into the laser cavity as the perturbation signal to act on the optical field. Due to the extreme sensitivity of the semiconductor laser to the external perturbation laser, the optical feedback method and the optical injection method can obtain rich nonlinear dynamics. Among them, there are two main reasons for the extreme sensitivity of semiconductor lasers to external optical perturbations [30]. First, for edge-emitting semiconductor lasers, due to the low reflectivity of the emitting side of the active area (usually 10%–30%), external laser light can easily enter the active cavity and generate disturbances. For the surface-emitting semiconductor laser, although it has a high reflectivity at the emitting side, the very small number of photons in the

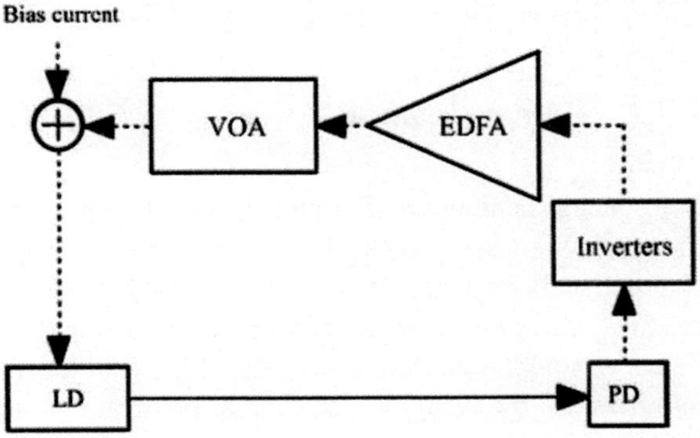

Fig. 1.4 Schematic diagram of the optoelectronic feedback structure

cavity makes it equally sensitive to weakly disturbing light. Second, semiconductor lasers usually have a high line width enhancement factor, which makes the carrier and field phase couple in the active region of the laser, aggravating the nonlinear state of semiconductor lasers. In the photoelectric feedback method, the output light is converted into an electrical signal to modulate the current drive to influence the laser carrier density dynamics, which in turn leads to dynamic changes in the optical field phase and photon number density through the line width enhancement factor and gain, respectively, and this method can also produce a nonlinear state similar to the optical feedback method and the optical injection method [31–34]. However, in order to respond to the relaxation oscillations of the laser and to obtain complex nonlinear states, it requires the use of large bandwidth photodetectors and related electronic devices. For both optical feedback and optical injection methods of chaotic laser generation, only one semiconductor laser is required for the optical feedback method, while two semiconductor lasers are required for the optical injection method, and the output wavelength of the laser needs to be precisely adjusted. In addition, the optical feedback method generates chaotic laser due to the dynamic competition between the internal relaxation oscillation frequency and the external cavity frequency, where the number of external cavity modes is large and varies by the optical feedback. In the light injection method, the injected light comes from the output light of other lasers and is relatively stable. Therefore, compared with the optical injection method, the optical feedback method is easier to generate high-dimensional chaotic lasers, and the chaotic spectrum generated is smoother, and the structure of this method is simple, easy to operate, and less expensive. In applications such as chaos-based secure communication, random number generation, and fiber optic sensing, optical feedback is the most common method.

1.3 Generation of Broadband Chaotic Lasers

1.3.1 Research Status

The spectral bandwidth of a typical chaotic laser is only a few gigahertz (GHz), and the energy is mainly concentrated around the relaxation oscillation frequency, so chaotic lasers have problems such as uneven spectrum and narrow bandwidth. The bandwidth of chaotic lasers has a significant impact on their application in various fields. Among them, the narrow bandwidth will limit the generation rate of random numbers and the resolution of chaotic lidar, and the uneven spectrum will limit the energy of low-frequency components in circuit acquisition and processing [35–37]. In the field of chaotic secure optical communication, the modulation format of the communication system used in conventional chaotic laser is mainly switch keying and differential phase shift keying, Therefore, high-speed binary signal encryption requires broadband chaotic signals, and the bandwidth of chaotic lasers limits the transmission rate and distance of communication systems [38]. In order to solve the problems of uneven spectrum and narrow bandwidth of chaotic lasers, and to promote chaotic lasers to practical applications, Researchers have proposed several methods to achieve bandwidth enhancement and spectral shaping of chaotic excitations. The methods include optical injection, chaotic excitation injection, ring cavity delayed self-interference, fiber oscillation ring, optical external aberration [23, 35, 36, 39–47], etc. In 2003, Professor Uchida of Saitama University, Japan, and others injected a chaotic laser generated by external optical feedback into another semiconductor laser, a semiconductor laser with a relaxation oscillation frequency of only 6.4 GHz can generate chaotic lasers with a bandwidth of up to 22 GHz, The experimental schematic and power spectrum are shown in Fig. 1.5. Due to the limitation of instrument bandwidth, Prof. Uchida et al. further simulated, He found that the generation of high-frequency broadband chaotic signals is mainly influenced by two factors: one is the high-frequency oscillation caused by the beat-frequency effect between the master and slave lasers; the other is that the chaotic laser injection increases the chaotic spectrum energy in the high-frequency range and makes the spectrum flatter. It is found that the range of the chaotic spectrum can be further changed by choosing the appropriate injection intensity and optical detuning frequency parameters [35].

In 2012, Prof. Wei Pan of Southwest Jiaotong University proposed a semiconductor laser cascade scheme, which consists of an external cavity feedback master laser, an independent intermediate laser and an independent slave laser. As shown in Fig. 1.6. The first two lasers are used for time delay feature suppression, and the last two lasers are used for bandwidth enhancement. By adjusting the frequency detuning and injection intensity, the experimentally generated chaotic bandwidth is 14.49 GHz, which is limited by the instrument bandwidth. Further, after the Lang-Kobayashi (L-K) equation simulation study, the 80% bandwidth of the chaotic signal reaches 35.34 GHz [23].

Prof. Zhang Mingjiang and Prof. Wang Anbang of Taiyuan University of Technology have made research on the bandwidth enhancement of chaotic lasers. In 2008,

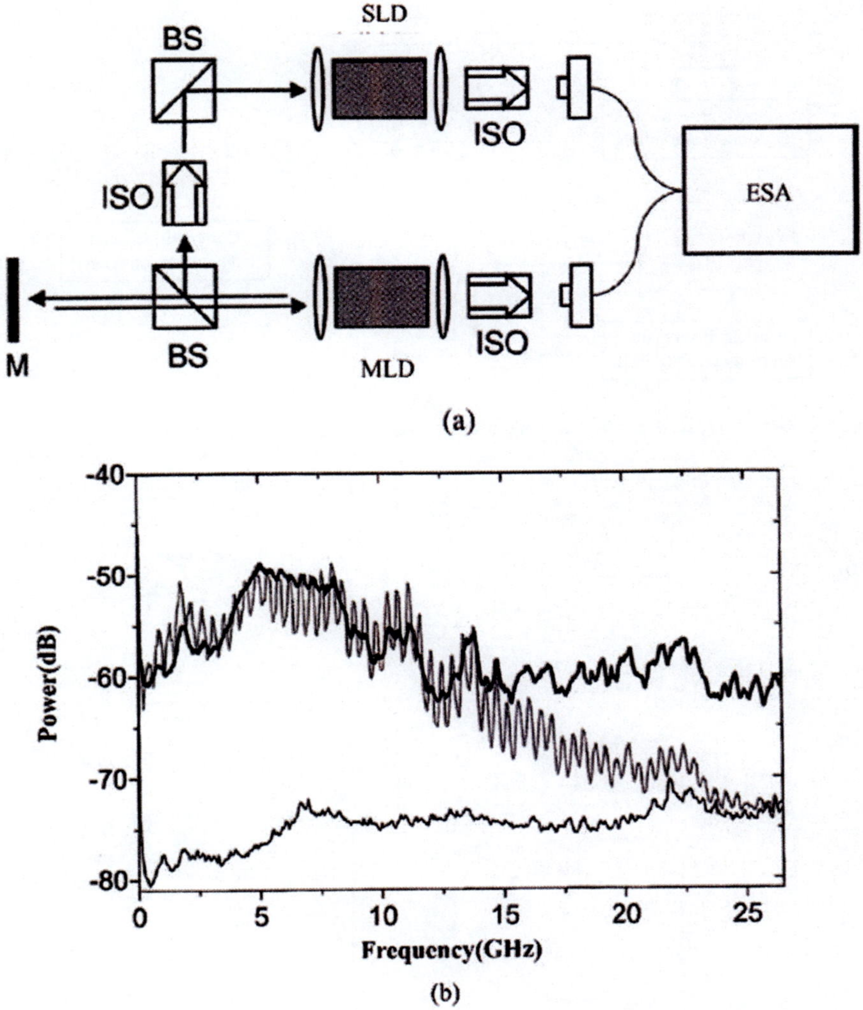

Fig. 1.5 Schematic diagram (**a**) and power spectrum (**b**) of chaotic laser injection to generate broadband chaotic laser

they proposed a semiconductor laser with continuous light injection into the external cavity with optical feedback, as shown in Fig. 1.7a. The bandwidth of chaotic signal is enhanced from 6.2 to 16.8 GHz by controlling the amount of frequency detuning and the intensity of light injection. It is shown that the bandwidth enhancement is due to the enhancement of high-frequency periodic oscillations and relaxation oscillations during unstable and stable locked injection, and that the high-frequency periodic oscillations due to positive detuning injection are more suitable for obtaining bandwidth-enhanced chaotic signals [39].

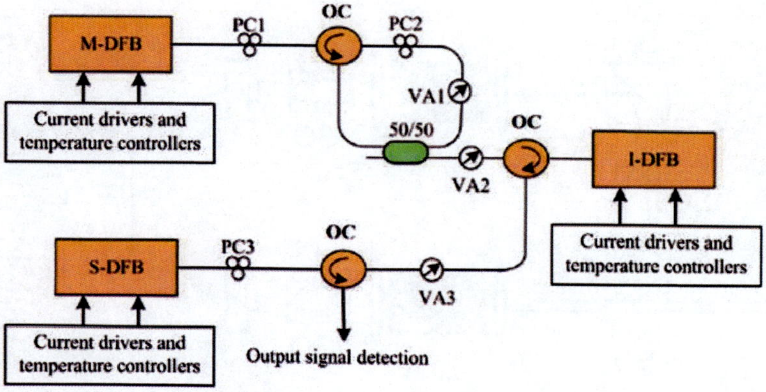

Fig. 1.6 Schematic diagram of a cascade laser generating broadband chaotic laser

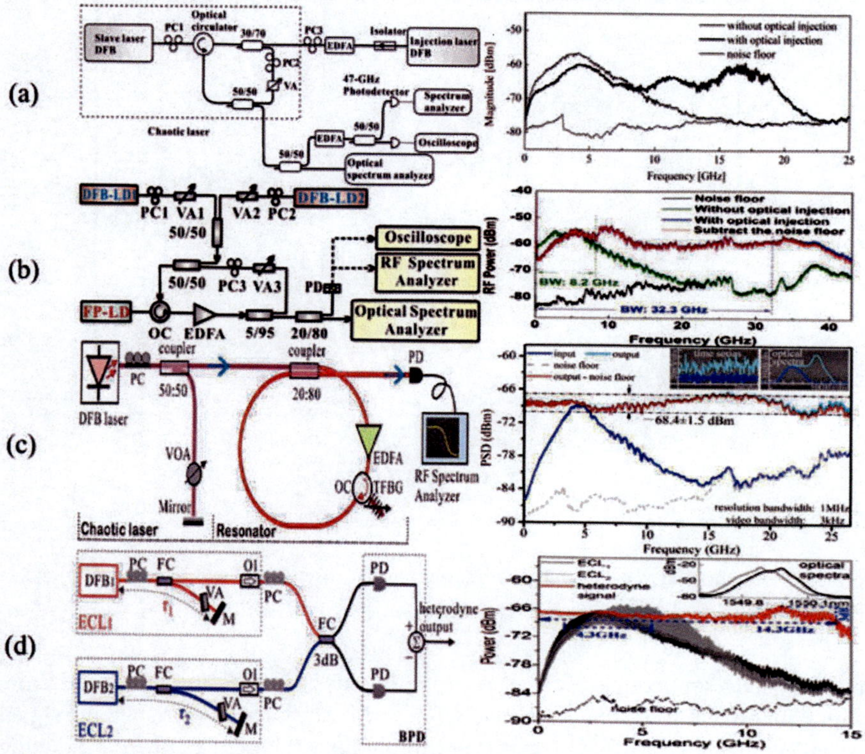

Fig. 1.7 Schematic and spectrum of multi-mode generation of broadband chaotic laser at Taiyuan University of Technology

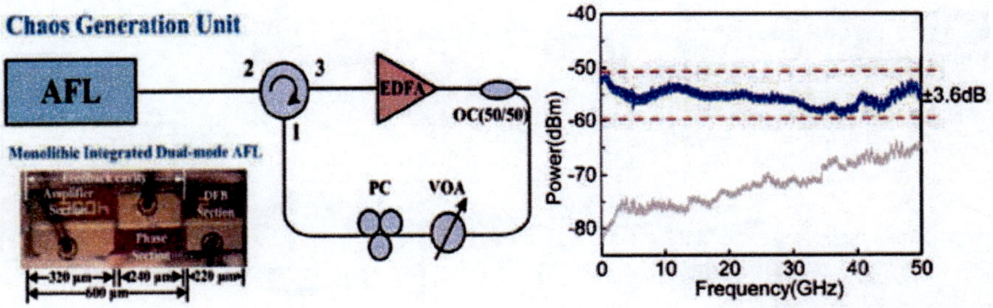

Fig. 1.8 Schematic and spectrum of the broadband chaotic laser generated by the applied perturbation of the AFL chip

In 2013, we proposed to obtain a broadband chaotic laser with a bandwidth larger than 26.5 GHz and a flatness of ± 1.5 dB by chaotic laser injection into a fiber ring resonant cavity. Among them, the optical amplifier and optical filter are added to the optical resonant cavity to cause the self-tap frequency effect between the injected light and its own delay, which achieves the enhancement of chaotic bandwidth and spectrum optimization [46]. In 2015, a chaotic laser with a flatness of 3 dB and a bandwidth of 14 GHz, i.e., white chaos, was generated by the optical outlier of a semiconductor laser with optical feedback from two external cavities, as shown in Fig. 1.6d. The optical external difference method excites the laser beat effect not only to eliminate the relaxation oscillation peak of the laser, but also to obtain a chaotic laser with high bandwidth and flat spectrum by adjusting the frequency detuning between the two semiconductor lasers [47].

In 2017, Lingjuan Zhao et al. at the Institute of Semiconductor Research, Chinese Academy of Sciences proposed an amplified feed-back semiconductor laser (AFL) chip plus fiber ring perturbation to generate chaotic laser, as shown in Fig. 1.8. By adjusting the drive current of the amplification area and distributed feedback area of the AFL to generate a dual-wavelength laser, the external optical feedback perturbs the AFL chip to generate a chaotic state, and the two modes of chaotic oscillations are coupled with each other to generate a broadband chaotic laser with a spectral range greater than 50 GHz and a flatness of ± 3.6 dB [36].

1.3.2 Dual-Wavelength Combined Optical Feedback Method

The authors' team enhanced the bandwidth of the chaotic laser by injecting dual-wavelength external light into the F–P semiconductor laser. The schematic diagram of the experimental system is shown in Fig. 1.9. The F–P semiconductor laser with a fiber feedback loop is used to generate a multi-wavelength chaotic laser with an erbium-doped fiber amplifier (EDFA) and a variable optical attenuator (VOA3) to control the intensity of the feedback light, and a polarization controller (PC3) to

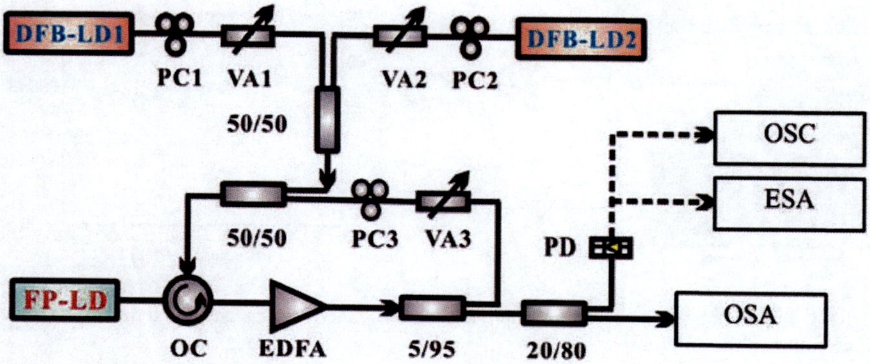

Fig. 1.9 Schematic diagram of broadband chaotic laser system generated by dual-wavelength external light injection into an F–P semiconductor laser

adjust the polarization state of the feedback light. Two distributed feedback semiconductor lasers (hereafter referred to as DFB lasers) (DFB laser 1 and DFB laser 2) are used as external injection sources and are injected into the F–P semiconductor laser through a 50:50 fiber coupler. The output wavelengths of the DFB laser 1 and DFB laser 2 are regulated by two precision temperature controllers respectively. The polarization state and power of the injected light were adjusted by PC1, PC2, VOA1 and VOA2 respectively, and a broadband dual-wavelength broadband chaotic laser was generated with the appropriate injection power and optical frequency detuning.

Measurement of the spectrum of the output chaotic excitations using a spectral analyzer (AQ6370B). An ultrafast photodetector with a bandwidth of 50 GHz was used to convert the chaotic laser into an electrical signal, and a spectrum analyzer (AgilentE4447A) with a bandwidth of 42.98 GHz was used to measure its spectral characteristics.

In the experiment, the current bias of the F–P semiconductor laser was set at 1.27 times the current threshold. Due to the large fluctuation of the chaotic oscillation energy distribution, the traditional definition of − 3 dB frequency point is no longer suitable for the calibration of the chaotic bandwidth, and researchers generally use 80% of the full bandwidth energy of the chaotic spectrum or the total bandwidth within − 20 dB of the highest power point. The green curve in Fig. 1.10 is the spectrum of the maximum bandwidth of the chaotic laser obtained when only optical feedback is available, and the bandwidth of the chaotic laser is 8.8 GHz at this time. The blue curve shows the spectrum of the broadband chaotic laser obtained by choosing the appropriate injection power and optical frequency detuning, with a bandwidth of 32.3 GHz, which is about 4 times more than that of the original chaotic laser. The total injected power at this time is − 2.11 dBm (about 50% of each), and the optical frequency detuning is 25.4 GHz and 32.9 GHz respectively. In addition, it is very important that the power spectrum of chaotic oscillations is flatter, with power fluctuations within ± 3 dBm in the frequency range of 2.5–38.5 GHz, and the energy distribution of the resulting chaotic spectrum is more uniform and the profile

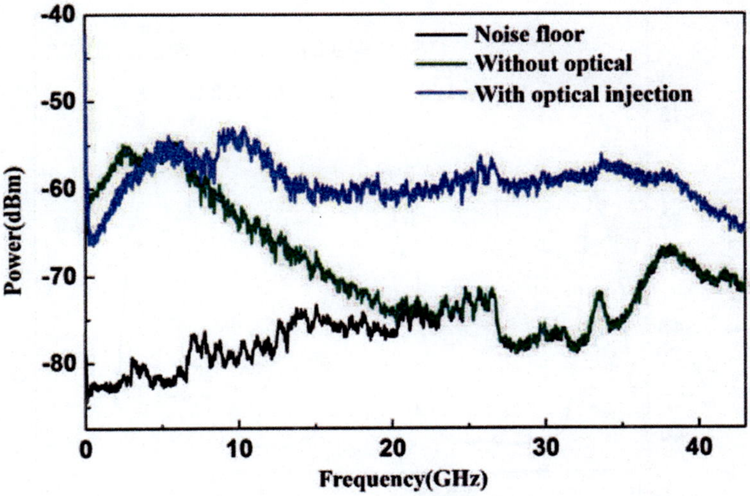

Fig. 1.10 Spectra of experimentally obtained broadband chaotic laser

is flatter. This is essential for high-speed chaotic and confidential communication and high-speed random number generation.

The spectra of the generated broadband chaotic laser are shown in Fig. 1.11. It is obvious that the F–P semiconductor laser is in an un-locked state, and the output spectrum contains multiple longitudinal mode outputs. Due to the optical feedback, each longitudinal mode is red-shifted and new optical frequency components are generated, and the spectrum is broadened. For the two longitudinal modes which are close to the injected wavelength, they interact with the injected wavelength nonlinear and generate more new optical frequency components, and the spectral broadening is wider than that in the case of pure feedback. The wavelength of the main laser 1 (MLD1: DFB laser 1) is 1549.170 nm, and the wavelength of the F–P semiconductor laser with similar mode after red shift is 1549.374 nm, and the wavelength detuning of both is 0.204 nm, and the corresponding optical frequency detuning is 25.4 GHz. The wavelength of MLD2 (DFB laser 2) is 1550.234 nm, and the mode wavelength of F–P semiconductor laser is 1550.498 nm after redshift, and the wavelength detuning of both is 0.264 nm, and the corresponding optical frequency detuning is 32.9 GHz. The frequency beat of each of the two injected lightwave modes interacts with the injected lightwave, resulting in a significant periodic oscillation spike in the spectrum of the chaotic laser at 25.4 and 32.9 GHz. The broadband high-frequency periodic oscillations are superimposed and coupled with the original chaotic oscillations, making the bandwidth of the chaotic laser wider and flatter due to the beat-frequency effect between multiple modes, especially due to the beat-frequency effect between the dual-wavelength injected laser and the original chaotic laser light field.

The waveform timing diagram of the above broadband dual-wavelength chaotic laser acquired using a real-time oscilloscope with 6 GHz bandwidth is shown in Fig. 1.12. From Fig. 1.10, we can see that the bandwidth of this chaotic laser is as

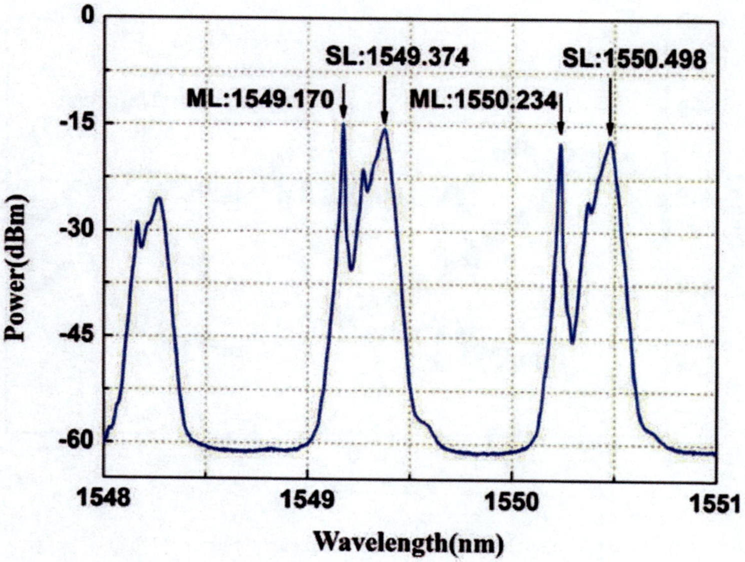

Fig. 1.11 Spectrogram of dual-wavelength broadband chaotic laser

high as 40 GHz. Therefore, due to the bandwidth limitation of the oscilloscope, the acquired waveform is equivalent to a low-pass filter with a bandwidth of 6 GHz, which can no longer truly reflect the true bandwidth of this broadband chaotic laser, and the result can only qualitatively analyze whether the F–P semiconductor laser is in a chaotic oscillation state.

The autocorrelation curves of the obtained broadband dual-wavelength chaotic laser are shown in Fig. 1.13. As can be seen from the figure, this broadband chaotic signal has an excellent δ-function characteristic, showing a near-ideal peg shape, which is a crucial characteristic for chaotic laser ranging and optical time-domain reflectometry applications. It can also be seen that there are two symmetrically distributed side flaps on both sides of the center with a time delay of 218.49 ns, corresponding to a distance of 43.698 m, which reflects the feedback cavity length information of the laser, which needs to be suppressed in chaotic and confidential communications to avoid eavesdroppers from obtaining hardware information of the transmitter. However, a hint from another point of view: the optical feedback method can be used to obtain information about the length of the fiber by measuring the autocorrelation curve of the generated chaotic laser.

The effect of the injected power and optical frequency detuning on the bandwidth of the resulting chaotic laser was then further investigated. Figure 1.14a shows the effect of the amount of optical frequency detuning on the bandwidth of the chaotic laser for a dual wavelength injected laser. The green curve shows the trend of the effect of the injected optical detuning of DFB laser 1 on the bandwidth of the chaotic laser, where the injected optical detuning of DFB laser 2 is fixed at 17.5 GHz and the total injected power is − 3.42 dBm. The blue curve depicts the effect of the injected

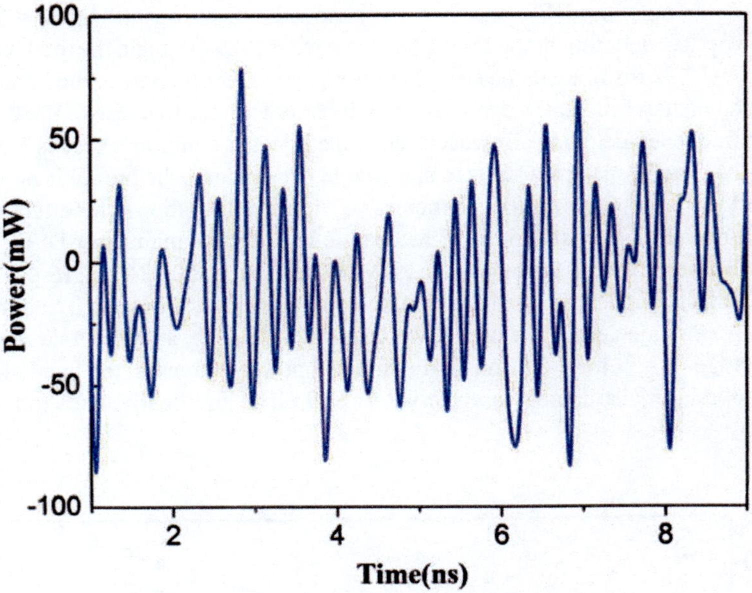

Fig. 1.12 Waveform time series diagram of broadband dual-wavelength chaotic laser

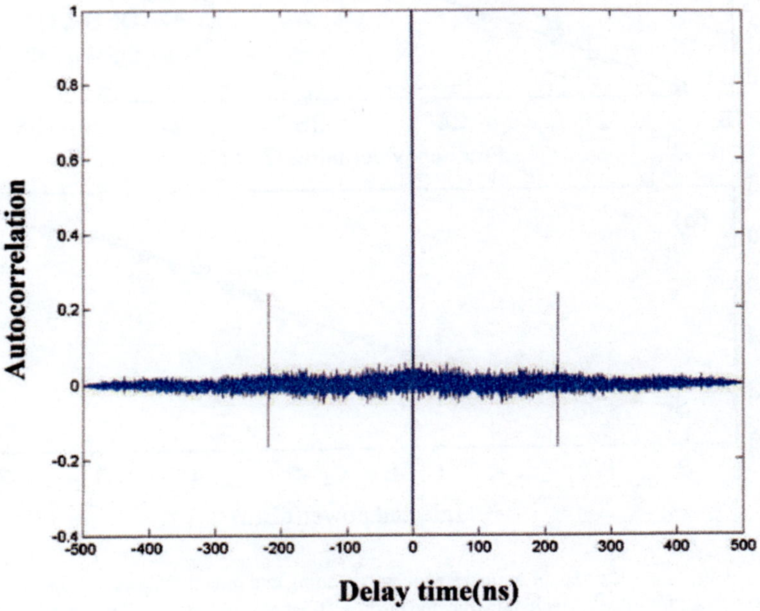

Fig. 1.13 Autocorrelation curve of broadband dual-wavelength chaotic laser

optical detuning of the DFB laser 2 on the bandwidth of the chaotic laser, when the injected optical detuning of the DFB laser 1 is fixed at 20.9 GHz and the total injected power is − 2.92 dBm. It can be seen that for a given injection power, the bandwidth of the chaotic laser increases approximately linearly with the increase of the injected optical frequency detuning. However, when the injected detuning exceeds a certain range, the chaotic spectrum starts to show large depressions, the profile is no longer flat, and the bandwidth no longer increases. Figure 1.14b shows the effect of the injected power on the bandwidth of the chaotic laser for the main laser DFB laser 1 and DFB laser 2 with wavelengths of 1549.170 nm and 1550.234 nm, respectively, and the injected optical frequency detuning corresponding to 25.4 and 32.9 GHz. At this point, the injected power of the two lasers is 50% each, and it can be seen that in the range of − 9 to − 2 dBm, as the injected power increases, the bandwidth of the chaotic laser also increases, from 9.2 to 32.9 GHz, and finally tends to a stable and slow change.

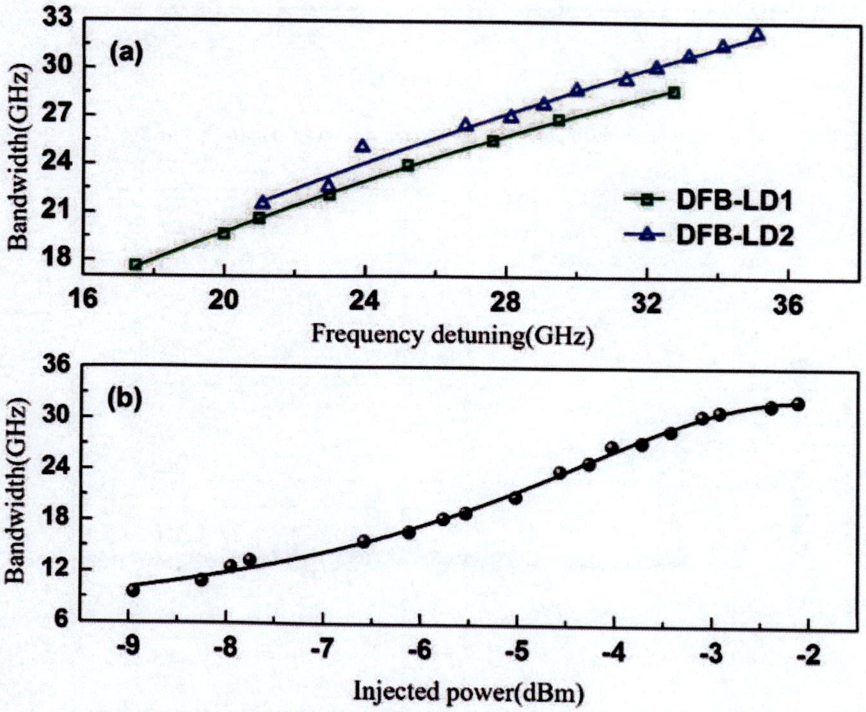

Fig. 1.14 Curves of the effect of optical frequency detuning and injection power on the bandwidth of a chaotic laser for a dual wavelength injected laser. **a** Effect of injected optical frequency detuning on the bandwidth of a chaotic laser; **b** The effect of injected power on the bandwidth of a chaotic laser

1.3.3 Mutual Injection Method

1. Mutual injection of F–P semiconductor lasers for the generation of multi-wavelength chaotic lasers

The experimental setup for mutual injection of two F–P semiconductor lasers to generate multi-wavelength chaotic lasers is shown in Fig. 1.15. At a current bias of 1.5 times the current threshold, the central wavelengths of F–P laser 1 and F–P laser 2 were 1545.72 nm and 1546.86 nm respectively, and the longitudinal mode spacing was 1.11 nm and 1.12 nm respectively, measured at room temperature. The laser output from the F–P semiconductor laser 1 is amplified by an erbium-doped fiber amplifier and divided into two channels. One way is injected into the F–P semiconductor laser 2 by an optical circulator after passing through a variable optical attenuator and polarization controller. The light output from the F–P semiconductor laser 2 is injected into the F–P semiconductor laser 1 by an optical circulator after passing through a variable optical attenuator and polarization controller respectively, thus forming a mutual injection system; the other part of the output chaotic laser is divided into two paths via a fiber coupler, The spectrum is measured by a spectrum analyzer, and the spectrum and waveform are measured by a spectrum analyzer and a real-time oscilloscope after conversion to electrical signals by a photodetector.

Figure 1.16 shows the spectra and spectrum of the chaotic laser obtained with the F–P semiconductor laser 1 in the presence of feedback only, when the bias current of the F–P semiconductor laser is set to 20 mA. The resulting chaotic laser spectrum has a center frequency of 4.83 GHz, an energy of 80%, a bandwidth of 6.5 GHz and a center wavelength of 1545.72 nm.

The spectrogram shows that only two wavelengths have comparable optical power, while the rest of the wavelengths have at least 5 dB lower optical power. The resulting chaotic laser is shown in Fig. 1.17. Figure 1.17a shows the spectrum of the generated multi-wavelength chaotic laser, which has eight wavelengths of equivalent optical

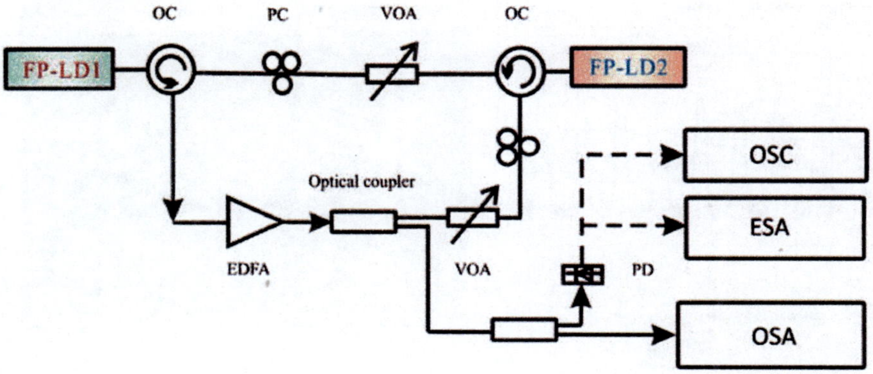

Fig. 1.15 Schematic diagram of the experimental setup for inter-injection of F–P semiconductor lasers to generate multi-wavelength chaotic lasers

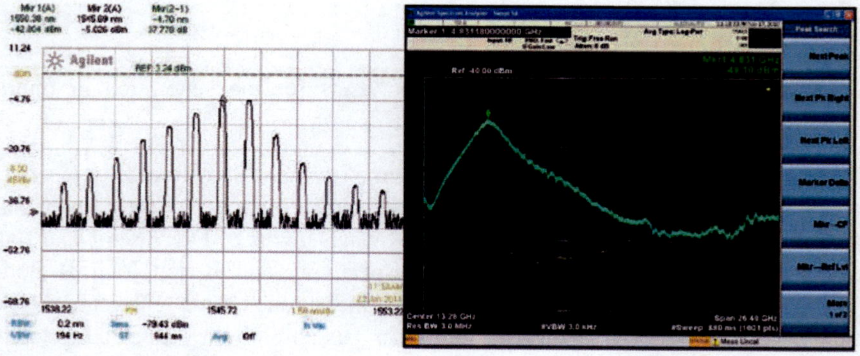

(a) (b)

Fig. 1.16 Chaotic laser generated by a single feedback F–P semiconductor laser. **a** Spectrogram; **b** spectrum diagram

power output due to mutual injection locking. The blue curve in Fig. 1.17b shows the spectrum of this chaotic laser with a bandwidth of 10.8 GHz and a much flatter spectrum profile. The yellow curve shows the noise floor of the spectrum analyzer. It can be seen that the bandwidth of the chaotic laser is enhanced by a factor of more than 1.5 by using mutual injection.

Figure 1.18 shows the time series diagram of the resulting multi-wavelength chaotic laser. The resulting waveform has a peak-to-peak value of 200 mV and nearly 30 chaotic pulse oscillations with a pulse width of approximately 150 ps in 8 ns with random fluctuations. Figure 1.19 shows the autocorrelation curve of the chaotic laser with a sharp spike shape at the sidelobe level (PSL) of − 11.8 dB.

The multi-wavelength chaotic laser generated in this experiment has important applications in chaotic laser wavelength division multiplexing communication

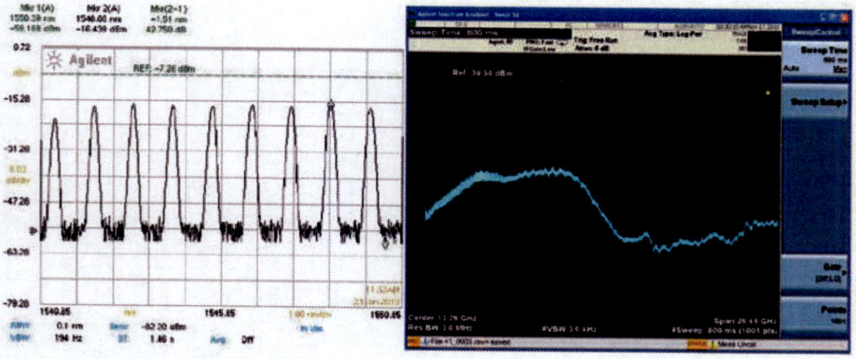

(a) (b)

Fig. 1.17 Chaotic laser generated by inter-injection of an F–P semiconductor laser. **a** Spectrogram; **b** spectrum diagram

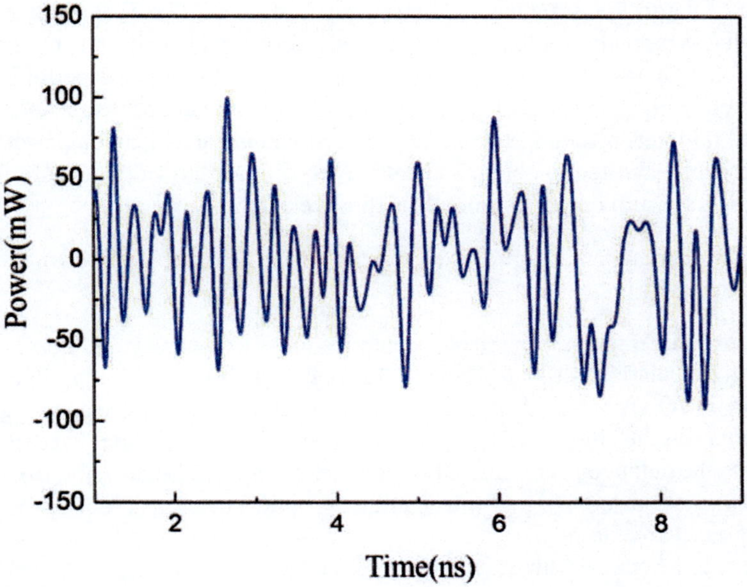

Fig. 1.18 Time series diagram of a multi-wavelength chaotic laser

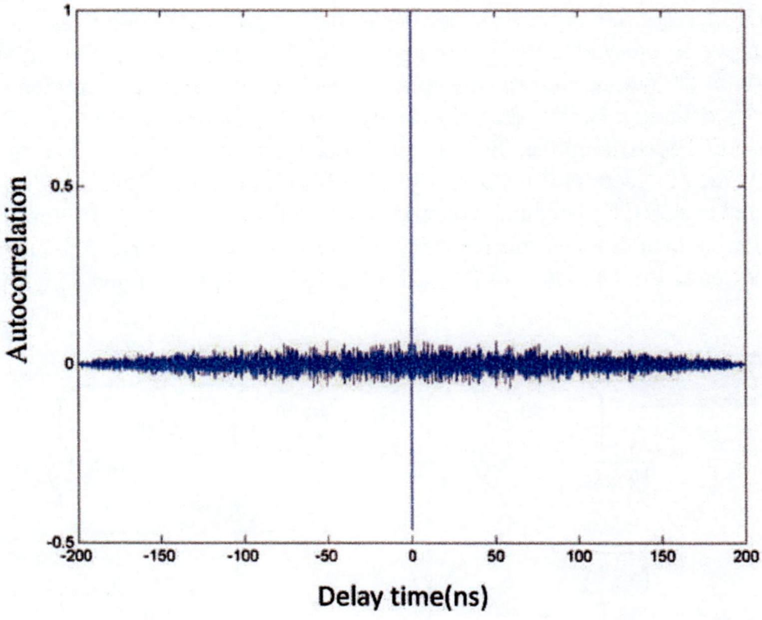

Fig. 1.19 Multi-wavelength chaotic laser autocorrelation curve

systems. Due to the differences in the parameters of the two F–P semiconductor lasers, it was difficult to optimize the detuning of each wavelength, and only a chaotic laser with a bandwidth of about 10 GHz was obtained. In further experiments, two F–P lasers with a wavelength separation of 0.8 nm and similar parameters can be selected for inter-injection, and the wavelength detuning and injection power can be appropriately adjusted to obtain a chaotic laser with a bandwidth of more than 20 GHz and a spectral output of more than 15 wavelengths.

2. Study of broadband chaotic laser generation based on inter-injection of semiconductor lasers

The experimental setup based on the interjection of semiconductor lasers is shown in Fig. 1.20. Polarization control controls the polarization state; the adjustable optical attenuator (VOA) controls the amount of inter-injected power, i.e. the coupling intensity, defined as the ratio of the optical power injected by the DFB laser 2 into the DFB laser 1 to the output power of the DFB laser 1 itself; optical isolator (OI, isolation ≥ 48 dB) prevents external light from entering the optical path and affecting the output state of the chaotic laser.

The specific optical path is: DFB laser 1 is connected to DFB laser 2 by two 90:10 couplers, The DFB laser 1 is connected to the DFB laser 2 via two 90:10 couplers, with the addition of PC and VOA in its intermediate optical path to regulate the output state of the chaotic laser, thus forming an optical inter-injection structure based on a semiconductor laser. A spectrometer (MS9740A, ANRITSU) with a resolution of 0.03 nm was connected to the 20% output of the fiber optic coupler to monitor the variation of the output power and central wavelength to obtain the variation of the frequency detuning. Highly sensitive erbium-doped optical fiber amplifier (CEFA-C-HG, KEOPSYS) amplifies 80% of the output side of an optical fiber coupler, which is then divided into two parts by a 50:50 splitting ratio fiber coupler: partly connected to a 50 GHz bandwidth photodetector (XPDV2120R-VF-FP, Finisar) and a 50 GHz bandwidth spectrum analyzer (FSW50, Rohde& Schwarz). The other part is connected to the same type of PD with a bandwidth of 50 GHz and a high-speed

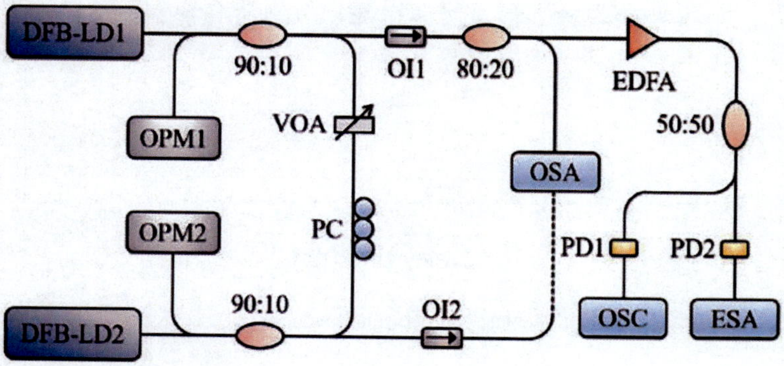

Fig. 1.20 Diagram of the experimental setup for broadband chaotic laser generation

real-time oscilloscope (MCM-Zi-A, LeCroy) with a bandwidth of 36 GHz to monitor the change in feedback and thus calculate the change in coupling strength. Thus, with the adjustment of the individual components in the optical path, the above-mentioned precision instruments can be used to achieve high bandwidth signal output tests.

The performance of the distributed feedback semiconductor laser chips used in the experiments is similar, so that the beat effect from cross-injection enables spectrum shaping and bandwidth enhancement. The DFB laser is a common commercial semiconductor laser manufactured by WTD, with no built-in isolator, FC/APC input and output, 14-pin butterfly package, DFB laser 1 and DFB laser 2 models E21238 and E21236 respectively. The laser is powered by a standard butterfly laser clamp (LDM-4980), the laser bias current and output power are controlled by a drive current source (LDX-3412, ILXLightwave) and the laser temperature and center wavelength are regulated by a temperature controller (LDT-5412B, ILXLightwave), which ensures high laser stability and low noise output.

The P-I curves for DFB laser 1 and DFB laser 2, as measured by an optical power meter (PM100D, Thorlabs), are shown in Fig. 1.21. The threshold current for both DFB laser 1 and DFB laser 2 is 7.2 mA. It is clear that the slope efficiencies of the two lasers are essentially the same, with a calculated slope efficiency of 0.173 W/A for DFB laser 1 and 0.168 W/A for DFB laser 2. It is therefore concluded that the performance of the two lasers is close to each other.

The output spectra of DFB laser 1 and DFB laser 2 was obtained by APEX high resolution spectrometer (resolution 0.04 pm) when the drive current source of

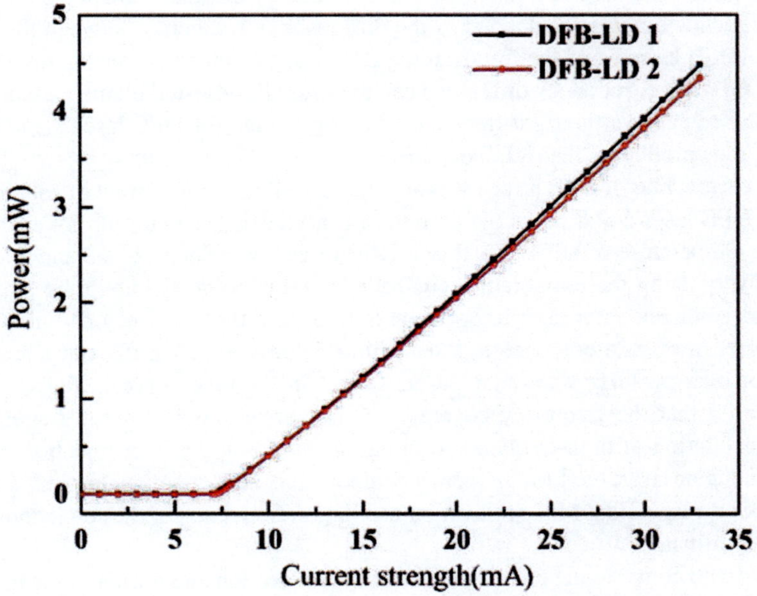

Fig. 1.21 P-I curves for DFB laser 1 and DFB laser 2

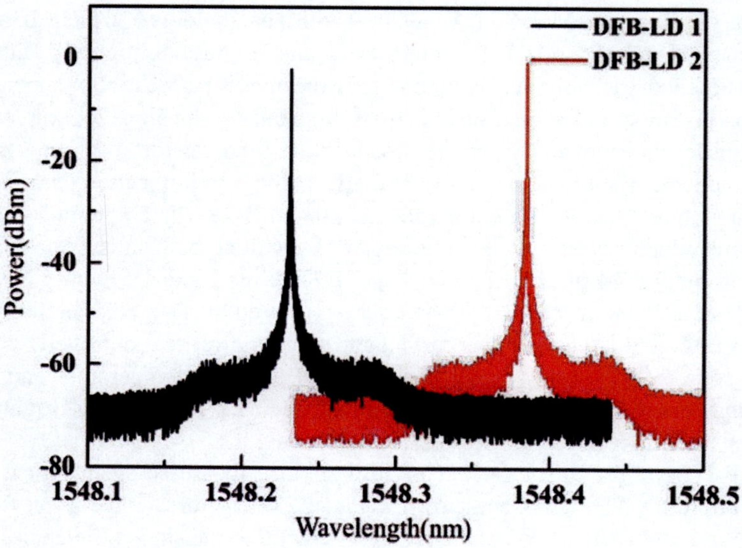

Fig. 1.22 Free running spectra of DFB laser 1 and DFB laser 2

DFB laser 1 and DFB laser 2 were set at 15 mA and the temperature controller at 25 °C, respectively, as shown in Fig. 1.22. The central wavelength of DFB laser 1 was measured to be 1548.231 nm with a − 3 dB line width of 4.64 MHz; the central wavelength of DFB laser 2 was 1548.384 nm with a − 3 dB line width of 8.679 MHz. The frequency detuning is defined as the difference in frequency between the free-running DFB laser 1 and the free-running DFB laser 2, which is calculated to be − 19.125 GHz. The frequency difference can therefore be adjusted by the temperature controller to achieve the right frequency detuning range for DFB laser 1 and DFB laser 2, thus producing the right beat effect.

The current bias of DFB laser 1 was adjusted to 1.4 times the current threshold and that of DFB laser 2–2.8 times the current threshold using the current drive source, and the temperature of DFB laser 1 and DFB laser 2 was set to 25 °C and 25.1 °C respectively using the temperature controller. At this point, the coupling strength based on semiconductor laser interjection is 1.635 and the frequency detuning is − 33.5 GHz. In this state, a chaotic laser with a flatness of ± 2.8 dB and a spectral range of over 50 GHz was obtained, as shown in the blue curve in Fig. 1.23. It is observed that the frequency coverage of the chaotic laser is greatly enhanced by the excitation of high-frequency oscillations based on the inter-injection of the semiconductor laser; the low-frequency component is also greatly enhanced, so that the peak corresponding to the relaxation oscillation frequency of the semiconductor laser is eliminated.

In order to compare the effects of optical inter-injection and normal optical feedback, the DFB laser 2 was replaced with an optical reflector to further investigate the effects of optical feedback-based chaotic lasers. The bias current of DFB laser

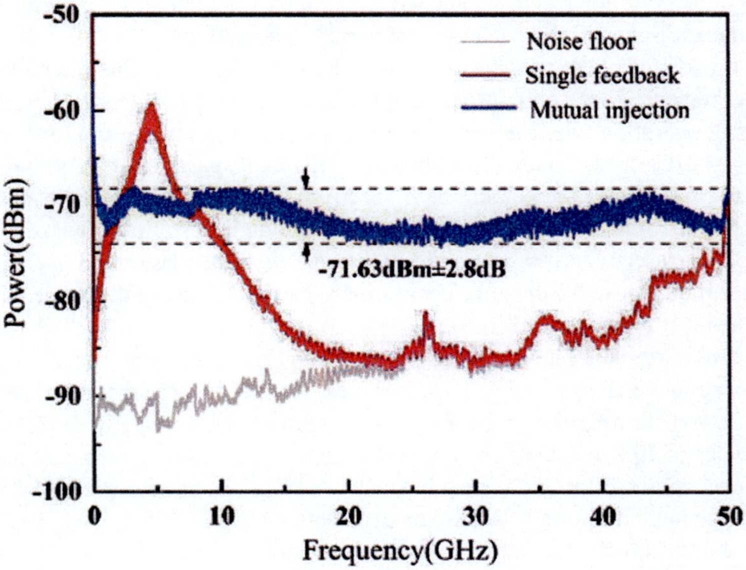

Fig. 1.23 Broadband chaotic laser spectrum

1 is still set to 1.4 times the threshold current and the feedback intensity is set to − 21 dB, i.e. the feedback of DFB laser 1. As a result, a chaotic laser based on optical feedback can be obtained, with the spectrum shown in red in Fig. 1.23 and the noise floor shown in grey in Fig. 1.23. It is clear that when generating chaotic lasers based on optical feedback, the relaxation oscillation frequency of the semiconductor laser dominates the energy of the chaotic laser. As a result, the starting position of the spectrum of the chaotic signal is basically close to the noise floor, and the spectrum also basically coincides with the noise when the frequency reaches 22 GHz. The low energy of the low and high frequency components of the chaotic signal greatly limits its bandwidth coverage, resulting in a restricted bandwidth and uneven spectrum for chaotic lasers, which will limit the practical application of chaotic signals.

As the energy distribution of chaotic signals is not uniform, i.e. the spectrum is not flat enough. If the − 3 dB bandwidth is used to measure the bandwidth of the spectrum, the low and high frequency components of the chaotic signals will be ignored, so the − 3 dB bandwidth is not suitable for this spectrum. In order to better describe the spectral bandwidth of a chaotic signal, we use the 80% bandwidth calculation method, which is defined as the width of the spectrum that occupies 80% of the total energy of the spectrum, starting from 0. The 80% bandwidth of the optical feedback-based chaotic laser in Fig. 1.23 is 6.0 GHz, while the 80% bandwidth of the inter-injection-based chaotic laser is increased to 38.6 GHz. It is clear that the bandwidth of the chaotic signal can be enhanced to about 6.42 times by inter-injection of semiconductor lasers. The bandwidth of chaotic signals has been significantly enhanced, which will be crucial for high-speed chaotic secure communications and high-speed random number generation.

The broadband chaotic laser spectrum is shown in Fig. 1.24. The red and purple curves correspond to the output spectra of DFB laser 1 and DFB laser 2, respectively, during free operation, where the center wavelengths of 1548.191 nm and 1548.459 nm and the − 20 dB line width of 0.075 nm and 0.073 nm, respectively, can be measured. The blue curve shows the chaotic laser spectrum of DFB laser 1 after interjection, with a − 20 dB line width spread of 0.537 nm, which is 7.16 times that of the free running DFB laser 1. As can be seen from the blue curve, the laser field of DFB laser 1 is red-shifted, due to the dynamic change in the carrier density of the semiconductor laser when it is inter-injected.

The time series and phase diagrams for chaotic laser generation based on optical inter-injection are shown in Fig. 1.25. It is clear from Fig. 1.25a that the time series shows a noise-like irregular random fluctuation with a large increase in amplitude and a peak value of 70.2 mV, so it is a typical chaotic state. In addition, the phase diagram is obtained from a series of transient intensities and, as can be seen in Fig. 1.25b, the phase diagram at this point is a complex distribution over a limited range, indicating that the output signal from the DFB laser 1 is chaotic.

In order to further verify that the DFB laser 1 was outputting a chaotic signal, the maximum Lyapunov exponent of 0.0366 was calculated using the collected time series data, and an exponent greater than zero is a chaotic signal. To further investigate the spectral details of the chaotic laser based on optical inter-injection, the output spectrum of the chaotic laser was tested using an APEX high resolution spectrometer (resolution 0.04 pm), as shown in Fig. 1.26, which corresponds to the blue curve in the spectrum of Fig. 1.24. It can be seen that there are two spikes in the spectrum

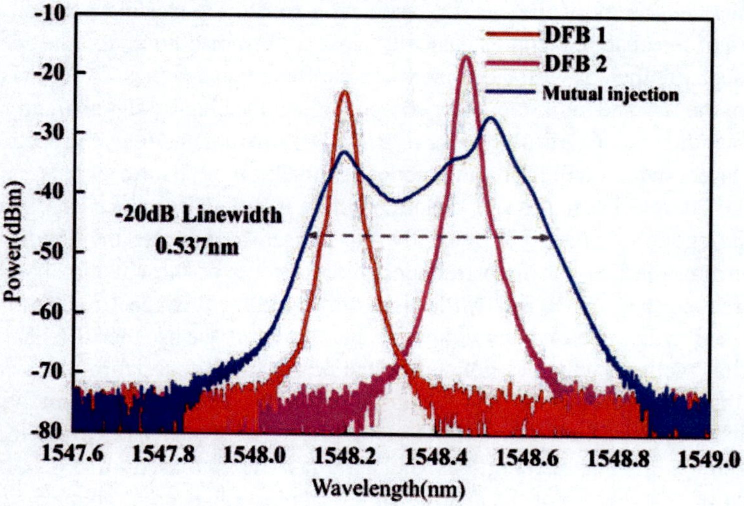

Fig. 1.24 Broadband chaotic laser spectra

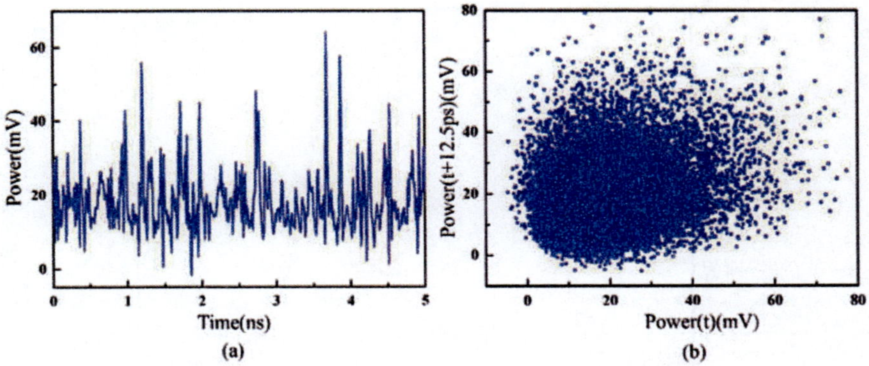

Fig. 1.25 Broadband chaotic laser timing diagram (**a**) and phase diagram (**b**)

of the chaotic laser, which correspond to the central wavelengths of the DFB laser 1 and DFB laser 2 in free-running mode, respectively, as can be seen in Fig. 1.24. The middle of the spectrum is lifted and broadened, but there is still a spectral sag and no flat broadening is achieved. This indicates that there is a large frequency detuning when the two lasers are injected into each other, thus producing a beat effect, which results in spectral lift and spectral broadening. For chaotic secure communication bandwidth and random number generation rate, which is mainly limited by the spectrum bandwidth, it is known from the above experimental results that the spectrum bandwidth of the system has been enhanced to 50 GHz, so as to meet the application of chaotic laser in this area. However, in the case of fiber optic sensing, which is largely dependent on the spectral line width of the chaotic laser, the frequency detuning of the laser can be further reduced to achieve further spectral optimization. The experimental system can therefore be adapted to different applications of the chaotic laser by adjusting different parameters.

The difference in the amplitude of the fluctuations in the power spectrum over the entire spectrum is defined as the flatness, and the spectrum bandwidth is still calculated using the 80% bandwidth method described previously, i.e. The width of the spectrum starts from zero and occupies 80% of the total energy of the spectrum. By holding the coupling strength constant, the effect of frequency detuning on bandwidth and flatness is investigated.

When the coupling strength is 1.635, the spectral bandwidth and flatness of the chaotic laser vary with the amount of frequency detuning were shown in Fig. 1.27. The blue and red curves represent the trend of the chaotic signal bandwidth and flatness respectively. The temperature of the DFB laser 1 is fixed at 25 °C and the temperature of the DFB laser 2 is regulated from 17.9 to 27.5 °C by means of a temperature controller. At this time, the center wavelength of DFB laser 1 is fixed at 1548.191 nm and the center wavelength of DFB laser 2 varies from 1547.738 to 1548.607 nm. The difference in center wavelength is calculated to be 0.416 to − 0.453 nm, so the corresponding frequency detuning is − 56.6 to 52.0 GHz.

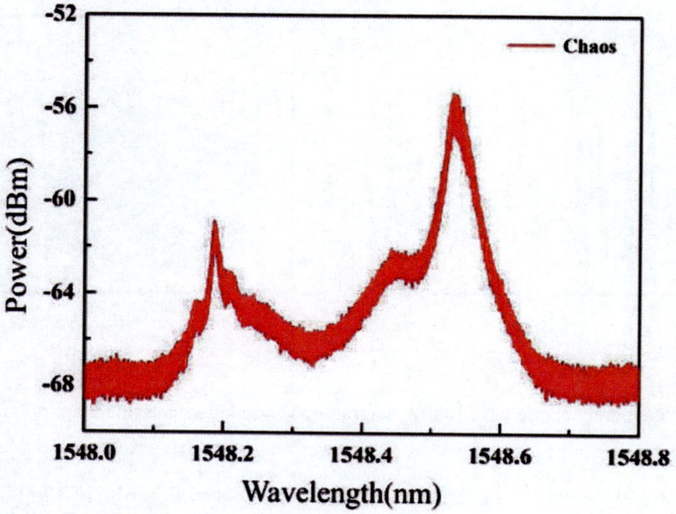

Fig. 1.26 Broadband chaotic laser high-resolution spectrogram

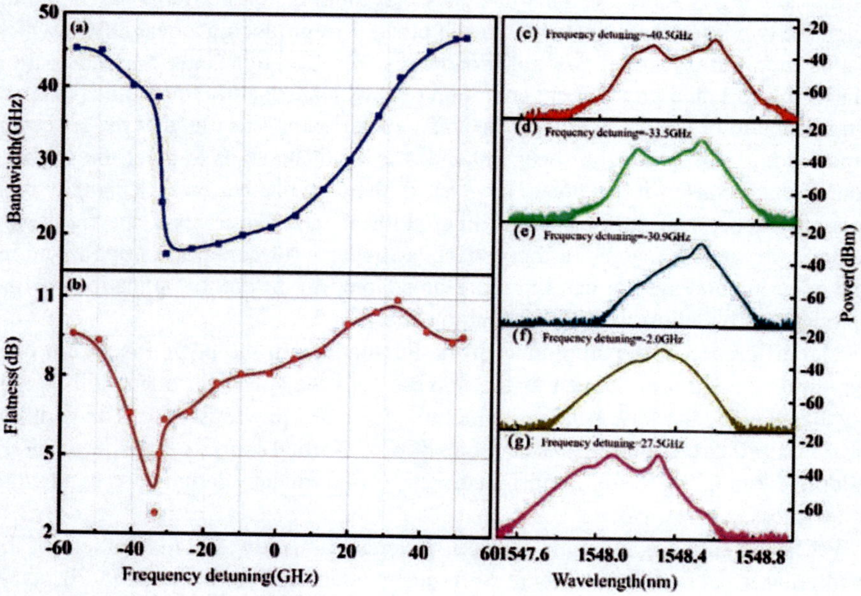

Fig. 1.27 Effect of frequency detuning on chaotic signals. **a** 80% bandwidth; **b** flatness; **c** frequency detuning = − 40.5 GHz; **d** frequency detuning = − 33.5 GHz; **e** frequency detuning = − 30.9 GHz; **f** frequency detuning = − 2.0 GHz; **g** frequency detuning = 27.5 GHz

When the frequency detuning is near zero, both the spectral bandwidth and flatness change slowly, and 80% of the bandwidth is small and flatness is poor, and the spectrum is shown in Fig. 1.27f. This is due to the uneven distribution of the energy in the power spectrum around the relaxation oscillation frequency due to the light-locked injection.

When the frequency detuning increases in the negative direction, the interacting two semiconductor lasers produce a beat frequency effect exciting a rich frequency component, as shown in Fig. 1.27c–e. As a result, the energy of the low-frequency component is enhanced, and then the energy of the low-frequency part is gradually transferred to the high-frequency part due to the beat-frequency effect. Therefore, the flatness shows a trend of decreasing and then increasing. During this process, the energy distribution gradually becomes more homogeneous, and when the frequency detuning is $-$ 33.5 GHz, the spectrum flatness is optimized to a minimum level of $\pm$ 2.8 dB.

When the amount of frequency detuning increases in the positive direction, as shown in Fig. 1.27g, the excitation of the high frequency components leads to an increase in 80% bandwidth and deterioration of flatness. As the frequency detuning increases further, the interaction between the two lasers becomes progressively weaker and therefore the high-frequency part of the energy diminishes. When the frequency detuning is close to $\pm$ 60 GHz, there is almost no interaction between the two lasers and the high frequency part tends to the noise floor, so that the 80% bandwidth and flatness show a weak variation.

To further analyze the influence of the spectral bandwidth and flatness of the chaotic signal, the influence of the coupling strength of the semiconductor laser was investigated. According to Fig. 1.27, when the frequency detuning is $-$ 33.5 GHz, the spectral flatness is optimized to the minimum level, so the frequency detuning is fixed at $-$ 33.5 GHz and the magnitude of the coupling strength is varied by controlling the inter-injection strength through an attenuator.

When the frequency detuning is $-$ 33.5 GHz, the trend of the spectral bandwidth and flatness of the chaotic laser with the coupling intensity is shown in Fig. 1.28. The bias current of the DFB laser 1 is fixed at 1.4 times its threshold current and the temperature is controlled at 25 °C. At the same time, the bias current of the DFB laser 2 was fixed at 2.8 times its threshold current and the temperature was controlled at 25.1 °C. As a result, the output power and frequency detuning of the two semiconductor lasers remain constant. By adjusting the VOA, the coupling intensity of the laser is varied between 0.468 and 1.89.

The blue curve in Fig. 1.28a shows the variation of the chaotic laser bandwidth, while the red curve in Fig. 1.28b shows the variation of the flatness. As shown in Fig. 1.28c the interaction between the two lasers is weak and the 80% bandwidth and flatness change slowly as the coupling intensity increases from 0.468 to 0.761. During the increase of the coupling strength from 0.761 to 1.635, the interaction between the two lasers becomes stronger, producing a four-wave mixing, as shown in Fig. 1.28d–f, with a uniform gain of the new frequency components in the spectrum. As a result, the energy in the low-frequency component is enhanced and further enhanced in the high-frequency component by the beat-frequency effect, resulting in enhanced

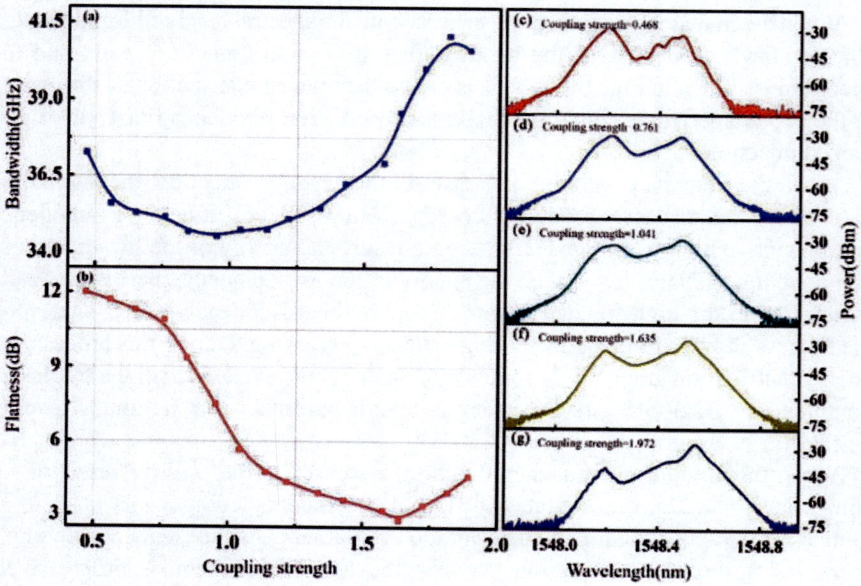

Fig. 1.28 Effect of coupling strength on chaotic signals. **a** 80% bandwidth; **b** flatness; **c** coupling strength = 0.468; **d** coupling strength = 0.761; **e** coupling strength = 1.041; **f** coupling strength = 1.635; **g** coupling strength = 1.972

spectral bandwidth and optimized flatness. It was found that the best flatness of ± 2.8 dB was achieved at a coupling strength of 1.635. When the coupling intensity increases to around 1.89, as shown in Fig. 1.28g, the energy of the DFB laser 2 dominates and the chaotic output state of the DFB laser 1 is suppressed, but the high frequency oscillations remain. As a result, the spectral flatness of the chaotic signal output from DFB laser 1 gradually deteriorates, while the 80% bandwidth increases accordingly.

1.4 Generation of Time-Delay-Free Chaotic Lasers

1.4.1 Research Status

For lasers with a fixed feedback surface and feedback cavity length structure using delayed time feedback, the chaotic lasers generated have obvious time delay characteristics, i.e. the chaotic signal has a certain periodicity, which is usually characterized by the autocorrelation function and mutual information technology. The information on time delay characteristics is extremely detrimental to the generation of high-speed physical random numbers using chaotic lasers as a source of physical entropy, reducing the randomness of the random numbers [48]. In the case of chaotic secure optical communications, the chaotic laser has time-delay information that can lead to security vulnerabilities and can introduce false alarms and false positives to chaotic radars and optical time-domain reflectors. On the one hand, the time delay characteristic signal of the chaotic laser is suppressed by selecting the appropriate feedback intensity and injection current from the chaotic laser source itself, which is the simplest suppression method available [49, 50]. At the same time, the time delay signature of a chaotic laser can be suppressed by using dual optical feedback, modulated multi-feedback, fiber Bragg grating feedback, polarization rotational feedback and filtered feedback with the appropriate choice of feedback strength [51–57]. However, by simply controlling the parameters of the semiconductor laser, whether simple or relatively complex feedback, these options are still very controversial as there is no clear mechanism for research [58]. In addition, two or three cascade-coupled semiconductor lasers can be used to output a time-delay-free chaotic laser, but they are extremely complex and require additional parameters to be optimized [23, 55, 59, 60]. On the other hand, depending on the application of chaotic lasers, other methods of suppressing the time delay characteristics are proposed. For example, in high-speed random number generation, the time delay feature is eliminated by performing an "iso-or" operation on the difference between two random sequences (which are uncorrelated) or continuously sampled 8-bit values extracted from chaotic fluctuations [61, 62]. Although this solution can suppress the delay characteristics to a certain extent, the high-speed logic devices required for the experiments are too expensive to be used in more applications, and these solutions also face problems such as electronic rate bottlenecks. For applications such as chaotic secure communications, possible time delay signatures can be hidden by integrating them into chaotic delay systems through pseudo-random binary sequences (as digital keys) [63]. However, this solution is only theoretically validated and still has many problems before it can be applied in practice.

In this book, we propose two schemes for suppressing the time-delay characteristics, the single-light injection combined with random scattering optical feedback method and the excited Brillouin scattering method, which have been verified and analyzed experimentally.

1.4.2 Single-Light-Injection Combined with Random Scattering Optical Feedback Method

This section presents a broadband acyclic chaotic signal generation method: Chaotic laser generation using external continuous light injection into an optical feedback semiconductor laser and self-injection of the chaotic laser backscattering signal in a single-mode fiber to eliminate the time delay feature; the bandwidth enhancement of the chaotic signal is then achieved by the self-coherence of the acyclic chaotic laser. The 3 dB bandwidth of the chaotic signal generated by this scheme is up to 13.6 GHz and its accuracy has been verified using an autocorrelation algorithm.

A schematic diagram of the experimental setup for acyclic chaotic signal generation is shown in Fig. 1.29. Backscattered signals from a single-mode fiber and continuous light injection from a master semiconductor laser (MLD) into a slave semiconductor laser (SLD) produce acyclic chaotic lasers. Then the bandwidth of the chaotic laser is enhanced by the self-coherence of the resulting chaotic laser, which increases the power of the low frequency signal and results in a broad and flat power spectrum. Self-injected loops for backward scattering signals: The output light from the SLD is injected into the single mode fiber (SMF) via PC1, coupler 1, optical circulator 1 (OC1), adjustable optical attenuator 2 (VOA2) and OC2, where the power and polarization state of the injected light are controlled by VOA2 and PC1 respectively. OI1 is placed at the end of the SMF and the reflection of the placed end surface forms a fixed feedback cavity. The backscattered light is injected into the SLD via OC2, EDFA2, adjustable optical attenuator 3 (VOA3), PC3, coupler 4, OC1, coupler 1 and PC1. The EDFA2 amplifies the scattered light signal and then controls its power and polarization state via VOA3 and PC3.

External continuous light injection loop: The output laser from the MLD is injected into the SLD via OI2, VOA1, PC2, Coupler 4, OC1, Coupler 1 and PC1. The external injection of continuous optical power and the polarization state is controlled by VOA1 and PC2 respectively, with OI2 preventing end-face reflections of the fiber from feeding back into the MLD, thus allowing it to operate in a chaotic state. Due to the instability of the backscattered light from single-mode fibers, the acyclic chaotic signal generated by self-injection is also unstable. Therefore, an excitation with a wavelength equal to the wavelength of the Stokes backward scattered light is injected into the SLD via the MLD output center, resulting in a stable chaotic laser output and a wider bandwidth for the chaotic signal due to the continuous external injection of light. If the central wavelength of the MLD output is not equal to the wavelength of the Stokes backscattered light, it is equivalent to multi-wavelength injection into a semiconductor laser, so the chaotic laser output from the SLD is not stable.

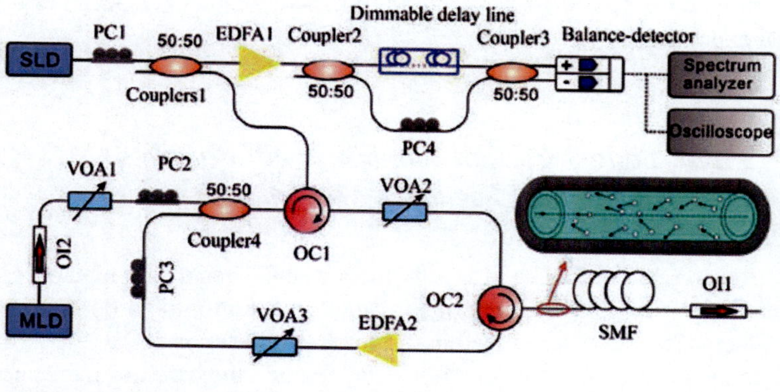

Fig. 1.29 Diagram of the experimental setup for acyclic chaotic signal generation

Self-adjacent loops: the output of acyclic chaotic laser is detected by an equilibrium detector after passing through the EDFA1 and a simple Mach–Zehnder interferometer consisting of two 50:50 couplers. One arm of the M-Z interferometer has PC4 and the other arm has a light tunable delay line (VODL), which is adjusted appropriately to make both arms equal in light range.

In the experiment, the final chaotic signal was detected by a balanced detector (bandwidth 40 GHz, DiscoveryDSC-R410) containing two photodetectors and a differential amplifier. The single mode fiber length is 10 km, the master and slave lasers are operating at 1.5 times threshold and the bias current is 24 mA. The waveforms and power spectra of the output signals were recorded with a real-time oscilloscope (bandwidth 6 GHz, LeCroySDA 806Zi-A) and a signal analyzer (AgilentN9010A) and observed with a spectrometer (YOKOGAWA AQ6370C).

Figure 1.30 shows the experimentally obtained waveform of an acyclic chaotic signal with a time duration of 500 μs. As shown in the figure, the resulting chaotic signal is a noise-like signal with randomly fluctuating amplitudes. In order to show more clearly the waveform characteristics of the experimentally obtained acyclic chaotic source, a detail of the time series from 0 to 39 ns is given in the upper right-hand illustration.

The experimentally obtained power spectrum of a broadband acyclic chaotic signal is shown in Fig. 1.31. The grey curve is the noise floor of the signal analysis, the blue curve is the experimentally obtained power spectrum, and the purple curve is the experimentally obtained power spectrum of the acyclic chaotic signal

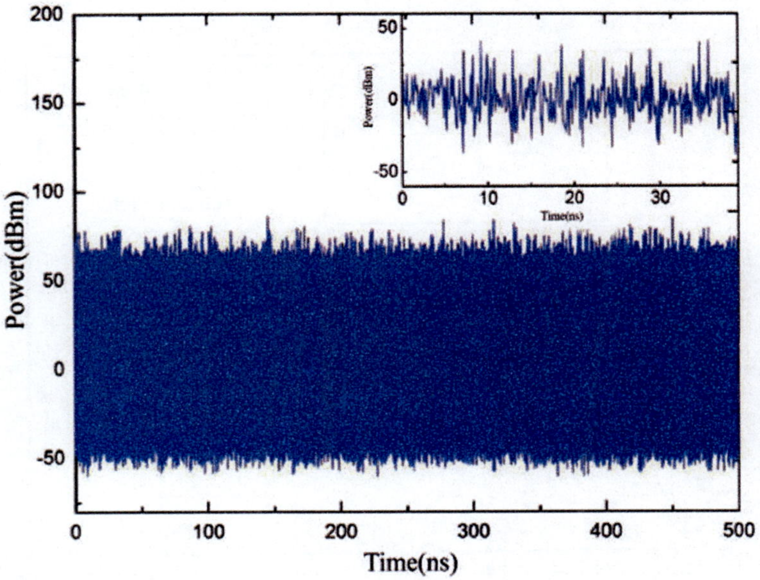

Fig. 1.30 Waveforms of experimentally obtained acyclic chaotic signals

after subtracting the noise of the signal analyzer. By adjusting the scattered light self-injection power to − 10.23 dBm, the external continuous light injection intensity to − 11.46 dBm and the frequency detuning to 10.7 GHz (Brillouin Stokes optical frequency shift), an acyclic chaotic laser with a 3 dB bandwidth up to 13.6 GHz was obtained as shown in Fig. 1.31. The self-injected scattered light in the experiment is backward Rayleigh and Brillouin scattered light provided by a single-mode fiber as a continuous scatterer with a length of 10 km. It can be seen that the relaxation oscillation peak of the semiconductor laser is masked and a broad and flat signal power spectrum is obtained. Furthermore, because of the self-coherence of the chaotic signal, the low frequency energy of the chaotic signal is greatly increased. In order to be able to clearly observe the acyclicity of the signal, the inset shows details of the spectrum with a bandwidth of 0.8 GHz. From 5.6 to 6.4 GHz, the power spectrum of the signal fluctuates randomly and without periodicity.

In order to analyze more clearly the effect of the suppression of chaotic time-delay characteristics, we introduced the autocorrelation function and the mutual information method. By comparing the values of the autocorrelation coefficient before and after Brillouin scattering and the peak of the mutual information curve, the suppression effect of the time-delay characteristics can be seen. The autocorrelation function can be defined as

$$C(\Delta t) = \frac{\langle [P(t+\Delta t) - \langle P(t)\rangle][P(t) - \langle P(t)\rangle]\rangle}{\sqrt{\langle [P(t) - \langle P(t)\rangle]^2\rangle \langle [P(t+\Delta t) - \langle P(t)\rangle]^2\rangle}} \tag{1.1}$$

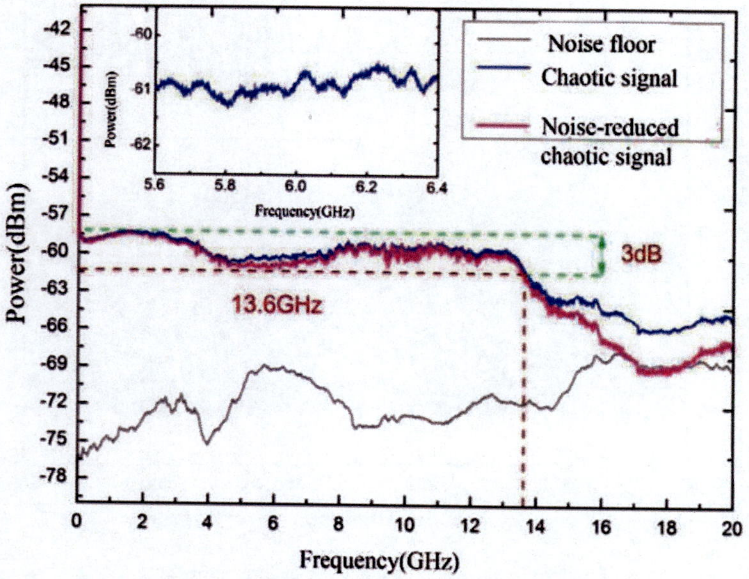

Fig. 1.31 Experimentally obtained power spectrum of a broadband acyclic chaotic signal

In this equation, $P(t) = |E(t)|^2$ represents the chaotic time series, and Δt is the delay time. The delay time corresponding to the feedback outer cavity can be extracted from the position of the peak of the autocorrelation curve of the chaotic laser signal.

Mutual information is the correlation between two sets of events and is a useful measure of information in information theory. Mutual information can be defined as

$$M(\Delta t) = \sum_{P(t), P(t+\Delta t)} \varphi(P(t), P(t+\Delta t)) \log \frac{\varphi(P(t), P(t+\Delta t))}{\varphi(P(t))\varphi(P(t+\Delta t))} \tag{1.2}$$

In this equation, $\varphi(P(t), P(t + \Delta t))$ represents the joint distribution probability density, and $\varphi(P(t))$ and $\varphi(P(t + \Delta t))$ represent the marginal distribution probability density, respectively. Thus, the suppression of the chaotic delay characteristics is evident from the peak position of the mutual information curve.

The experimentally obtained autocorrelation function for acyclic chaotic signals is shown in Fig. 1.32. The autocorrelation curve shown in the figure has no side flaps. Because the proposed chaotic signal generation is based on the self-injection of single-mode fiber backward Brillouin scattered and Rayleigh scattered light plus the injection of external continuous light, there is no fixed feedback cavity in the system, i.e. The chaotic signal generated is acyclic. The inset shows the details of the time and correlation peaks around 0 μs, giving a clearer picture of the autocorrelation curve characteristics of acyclic chaotic signals. It can be clearly seen that the autocorrelation curve has no side flaps and the noise floor is very low, because the autocorrelation curve noise of the chaotic signal is reduced by the autocorrelation.

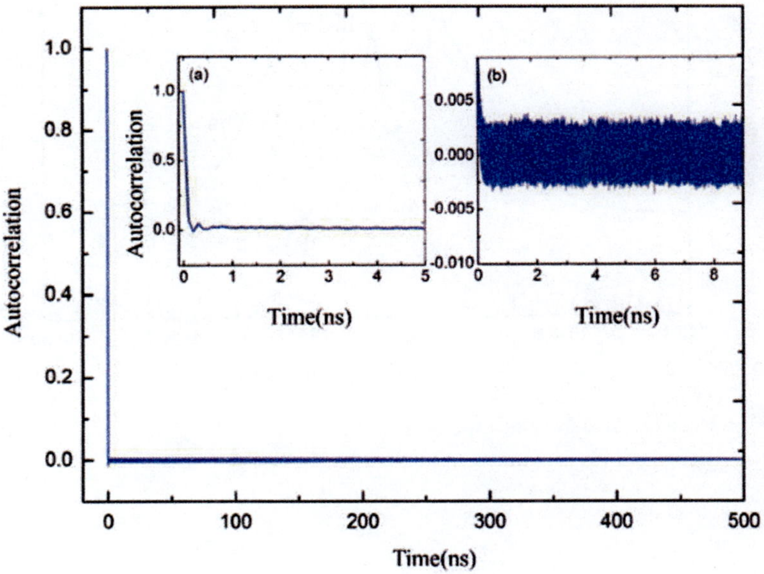

Fig. 1.32 Autocorrelation function for acyclic chaotic signals

When the SLD has no scattered light self-injection and external continuous light injection, its output laser is injected into the SMF, and its backscattered light spectrum is shown in the red curve in Fig. 1.33. At this time, the SLD output power is 0.62 mW. The main peak is Rayleigh scattered light and the side peaks are Brillouin scattered light. The central wavelength of the MLD is 1549.976 nm, which is the same as that of the Brillouin Stokes light. By injecting external continuous light, the resulting acyclic chaotic signal is more stable and, at the same time, the bandwidth is largely enhanced. In the figure, the blue curve shows the spectrum of a chaotic laser produced by self-injection and external continuous light injection using backward Rayleigh scattering and Brillouin scattered light. Due to the injection of backscattered light and external continuous light, the two peaks of the acyclic chaotic signal spectrum become 1549.876 nm and 1549.978 nm respectively.

Overall, we present a method for the experimental generation of broadband time-delay-free chaotic lasers, the use of single-mode fibers for backward Brillouin and Rayleigh scattered light self-injection into a semiconductor laser plus external continuous light injection into a semiconductor laser produces a chaotic laser with no time delay. A broadband chaotic laser with a 3 dB bandwidth of up to 13.6 GHz was obtained by combining the self-coherent effect of the chaotic laser.

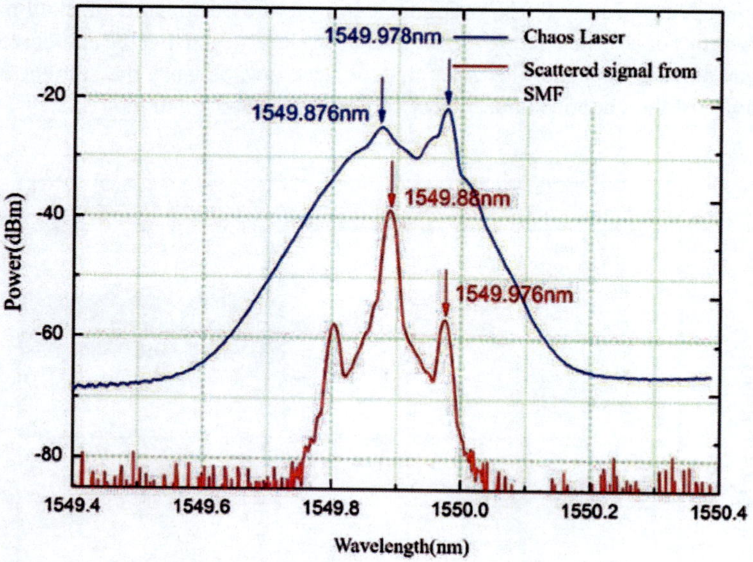

Fig. 1.33 Spectra obtained from the experiment

1.4.3 Stimulated Brillouin Scattering Method

Subsequent manipulation of the chaotic laser signal output directly from the outer cavity of the semiconductor laser can eliminate its time delay characteristics. For example, time delay features are suppressed by using optical delay self-interference of chaotic signals or electrical outlier methods [64, 65]. Here, we propose a relatively novel scheme to suppress the time delay feature using Brillouin backward scattering and present a method to suppress the chaotic laser time delay feature signal using the Brillouin scattering effect of optical fibers. The experimental results show that the time delay characteristics of the chaotic laser can be significantly suppressed by 0.5, 2, 4.5 or 6 km lengths of G.652 type fibers when the average power of the chaotic laser is injected into the single-mode fibers in the range of 400–1500 mW. It was also demonstrated that the G.655 type of fiber has the same suppression effect as the G.652 type of fiber. Furthermore, the suppression of the delay characteristics was very stable over the course of 50 measurements. In addition, it is strongly demonstrated that the suppression of chaotic laser time delay features is independent of the length of the chaotic sequence.

Figure 1.34 shows the experimental setup for chaotic laser time delay feature suppression, with the generated chaotic laser shown in the dashed box. The DFB laser has a threshold current and a central wavelength of 22 mA and 1550 nm respectively. The fiber optic feedback loop consists of an optical circulator, a 3 dB optical couple, an adjustable optical attenuator and a polarization controller. By adjusting the polarization state of the external feedback light, the intensity of the feedback and the bias current of the laser, the DFB laser produces a chaotic excitation. The generated chaotic laser light is first passed through an optical isolator and then amplified by an EDFA. The amplified chaotic laser is then injected into a single-mode fiber via OC2 with lengths of 0.5 km, 2 km, 4.5 km and 6 km in G.652 and G.655 respectively. At the same time, the end of the fiber is placed in a matching solution in order to suppress the resulting Fresnel reflections. The chaotic laser is scattered back through the fiber and output through the scattering end of OC2, which is then filtered by an adjustable bandpass filter to obtain the desired chaotic Stokes signal. The 50:50 optocoupler is then used to split the two circuits. One way, light is measured by a high-resolution spectrometer (OSA) with a maximum resolution of 5 MHz, model APEXAP2041B. The other way is converted into an electrical signal using a photodetector, and then the time series of the chaotic signal is analyzed by a real-time oscilloscope (OSC) with a bandwidth of 36 GHz and a sampling rate of 80 GS/s.

In the experiments, the operating current of the DFB laser was set to 1.5 times the threshold current (33 mA). Figure 1.35 depicts the spectrogram of a chaotic laser. The dark yellow line represents the spectrum of the chaotic laser produced directly by the outer cavity of the semiconductor laser. At this point, the − 3 dB linewidth of the chaotic laser spectrum is 1.4 GHz and the − 10 dB linewidth is 6.2 GHz. The red and purple lines represent the spectra of the backscattered light produced by a chaotic laser propagating through a 6 km and 0.5 km fiber, respectively. The light

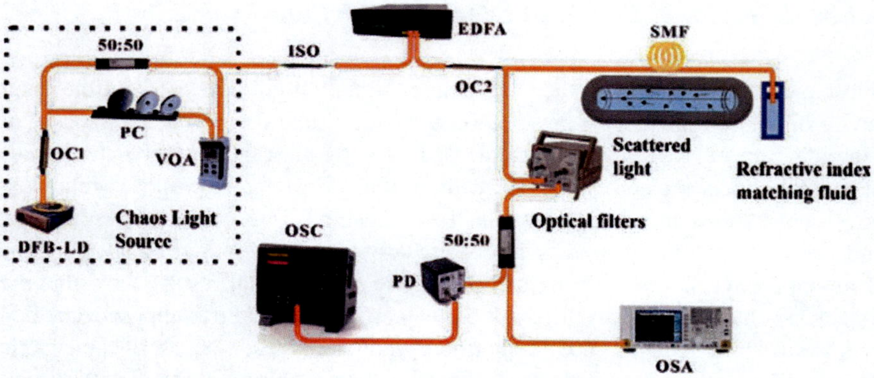

Fig. 1.34 Diagram of the experimental setup for chaotic laser time delay feature suppression

blue and dark blue lines represent the spectrum of Stokes light, which is filtered out using a tunable filter.

In the experiment, the chaotic excitation sequence generated directly from the outer cavity of the light-feeding semiconductor exciter is shown in Fig. 1.36a-I. The original chaotic laser was amplified to 500 mW using an erbium-doped fiber amplifier, and then the amplified chaotic laser was injected into the incident end of a 6 km long single-mode fiber (type G.652). After the Brillouin scattering effect of the optical fiber, the time sequence of the Stokes light signal filtered out by the filter is shown in Fig. 1.36b-I. At this chaotic time series, the Lyapunov index was calculated to be 0.2167, while the Lyapunov index for a chaotic laser generated directly from the

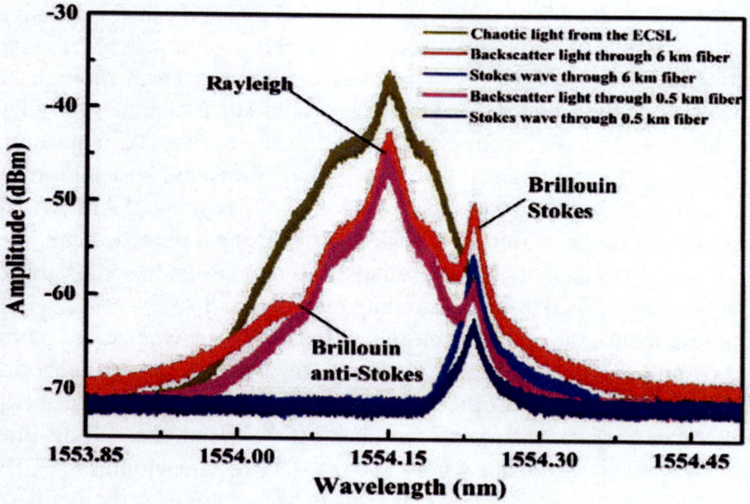

Fig. 1.35 Spectrograms of different chaotic laser signals

external cavity of an optical feedback semiconductor laser was 0.2213. Figure 1.36a-II and b-II show the corresponding autocorrelation curves before and after the effect of Brillouin scattering by the fiber, respectively. The chaotic laser signal output directly from the optical feedback semiconductor laser has significant peaks in the autocorrelation function curve at the external cavity feedback delay $\tau = 105$ ns and $2\tau = 210$ ns, with $C = 0.37$ and $C = 0.087$, respectively. However, after Brillouin scattering, the correlation peak of the chaotic laser drops significantly to $C = 0.026$ at a feedback delay of $\tau = 105$ ns, while at a delay of $2\tau = 210$ ns the correlation peak is already drowned in noise and is difficult to observe. Figure 1.36a-III and b-III show the corresponding mutual information curves before and after Brillouin scattering by the fiber respectively. It can also be seen that the peak of the mutual information curve is also evident at the outer cavity feedback delay $\tau = 105$ ns, reaching $M = 0.169$, and is also clearly observable at $2\tau = 210$ ns, with a value of $M = 0.079$. However, after Brillouin scattering, the corresponding peak drops from 0.169 to 0.0017 and is completely submerged in noise and difficult to observe. This shows that the method is effective in suppressing the time delay characteristics of chaotic lasers when a G.652 fiber of 6 km length is selected.

The chaotic delay characteristics are effectively suppressed probably because: the excited Brillouin backscattering itself is a non-linear process, and the chaotic pump

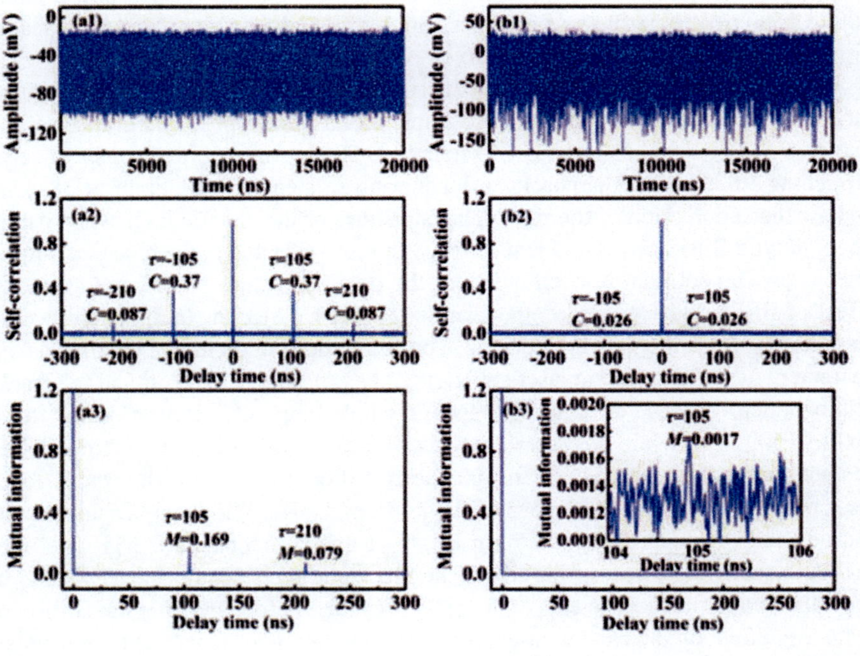

Fig. 1.36 Comparison of before and after Brillouin scattering by an optical feedback semiconductor laser. **a** Chaotic characteristics of the fiber before Brillouin scattering; **b** chaotic characteristics of the fiber after Brillouin scattering (injection power of 500 mW)

light injected into the fiber interferes with the backscattered Stokes light, which then generates acoustic waves through the electrostriction effect. The resulting acoustic wave acts as a moving Brillouin dynamic grating which backscatters the chaotic laser signal. Due to the large number of special gratings distributed along the fiber producing many scattering points, these random scattering points form the backward scattering chaotic laser, making the backward scattered light transmitted irregularly, and the weak period of the chaotic laser signal corresponding to the time delay is destroyed during the Brillouin scattering in the chaotic laser. As a result, the chaotic laser time delay characteristics are effectively suppressed in the process.

To verify the suppression effect of this method, we integrated the autocorrelation curve of the chaotic laser and formed a two-dimensional plot of the injection power from 400 to 1500 mW, as shown in Fig. 1.37. Figure 1.37a shows the case of a chaotic laser generated directly by an optical feedback semiconductor laser. It is clear from the figure that the chaotic laser has a distinct time delay signature at the external cavity feedback time delay $\tau = \pm 105$ ns for different injection powers. In our experiments, we have demonstrated that different types and lengths of fibers have good suppression of chaotic time delay signals. Figure 1.37b–d show the suppression of the chaotic laser time delay characteristics by Brillouin scattering when passing through a 0.5 km, 4.5 km and 6 km long G.655 type fiber respectively. It can be seen that the chaotic delay characteristics of the outer cavity feedback delay at $\tau = \pm$ 105 ns can be significantly suppressed in the whole range of injected power from 400 to 1500 mW, regardless of the fiber length of 0.5, 4.5 or 6 km, and are already drowned in noise and difficult to observe.

In addition, the effect of Brillouin scattering on the chaotic laser time delay signature was further analyzed for the G.652 and G.655 fiber types. The optical effective area A_{eff} of the G.652 and G.655 fibers are 80 μm^2 and 50 μm^2, respectively, producing different Brillouin backward scattering efficiencies [66]. Figure 1.38a and b show the suppression of the time delay signature of the chaotic laser when using fibers of type G.652 and G.655 respectively. In Fig. 1.38a and b, the blue and purple lines show the correlation coefficients of the delay signature before and after the effect of Brillouin scattering on the chaotic laser at $\tau = 105$ ns in the outer cavity, respectively, for a fiber length of 0.5 km. The correlation coefficient of the time delay characteristics of the chaotic laser signal directly generated by the optical feedback semiconductor laser is close to 0.38 for the whole range of injection power from 400 to 1500 mW, while the correlation coefficient of the chaotic laser time delay characteristics decreases to close to 0.03 due to Brillouin scattering when the chaotic laser passes through a G.652 fiber of 0.5 km length. The same suppression effect can be achieved when using a 0.5 km length of optical fiber type G.655, after the Brillouin scattering effect of the fiber. The black and red lines in Fig. 1.38a and b show the correlation coefficients of the delay signature before and after the Brillouin scattering effect on the chaotic laser at $\tau = 105$ ns for a fiber length of 6 km and an external cavity delay of $\tau = 105$ ns, respectively. As can be seen from the figure, the method also provides good suppression of the time delay signature of the chaotic laser for the entire range of injection power from 400 to 1500 mW for a fiber length

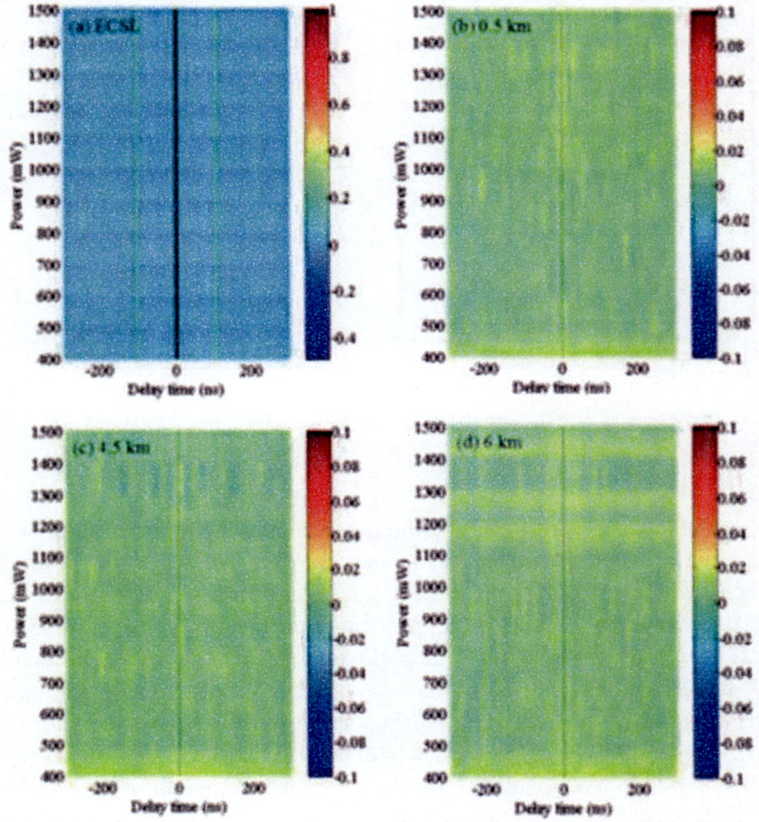

Fig. 1.37 Chaotic laser autocorrelation curves for injection powers of 400–1500 mW. G.655 fiber with fiber lengths of 0.5, 4.5 and 6 km before (**a**) and after (**b**)–(**d**) the Brillouin scattering effect

of 6 km. By comparing the two cases before and after Brillouin backscattering, the same suppression effect can be obtained by using different types and lengths of fibers.

The error bars in Fig. 1.38 further show that for different injected powers, through 50 measurements and calculating their standard deviations, after the Brillouin scattering effect of the fibers, it can be seen that there is some stability in the suppression effect of the time delay characteristics using the Brillouin scattering effect. Both G.652 and G.655 fiber types are still used here, with fiber lengths of 0.5 km and 6 km respectively. The correlation coefficient of the chaotic laser time delay signature remains almost constant at 0.38 for 50 measurements with an injected power of 600 mW and a 6 km long G.652 fiber, but drops to 0.012 after Brillouin scattering from the fiber. The same results can be obtained for other different injection powers as well as for other types and lengths of fibers. This proves that the method is very stable for the suppression of time delay features.

As mentioned above, the time delay signature of a chaotic laser produced by an optical feedback semiconductor laser is usually characterized by two parameters.

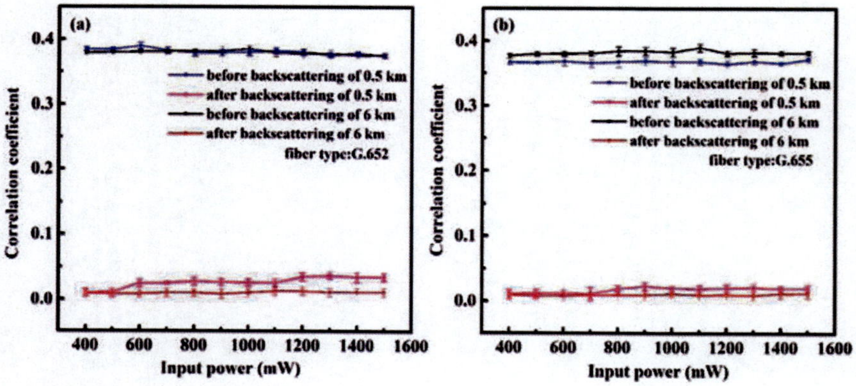

Fig. 1.38 Suppression of chaotic time-delay signature signals for two types of fibers, G.652 (**a**) and G.655 (**b**)

One is the delay value determined by the outer cavity length of the semiconductor laser alone. The other is the intensity of the delay signature directly described by the autocorrelation correlation coefficient. Intuitively, the intensity of the time delay characteristics of a chaotic laser is independent of the length of the chaotic sequence. Figure 1.39 represents the correlation coefficient of the outer cavity time delay at $\tau = 105$ ns versus the length of the chaotic sequence for different injection powers and fiber lengths. As can be seen from Fig. 1.39a, for a single-mode fiber of type G.655 with a length of 2 km, the injected power is 200 mW, 400 mW, 600 mW and 800 mW, respectively, and for the chaotic sequence length of 600–11,000 ns. The correlation coefficients of the chaotic laser time delay characteristics are almost independent of the chaotic sequence length after Brillouin scattering from the fiber. The correlation coefficients remained stable at 0.5, 2, 4.5 and 6 km as the length of the chaotic sequence increased from 600 to 11,000 ns. Therefore, the suppression of the chaotic laser time delay signature is independent of the length of the chaotic sequence of Brillouin scattering effects in the fiber.

This section focuses on the suppression of the time-delayed characteristic signal of chaotic lasers through the Brillouin scattering effect of optical fibers. The effect of Brillouin scattering on the time delay characteristics of chaotic lasers is analyzed in experiments with different types and lengths of fibers. The results show that the time delay characteristics of the chaotic laser can be significantly suppressed by 0.5, 2, 4.5 or 6 km lengths of G.652 fibers when the average power of the injected chaotic laser is in the range of 400–1500 mW. It has also been demonstrated that the G.655 type of fiber has the same suppression effect as the G.652 type of fiber at the above fiber lengths. The stability of the time delay suppression was also very good over 50 measurements. Therefore, the Brillouin scattering effect of optical fibers can be used to effectively suppress the time delay characteristics of optical feedback semiconductor lasers.

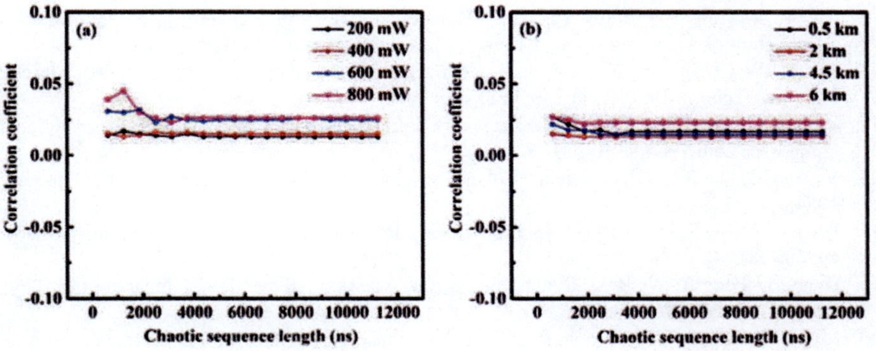

Fig. 1.39 Correlation coefficient versus chaotic sequence length for external cavity time delay at $\tau = 105$ ns. **a** Injection power of 200, 400, 600 and 800 mW; **b** Fiber length of 0.5 km, 2 km, 4.5 km and 6 km respectively

At the same time, the Brillouin scattering method uses standard single-mode fibers to suppress the time delay signature of chaotic lasers, and is entirely in the optical domain while the time delay signature of chaotic lasers is suppressed, in contrast to methods such as optical delay self-interference and electrical outlier, which use electronic components to achieve the suppression of chaotic laser time delay signatures [65, 66]. In the process of chaotic laser time delay signature signal suppression using methods such as optical delay self-interference and electrical external aberration, the use of electronic components can cut off the RF spectrum of the chaotic signal and lose the high frequency signal. However, the Brillouin scattering method for optical fibers can avoid these problems.

References

1. Lorenz EN. Deterministic nonperiodic flow. J Atmos Sci. 2004;20(2):130–41.
2. Li B, Jiang WS. Chaos optimization methods and its applications. Control Theory Appl. 1997;4:613–5.
3. Huang Z. Chaos and butterfly effect. Rev World Inven. 2002;3:32–32.
4. Wang YC. Generation and applications of chaotic laser. Laser Optoelectr Prog. 2009;46(4): 13–21.
5. Wang YC. Basic and applied research on chaotic lasers. In: The fourth joint set of twelve western provinces (regions) and municipal physics societies; 2008.
6. Lang R, Kobayashi K. External optical feedback effects on semiconductor injection laser properties. IEEE J Quantum Electron. 1980;16(3):347–55.
7. Chow WW. Theory of line narrowing and frequency selection in all injection locked laser. IEEE J Quantum Electron. 1983;19(2):243–9.
8. Zhang HY, Pan ZQ, Yang JQ, et al. Semiconductor lasers with tunable narrow-linewidth single-longitudinal-mode external cavity. Physics. 1995;24(7):429–32.
9. Gauthier DJ, Narum P, Boyd RW. Observation of deterministic chaos in a phase-conjugate mirror. Phys Rev Lett. 1987;58(16):1640–3.

10. Roy R, Murphy TW, Maier TD, et al. Dynamical control of a chaotic laser: experimental stabilization of a globally coupled system. Phys Rev Lett. 1992;68(9):1259–62.
11. Tkach RW, Chraplyvy AR. Regimes of feedback effects in 1.5-pm distributed feedback lasers. J Lightw Technol. 1986;4(11):1655–61.
12. Weiss CO, Telle HR, Klische W. Chaos in a solid-state laser with a periodically modulated pump. Opt Lett. 1984;9(12):561–3.
13. Valley GC, Dunning GJ. Observation of optical chaos in a phase-conjugate resonator. Opt Lett. 1984;9(11):513–5.
14. Pecora LM, Carroll TL. Synchronization in chaotic systems. Controlling Chaos. 1996;6(08):142–5.
15. Huang L, Feng R, Wang M. Synchronization of chaotic systems via nonlinear control. Phys Lett A. 2004;320(4):271–5.
16. Mukai T, Otsuka K. New route to optical chaos: successive-subharmonic-oscillation cascade in a semiconductor laser coupled to an external cavity. Phys Rev Lett. 1985;55(17):1711–4.
17. Erneux T, Gavrielides A, Sciamanna M. Stable microwave oscillations due to external-cavity-mode beating in laser diodes subject to optical feedback. Phys Rev A. 2002;66(3):033809.
18. Fischer APA, Yousefi M, Lenstra D, et al. Filtered optical feedback induced frequency dynamics in semiconductor lasers. Phys Rev Lett. 2004;92(2):023901.
19. Guo CZ, Liu P. Stability of the coherent light injection locking in semiconductor lasers, the related instability phenomena and their routes to chaos. Acta Phys Sinica. 1990;39(11):1730–8.
20. Hong Y, Rees P, Spencer PS, et al. Polarization-resolved chaos in a vertical cavity surface emitting laser subject to optical injection. Quantum Electr Laser Sci Conf; 2002.
21. Murakami A. Phase locking and chaos synchronization in injection-locked semiconductor lasers. IEEE J Quantum Electron. 2003;39(3):438–47.
22. Wieczorek S, Krauskopf B, Simpson TB, et al. The dynamical complexity of optically injected semiconductor lasers. Phys Rep. 2012;416(1–2):1–128.
23. Li N, Pan W, Xiang S, et al. Loss of time delay signature in broadband cascade-coupled semiconductor lasers. IEEE Photonics Technol Lett. 2012;24(23):2187–90.
24. Yuan G, Zhang X, Wang Z. Chaos generation in a semiconductor ring laser with an optical injection. Optik-Int J Light Electron Opt. 2013;124(22):5715–8.
25. Yuan G, Zhang X, Wang Z. Generation and synchronization of feedback-induced chaos in semiconductor ring lasers by injection-locking. Optik-Int J Light Electron Opt. 2014;125(8):1950–3.
26. Tang S, Liu JM. Chaotic pulsing and quasi-periodic route to chaos in a semiconductor laser with delayed opto-electronic feedback. IEEE J Quantum Electron. 2001;37(3):329–36.
27. Lin FY, Liu JM. Nonlinear dynamics of a semiconductor laser with delayed negative optoelectronic feedback. IEEE J Quantum Electron. 2003;39(4):562–8.
28. Yan SL. Controlling chaos in a semiconductor laser via photoelectric delayed negative-feedback. Acta Phys Sinica. 2008;57(4):2100–6.
29. Yan SL. Control of chaos in a semiconductor laser using photoelectric nonlinear feedback. In: Ninth international conference on natural computation; 2014.
30. Wang AB. Broadband chaos generation and chaos OTDR. Taiyuan: Taiyuan University of Technology; 2014.
31. Lee CH, Shin SY. Self pulsing, spectral bistability, and chaos in a semiconductor laser diode with optoelectronic feedback. Appl Phys Lett. 1993;62(9):922–4.
32. Lin FY, Tsai MC. Chaotic communication in radio-over-fiber transmission based on optoelectronic feedback semiconductor lasers. Opt Express. 2007;15(2):302–11.
33. Tang S, Liu JM. Chaos synchronization in semiconductor lasers with optoelectronic feedback. IEEE J Quantum Electron. 2003;39(6):708–15.
34. Lin FY, Liu JM. Nonlinear dynamical characteristics of an optically injected semiconductor laser subject to optoelectronic feedback. Opt Commun. 2003;221(1):173–80.
35. Uchida A, Okumura H, Aida H, et al. Fast random bit generation with bandwidth-enhanced chaos in semiconductor lasers. Opt Express. 2010;18(6):5512–24.
36. Zhang LM, Pan BW, Chen GC, et al. 640-Gbit/s fast physical random number generation using a broadband chaotic semiconductor laser. Sci Rep. 2017;8:45900.

37. Lin FY, Liu JM. Chaotic lidar. IEEE J Sel Top Quantum Electron. 2004;10(5):991–7.
38. Ke JX, Yi LL, Xia GQ, et al. Chaotic optical communications over 100-km fiber transmission at 30-Gb/s bit rate. Opt Lett. 2018;43(6):1323
39. Wang AB, Wang YC, He H. Enhancing the bandwidth of the optical chaotic signal generated by a semiconductor laser with optical feedback. IEEE Photonics Technol Lett. 2008;20(19):1633–5.
40. Zhang MJ, Liu TG, Li P, et al. Generation of broadband chaotic laser using dual-wavelength optically injected Fabry-Pérot laser diode with optical feedback. IEEE Photonics Technol Lett. 2011;23(24):1872–4.
41. Feng Y, Yang YB, Wang AB, et al. Generation of 27 GHz flat broadband chaotic laser with semiconductor laser loop. Acta Phys Sinica. 2011;60(6):325–9.
42. Hong Y, Spencer PS, Shore KA. Flat broadband chaos in vertical-cavity surface-emitting lasers subject to chaotic optical injection. IEEE J Quantum Electron. 2012;48(12):1536–41.
43. Xiang SY, Pan W, Luo B, et al. Wideband unpredictability-enhanced chaotic semiconductor lasers with dual-chaotic optical injections. IEEE J Quantum Electron. 2012;48(8):1069–76.
44. Hong Y, Chen X, Spencer PS, et al. Enhanced flat broadband optical chaos using low-cost VCSEL and fiber ring resonator. IEEE J Quantum Electron. 2015;51(3):1–6.
45. Wang LY, Zhong ZQ, Wu ZM, et al. Bandwidth enhancement and time-delay signature suppression of chaotic signal from an optical feedback semiconductor laser by using cross phase modulation in a highly nonlinear fiber loop mirror. Semicond Lasers Appl VII. 2016;10017:82–7.
46. Wang AB, Wang YC, Yang YB, et al. Generation of flat-spectrum wideband chaos by fiber ring resonator. Appl Phys Lett. 2013;102(3):031112.
47. Wang AB, Wang BJ, Li L, et al. Optical heterodyne generation of high-dimensional and broadband white chaos. IEEE J Sel Top Quantum Electron. 2015;21(6):1–10.
48. Syvridis D, Argyris A, Bogris A, et al. Integrated devices for optical chaos generation and communication applications. IEEE J Quantum Electr. 2009;45(11):0–1428.
49. Rontani D, Locquet A, Sciamanna M, et al. Loss of time-delay signature in the chaotic output of a semiconductor laser with optical feedback. Opt Lett. 2007;32(20):2960–2.
50. Wu JG, Xia GQ, Tang X, et al. Time delay signature concealment of optical feedback induced chaos in an external cavity semiconductor laser. Opt Express. 2010;18(7):6661–6.
51. Lee MW, Rees P, Shore KA, et al. Dynamical characterisation of laser diode subject to double optical feedback for chaotic optical communications. Optoelectr IEE Proc. 2005;152(2):97–102.
52. Xia GQ, Wu JG, Wu ZM. Suppression of time delay signatures of chaotic output in a semiconductor laser with double optical feedback. Opt Express. 2009;17(22):20124–33.
53. Shahverdiev EM, Shore KA. Impact of modulated multiple optical feedback time delays on laser diode chaos synchronization. Opt Commun. 2009;282(17):3568–72.
54. Li SS, Liu Q, Chan SC. Distributed feedbacks for time-delay signature suppression of chaos generated from a semiconductor laser. IEEE Photonics J. 2012;4(5):1930–5.
55. Li SS, Chan SC. Chaotic time-delay signature suppression in a semiconductor laser with frequency-detuned grating feedback. IEEE J Sel Top Quantum Electron. 2015;21(6):541–52.
56. Wu JG, Xia GQ, Cao LP, et al. Experimental investigations on the external cavity time signature in chaotic output of an incoherent optical feedback external cavity semiconductor laser. Opt Commun. 2009;282(15):3153–6.
57. Wu Y, Wang B, Zhang J, et al. Suppression of time delay signature in chaotic semiconductor lasers with filtered optical feedback. Math Probl Eng. 2013;2013(4):1–7.
58. Wu Y, Wang YC, Li P, et al. Can fixed time delay signature be concealed in chaotic semiconductor laser with optical feedback. IEEE J Quantum Electron. 2012;48(11):1371–9.
59. Wu JG, Wu ZM, Xia GQ, et al. Evolution of time delay signature of chaos generated in a mutually delay-coupled semiconductor lasers system. Opt Express. 2012;20(2):1741–53.
60. Hong Y, Quirce A, Wang B, et al. Concealment of chaos time-delay signature in three-cascaded vertical-cavity surface-emitting lasers. IEEE J Quantum Electron. 2016;52(8):1–8.

61. Uchida A, Amano K, Inoue M, et al. Fast physical random bit generation with chaotic semiconductor lasers. Nat Photonics. 2008;2(12):728–32.
62. Reidler I, Aviad Y, Rosenbluh M, et al. Ultrahigh-speed random number generation based on a chaotic semiconductor laser. Phys Rev Lett. 2009;103(2):024102.
63. Nguimdo RM, Colet P, Larger L, et al. Digital key for chaos communication performing time delay concealment. Phys Rev Lett. 2011;107(3):034103.
64. Wang A, Yang Y, Wang B, et al. Generation of wideband chaos with suppressed time-delay signature by delayed self-interference. Opt Express. 2013;21(7):8701–10.
65. Cheng CH, Chen YC, Lin FY. Chaos time delay signature suppression and bandwidth enhancement by electrical heterodyning. Opt Express. 2015;23(3):2308–19.
66. Kobyakov A, Sauer M, Chowdhury D. Stimulated Brillouin scattering in optical fibers. Adv Opt Photonics. 2009;2:1–59.

Chapter 2
Photonic Integrated Chaotic Lasers

At present, chaotic laser generation systems are mostly built by semiconductor lasers combined with a variety of external discrete optical elements. The system has some disadvantages, such as large volume, vulnerable to the external environment and unstable output. It is not conducive to the further commercial application of chaotic laser. Combined with the advantages of the photonic integrated chip, photonic integrated chaotic semiconductor laser arises at the historic moment, which has the advantages of small size and stable performance, and is the key to the practicality and marketization of chaotic laser.

2.1 Research Overview

In order to promote the application of chaotic laser, scholars at home and abroad have carried out a series of research and development of the integrated chaotic laser, and made a lot of progress. According to the existing technical scheme, photon integrated chaotic laser is mainly divided into monolithic integrated chaotic laser and hybrid integrated chaotic laser.

2.1.1 Monolithic Integrated Chaotic Semiconductor Lasers

Monolithic integrated chaotic semiconductor lasers are fabricated by growing lasers, disturbance elements and other related functional elements on the same substrate material.

M. Zhang et al., *Novel Optical Fiber Sensing Technology and Systems*, Progress in Optical Science and Photonics 28, https://doi.org/10.1007/978-981-99-7149-7_2

1. **Single cavity four section structure**

In 2008, A. Argyris from Optical Communications Laboratory, Ministry of Information and Communication, University of Athens, Greece, and M. Hamacher from Fraunhofer Telecommunications Institute, Heinrich-Hertz Research Institute, Germany, proposed a new type of photonic integrated chip [1]. As shown in Fig. 2.1, the chip consists of the DFB laser region, gain/absorption region, phase region and passive waveguides coated with high reflective film at the end [2–6]. Among them, the high reflective film plated at the end of the passive waveguide provides single-cavity feedback for the DFB laser. The gain/absorption region and phase region can control and adjust the intensity and phase of the feedback light respectively, so that it can produce high-dimensional controllable broadband chaotic signals.

2. **Air gap multi-feedback structure**

In 2009, V. Z. Tronciu, Vilstras College, Berlin, C. Mirasso and P. Colet, University of the Balearic Islands, Spain, M. Hamacher of Fraunhofer Telecommunications Institute of Heinrich-Hertz Research Institute in Germany and V. Annovazzi-Lodi of Tipavia University in Italy proposed a multi-feedback photonic integrated chaotic semiconductor laser chip with the air gap. As shown in Fig. 2.2, it includes a DFB laser region, two phase regions, an air gap, and a passive waveguide part. Among them, the two sides of the air gap and the DFB laser region coated with highly reflective film form three-cavity feedback, and the two-phase regions can control the

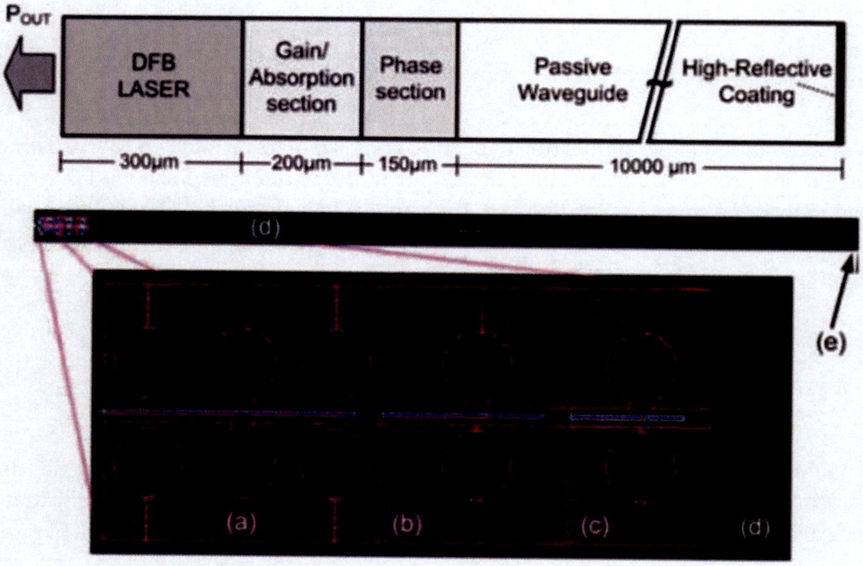

Fig. 2.1 Schematic diagram of the four-segment chip structure consisting of the DFB laser region, gain/absorption region, phase region and passive waveguide. (a) DFB laser region; (b) gain/absorption region; (c) phase region; (d) passive waveguide

feedback phase. In 2010, they combined with M. Benedetti and V. Vercesi of the University of Tipavia in Italy to simulate and test the output characteristics of the integrated chaotic semiconductor laser [7].

3. **Ring waveguide structure**

In 2011, S. Sunada of NTT Company of Japan and A. Uchida of Saitama University developed monolithic integrated chaotic semiconductor lasers with annular passive optical waveguides [8]. As shown in Fig. 2.3, the chip consists of a DFB laser, two semiconductor optical amplifiers (SOA), a photodetector and a section of annular optical waveguides [9]. Among them, SOA can control the light intensity and phase of the feedback signal at the same time, and the chaotic laser can be generated by adjusting the injection current of two SOA.

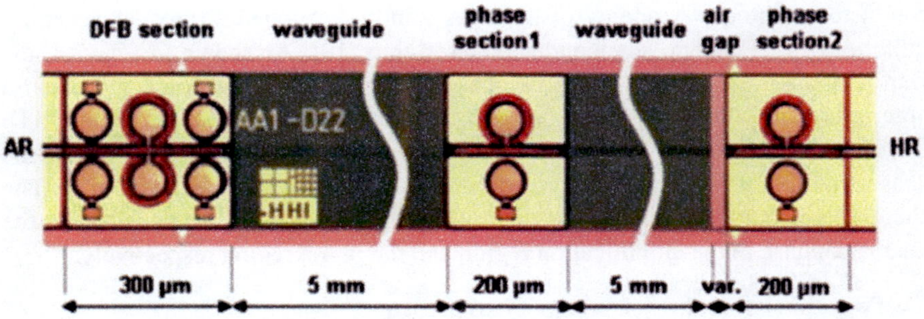

Fig. 2.2 Schematic diagram of a multi-feedback integrated chaotic semiconductor laser chip with an air gap

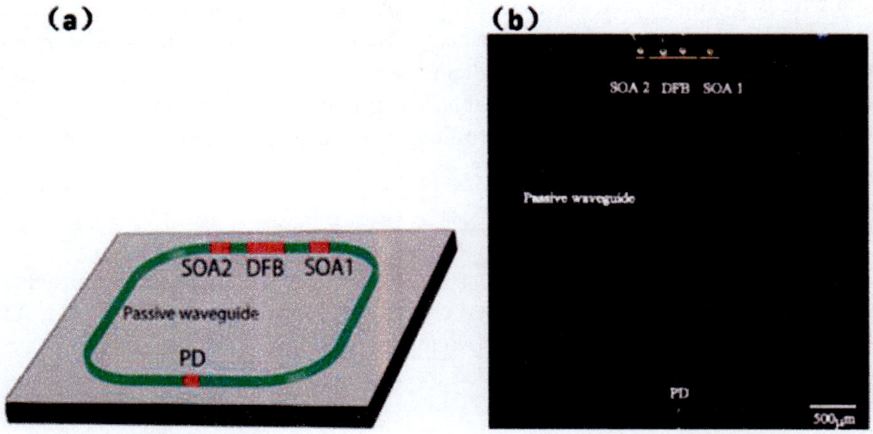

Fig. 2.3 Monolithic integrated chaotic semiconductor laser with annular optical waveguide structure **a** schematic structure; **b** physical view

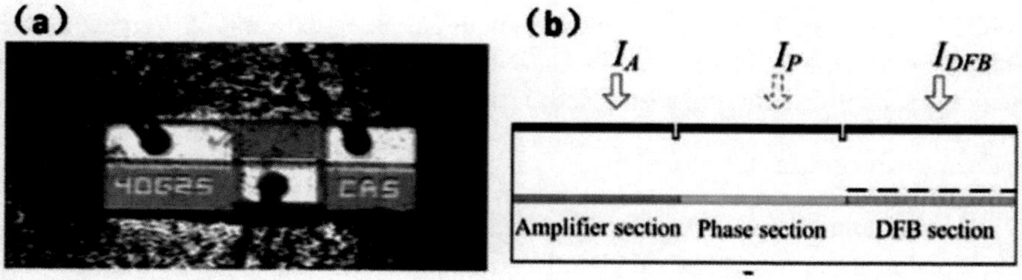

Fig. 2.4 Three-section monolithic integrated semiconductor laser chip consisting of DFB laser area, phase area and amplification area **a** physical view; **b** schematic diagram

4. **Single-cavity three-segment structure**

In 2013, Wu Jiagui of Southwest University and Zhao Lingjuan of Institute of Semiconductors, Chinese Academy of Sciences jointly developed a three-stage monolithic integrated chaotic semiconductor laser chip [10]. As shown in Fig. 2.4, the chip contains a DFB laser region, a phase region, an amplification region, and a high reflection film is coated on one end to form an optical feedback cavity [11]. Quantum well fusion (QWI) technology is used to reduce the absorption loss in the phase region, and high reflectivity is produced by coating high reflective film. The chaotic laser output is realized by controlling the feedback light intensity and the feedback phase in the amplification region and the phase region respectively.

5. **Two-dimensional external cavity structure**

In 2014, S. Sunada and M. Adachi from Kanezawa University, T. Fukushima from Okayama Prefectural University, S. Shinohara from NTT and T. Hirayama from Toyo University and others have jointly developed a chaotic semiconductor laser chip with two-dimensional external cavity structure [12]. As shown in Fig. 2.5, the chip includes a laser part and a two-dimensional external cavity part, the size of which is less than 230 μm × 1 mm. Among them, the two-dimensional external cavity can make the laser produce multiple feedback to produce a larger optical delay, and the chaotic laser is generated by controlling the feedback intensity by injecting the current into the external cavity.

6. **Mutual injection coupling structure**

In 2014, Sun Changzheng of Tsinghua University developed an ultra-short delay time intercoupled monolithic integrated chaotic semiconductor laser chip [13]. As shown in Fig. 2.6, the chip consists of two mutually coupled DFB lasers and a phase region. Chaotic signals can be generated by changing the phase region current and the laser bias current.

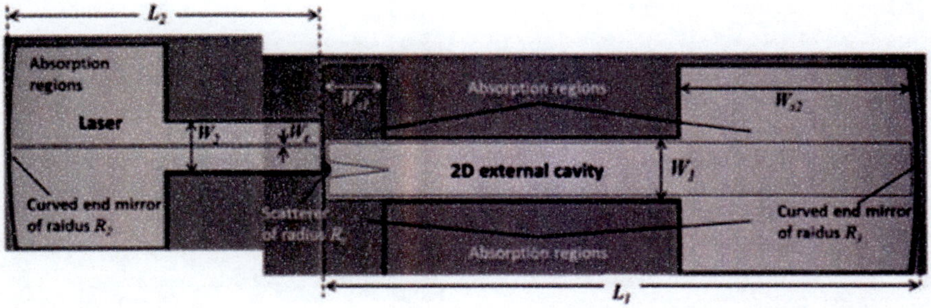

Fig. 2.5 Chaotic laser chip with two-dimensional external cavity structure

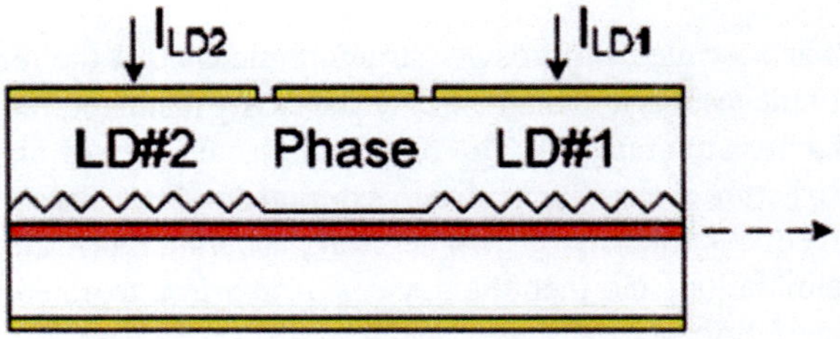

Fig. 2.6 Schematic diagram of the structure of a mutually coupled monolithic integrated chaotic laser chip

2.1.2 *Hybrid Integrated Chaotic Semiconductor Lasers*

Hybrid integrated chaotic semiconductor lasers are fabricated by combining discrete laser chips, disturbance elements and other related functional components on the same substrate. A hybrid integrated chaotic semiconductor laser has been developed by Taiyuan University of Technology and Institute of Semiconductors, Chinese Academy of Science [14]. The structure diagram and physical diagram of the integrated chaotic semiconductor laser are shown in Fig. 2.7, which adopts the coupling form of the DFB laser chip, collimating lens, reflector, focusing lens and optical fiber module, wherein the reflector provides single cavity feedback to the DFB laser chip, and the collimating lens and focusing lens shape the optical path to improve the coupling efficiency. The beam is output by the tail fiber of the optical fiber module. Chaotic lasers with sequential peaks of 30 mV, spectrum width greater than 5 GHz and no time delay information can be generated, which will be described in detail in Sect. 2.2.

2.2 Hybrid Integrated Chaotic Semiconductor Lasers

Hybrid integration has the advantages of simple structure, easy integration and low cost, so it is indispensable in the development of integrated chaotic semiconductor lasers. In general, the short cavity optical feedback structure is easy to integrate [15].

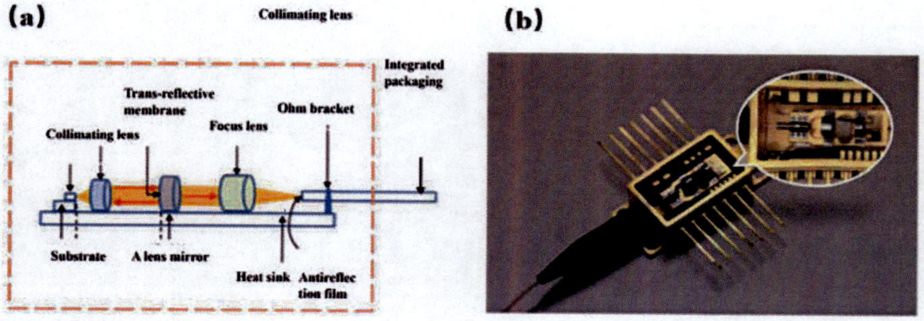

Fig. 2.7 Hybrid integrated chaotic semiconductor laser **a** schematic diagram of the structure; **b** physical drawing

The so-called short cavity structure means that the relaxation oscillation frequency of the laser is less than the external cavity resonant frequency, otherwise it is the long cavity structure [5]. The output characteristics of chaotic lasers with short cavity structure are very sensitive to external conditions such as feedback cavity length and feedback intensity [16]. Therefore, accurate fabrication parameters are obtained by simulation, and then the devices of hybrid integrated chaotic semiconductor lasers are fabricated.

2.2.1 Parameter Extraction and System Simulation

The internal parameters of semiconductor laser chips, such as linewidth enhancement factor, carrier lifetime and photon lifetime, have a great influence on the dynamic characteristics of semiconductor lasers, and different lasers need different external conditions to generate chaos. In general, the typical internal parameters of the laser [17–19] are used in the simulation rate equation, and the 44 results only represent the typical results, which cannot provide accurate parameters for the fabrication of hybrid integrated chaotic semiconductor lasers. Therefore, we should first extract the real values of the internal parameters of the semiconductor laser chip, and then theoretically simulate the dynamic characteristics of hybrid integrated chaotic semiconductor lasers to guide the practical fabrication of hybrid integrated chaotic semiconductor lasers.

1. **Model design**

Figure 2.8 shows a schematic diagram of a short-cavity hybrid integrated chaotic semiconductor laser based on optical feedback. The semiconductor laser chip uses a DFB laser chip, and the reflector partially transmits and reflects light. Part of the laser emitted by the semiconductor laser chip goes back to the chip through the reflection of the reflector and disturbs it to produce a chaotic laser, while the other part of the laser is coupled into the optical fiber through the transmission of the reflector. The chaotic laser generated is finally output by the tail fiber.

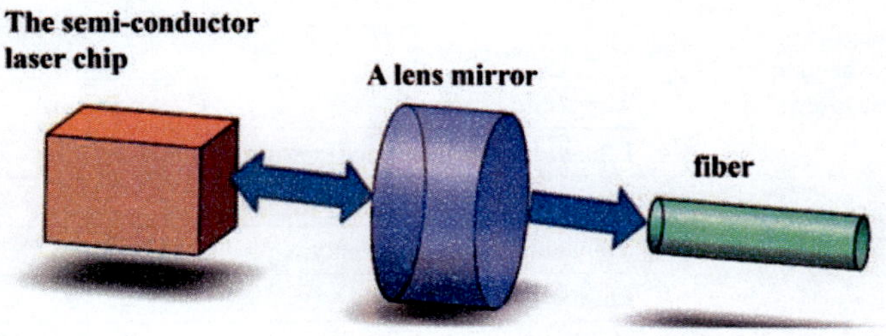

Fig. 2.8 Schematic diagram of a short-cavity hybrid integrated chaotic semiconductor laser based on optical feedback

Before fabrication, the intensity reflectivity of the reflector and the optimal length of the external feedback cavity need to be determined by theoretical simulation in order to ensure that the hybrid integrated chaotic semiconductor laser can output a stable chaotic laser. Combined with the rate equation of semiconductor laser, the rate equation of hybrid integrated chaotic semiconductor laser based on external optical feedback short cavity can be derived, that is, the famous L-K equation [20, 21]· The equation reflects the dynamic changes of photon density S(t), carrier density N(t) and phase φ(t) with time in the active cavity of the semiconductor laser:

$$\frac{dS(t)}{dt} = \left[\frac{\Gamma \times g_0[N(t) - N_0]}{1 + \varepsilon S(t)} - \frac{1}{\tau_p}\right] \times S(t) + \frac{2K_{ap}}{\tau_{in}} \times \sqrt{S(t - \tau_t)S(t)} \times \cos\theta(\mathrm{t}) + \frac{\Gamma \times \beta_{sp} \times N(t)}{\tau_c} \tag{2.1}$$

$$\frac{dN(t)}{dt} = \frac{I(t)}{q \times V_{act}} - \frac{N(t)}{\tau_c} - \frac{g_0 \times [N(t) - N_0]}{1 + \varepsilon E^2(t)} \times E^2(t) \tag{2.2}$$

$$\frac{d\phi(t)}{dt} = \frac{\alpha}{2}\left[\frac{\Gamma g_0[N(t) - N_0]}{1 + \varepsilon S(t)} - \frac{1}{\tau_p}\right] - \frac{K_{ap}}{\tau_{in}} \times \sqrt{\frac{S(t - \tau_t)}{S(t)}} \times \sin\theta(t) \tag{2.3}$$

$$Kap = (1 - R_2) \times \sqrt{\frac{R_3}{R_2}} \tag{2.4}$$

$$\theta(t) = \omega\tau_t + \varphi(\mathrm{t}) - \varphi(t - \tau_t) \tag{2.5}$$

Equation (2.4) indicates the feedback intensity, where $R_2 = 0.3$ represents the intensity reflectivity of the output surface of the chaotic semiconductor laser chip, and R_3 represents the intensity reflectivity of the coating on the reflector, which is a variable parameter. $\tau_{in} = 7.1 \times 10^{-12}$ s in the sum of Eqs. (2.1) and (2.3) represents the round trip time of light in the active region of the chaotic semiconductor laser chip; $\theta(t) = \omega\tau_t + \varphi(t) - \varphi(t - \tau_t)$ in Eqs. (2.1) and (2.3) represents the feedback light phase, where ω is the angular frequency of the static laser, φ(t) represents the electric field phase, and $\tau_t = (2n_{ext}L)/C$ is the external cavity feedback delay time, C

Table 2.1 Internal parameters of the chaotic semiconductor laser chip

Internal parameters	Units
Threshold current I_{th}	mA
Linewidth enhancement factor α	
Threshold carrier density N_{th}	m^{-3}
Transparent carrier density N_0	m^{-3}
Photon lifetime τ_p	s
Carrier lifetime τ_c	s
Optical confinement factor Γ	
Differential gain g_0	m^3/s
Gain saturation coefficient ε	m^3

= 3.0 × 10^8 m/s represents the speed of light in vacuum, L represents the length of the external feedback cavity, and next represents the refractive index of the external cavity. In Eq. (2.2), $V_{act} = 4.8 \times 10^{-18}$ m^3 is the active region volume of chaotic semiconductor laser chip, $q = 1.602 \times 10^{-19}$ C is the electron charge, and I(t) is the bias current.

It can be seen from the above equations that the rate equation of hybrid integrated the chaotic semiconductor lasers contains many internal parameters of semiconductor lasers, which need to be extracted as shown in Table 2.1.

It is worth noting that the internal parameters of the semiconductor laser chip are closely related to the relaxation oscillation frequency, damping coefficient and threshold current. The internal parameters of the semiconductor laser chip satisfy the following formula derivation [22–25].

When the bias current is equal to the threshold current, the rate Eqs. (2.1) and (2.2) indicate that the semiconductor laser is in a steady state, dN(t)/dt and $dS(t)/dt$ are equal to zero, and the electric field amplitude E(t) is approximately zero, then the formula (2.2) can be approximately equal to

$$0 = \frac{I_{th}}{qV_{act}} - \frac{N_{th}}{\tau_c} \tag{2.6}$$

When the bias current is equal to the threshold current, the spontaneous emission is very small and can be ignored.

$$0 = \left[\frac{\Gamma g_0[N(t) - N_0]}{1 + \varepsilon S(t)} - \frac{1}{\tau_p}\right] \times S(t) + \frac{\Gamma \beta_{sp} N(t)}{\tau_c} \tag{2.7}$$

It is also because at the threshold current, εS(t) is very small and can be ignored, and Eq. (2.7) can be simplified.

$$0 = \Gamma g_0[N_{th} - N_0] - \frac{1}{\tau_p} \tag{2.8}$$

From the above relationship, we can see that the internal parameters of the semiconductor laser are interrelated, and the other internal parameters can be calculated by the linear relationship from several known parameters. The internal parameters of the semiconductor laser satisfy the following formula:

$$2\pi f_r = \left[\frac{\Gamma g_0[I(t) - I_{th}]}{qV_{act}}\right]^{\frac{1}{2}} \tag{2.9}$$

$$\upsilon_o = \frac{1}{\tau_c} + Kf_{ro}^2 \tag{2.10}$$

$$K = 4\pi^2(\tau_p + \tau_n) \tag{2.11}$$

$$\tau_n = (2\pi f_c)/(2\pi f_{ro})^2 \tag{2.12}$$

f_r, v_o and f_c represent the relaxation oscillation frequency, damping coefficient and chirp frequency of semiconductor lasers at threshold current, respectively. According to the above relationship, the P-I characteristic curve and frequency response curve of the chip in steady state can be measured by experiments, and the internal parameters of the semiconductor laser chip can be solved indirectly by linear equation.

2. **P-I characteristic of the semiconductor laser chip**

Since the semiconductor laser satisfies the formula (2.6) and the formula (2.8) when the bias current is equal to the threshold current, the following relationship can be derived:

$$I_{th} = \frac{qV_{act}N_{th}}{\tau_c} = \frac{qV_{act}}{\tau_c} \times \left(N_0 + \frac{1}{\Gamma g_0 \tau_p}\right) \tag{2.13}$$

The steady-state photon Eq. (2.1) is substituted into the carrier Eq. (2.2), and the term can be simplified and shifted.

$$S(t) = \frac{\Gamma I \tau_P}{qV_{act}} - \frac{\tau_p}{\tau_c}(1 - \beta_{sp})\Gamma N(t) \tag{2.14}$$

The equation related to the carrier density can be obtained by shifting the term of the carrier Eq. (2.2).

$$N(t) = \frac{N_{th}}{1 + \frac{\beta_{sp}}{g_0 \tau_c}\frac{1+\varepsilon S(t)}{S(t)}} \tag{2.15}$$

Equation (2.15) can be replaced by the Eq. (2.14), and the Eq. (2.13) can be obtained.

$$\frac{qV_{act}\beta_{sp}}{g_0\tau_c\tau_p\Gamma} + \frac{qV_{act}}{\tau_p}\left(1 + \frac{\beta_{sp}\varepsilon}{g_0\tau_c}\right)\frac{S}{\Gamma} = \left(1 + \frac{\beta_{sp}\varepsilon}{g_0\tau_c}\right)I + \frac{\Gamma}{S}\frac{\beta_{sp}\varepsilon}{g_0\tau_c}I - (1 - \beta_{sp})I_{th} \tag{2.16}$$

The leakage current Is of a semiconductor laser is:

$$I_s = \frac{qV_{act}\beta_{sp}}{g_0\tau_c\tau_p\Gamma} \tag{2.17}$$

Substituting Eq. (2.17) into Eq. (2.16), we can get

$$I_s + \frac{qV_{act}}{\tau_p}\left(1 + \frac{\beta_{sp}\varepsilon}{g_0\tau_c}\right)\frac{S}{\Gamma} = \left(1 + \frac{\beta_{sp}\varepsilon}{g_0\tau_c}\right)I + \frac{\Gamma}{S}\frac{1}{\frac{qV_{act}}{\tau_c}}I_sI - (1 - \beta_{sp})I_{th} \tag{2.18}$$

Since the β_{sp} and $(\beta_{sp\varepsilon})/(g_0\tau_c)$ of the semiconductor laser is far less than 1 at the threshold current, Eq. (2.18) can be simplified as

$$(FP)^2 - (I - I_s - I_{th})FP - I_sI = 0 \tag{2.19}$$

In the formula, $F = (2q\lambda)/\eta hc$, where Q is the electron charge; λ is the working wavelength of the semiconductor laser, which is 1550 nm in this chapter; η is the differential quantum efficiency of the semiconductor laser; h is the Planck constant, $h = 6.626 \times 10^{-34}$ J s, c is the speed of light propagating in vacuum, $c = 3.0 \times 10^8$ m/s.

From the Eq. (2.19), the output power P of the semiconductor laser can be expressed as

$$P = \frac{I - I_{th} - I_s + \sqrt{(I - I_{th} - I_s)2 + 4I_sI}}{2F} \tag{2.20}$$

Equation (2.20) is the linear equation satisfied by the P-I curve of the semiconductor laser in a steady state. From the P-I curve, the threshold current, differential quantum efficiency and leakage current of the semiconductor laser can be obtained, and the threshold current and leakage current satisfy the formula (2.13) and Eq. (2.17), respectively.

Figure 2.9 shows the experimental device for semiconductor laser to measure the P-I characteristic curve. Part a is a free moving platform, which is used to fix and move the chip; part b is the image of the chip, and the bias current is loaded by the probe. The right half in Fig. 2.9 is a low noise DC source and an optical power meter, which are used to provide DC current to the chip and to record the output power of the chip, respectively.

By changing the bias current, we test the P-I characteristic curves of the two chips, and the experimental results are shown in Fig. 2.10.

From Fig. 2.10, it can be concluded that the threshold current of semiconductor laser chip 1 is 11.3 mA, and that of chip 2 is 11.4 mA. Comparing with Fig. 2.10a and

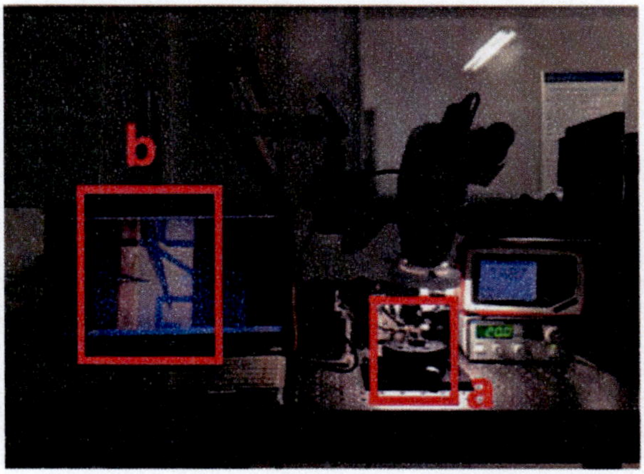

Fig. 2.9 Experimental setup for measuring the P-I characteristic curves

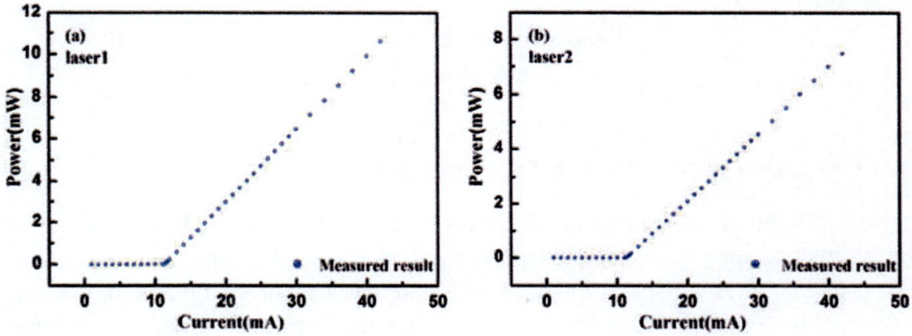

Fig. 2.10 Experimental test results for P-I characteristic curves of two semiconductor laser chips **a** semiconductor laser chip 1; **b** semiconductor laser chip 2

b, it can be seen that when the bias current of semiconductor laser chip 2 is greater than the threshold current, slope of the P-I curve of semiconductor laser chip 2 is larger than that of semiconductor laser chip 1. This means that the external differential gain efficiency of semiconductor laser chip 2 is larger than that of semiconductor laser chip 1. The experimental test data are fitted by the least square method according to the Eq. (2.20), and the fitting results are shown in Fig. 2.11.

Figure 2.11a shows the fitting result of semiconductor laser chip 1, and Fig. 2.11b shows the fitting result of semiconductor laser chip 2. The blue curve in the picture is the experimental test curve, and the red curve is the fitting curve. The fitting curve is in good agreement with the experimental curve. The fitted parameters are shown in Table 2.2.

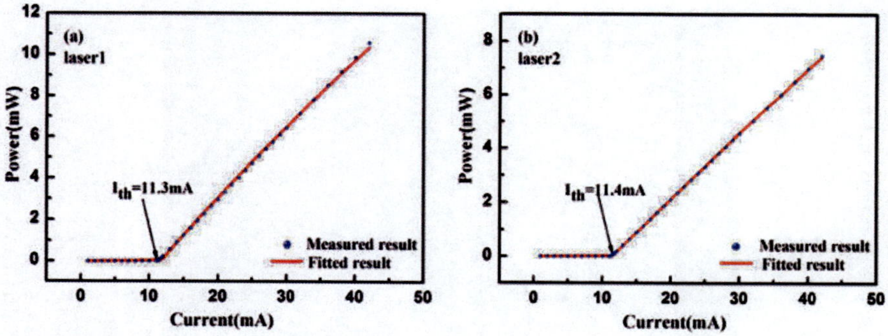

Fig. 2.11 Fitting results for the P-I characteristic curves of two semiconductor laser chips **a** semiconductor laser chip 1; **b** semiconductor laser chip 2

Table 2.2 P-I characteristic curve fitting results for semiconductor laser chips

Fit parameters	Chip 1	Chip 2	Units
Threshold current I_{th}	11.3	11.4	mA
Leakage current I_s	1.27×10^{-6}	0.367×10^{-6}	A
Intermediate parameter F	4.126×10^{23}	2.894×10^{23}	

3. **Small signal power testing for the semiconductor laser chip**

Intensity modulation response is one of the most important characteristics to reflect the modulation speed of semiconductor laser in optical wave communication system. The principle of small signal modulation frequency response is that when the bias current of the semiconductor laser is greater than the threshold current, at the same time, DC bias current and continuous sine wave modulation current signal are added to the semiconductor laser. The output power signal of the semiconductor laser will change with the frequency and power of the continuous sine wave modulated current signal. When the DC bias current of the semiconductor laser changes, the output power of the semiconductor laser will have different frequency response characteristics. Figure 2.12 is a schematic diagram of small signal current modulation of semiconductor lasers. The commonly used instruments for testing the frequency response curve of semiconductor lasers are microwave vector network analyzer and optical wave element analyzer. The microwave vector network analyzer is used here, which is collectively referred to as the vector network analyzer.

From the rate Eqs. (2.1) and (2.2), the steady-state solution can be expressed as Eqs. (2.7) and (2.21) when the bias current is very small. When a small signal modulation is added to the semiconductor laser and the semiconductor laser works greater than the threshold current, the spontaneous emission signal of the semiconductor laser can be ignored, and the photon density of the semiconductor laser can be expressed by the Eq. (2.22). The details are as follows:

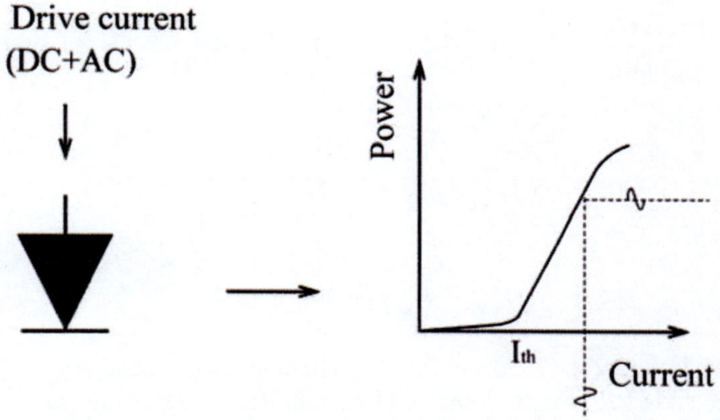

Fig. 2.12 Schematic diagram of small signal current modulation for semiconductor lasers

$$0 = \frac{I(t)}{qV_{act}} - \frac{N(t)}{\tau_c} - \frac{g_0[N(t) - N_0]}{1 + \varepsilon S(t)} S(t) \tag{2.21}$$

$$S(t) = \frac{\tau_p \Gamma}{qV_{act}} (I(t) - I_{th}) \tag{2.22}$$

The bias current of the semiconductor laser is set be $I_0(I_0 > I_{th})$, the photon density of the semiconductor laser S_0, and the carrier density N_0. $\Delta I(t)$ represents the small signal modulation current, then the total injection current of the semiconductor laser is expressed by the Eq. (2.23), the total photon density is expressed by the Eq. (2.24), and the total carrier density is expressed by the Eq. (2.25), as follows:

$$I(t) = I_0 - \Delta I(t) \tag{2.23}$$

$$S(t) = S_0 - \Delta S(t) \tag{2.24}$$

$$N(t) = N_0 - \Delta N(t) \tag{2.25}$$

According to the linear rate equation, the Taylor series expansion of the optical gain of a semiconductor laser can be expressed as follows:

$$g(N, S) = g(N_0, S_0) + \frac{\partial g}{\partial N} \Delta N(t) + \frac{\partial g}{\partial S} \Delta S(t) \ldots \tag{2.26}$$

Then substitute Eqs. (2.23) to (2.26) into Eqs. (2.1) and (2.2), simplify and shift the terms to obtain the following relations:

$$\frac{d\Delta S}{dt} = \left[\Gamma g_0 + \Gamma \frac{\partial g}{\partial S} S_0 - \frac{1}{\tau_p}\right] \Delta S(t) + \left(\Gamma \frac{\partial g}{\partial N} S_0 + \frac{\Gamma \beta_{sp}}{\tau_c}\right) \Delta N(t) \tag{2.27}$$

$$\frac{d\Delta N}{dt} = \frac{\Delta I(t)}{qV_{act}} - \left(\frac{1}{\tau_c} + \frac{\partial g}{\partial N}S_0\right)\Delta N(t) - \left(g_0 + \frac{\partial g}{\partial S}S\right) \times \Delta S(t) \tag{2.28}$$

Moreover:

$$\Gamma g_0 - \frac{1}{\tau_p} = -\frac{\beta_{sp}N_0}{S_0\tau_c} \approx 0 \tag{2.29}$$

$$\Gamma\frac{\partial g}{\partial S}S_0 \approx 0 \tag{2.30}$$

Then the Eqs. (2.27), (2.29) and (2.30) can be combined to obtain Eq. (2.31), from Eqs. (2.28) and (2.29) can be combined to obtain Eq. (2.32), as follows:

$$\frac{d\Delta S}{dt} = \left(\Gamma\frac{\partial g}{\partial N}S_0\right)\Delta N(t) \tag{2.31}$$

$$\frac{d\Delta N}{dt} = \frac{\Delta I(t)}{qV_{act}} - \left(\frac{1}{\tau_c} + \frac{\partial g}{\partial N}S_0\right)\Delta N(t) - \frac{1}{\Gamma\tau_P} \times \Delta S(t) \tag{2.32}$$

The bias current, photon density and carrier density of a semiconductor laser are composed of amplitude (real part) and frequency (imaginary part), which are expressed as Eqs. (2.33), (2.34) and (2.35), respectively, where Ω denotes the angular frequency:

$$\Delta I(t) = \Re\{\Delta I(t)\exp(j\Omega t)\} \tag{2.33}$$

$$\Delta S(t) = \Re\{\Delta S(t)\exp(j\Omega t)\} \tag{2.34}$$

$$\Delta N(t) = \Re\{\Delta N(t)\exp(j\Omega t)\} \tag{2.35}$$

and (2.31) and (2.32) can be combined to obtain the following equation:

$$j\Omega\Delta S = \left(\Gamma\frac{\partial g}{\partial N}S_0\right)\Delta N(t) \tag{2.36}$$

$$j\Omega\Delta N = \frac{\Delta I(t)}{qV_{act}} - \left(\frac{1}{\tau_c} + \frac{\partial g}{\partial N}S_0\right)\Delta N(t) - \frac{1}{\Gamma\tau_P}\Delta S(t) \tag{2.37}$$

also because

$$f_{ro} = \left(\frac{\frac{\partial g}{\partial N}S_0}{\tau_p}\right)^{\frac{1}{2}} = \left(\frac{\Gamma\frac{\partial g}{\partial N}(I_t - I_{th})}{qV_{act}}\right)^{\frac{1}{2}} = \left[\frac{\Gamma g_0[I(t) - I_{th}]}{qV_{act}}\right]^{\frac{1}{2}} \tag{2.38}$$

$$\nu = \frac{1}{2}\left(\frac{1}{\tau_c} + \frac{\partial g}{\partial N} S_0\right) = \frac{1}{2}\left(\frac{1}{\tau_c} + \frac{g_0 \Gamma \tau_p}{q V_{act}}(I_t - I_{th})\right) = \frac{1}{2}\left(\frac{1}{\tau_c} + \tau_p \times f_{ro}^2\right) \tag{2.39}$$

$$\frac{1}{\tau_c} + \frac{\partial g}{\partial N} S_0 = 2v \tag{2.40}$$

In Eq. (2.38), f_{ro} represents the relaxation oscillation frequency of the semiconductor laser, and in Eq. (2.39), v represents the damping factor of the semiconductor laser. The output and input current of the semiconductor laser can be obtained by combining the above equations.

$$\frac{\Delta S(t)}{S_0} = \frac{4\pi^2 f_{ro}^2}{jv\Omega + 4\pi^2 f_{ro}^2 - \Omega^2} \frac{\Delta I(t)}{I_0} \tag{2.41}$$

Simplifying the imaginary part j, the transfer function of the semiconductor laser under small signal modulation is Eq. (2.42), where P1 is the small signal output to input ratio, which is the dependent variable:

$$|P_1| = \left(\frac{f_{ro}^2}{\sqrt{(v_1\Omega)^2 + ((2\pi f_{ro})^2 - \Omega^2)^2}}\right)^2 \tag{2.42}$$

Since $\Omega = 2\pi f$, where f denotes the modulation frequency of the sinusoidal modulated signal, which is the independent variable, the relationship between the output power of the semiconductor laser and the input signal when the power is in the logarithmic domain can be expressed as

$$|P_1| = 20\,lg\left(\frac{f_{ro1}^2}{\sqrt{\left(\frac{v_1 f}{2\pi}\right)^2 + (f_{ro1}^2 - f^2)^2}}\right) \tag{2.43}$$

It is a more traditional method to test the frequency response curve of a semiconductor laser chip with a vector network analyzer, but because the connection between the chip and the instrument requires a chip fixture, which introduces parasitic parameters and makes the test results inaccurate, the vector network analyzer should be calibrated before use. Figure 2.13 shows the experimental test setup for the frequency response of a semiconductor laser chip. The fiber and erbium-doped fiber amplifier in the figure will be used to measure the frequency response curve of the chaotic semiconductor laser chip in the fiber, and the AC signal power provided by the vector network analyzer is – 20 dBm during the experiment. A low-noise driving DC current source (ILX Lightwave, LDX3412) and a vector network analyzer (ROHDE&SCHWARZ, ZVA24, bandwidth 10 MHz–24 GHz) provide DC current and AC signal to the semiconductor laser chip, respectively. The DC current and AC signal are used to supply the drive current to the semiconductor laser chip through the vector network analyzer at the same time. The current source and the vector network analyzer are connected to the semiconductor laser chip through a high frequency probe (Cascade Microtech, ACP40-GS-200, frequency coverage from DC to 40 GHz). When DC and AC drives are provided to the semiconductor laser chip, the chip emits a continuous optical signal, which is converted into an electrical signal by a photodetector (Finisar, XPDV2120RA, 50 GHz bandwidth). The frequency response curves of semiconductor laser chip 1 and semiconductor laser chip 2 were tested separately, as shown in Fig. 2.14.

Figure 2.14a shows the frequency response curves of the semiconductor laser chip 1 when the bias currents are 13 mA, 16 mA, 22 mA, and 28 mA, respectively. The peak on the curve is caused by the modulation frequency value close to the relaxation oscillation frequency of the semiconductor laser chip. From the figure, it can be seen that the relaxation oscillation frequency of chip 1 increases with the increase of bias

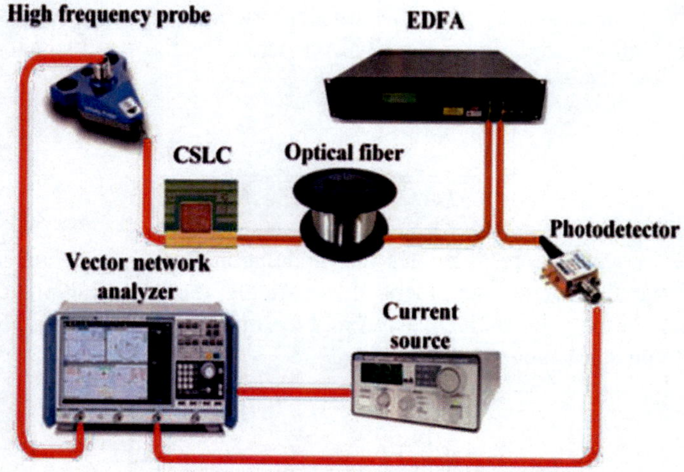

Fig. 2.13 Experimental setup for measuring frequency response

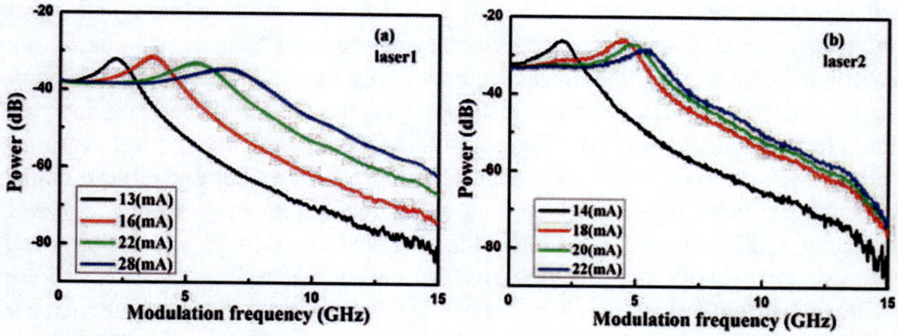

Fig. 2.14 Frequency response curves of semiconductor laser chips at different bias currents. **a** Semiconductor laser chip 1; **b** semiconductor laser chip 2

current. Figure 2.14b shows the frequency response curves of the semiconductor laser chip 2 when the bias currents are 14 mA, 18 mA, 20 mA and 22 mA, respectively. Comparing the frequency response curves of chip 1 and chip 2, the 3 dB bandwidth is narrower (the 3 dB bandwidth is the frequency point where the frequency responsiveness of the laser changes up to 3 dB relative to the low frequency responsiveness of the laser), and the high frequency response curve of semiconductor laser chip 1 is flatter and smoother, which means that the high frequency response performance of semiconductor laser chip 1 is better. In order to eliminate the problem of uneven frequency response caused by electronic components in the experiment, we used the deduction method in the process of extracting parameters. The subtraction method is to subtract the frequency response curve of the semiconductor laser chip with a larger bias current from the frequency response curve with a smaller bias current, and the process of subtraction can eliminate the systematic errors in the system that do not change with the change of the bias current. In the previous derivation, the transfer function of the frequency response satisfies the Eq. (2.43), and the transfer function after the phase subtraction can be derived as

$$|P_2 - P_1| = 20lg\left(\frac{f_{ro2}^2\sqrt{\left(\frac{v_1 f}{2\pi}\right)^2 + (f_{ro1}^2 - f^2)^2}}{f_{ro1}^2\sqrt{\left(\frac{v_2 f}{2\pi}\right)^2 + (f_{ro2}^2 - f^2)^2}}\right) \tag{2.44}$$

The results of the frequency response curve fitting after subtraction are shown in Fig. 2.15, in which the blue curve represents the test result and the red curve is the fitting result. Figure 2.15a describes the fitting results of the semiconductor laser chip 1 using the subtraction method. The three curves in the figure are the frequency response curves of the bias current at 16 mA, 22 mA, 28 mA minus the bias current of 13 mA; fig. 2.15b describes the fitting results of the semiconductor laser chip 2 using the subtraction method. The three curves in the figure are the frequency

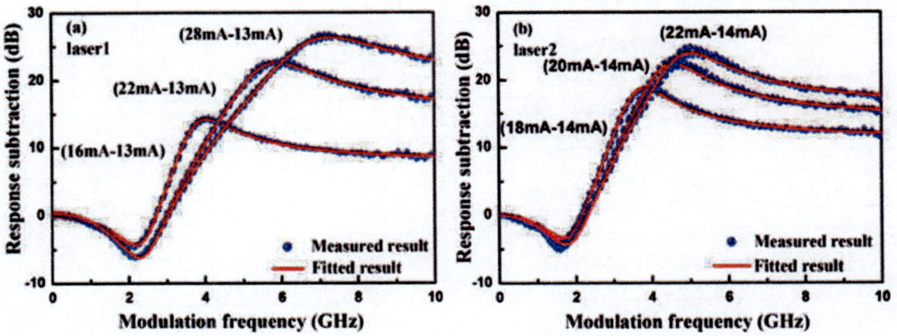

Fig. 2.15 Fitting results for the frequency response curve of a semiconductor laser chip. **a** The fitting of the frequency response curve of semiconductor laser chip 1 at a higher bias current minus 13 mA; **b** the fitting of the frequency response curve of semiconductor laser chip 2 at a higher bias current minus 14 mA

response curves of the bias current at 18 mA, 20 mA, 22 mA minus the bias current of 14 mA. It can be seen from the diagram that the system error is eliminated after the deduction method is used. The test results of the curve are in good agreement with the fitting results, which can ensure that the internal parameters of the extracted chip are accurate. It is worth noting that the relaxation oscillation frequency f_{ro} of the semiconductor laser chip obtained by fitting Eq. (2.44) is an approximate value, and the exact value of the relaxation oscillation frequency f_r satisfies the following equation:

$$f_r = \frac{1}{2\pi}\sqrt{f_{ro}^2 - 0.5v^2} \tag{2.45}$$

By fitting, the parameters of the semiconductor laser chip 1 are shown in Table 2.3, and the parameters of the semiconductor laser chip 2 are shown in Table 2.4.

The squared f_{2r} of the relaxation oscillation frequency of the semiconductor laser chip and the damping coefficient v are linearly related, by recording the relaxation oscillation frequency f_r of the semiconductor laser chip at different bias currents and the corresponding damping coefficient v, a curve can be obtained, and a linear fit to the curve can give the slope K. This K value is the K-factor [23], from Eq. (2.11) it can be seen that K and the semiconductor laser Based on the above experimental results, the fitted K results are shown in Fig. 2.16. The experimental results of the two semiconductor laser chips in the figure are basically a straight line, which basically matches the fitted straight line, with K = 0.3 ns for semiconductor laser chip 1 and K = 0.5 ns for semiconductor laser chip 2.

4. **Modulation response of the chaotic semiconductor laser chip in fiber optic systems**

The light wave signal generated by the DFB laser is not strictly monochromatic, and the spectrum of this output signal has a central frequency and a certain line

Table 2.3 Fitting results of frequency response curve of semiconductor laser chip 1

Fit parameters	Bias current					
	13 mA	16 mA	22 mA	28 mA	34 mA	40 mA
f_{ro} (× 10^9)	2.4148	3.815476	5.643	6.942	8.032	8.9
f_r (× 10^9)	2.3	3.708	5.474	6.681	7.673	8.463
v (× 10^9)	6.53	7.981481	12.182	16.751	21.092	24.474

Table 2.4 Fitting results of frequency response curve of semiconductor laser chip 2

Fit parameters	Bias current					
	14 mA	20 mA	21 mA	22 mA	23 mA	24 mA
f_{ro} (× 10^9)	2.23	4.471621	4.68572	4.93644	5.13784	5.367319
f_r (× 10^9)	2.127	4.38665	4.58	4.817677	5	5.2
v (× 10^9)	6.16357	7.709192	8.78119	9.563708	10.71857	11.7366

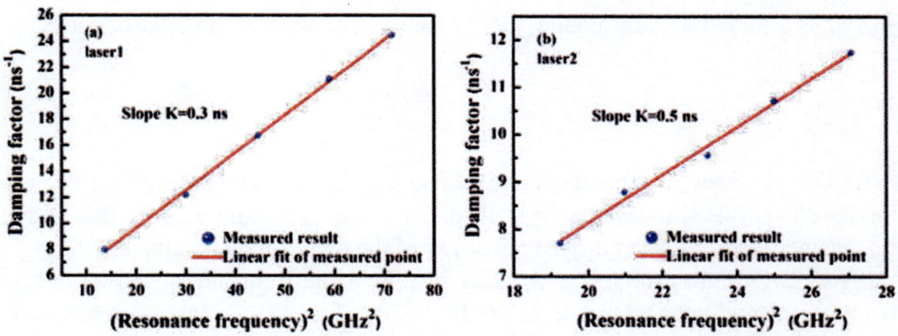

Fig. 2.16 Plot of the relationship between the damping coefficient and the squared relaxation oscillations of the semiconductor laser. **a** Fitting results for a slope K-factor equal to 0.3 ns for semiconductor laser chip 1; **b** fitting results for a slope K-factor equal to 0.5 ns for semiconductor laser chip 2

width. When a semiconductor laser is modulated with an injected current, the center frequency of the output light wave signal varies with the output power, which is called frequency chirp. Because of the frequency chirp characteristic, the linewidth of the light wave becomes wider when the DFB laser is modulated. When the chirp signal is transmitted through the dispersive fiber, the signal carried by the light wave will be distorted and the output power through the fiber will be affected, and there is a linear relationship between the chirp frequency of the chip and the internal parameters of the laser, so it is necessary to study the characteristics of the semiconductor laser in the fiber system by the small signal intensity modulation response.

The direct intensity modulated signal characteristics of the input fiber are

$$\Delta S_{\text{in}}(j\omega) = \left(\frac{\tau_p}{q}\right) H(j\omega)\Delta I(j\omega) \tag{2.46}$$

where H(jω) is equal to the transfer function in Eq. (2.41), then the output fiber modulation response is

$$\Delta S_{out}(j\omega) = \left[\cos(F_D\omega^2) - \alpha\left(1 - j\frac{\omega_c}{\omega}\right)\sin(F_D\omega^2)\right]\left(\frac{\tau_p}{q}\right)H(j\omega)\Delta I(j\omega) \tag{2.47}$$

Then, the transfer function of the modulated response in the fiber is given by Eqs. (2.46) and (2.47)

$$\begin{aligned} H_{fiber,dB} &= \Delta P_{fiber,dB} - \Delta P_{LD,dB} \\ &= 20\ \lg\left[\cos F_D\omega^2 - \alpha\left(1 - j\frac{\omega_c}{\omega}\right)\sin F_D\omega^2\right]\left(\frac{\tau_p}{q}\right)H(j\omega)\Delta I(j\omega) \\ &= 20\ \lg\sqrt{\cos^2 F_D(2\pi f)^2 + \alpha^2\left(1 + \frac{f_c^2}{f^2}\right)\sin^2 F_D(2\pi f)^2 - \alpha\sin 2F_D(2\pi f)^2} \end{aligned} \tag{2.48}$$

where f is the modulation frequency, f_c is the chirp frequency of the laser chip,

$$F_D = \frac{\lambda^2 D L_{fibre}}{4\pi c} \tag{2.49}$$

where D is the fiber dispersion coefficient, L_{fibre} is the fiber length, the fiber dispersion coefficient used in this paper is 17 ps/km, L_{fibre} is 50 km, and the line width enhancement factor α and chirp frequency f_c of the chip can be obtained by fitting.

Figure 2.17 shows the test and fitting results of the frequency response of the chaotic semiconductor laser chip in the fiber. The blue curve is the test result, and the red curve is the fitting result. Figure 2.17a shows the fitting result of chaotic semiconductor laser chip 1, and Fig. 2.17b shows the fitting result of chaotic semiconductor laser chip 2. In order to eliminate systematic errors in the experimental system, the same deduction method is used here. According to the experimental results, the frequency response of the chip at the same bias current is subtracted from the response of the chaotic semiconductor laser chip 1 at a bias current of 16 mA in the fiber, and the final curve obtained satisfies the Eq. (2.45). Similarly, by subtracting the frequency response of the chip at the same bias current from the response of the chaotic semiconductor laser chip 2 at 17 mA in the fiber, the final curve obtained satisfies Eq. (2.47), and the chirp frequency and linewidth enhancement factor of chip 2 at 17 mA can be obtained by fitting the results.

5. **Chip internal parameters results and verification**

The direct parameters of the two chips were obtained by testing the P-I curve and frequency response curve of the semiconductor laser chip, as shown in Table 2.5. The extracted parameters were calculated according to the following formulas.

$$N_{th} = \frac{\tau_c I_{th}}{q V_{act}} \tag{2.50}$$

$$0 = \Gamma g_0 [N_{th} - N_0] - \frac{1}{\tau_p} \tag{2.51}$$

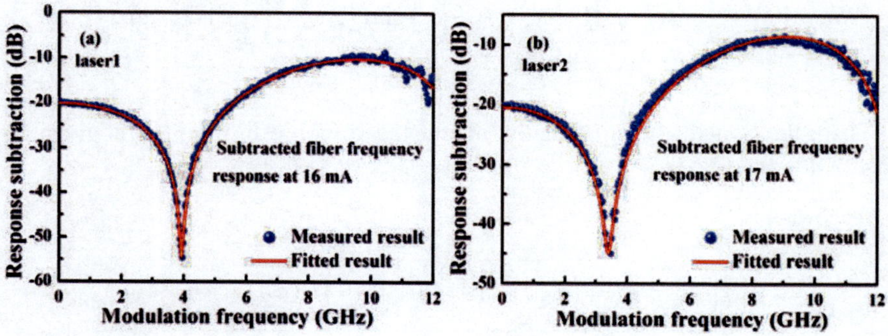

Fig. 2.17 The fitting of frequency response curves for chaotic semiconductor laser chips in optical fibers. **a** Chaotic semiconductor laser chip 1; **b** chaotic semiconductor laser chip 2

$$(2\pi f_r)^2 = \frac{\Gamma g_0[I_t - I_{th}]}{qV_{act}} \tag{2.52}$$

$$\nu_o = \frac{1}{\tau_c} + Kf_r^2 \tag{2.53}$$

$$K = 4\pi^2(\tau_p + \tau_n) \tag{2.54}$$

$$\tau_n = 2\pi f_c/(2\pi f_{ro})^2 \tag{2.55}$$

$$I_{th} = \frac{qV_{act}N_{th}}{\tau_c} = \frac{qV_{act}}{\tau_c}\left(N_0 + \frac{1}{\Gamma g_0 \tau_p}\right) \tag{2.56}$$

$$I_s = \frac{qV_{act}\beta_{sp}}{g_0\tau_c\tau_p\Gamma} \tag{2.57}$$

The internal parameters of the obtained semiconductor laser chip are shown in Table 2.6.

Table 2.5 The fitting results of semiconductor laser chip

Fit parameters	Chip 1	Chip 2	Units
Threshold current I_{th}	11.3	11.4	mA
Leakage current I_s	1.27×10^{-6}	0.367×10^{-6}	A
Intermediate parameter F	4.126×10^{23}	2.894×10^{23}	
Relaxation oscillation frequency f_r	2.3×10^9	2.127×10^9	Hz
Damping factor v	6.53×10^9	6.16357×10^9	
Linewidth enhancement factor α	3.0	3.8558	
Chirped frequency f_c	0.08×10^9	0.212×10^9	

Table 2.6 Extracted results of internal parameters of the semiconductor laser chip

Fit parameters	Chip 1	Chip 2	Units
Threshold current I_{th}	11.3	11.4	mA
Photon lifetime τ_p	5.2×10^{-12}	5.2×10^{-12}	s
Carrier lifetime τ_c	0.2×10^{-9}	0.256×10^{-9}	s
Optical confinement factor Γ	0.01	0.01	Hz
Differential gain g_0	9.434×10^{-12}	5.28×10^{-12}	m^3/s
Linewidth enhancement factor α	3.0	3.8558	
Gain saturation coefficient ε	0.08×10^9	3.938×10^{-23}	m^3
Threshold carrier density N_{th}	2.943×10^{24}	3.8×10^{24}	m^{-3}
Transparent carrier density N_0	0.905×10^{24}	0.16×10^{24}	m^{-3}

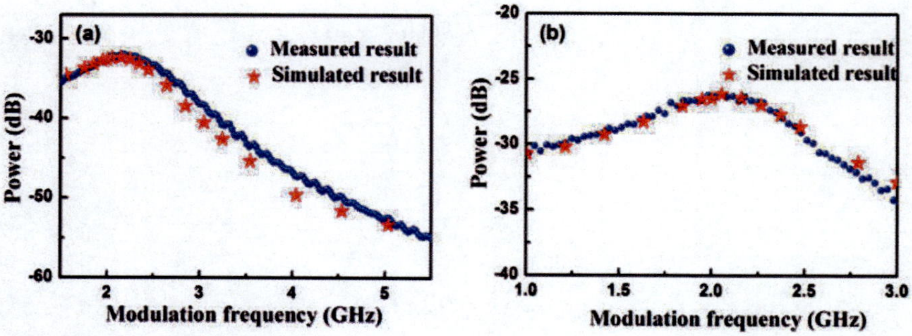

Fig. 2.18 Comparison of the simulated frequency response curve of the semiconductor laser chip at 13 mA (red curve) and the experimental data at the same current (blue curve). **a** Experimental results for chip 1; **b** experimental results for chip 2

In order to verify the accuracy of the extracted parameters, the frequency response curves of semiconductor laser chip 1 and semiconductor laser chip 2 at bias currents of 13 mA and 14 mA, respectively, were simulated by substituting the extracted parameters into the rate equation, and compared with the experimental results at the same bias current.

As can be seen in Fig. 2.18, the experimental results are in good agreement with the simulation results. Next, the extracted internal parameters are substituted into the simulated rate equation to investigate the dynamic characteristics of the designed hybrid integrated chaotic semiconductor laser structure.

6. **Path of hybrid integrated chaotic semiconductor lasers into chaos**

By studying the evolution of the oscillations of the system as the feedback intensity varies from small to large, the path of the system into chaos can be obtained, i.e. by observing the bifurcation diagram of the system [26], which in this section is a sample of the extreme values taken from the output of the electric field intensity. Figure 2.19a is the bifurcation diagram of the chaotic semiconductor laser chip 1, from which it can be seen that with the increase of the feedback intensity, chip 1 goes through the steady state (A), single periodic oscillation (B), doubled period oscillation (C), four periodic oscillation (D), and finally enters the chaotic state (E), from which it can be seen that the chaotic path of chip 1 is the multi-fold period path; Fig. 2.19b is the bifurcation diagram of the chaotic Fig. 2.19b is the bifurcation diagram of the chaotic semiconductor laser chip 2, from which it can be seen that with the increase of feedback strength, chip 2 goes through the steady state (A), single periodic oscillation (B), chaotic state (C), and doubled period oscillation (D).

7. **Characteristics in the time and frequency domains**

Figure 2.20 depicts the time domain and frequency domain characteristic curves of the chaotic semiconductor laser chip 1, from which it can be seen that the chaotic path of chip 1 is a multiperiodic path. Figure 2.20a shows that the system is in a steady state when the feedback strength is 0.015. Figure 2.20a-I is the time series

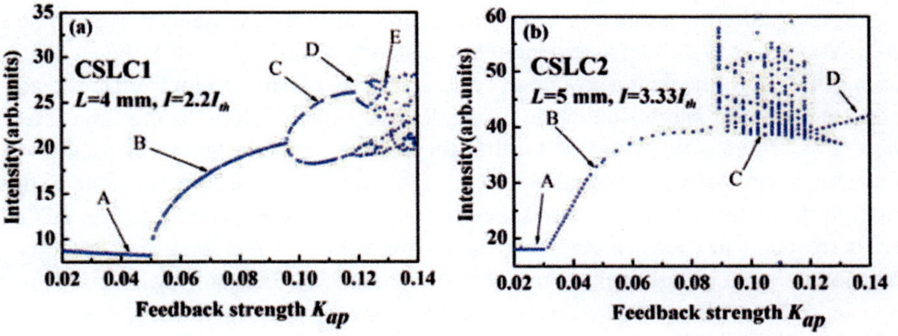

Fig. 2.19 Bifurcation diagram for chaotic semiconductor lasers. **a** Bifurcation diagram of chaotic semiconductor laser chip 1, (A) steady state, (B) single periodic oscillation, (C) doubled period oscillation, (D) four periodic oscillation, (E) chaotic state; **b** bifurcation diagram of chaotic semiconductor laser chip 2, (A) steady state, (B) single periodic oscillation, (C) chaotic state, (D) doubled period oscillation

diagram of the steady state, from which it can be seen that the output of the laser is constant at this time, and the laser is damped oscillation at this time; Fig. 2.20a-II is the spectral diagram of the steady state, and it can be seen that the spectrum has only one peak; Fig. 2.20a-III is the spectral diagram of the steady state, and it can be seen that the spectral energy is very low at this time, and there is only a little elevation at the relaxation oscillation frequency.

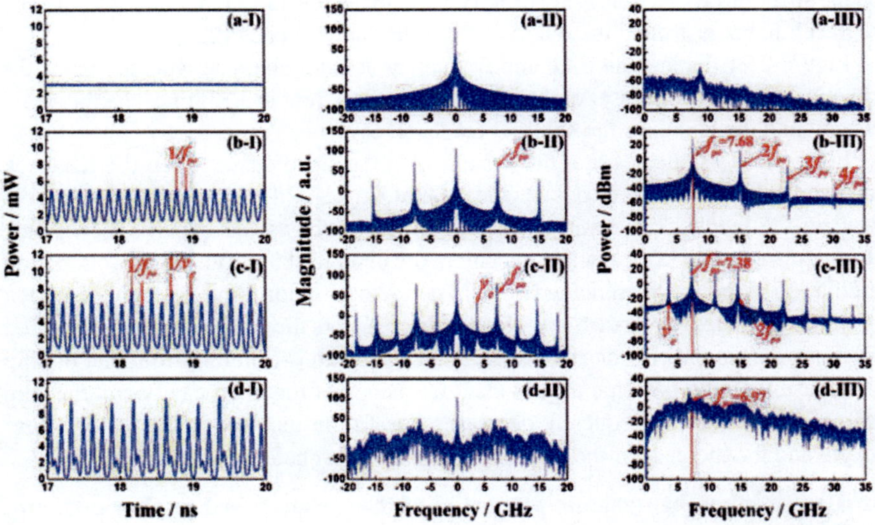

Fig. 2.20 Time series (I), optical spectra (II), RF spectra (III) of chaotic semiconductor laser chip 1. **a** Steady state, kap = 0.015; **b** single period, kap = 0.07; **c** double period, kap = 0.11; **d** chaotic state, kap = 0.16. All states above are the dynamic characteristics of the system at a bias current of 2.2 times the threshold current and an external feedback cavity length of 4 mm

Figure 2.20b shows that some damped oscillations of the laser become undamped oscillations due to the enhancement of the feedback perturbation when the feedback strength is 0.07. The increase of new oscillations changes the system from the steady state to a state of single-fold periodic oscillation. Figure 2.20b-I is the time series diagram of single-fold period, at this time the output state of the laser is an oscillation waveform with the period equal to $1/f_{po}$, and there is only one extreme value of the laser output at this time; and the four peak frequencies on the spectrum of Fig. 2.20b-III correspond to the relaxation oscillation frequency of the laser and its higher harmonics, and the higher harmonic frequencies are $2f_{po}$, $3f_{po}$ and $4f_{po}$ respectively.

Figure 2.20c shows that when the feedback intensity equals to 0.11, the system further bifurcates from the single period oscillation to the double period oscillation state. From the time series diagram 2.20c-I of this state, it can be seen that the time series output period of the laser at this time is $1/v_e$ with two different great values, and the period interval of these two great values is $1/f_{po}$; from the spectral diagram 2.20c-II and c-III, it can also be seen that harmonics appear between f_{po} and $2f_{po}$, $f_{po} = 2v_e$, which means that at this time the system is in a two-fold period oscillation state [27, 28].

And when the feedback strength continues to increase to 0.16, the system appears chaotic, as shown in Fig. 2.20d. Figure 2.20d-I is the time series diagram of chaos at this time, the output of the laser is not regular, and there are several extreme value points; and the spectra of Fig. 2.20d-II and d-III are continuous, which proves that the system has entered the chaotic state at this time. From Fig. 2.20, it can be found that the relaxation oscillation frequency of the chaotic semiconductor laser chip 1 gradually decreases as the feedback intensity increases, which indicates that the peak on the spectrum at this time is not exactly equal to the relaxation oscillation frequency of the chip, but is around the relaxation oscillation frequency [28].

Figure 2.21 depicts the time and frequency domain characteristic curves of the chaotic semiconductor laser chip 2, when the bias current is 3.33 times the threshold current and the length of the external feedback cavity is 5 mm.

Figure 2.21a indicates that the system is in the steady state when the feedback intensity is 0.02; Fig. 2.21b depicts the state of single-fold period oscillation of chip 2 when the feedback intensity is 0.06; Fig. 2.21c shows that when the feedback strength is equal to 0.11, the system enters the chaotic state from single-fold period oscillation without experiencing two-fold period oscillation. And when the feedback strength continues to increase to 0.15, the system enters the state of single-fold period oscillation from the chaotic state once again. The path is different from that of chip 1 because the ratio between the resonant frequency of the outer cavity and the chip relaxation oscillation frequency decreases due to the increase of the outer cavity length and the increase of the chip bias current when chaos is generated.

8. **Determining the dynamic properties of the system based on the maximum Lyapunov index**

The maximum Lyapunov exponent is one of the most important measures for determining whether a system is in a chaotic state [29]. The maximum Lyapunov exponent describes the dispersion of the two nearest oscillatory trajectories in space. If two different initial points are separated from each other after iteration, the maximum

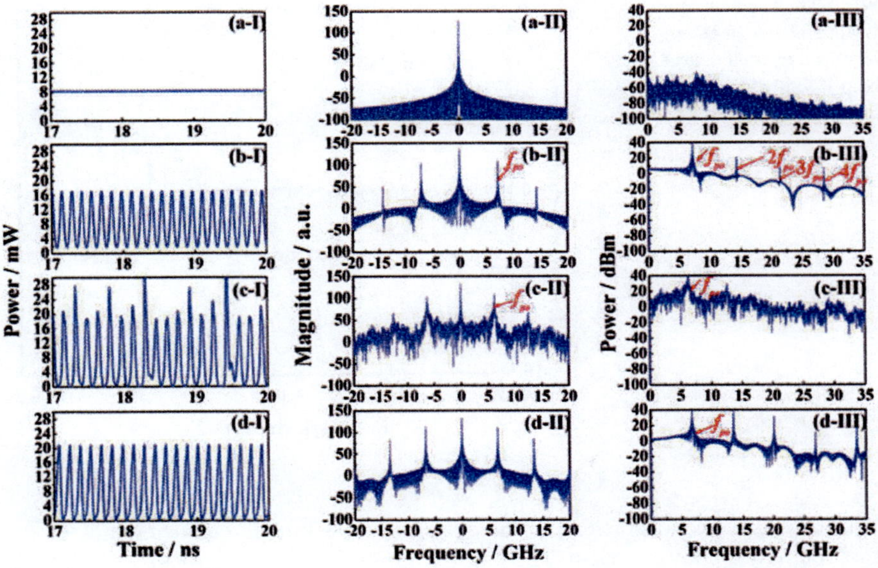

Fig. 2.21 Time series (I), optical spectra (II), RF spectra (III) of chaotic semiconductor laser chip 2. **a** Steady state, kap = 0.02; **b** single period, kap = 0.06; **c** chaotic state, kap = 0.11; **d** single period, kap = 0.15. All the above states are the dynamic characteristics of the system at a bias current of 3.33 times the threshold current and an external feedback cavity length of 5 mm

Lyapunov exponent of the system is positive, indicating that the system is in a chaotic state; if two different initial points are converged after iteration, the maximum Lyapunov exponent of the system is negative or near zero, indicating that the system is in a steady state or a doubly periodic oscillation.

Figure 2.22 depicts the variation of the maximum Lyapunov index of the system with the feedback intensity for a feedback cavity length of 4 mm and a bias current of 2.2 times the threshold for the chaotic semiconductor laser chip 1. From the figure, it can be seen that, when the feedback strength is less than 0.04, the maximum Lyapunov exponent of the system is much less than zero, the oscillation of the system is damped oscillation, and the system is in the steady state; as the feedback strength increases from 0.04 to 0.12, the maximum Lyapunov exponent of the system is near zero, and some oscillations in the system change from damped oscillation to non-damped oscillation, and the system enters the doubling cycle from the steady state; when the feedback strength increases from 0.12 When the feedback strength continues to increase from 0.12, new non-damped oscillations appear in the system, and the maximum Lyapunov exponent is much larger than zero at this time, which indicates that the chaotic state of the system appears at this time.

In order to obtain the specific fabrication parameters of the chip to generate chaos through theoretical study, and to provide theoretical guidance for the fabrication of hybrid integrated chaotic semiconductor lasers later, we studied the dynamic characteristics of the chaotic semiconductor laser chip 1 at different external cavity lengths with bias currents ranging from 15 to 41 mA and feedback intensity increasing

Fig. 2.22 The maximum Lyapunov exponent versus feedback intensity curve for chip 1 with a feedback cavity length of 4 mm and a bias current of 2.2 I_{th}

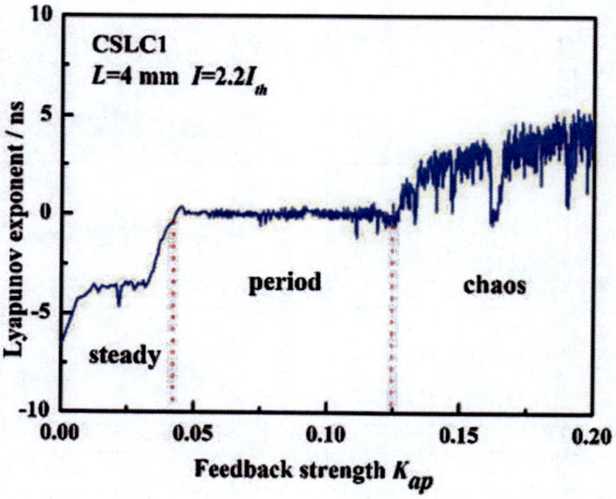

from 0 to 0.2, and characterized the system by the kinetic characteristics of the maximum Lyapunov exponent. The state of the system is characterized by the kinetic characteristics of the maximum Lyapunov exponent, as shown in Fig. 2.23.

Figure 2.23 shows the dynamic characteristics of the chaotic semiconductor laser chip 1, and Fig. 2.23a–f show the different conditions of the external cavity length. The different colors in the figure represent the different state characteristics of the system, red and yellow indicate that the value of the maximum Lyapunov exponent is greater than zero, and the system is chaotic; green indicates that the value of the maximum Lyapunov exponent is near zero, and the system is oscillating at times the period; blue indicates that the value of the maximum Lyapunov exponent is much less than zero, and the system is in the steady state. From Fig. 2.23a and b, we can see that when the feedback cavity length is equal to 2 and 3 mm, the maximum Lyapunov index of the system is much less than zero, which is indicated in blue, which means that the oscillation of the system is damped oscillation, and the different oscillation trajectories in the system will be clustered together, and the system is the steady state at this time.

Figure 2.23c depicts the dynamic characteristics of the system for an external feedback cavity length of 4 mm. As can be seen from the figure, when the feedback strength is less than 0.12, the maximum Lyapunov exponent of the system is negative. When the feedback strength is less than 0.04, the maximum Lyapunov exponent of the system is much less than zero, and the system is the steady state at this time, which is indicated by the blue color. When the feedback strength is 0.04–0.12, there is a non-damped oscillation in the system, and the maximum Lyapunov exponent of the system is near zero, which indicates that the system has entered the state of doubly periodic oscillation from the steady state, which is indicated by green. When the feedback intensity continues to increase, at 0.12–0.20, and the bias current is in the range of 22–41 mA, the maximum Lyapunov exponent of the system appears to be greater than zero, indicating that the system is prone to chaotic lasers at this time, which is indicated by red and yellow. And it can be seen that the system generates

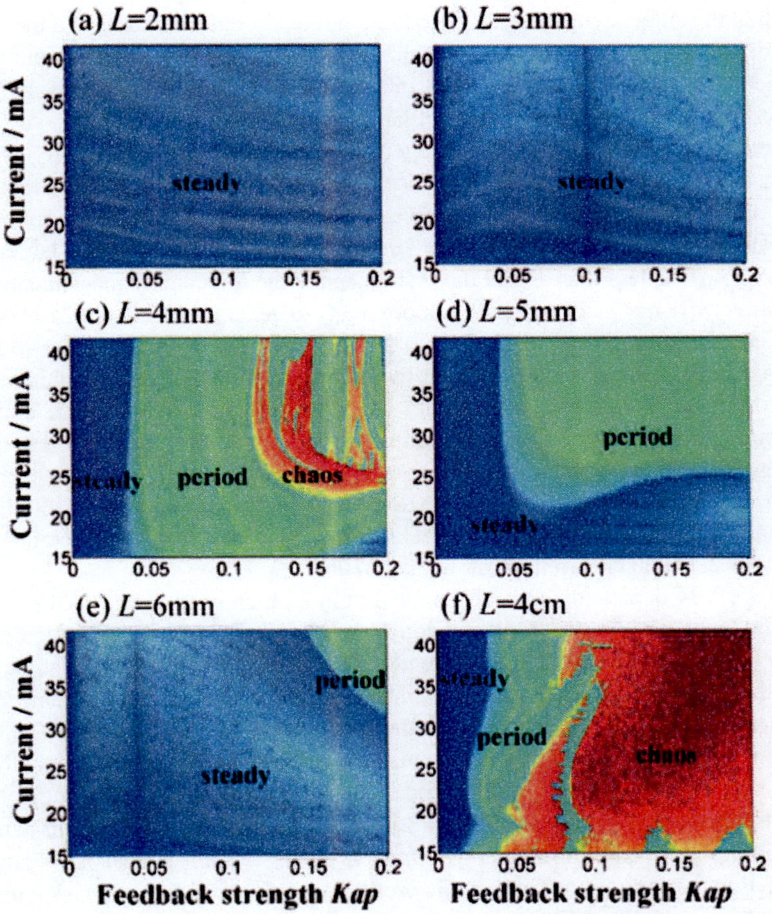

Fig. 2.23 Diagram of the dynamic characteristics of the chaotic semiconductor laser chip 1. **a** L = 2 mm; **b** L = 3 mm; **c** L = 4 mm; **d** L = 5 mm; **e** L = 6 mm; **f** L = 4 cm

chaos in the interval of 22–25 mA, when the feedback intensity is greater than 0.12. When the bias current is greater than 25 mA and the feedback intensity is 0.14–0.16, the red color is the deepest, which indicates that the maximum Lyapunov index value of the system is larger, and the system is prone to produce chaotic laser with the higher dimension.

When the feedback cavity length increases to 5 mm, the chaotic state of the system disappears, and the steady state and doubly periodic oscillation state increases 64 plus, as shown in Fig. 2.23d. When the feedback strength is less than 0.04, or the current is less than 20 mA, the system is in the steady state, indicated by blue; when the feedback strength is greater than 0.04 and the bias current is greater than 20 mA, the system is in the doubly-periodic oscillation state, indicated by green. When the length of the outer cavity increases to 6 mm, the steady state increases and the times-periodic oscillation state decreases, as shown in Fig. 2.23e. In this figure, only when the feedback strength is greater than 0.15 and the bias current is greater than 30 mA, the system is in a doubly-periodic oscillation state.

We also studied the dynamic characteristics of chip 1 in the long cavity case, as shown in Fig. 2.23f. The feedback cavity length is 4 cm and the external feedback frequency is $f_{\text{ext}} = 3.7$ GHz. when the bias current of chip 1 is 15 mA, the relaxation oscillation frequency of chip 1 is $f_{\text{r}} = 3.7$ GHz. when the bias current is larger than 15 mA, $f_{\text{ext}}/f_{\text{r}}$ is less than 1, and the system enters the long cavity mechanism from the short cavity mechanism, which is more likely to generate the chaotic laser. When the feedback strength is greater than 0.05, the system is prone to generate high-dimensional chaotic signals. However, the hybrid integrated chaotic semiconductor laser designed in this section needs to be packaged into a commercial butterfly housing, and it is difficult to realize the final package when the feedback cavity is longer than 1 cm. Based on the numerical simulation results, it is easy to conclude that the optimal fabrication parameters for the chaotic signal generation of the chaotic semiconductor laser chip 1 are: feedback cavity length of 4 mm, feedback intensity of 0.12–0.2, and bias current of more than 22 mA.

Figure 2.24 depicts the plot of the maximum Lyapunov exponent of the system with the feedback intensity for the chaotic semiconductor laser chip 2 at a feedback cavity length of 5 mm and a bias current of 3.33 times the threshold. It can be seen from the graph that the interval range of the chaotic laser generated by chip 2 and chip 1 are completely different. When the feedback intensity is less than 0.03, the maximum Lyapunov index of the system is much less than zero, and the oscillation in the system is damped oscillation, and the system is the steady state at this time; when the feedback intensity is 0.03–0.08 and 0.12–0.20, the system has non-damped oscillation, the maximum Lyapunov exponent of the system is near zero, and the system is in the state of doubly periodic oscillation; only when the feedback intensity is in the range of 0.08–0.12, the maximum Lyapunov exponent of the system is much larger than zero, and the dynamics of the system is chaotic at this time. The dynamic characteristics of the chaotic semiconductor laser chip 2 are shown in Fig. 2.25. From Fig. 2.25a and b, we can see that the maximum Lyapunov exponent of the system is much less than zero when the feedback cavity length is equal to 2 and 3 mm, which is indicated by the blue color, which means the system is in the steady state at this time. Figure 2.25c depicts the dynamic characteristics of the system when the length of the external feedback cavity is 4 mm, from the figure, it can be seen that the system is in the steady state when the feedback intensity is 0–0.03 and 0.16–0.20; when the feedback intensity is 0.03–0.16, the system is in the state of doubly periodic oscillation and there is no chaotic state in the system, which is completely different from the dynamic characteristics of chip 1 when the length of the external cavity is 4 mm. This is completely different from the dynamics of chip 1 when the outer cavity is 4 mm long.

With the increase of the feedback cavity length to 5 mm, a new non-damped oscillation appears in the system. From Fig. 2.25d, we can see that when the feedback strength is less than 0.03, the oscillation in the system is damped oscillation, and the

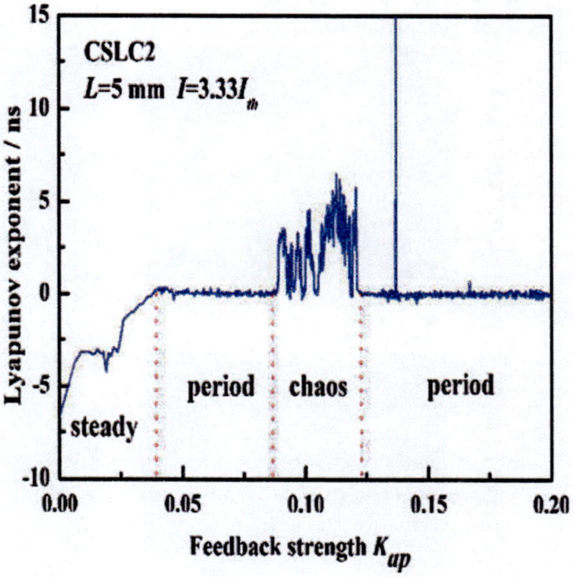

Fig. 2.24 The maximum Lyapunov index of chip 2 as a function of feedback intensity for a feedback cavity L of 5 mm and a bias current of 3.33 I_{th}

system is still steady state; with the increase of feedback strength, when the feedback strength is 0.03–0.07, the system enters the doubly-periodic oscillation state; with further enhancement of feedback, when the feedback strength is 0.07–0.12 and the bias current is 22–39 mA, the maximum Lyapunov When the feedback strength is 0.07–0.12 and the bias current is 22–39 mA, the maximum Lyapunov exponent is greater than zero, and the system has a new non-damped oscillation and a chaotic state. As can be seen from the figure, the outer cavity length is 5 mm and the feedback strength is 0.07–0.12, Chip 2 is prone to chaotic laser, which is not the same as Chip 1. When the outer cavity length increases to 6 mm, the steady state increases and the times-period oscillation state decreases, as shown in Fig. 2.25e. In the figure, only when the feedback strength is greater than 0.15 and the bias current is greater than 32 mA, the system has a doubly-periodic oscillation state. This state is almost identical to the dynamic characteristics of chip 1. In the long cavity case, the dynamic characteristics of chip 2 are shown in Fig. 2.25f, where the feedback cavity length is 4 cm and the external feedback frequency is $f_{ext} = 3.7$ GHz. From the figure, it can be seen that the system is prone to generate high-dimensional chaotic signals when the feedback strength is greater than 0.04. According to the numerical simulation results, the optimal production parameters for chaotic signals generated by the chaotic semiconductor laser chip 2 are: feedback cavity length of 5 mm, feedback intensity of 0.07–0.12, and bias current of more than 22 mA. From the above study, it is known that the chaotic dimension of the short-cavity optical feedback structure is low, and the external conditions of the chaotic laser generated by different chips are different due to the different internal parameters; Based on the maximum Lyapunov exponent of the calculated system, the optimal fabrication parameters of the two chips for chaos generation can be determined, the optimal manufacturing parameters for

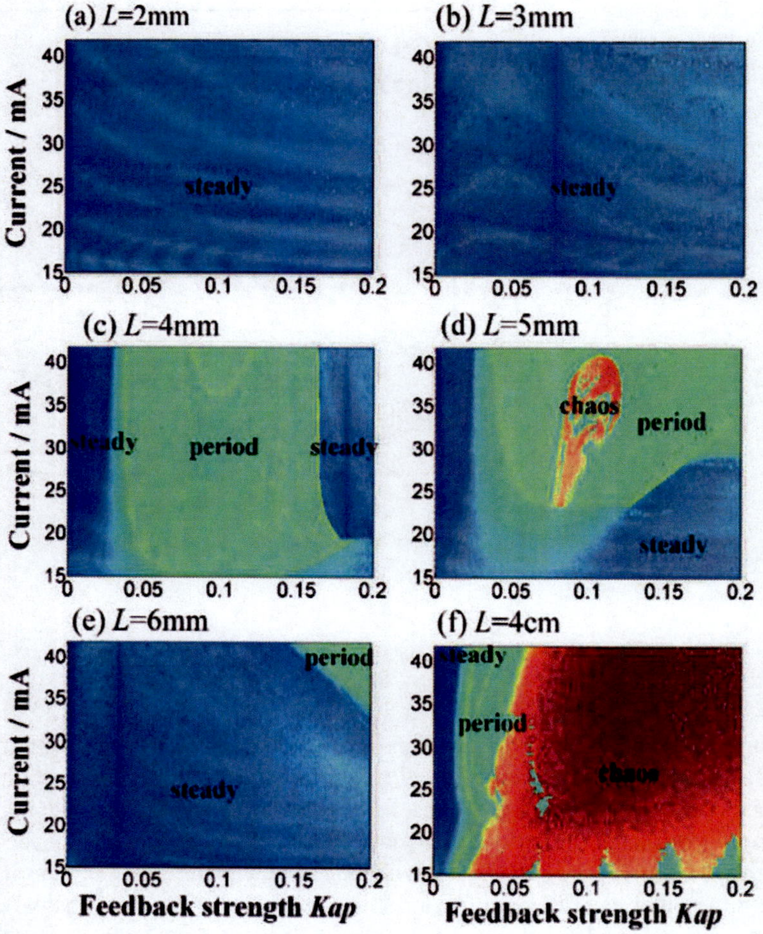

Fig. 2.25 Dynamic characteristics diagram of the chaotic semiconductor laser chip 2. **a** $L = 2$ mm; **b** $L = 3$ mm; **c** $L = 4$ mm; **d** $L = 5$ mm; **e** $L = 6$ mm; **f** $L = 4$ cm

the chaotic semiconductor laser chip 1 are a 4 mm external feedback cavity length and a feedback intensity of 0.12–0.20; the optimal manufacturing parameters for the chaotic semiconductor laser chip 2 are a 5 mm external feedback cavity length and a feedback intensity of 0.07–0.12, which are useful for the fabrication of the hybrid integrated chaotic semiconductor laser. This is an important guideline for the fabrication of the hybrid integrated chaotic semiconductor laser, and at the same time, the differences between the system simulation and the actual fabrication of the device should be considered, and be combined.

2.2.2 The Development of Devices

1. Structural design

Based on the existing production process of butterfly-packaged optical transmitter devices, the authors' team designed two hybrid integrated chaotic semiconductor laser structures: the in-coupled structure and the out-coupled structure. It is worth mentioning that the feedback cavity length, i.e., the distance between the laser chip and the transmissive mirror, is fixed for both structures due to the processing limitations of the existing butterfly package semiconductor lasers. The feedback cavity length is 2 mm for the in-coupled structure and 8.25 mm for the out-coupled structure, which are described in detail below.

Figure 2.26 shows a schematic diagram of the coupling structure inside the hybrid integrated laser, where a distributed feedback semiconductor laser chip, a collimating lens, a transmission and reflection mirror, a focusing lens and a section of optical fiber are arranged sequentially and coaxially coupled. The distributed feedback semiconductor laser chip is bonded to the chip carrier with conductive adhesive for power supply. In addition, in order to monitor the temperature of the distributed feedback semiconductor laser chip, a thermistor is bonded to the chip carrier and placed close to the distributed feedback semiconductor laser chip. The chip carrier, collimating lens, transverse mirror, focusing lens, and optical fiber are fixed on the heat sink, where the optical fiber is fixed on the heat sink by Ω-supports [30–33]. The entire heat sink is welded to the semiconductor cooler, and all the above devices are packaged in a common commercial 14-pin butterfly laser housing, which is filled with nitrogen gas to improve the lifetime and stability of the device.

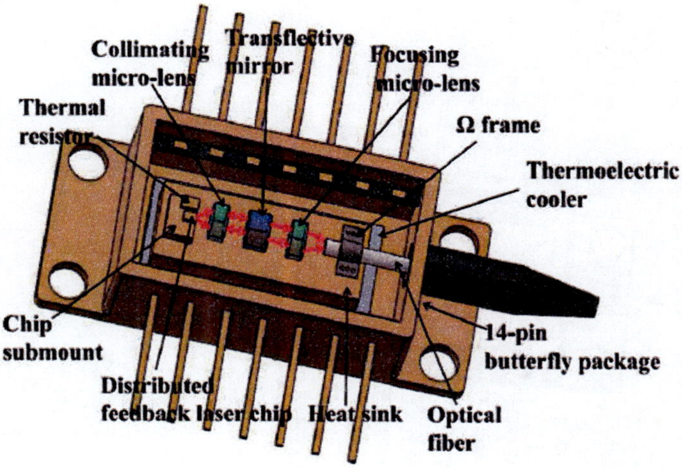

Fig. 2.26 Schematic diagram of the internal coupling structure of a hybrid integrated laser

Figure 2.27 shows the schematic diagram of the external coupling structure of the hybrid integrated laser, from which it can be seen that this structure is similar to the internal coupling structure, which also integrates a semiconductor laser chip, a thermistor, a chip carrier, a heat sink, a semiconductor cooler, and other components in a 14-foot butterfly-shaped metal housing. The laser beam is then coupled to the fiber through the collimation of the collimating lens and the focusing of the focusing lens. The difference is that in the internal coupling structure, the coupling of the optical fiber is done by reaching into the inner shell, while in the external coupling structure, the coupling of the optical fiber is done in the casing at the exit of the shell, as can be seen in Fig. 2.27, where the collimating lens is fixed on the heat sink, while the transverse mirror, focusing lens and optical fiber are all coupled in the casing. In the above two structural designs, three kinds of reflective mirrors with reflectivity of 20, 10 and 5% are used. The coupling efficiency between the collimating lens and the distributed feedback semiconductor laser chip is 90%, and the coupling efficiency between the collimating lens and the trans-reflective mirror is parallel light, so the coupling efficiency can be regarded as 100%. Therefore, according to the principle of optical reversibility, the feedback rates of the three reflectance trans-reflective mirrors for the chip are about 16.2%, 8.1% and 4.05%, respectively. The feedback rates of the latter two cases are in the theoretical range of − 40 and − 10 dB for chaotic laser generation, while the feedback rate of the first case is larger than the upper limit of the theoretical range, because the coupling efficiency between the collimating lens and the trans-reflective mirror may not reach the ideal situation in actual operation or some other factors may cause the loss in the actual optical path.

2. **Device fabrication**

The integrated package of the hybrid integrated chaotic semiconductor laser adopts the current mature semiconductor device packaging process, including SMD, gold

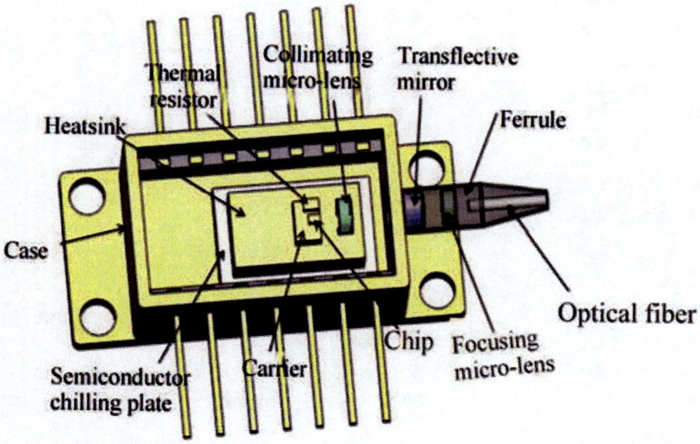

Fig. 2.27 Schematic diagram of the external coupling structure of a hybrid integrated laser

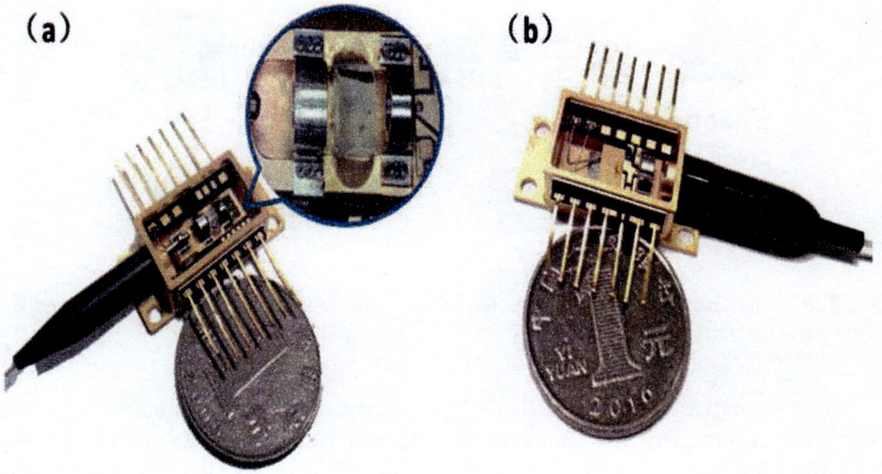

Fig. 2.28 Physical view of a hybrid integrated chaotic semiconductor laser. **a** Internally coupled structure; **b** externally coupled structure

wire ball welding, heat welding, coupling alignment, laser welding, etc. Finally, the hybrid integrated chaotic semiconductor laser with both internal and external coupling structures is fabricated (Fig. 2.28).

3. **The experiment equipment for output characteristics testing**

The authors also conducted experimental tests on the output characteristics of the hybrid integrated chaotic semiconductor laser, and the experimental setup is shown in Fig. 2.29. The integrated chaotic semiconductor laser is driven by a precision current source (ILX Lightwave, LDX3412) to provide bias current, and a temperature controller (ILX Lightwave, LDT-5412B) to regulate the operating temperature of the chip. The output signal of the integrated chaotic semiconductor laser first passes through an optical isolator (ISO, isolation degree > 50 dB) to prevent the influence of optical feedback from the detection path, and then is divided into two paths via an optical coupler (splitting ratio 80:20), of which 20% goes to a high-resolution spectrum analyzer (APEX, AP2041B, wavelength resolution up to 0.04 p.m.) to observe the spectrum. The amplified optical signal is further divided into two paths by an optical coupler (50:50 splitting ratio), one of which passes through a high-speed photodetector (Finisar, XPDV2120RA, 50 GHz rate). The converted electrical signal goes to a high-bandwidth real-time oscilloscope (TELEDYNELECOY, 10-36Zi, 36 GHz bandwidth, 80 GHz sampling rate) to monitor the time series; the other passes through the same type of high-speed photodetector, and the converted electrical signal goes to a spectrum analyzer (Rohde & Schwarz, FSW-26, 26.5 GHz bandwidth) to monitor the time series. 5 GHz bandwidth) to observe the spectrum. In Fig. 2.29, the solid line shows the optical signal path and the dashed line shows the electrical signal path.

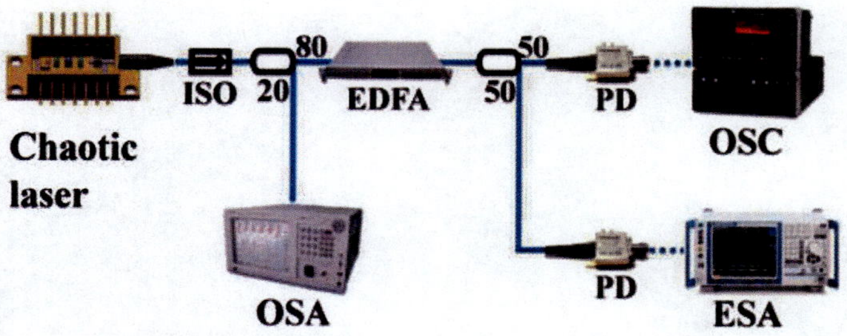

Fig. 2.29 Experimental setup for testing the output characteristics of hybrid integrated chaotic semiconductor lasers

After the experimental tests, the hybrid integrated chaotic semiconductor laser with a 20% reflectivity trans-reflector and a feedback cavity of 8.25 mm (R = 20%, L = 8.25 mm) and a 5% reflectivity and 2 mm feedback cavity (R = 5%, L = 2 mm) can output a more ideal chaotic laser signal, and their spectral bandwidths are above 4.5 GHz. It is also found that the chaotic output state of the latter does not change with the change of bias current, and the experimental results are more satisfactory, which are described below.

4. **Output characteristics of the laser with a 20% reflectivity trans-reflector and a feedback cavity of 8.25 mm (R = 20%, L = 8.25 mm)**

Two typical output states of this laser are found: a single-fold oscillation state with a bias current of 20.9 mA and an operating temperature of 11.9 °C and a chaotic state with a bias current of 20.9 mA and an operating temperature of 12.4 °C. Figure 2.30 shows the P-I characteristic curves of the laser in both cases. From Fig. 2.30a, it can be seen that the threshold current of the integrated chaotic semiconductor laser at an operating temperature of 11 °C. The threshold current of the integrated chaotic semiconductor laser at an operating temperature of 9 °C is 10.1 m A, with a slope efficiency of 0.1 W/A. The power of the integrated chaotic semiconductor laser is 3 mW at a bias current of 40 mA. As can be seen from Fig. 2.30b, the threshold current of the integrated chaotic semiconductor laser at an operating temperature of 12.4 °C is 10.4 mA, and the slope efficiency is 0.098 W/A. The power of the integrated chaotic semiconductor laser is 2.94 mW at a bias current of 40 mA. That is to say, corresponding to the operating temperatures of 11.9 °C and 12.4 °C, the bias current of the laser which is 20.9 mA is 2.07 times and 2.01 times of the laser threshold current respectively.

Figure 2.31 shows the time series, spectrum, spectra and autocorrelation curves of the laser at a bias current of 20.9 mA and an operating temperature of 11.9 °C. It can be seen from Fig. 2.31a that the time series peaks are around 25 mV. The fluctuation of the time series line shows obvious periodicity, and the interval between two adjacent spikes is 0.19 ns, then the minimum positive period is 0.19 ns, and the frequency of change is 5.26 GHz. It is obvious from Fig. 2.31b that a spike of intensity around − 55 dBm appears at a frequency of 5.2 GHz, which exceeds 27 dB, the intrinsic noise of the system. This indicates the relaxation oscillation frequency is about 5.2 GHz

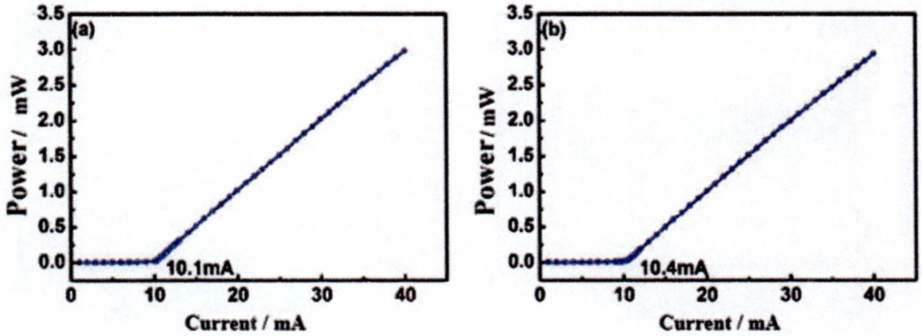

Fig. 2.30 P-I characteristic curve of the laser. **a** Operating at 11.9 °C, **b** operating at 12.4 °C

at an applied current of 20.9 mA and an operating temperature of 11.9 °C. At this point the 80% bandwidth of the output spectrum is 3.6 GHz. The − 20 dB linewidth of the spectrum in Fig. 2.31c is 0.4349 GHz, at 1549.02 nm wavelength and the left and right symmetrical distribution wavelengths are 1548.94, 1548.98, 1549.06 and 1549.11 nm, and the frequency difference between the two adjacent spikes is around 5.2 GHz. The autocorrelation curve in Fig. 2.31d shows a rhombic distribution and consists of a series of spikes with the interval of 0.19 ns between two adjacent spikes, which corresponds to the frequency of 5.26 GHz. Therefore, it is judged that the laser is in a single period oscillation with a bias current of 20.9 mA and an operating temperature of 11.9 °C, and the period is about 0.19 ns. The output energy is mainly concentrated in the relaxation oscillation near the frequency of 5.2 GHz, and the time series shows periodic fluctuations.

Figure 2.32 shows the time series, spectrum, spectra and autocorrelation curves of the laser at a bias current of 20.9 mA and an operating temperature of 12.4 °C. From Fig. 2.32a, we can see that the time series peak to the peak value of the laser is around 10 mV, and the time series fluctuation is non-periodic, which is more random compared with the operating temperature of 11.9 °C. From Fig. 2.32b, it can be seen that the spectrum is flatter and the bandwidth increases compared with 11.9 °C, and the output energy distribution tends to be dispersed, and the 80% bandwidth of the spectrum is 6.5 GHz at this time. Figure 2.32c, The − 20 dB linewidth of its spectrum is 1.1346 GHz, which is 1.5 times of the 0.4349 GHz, the − 20 dB linewidth of the spectrum of 11.9 °C. The central wavelength is 1549.09 nm, which increases compared to the central wavelength of the spectrum at 11.9 °C. This is due to the increase in temperature which leads to the decrease in the forbidden bandwidth of the semiconductor material and thus the increase in wavelength. Figure 2.32d shows a peg-shaped distribution of the autocorrelation curve at this time, which is a typical chaotic state distribution characteristic. Therefore, it is judged that the laser is in a chaotic output state when the bias current is 20.9 mA and the operating temperature is 12.4 °C, and the time series shows random fluctuations, the spectrum is broadened compared with 11.9 °C, and the spectrum tends to be flat, and the energy output is more dispersed.

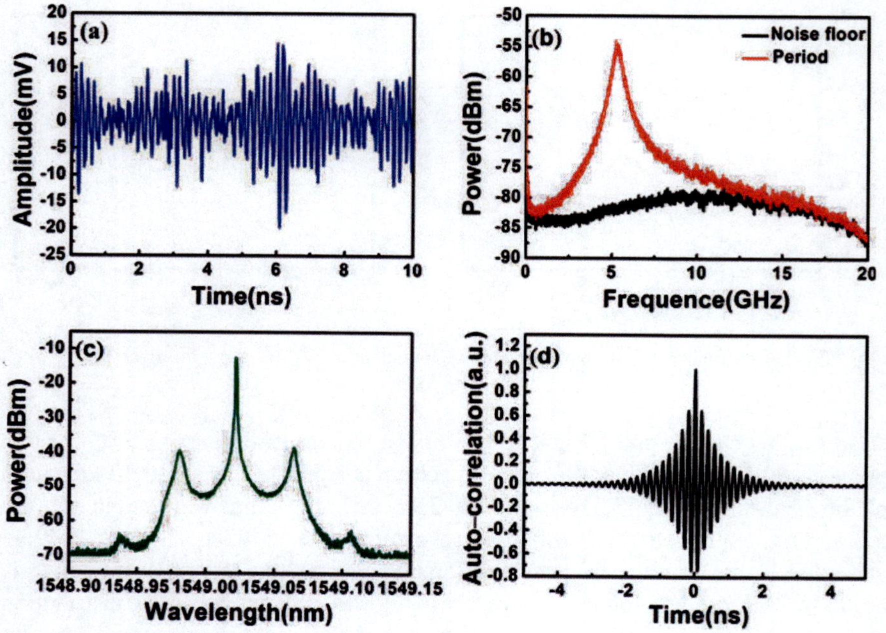

Fig. 2.31 **a** Time series **b** RF spectra **c** optical spectra **d** autocorrelation plots of the laser at an operating temperature of 11.9 °C for a bias current of 20.9 mA

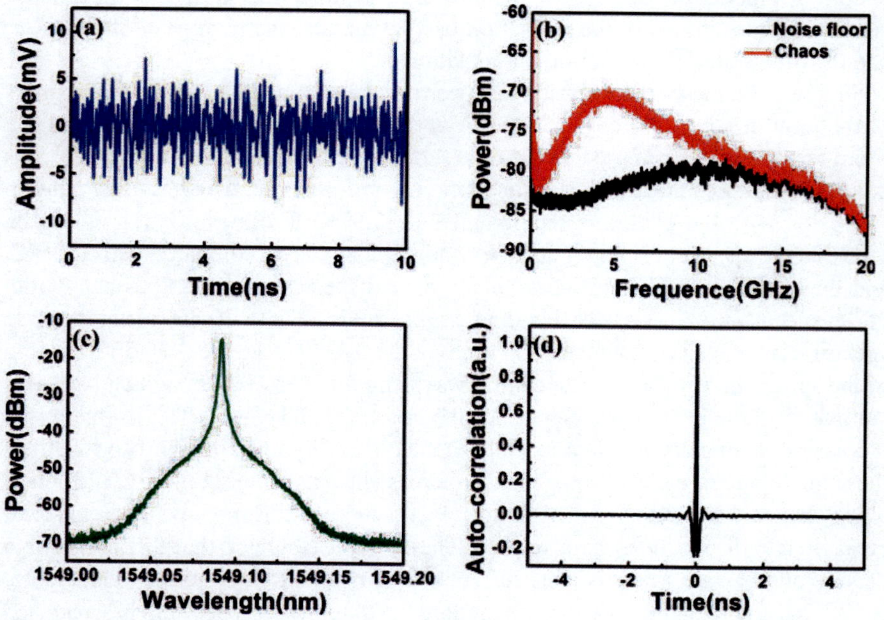

Fig. 2.32 **a** Time series **b** RF spectra **c** optical spectra **d** autocorrelation plots of the laser at an operating temperature of 12.4 °C for a bias current of 20.9 mA

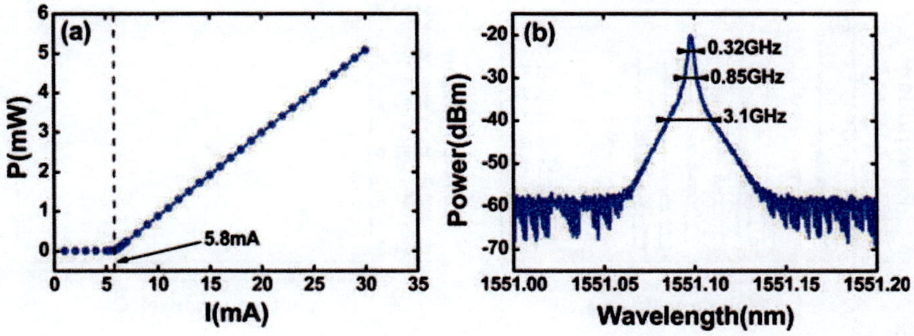

Fig. 2.33 **a** P-I characteristic curve and **b** typical spectrogram of a hybrid integrated chaotic semiconductor laser at an operating temperature of 18.5 °C

5. **Output characteristics of the laser with a 5% reflectivity trans-reflector and a feedback cavity of 2 mm (R = 5%, L = 2 mm)**

The output characteristics of the hybrid integrated chaotic semiconductor laser with a 5% reflectivity trans-reflective mirror and a feedback cavity length of 2 mm are described below.

Figure 2.33 shows the laser's P-I characteristic and optical spectrum at 18.5 °C and the bias current is 7.8 mA, which is 1.345 times of threshold current. As shown in Fig. 2.33a, the threshold current of this integrated chaotic semiconductor laser is 5.8 mA at this temperature and a slope efficiency of 0.21 W/A. As shown in Fig. 2.33b, at an operating temperature of 18.5 °C, the central wavelength of the spectrum of the integrated chaotic semiconductor laser is about 1551.10 nm. The spectrum's − 3 dB, − 10 dB and − 20 dB linewidth is 32 GHz, 0.85 GHz and 3.1 GHz, respectively, and the spectra have no edge mode at this time.

Figure 2.34 shows a graph of the variation of the spectrum with temperature for a hybrid integrated chaotic semiconductor laser. Figure 2.34a shows the spectral curve at the operating temperature from 19.7 to 35.7 °C, it can be seen that the center wavelength of the laser spectrum increases with the increase of temperature, in order to see the change of the center wavelength of the spectrum with the temperature more intuitively, Fig. 2.34b shows the line graph of the center wavelength at different temperatures, from which it can be seen that the center 73 wavelength of the laser spectrum increases linearly with the temperature The slope is 0.11 nm/°C.

Figure 2.35 shows a typical output state of a hybrid integrated chaotic semiconductor laser, where the applied bias current is 7.8 mA and the temperature controller controls the temperature at 18.5 °C. Figure 2.35a shows the time series diagram of the integrated chaotic semiconductor laser. It can be seen that the time series peak value is around 5 mV, and the time series curve shows a certain amplitude of vibration and significant fluctuations at the sub-nanosecond level, with random fluctuations and irregularities. In Fig. 2.35b, the upper red curve is the spectrum of the integrated chaotic semiconductor laser at this time, and the black curve is the system noise floor. It can be found that at a frequency of about 2.3 GHz, the laser spectrum shows

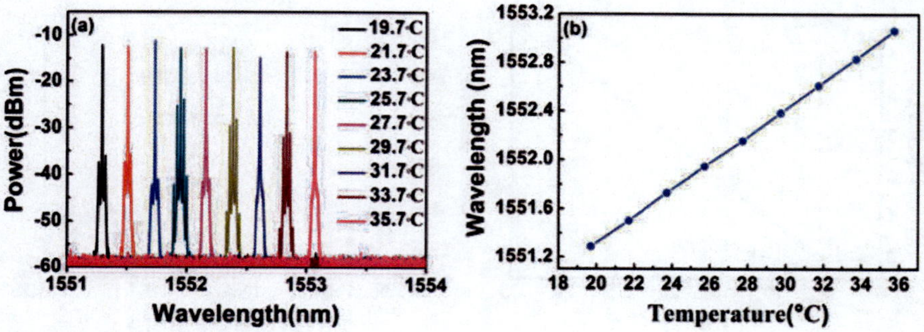

Fig. 2.34 Temperature dependence of a hybrid integrated chaotic semiconductor laser. **a** Spectra at different temperatures; **b** central wavelengths at different temperatures

a relatively gentle peak of – 76 dBm, which is caused by the relaxation oscillation of the chip itself, and the relaxation oscillation frequency of the laser can be judged to be around 2.3 GHz at this time. The peak exceeds the noise floor by about 24 dB, and the energy distribution is more dispersed at this time, and the spectrum is continuously raised in a wider frequency range up to about 9 GHz, and the 80% bandwidth of the spectrum is about 5 GHz at this time. At this time, the phase diagram data point distribution is scattered, showing the typical chaotic attractor characteristics, complex, ergodic and bounded, as shown in Fig. 2.35c. Figure 2.35d shows the autocorrelation curve at this point, and the image shows a peg-shaped distribution, which is also typical of the chaotic state distribution. The maximum Lyapunov exponent is 0.049935, which also indicates that the laser is in a chaotic output state.

At this operating temperature, the authors varied the applied bias current and found a wider range of chaotic states, as shown in Fig. 2.36a for several typical spectral plots of the integrated chaotic semiconductor laser at T = 18.5 °C with different bias currents, all of which have a wide range of continuous boosts with respect to the noise floor. Figure 2.36b shows the 80% bandwidth of the spectrum of the integrated chaotic semiconductor laser at this temperature with bias current, which has been normalized to the threshold current. The figure shows that with the increase of bias current, the 80% bandwidth of the spectrum basically shows a trend of first increasing and then slowly decreasing, with a small kink only at the peak. When the bias current is 1.55 times the threshold current, i.e., about 9 mA, the 80% bandwidth of the spectrum reaches the maximum value of 5.9 GHz. When the normalized bias current is between 1.3 and 2.2, the 80% bandwidth of the output spectrum is more than 4.5 GHz. Figure 2.36c shows the peak power variation of the spectrum at different bias currents, and it can be seen that the peak power of the spectrum shows a trend of first decreasing and then slowly increasing when the bias current is continuously increasing. Similarly, the peak power of the spectrum reaches the lowest when the bias current is 1.55 times the threshold current, i.e., about 9 mA. It can be judged that: with the change of bias current, the bandwidth of 80% of the spectrum and the peak power of the spectrum are inversely proportional. The blue points in the figure are the experimental data points and the red curve is the fitted

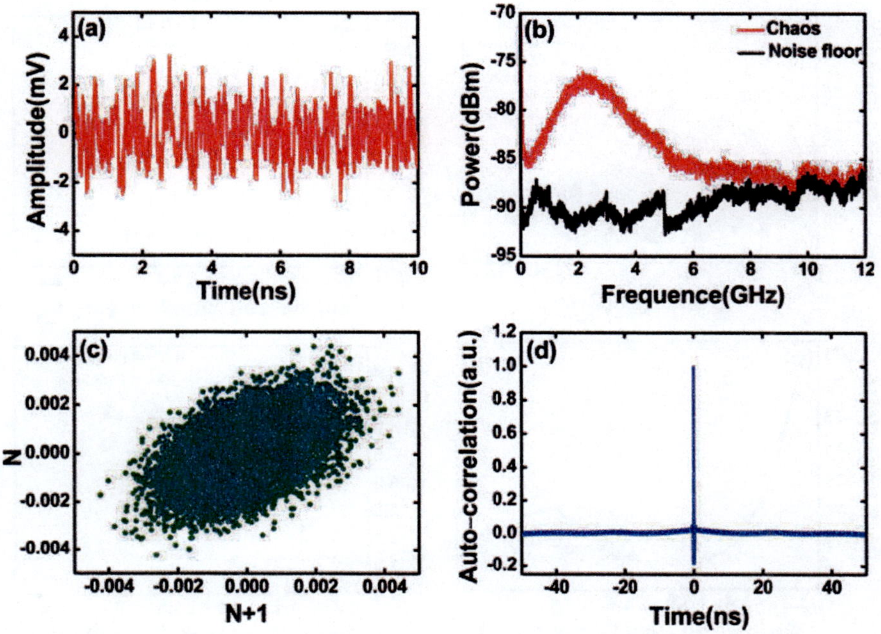

Fig. 2.35 Typical output states of a hybrid integrated chaotic semiconductor laser. **a** Time series **b** RF spectrum **c** phase diagram **d** autocorrelation curve

curve, the experimental data points are evenly scattered on both sides of the fitted curve. This is due to the increase in the bias current, which leads to the increase in the relaxation oscillation frequency of the laser, which is reflected in the spectrum as a continuous increase in the peak frequency.

In brief, at T = 18.5 °C, we found the chaotic state at multiple bias current whose 80% bandwidth is more than 4.5 GHz. As the current increases, the 80% bandwidth of the spectrum shows a trend of first increasing and then slowly decreasing. This is because the spectral bandwidth of the chaotic state is related to the relaxation oscillation frequency of the laser chip within a certain current range, and the magnitude of the relaxation oscillation frequency is proportional to the square root of the difference between the applied bias current and its own threshold current. Therefore, in Fig. 2.36b, the 80% bandwidth of the spectrum initially increases as the current increases, and then as the bias current continues to increase, the output energy of the laser begins to concentrate at the peak of the spectrum, and the randomness decreases and the periodicity increases, but the output is still in a chaotic state. As reflected in the figure, the 80% bandwidth of the spectrum increases to the peak and then tends to decrease slowly, but the lowest value is still around 4.5 GHz. The chaotic output state of this hybrid integrated chaotic semiconductor laser does not change with the bias current, which is very favorable for the application.

In addition, when the bias current is fixed and the operating temperature of the chip is changed, the output of this hybrid integrated chaotic semiconductor laser is

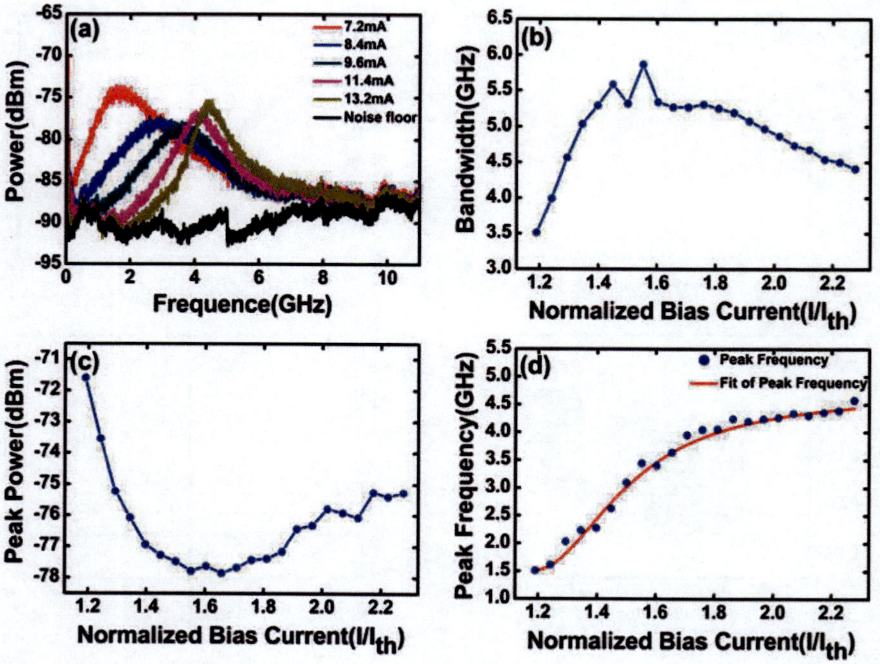

Fig. 2.36 **a** RF spectra of the hybrid integrated chaotic semiconductor laser at bias currents of 7.2 mA, 8.4 mA, 9.6 mA, 11.4 mA and 13.2 mA, respectively. **b** 80% bandwidth of the spectra, **c** peak power and **d** peak frequency of the laser at different normalized bias currents

found to be typical of a multiple period state. Figure 2.37 shows the time series, spectrum, spectra and autocorrelation curves of the laser at a bias current of 7.8 mA and a temperature of 18.8 °C. From Fig. 2.37a, the time series of the integrated chaotic semiconductor laser, it can be seen that the time series peaks at this time are around 11 mV and have a clear periodicity. The minimum positive period is about 0.408 ns, which corresponds to a frequency of about 2.45 GHz, as 24.5 cycles can be found in the 10 ns visual range of the time series. From the spectrum shown in Fig. 2.37b, it is obvious that a spike with an intensity of about − 54.6 dBm at a frequency of 2.46 GHz and a spike with an intensity of about − 76 dBm at a frequency of 4.92 GHz, indicating that the relaxation oscillation frequency of the laser is about 2.46 GHz at a bias current of 7.8 mA and a temperature of 18.8 °C. The output energy at this time is concentrated around the relaxation oscillation frequency. As seen in Fig. 2.37c, the central wavelength of the spectrum is about 1551.22 nm, and two pairs of spikes appear symmetrically on both sides of the main peak of the spectrum. The frequency difference between the two adjacent spikes is about 2.47 GHz, which is close to 2.46 GHz. It is obvious that when the bias current is 7.8 mA and the operating temperature is 18.8 °C, the laser is in a single period state, and the energy is mostly concentrated near the frequency of 2.46 GHz. The time series is fluctuating periodically.

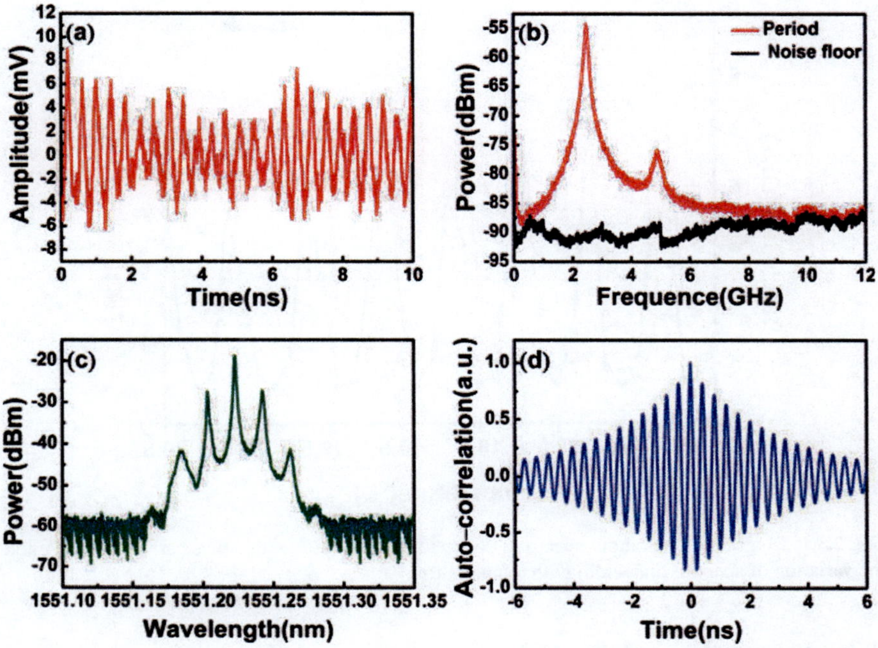

Fig. 2.37 Output state of an integrated chaotic semiconductor laser with a bias current of 7.8 mA at a temperature of 18.8 °C. **a** Time series **b** RF spectrum **c** optical spectrum **d** autocorrelation curve

In fact, when the bias current of the integrated chaotic semiconductor laser is kept constant and the operating temperature is varied, the output characteristics of the laser are found to vary periodically between the doubly periodic and chaotic states. The variation of the output spectral bandwidth and the spectral linewidth with temperature for a bias current of 8 mA is shown in Fig. 2.38.

The black dotted line in the figure indicates the spectral bandwidth, and the red square line indicates the spectral linewidth, where the spectral bandwidth is 80% bandwidth and the spectral linewidth is − 20 dB linewidth. As can be seen from the graph, with the change of temperature, the spectral bandwidth and spectral linewidth both show periodic changes, and the trend of the spectral bandwidth and spectral linewidth is basically the same, within the visible range of 3.5 °C change, 7 cycles can be found, the average minimum half-cycle is 0.25 °C, which is basically the same as the above experimental chaotic state and the single cycle state 0.3 °C change (the temperature controller The accuracy of temperature control is 0.1 °C).

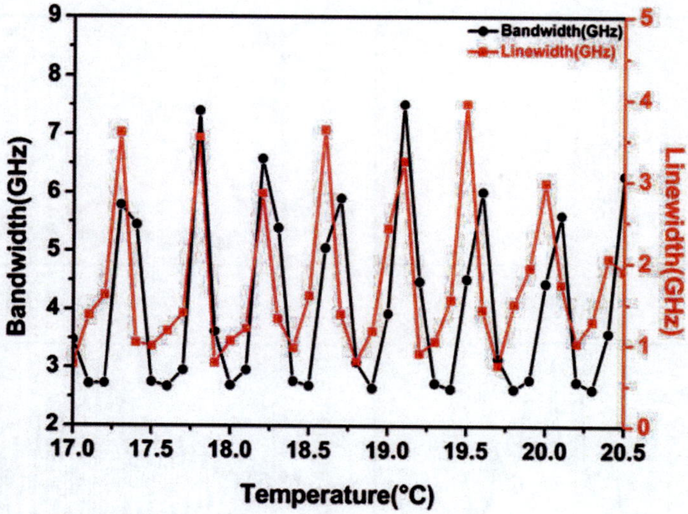

Fig. 2.38 Diagram of the output spectrum bandwidth of the laser with a bias current of 8 mA and the variation of spectral line width with temperature

2.3 Monolithic Integrated Chaotic Semiconductor Lasers

2.3.1 Structure Design of a Monolithic Integrated Chaotic Semiconductor Laser Chip

As a kind of photonic integrated chip, the monolithic integrated chaotic semiconductor laser has the advantages of compact structure, stable output, and suitable for mass production, etc. Through special design and fabrication, it can be controlled parametrically to output signals with different kinetic states. In the following, we present the chip structures of various time-delay-free and broadband monolithic integrated chaotic semiconductor lasers designed by our team.

1. **Time-delay-free, flat-broadband-spectrum monolithic integrated chaotic semiconductor lasers**

The Erbium-doped passive optical waveguide is used as a continuous scatterer to form a continuous distributed feedback cavity, and a non-isolated bi-directional amplified semiconductor optical amplifier chip is used to control the magnitude of the optical power injected into the left and right distributed feedback semiconductor laser chips and the feedback intensity of the passive optical waveguide to the left distributed feedback semiconductor laser chip, in order to solve the problems of chaotic laser light generated by semiconductor lasers with dark time delay characteristics, narrow signal bandwidth, uneven spectrum and unstable output. The structure of the designed time-delay-free, spectrally flat, broadband photonically integrated chaotic semiconductor laser is shown in Fig. 2.39.

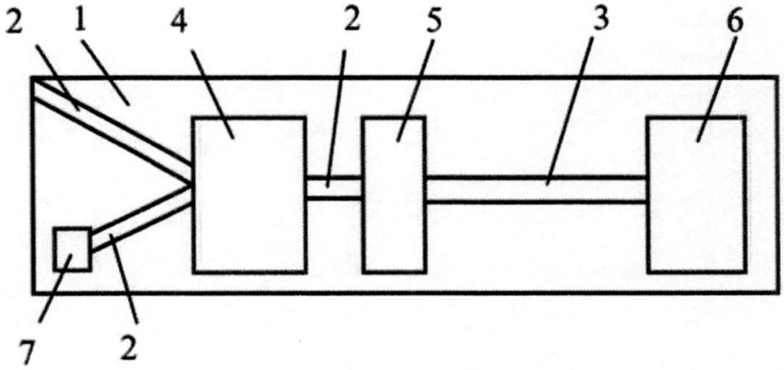

Fig. 2.39 Schematic diagram of a time-delay-free, spectrally flat, broadband photonically integrated chaotic semiconductor laser. 1—substrate, 2—optical waveguide, 3—erbium-doped passive optical waveguide, 4—left distributed feedback semiconductor laser chip, 5—semiconductor optical amplifier chip without isolated bidirectional amplification, 6—right distributed feedback semiconductor laser chip, 7—high-speed photodetector chip

2. Monolithic integrated laser chip for chaotic laser generation with highly scattered doped optical waveguide feedback

The monolithically integrated laser chip with high scattering doped light waveguide feedback for chaotic laser generation uses high scattering doped light waveguide random feedback for chaotic laser generation, which eliminates the time delay characteristics of the chaotic laser, and adopts the monolithic integrated structure for more compact structure and better stability.

The chip includes a DFB laser structure, a non-doped optical waveguide structure and a highly scattered doped optical waveguide structure, where the DFB laser structure and the highly scattered doped optical waveguide structure are distributed at the two ends of the non-doped optical waveguide structure, respectively. The bandgap wavelength of the semiconductor material for the DFB laser structure is 1.55 μm or greater than 1.55 μm, and the bandgap wavelength of the semiconductor material for the non-doped optical waveguide structure and the highly scattered doped optical waveguide structure is 1.45 μm or less than 1.45 μm. The active region material used in the DFB laser structure is the strain quantum well material. The doping elements of the high scattering doped optical waveguide structure can be silicon, iron, boron, and the sub-devices of the DFB laser structure, non-doped optical waveguide structure and high scattering doped optical waveguide structure are integrated on the same InP substrate using either quantum well mixing or selective region epitaxy. The only difference between the highly scattering doped and undoped optical waveguide structures is whether the active layer of the waveguide is doped or not. Figure 2.40 shows the structure of the designed monolithic integrated laser chip with high scattering doped optical waveguide feedback to generate chaotic laser, and Fig. 2.41 shows the specific growth diagram of the laser chip.

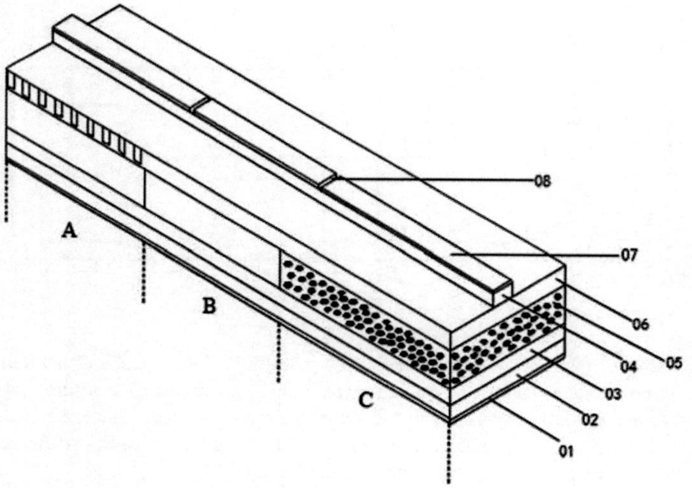

Fig. 2.40 Schematic diagram of the structure of a monolithic integrated laser chip with highly scattered doped optical waveguide feedback to generate chaotic light. A—DFB laser structure, B—undoped optical waveguide structure, C—high scattering doped optical waveguide structure

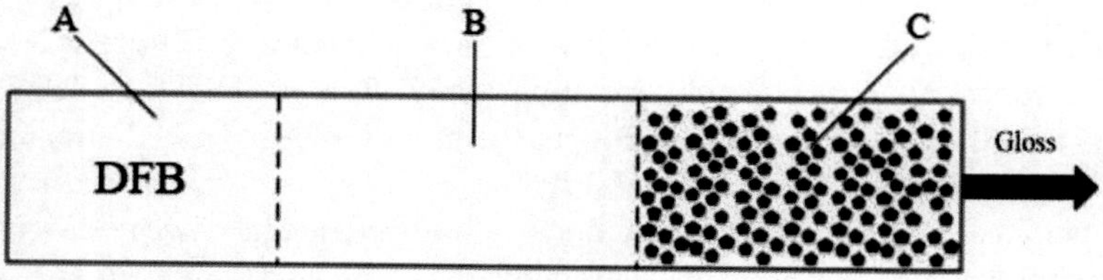

Fig. 2.41 Schematic diagram of the specific growth of the laser chip. 1—N^+ electrode layer, 02—substrate, 03—lower confinement layer, 04—waveguide layer, 05—active layer, 06—upper confinement layer, 07—P^+ electrode layer, 08—isolation trench

3. **InP-based monolithic integrated chaotic semiconductor lasers based on randomly distributed Bragg reflection grating**

The schematic structure of the designed InP-based monolithic integrated chaotic semiconductor laser chip based on a randomly distributed Bragg reflection grating is given in Fig. 2.42. The laser chip mainly consists of the following parts: left DFB laser, bi-directionally amplified semiconductor optical amplifier, randomly distributed Bragg reflection grating and right DFB laser. The random distributed Bragg reflection grating part is etched with a random distributed Bragg reflection grating layer using the phase mask method; the InGaAsP upper limiting layer is etched with a distributed feedback Bragg grating in the area corresponding to both the left and right DFB lasers. The length of the left DFB laser is 500 μm, the length of the bi-directional amplifier is 200 μm, the length of the randomly distributed Bragg reflection grating is 4–10 mm, and the length of the right DFB laser is 500 μm. The

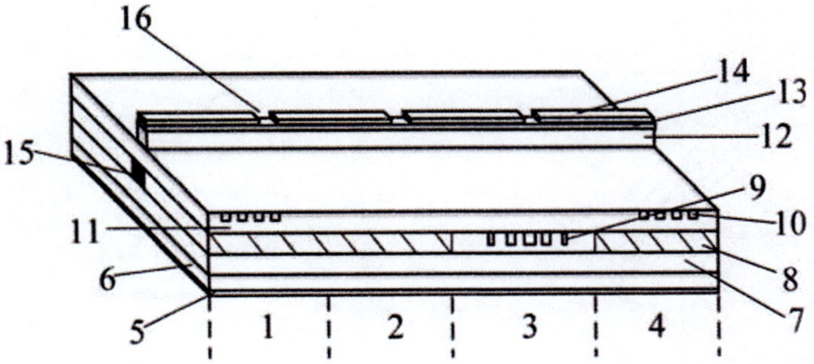

Fig. 2.42 Schematic of the structure of an InP-based monolithic integrated chaotic semiconductor laser chip based on randomly distributed Bragg reflection gratings. 1—Left DFB laser, 2—bidirectional semiconductor optical amplifier, 3—randomly distributed Bragg reflection grating section, 4—right DFB laser, 5—N^+ electrode layer, 6—N-type substrate, 7—InGaAsP lower confinement layer, 8—undoped InGaAsP multi-quantum well active region layer, 9—randomly distributed Bragg reflection grating layer 10—distributed feedback Bragg grating, 11—InGaAsP upper confinement layer, 12—P-type heavily doped InP cover layer, 13—P-type heavily doped InGaAs contact layer, 14—P^+ electrode layer, 15—outlet, 16—isolation trench

refractive index difference exists between the randomly distributed Bragg reflection grating, the bi-directional amplifier and the right DFB laser, which plays the role of optical coupling and optical random feedback. The randomly distributed Bragg reflection grating provides random optical feedback to the left and right DFB lasers, resulting in a chaotic laser with no time delay characteristics. The materials of the distributed feedback Bragg grating are InP and InGaAsP with thicknesses of 50–200 nm, and the Bragg grating period is 290 nm, corresponding to the excitation peak in the 1550 nm band. There is a parameter mismatch between the left and right DFB lasers, and the frequency difference between the center wavelengths of the two is 10–15 GHz, and the output power deviation between the two is less than 70%.

4. **Random scattering optical feedback InP-based monolithic integrated chaotic semiconductor laser**

Figure 2.43 shows the structure of the InP-based monolithic integrated chaotic semiconductor laser chip with random scattered optical feedback, from which it can be seen that the laser chip mainly consists of the following structures: left DFB laser, bi-directional amplified semiconductor optical amplifier, left passive optical waveguide, doped passive optical waveguide, right passive optical waveguide and right DFB laser. The left passive optical waveguide, doped passive optical waveguide and right passive optical waveguide are epitaxially grown on the lower limiting layer of InGaAsP at the same time, and then doped with impurities in the doped passive optical waveguide part, the impurities can be gain medium erbium particles or zinc particles. The doped passive optical waveguide part is doped with a certain concentration of impurities, and the layer where the impurities are located corresponds to the undoped InGaAsP multi-quantum well active region layer; the impurities can generate strong random backscattered light per unit length when continuous light

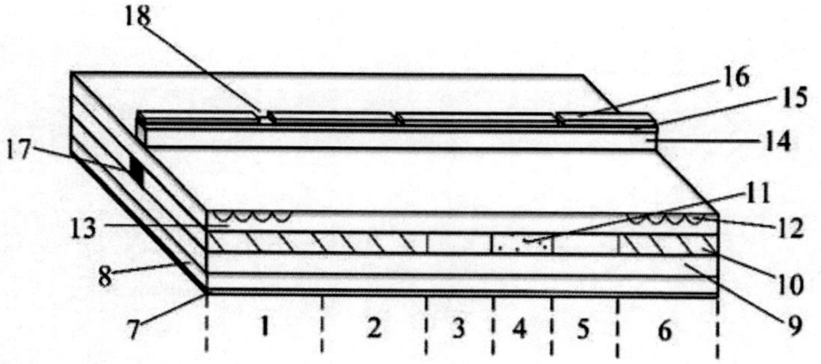

Fig. 2.43 Schematic of the structure of an InP-based monolithic integrated chaotic semiconductor laser chip with random scattered optical feedback. 1—Left DFB laser, 2—bidirectionally amplified semiconductor optical amplifier, 3—left passive optical waveguide section, 4—doped passive optical waveguide section, 5—right passive optical waveguide section, 6—right DFB laser, 7—N^+ electrode layer, 8—N-type substrate, 9—InGaAsP lower confinement layer, 10—undoped InGaAsP multi-quantum well active region layer, 11—impurities, 12—distributed feedback Bragg grating, 13—InGaAsP upper confinement layer, 14—P-type heavily doped InP cover layer, 15—P-type heavily doped InGaAs contact layer, 16—P^+ electrode layer, 17—outlet, 18—isolation trench

passes through, providing random optical feedback perturbation to the left and right DFB lasers. There is a parametric mismatch between the left and right DFB lasers, with the frequency difference between the two central wavelengths being 10–15 GHz and the output power deviation between the two being less than 70%. The left DFB laser is connected to the left side of the bi-directionally amplified semiconductor optical amplifier on the right side of the integrated chip, and the right side of the bi-directionally amplified semiconductor optical amplifier is connected to the left side of the doped passive optical waveguide section after a left passive optical waveguide section, and the right side of the doped passive optical waveguide section is connected to the left side of the right DFB laser after a right passive optical waveguide section. The optical inter-injection perturbation process is realized.

5. **Monolithic integrated chaotic lasers based on random grating feedback**

The structure of the monolithic integrated chaotic laser chip based on random grating feedback is shown in Fig. 2.44, which mainly includes a DFB laser region and a random feedback region. The random feedback area provides random multi-feedback to the light emitted from the DFB laser area, and the random feedback grating is produced in the active layer corresponding to the random feedback area. The structure uses a random grating feedback structure to generate a chaotic laser, completely eliminating the time delay characteristics of a single-cavity optical feedback chaotic laser.

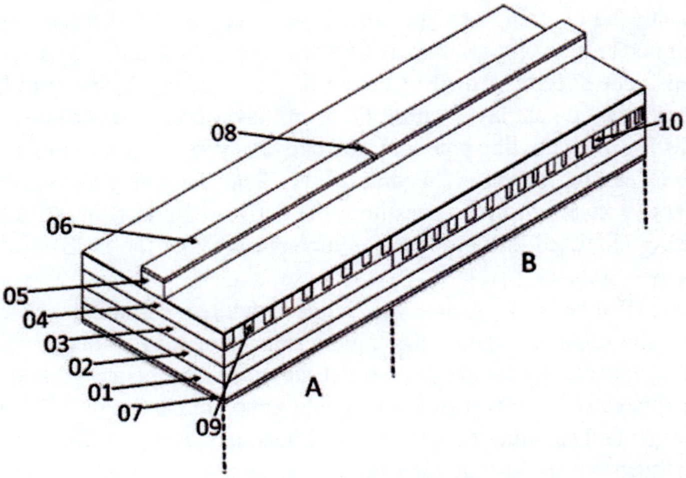

Fig. 2.44 Schematic diagram of a monolithic integrated chaotic laser chip based on random grating feedback. 1—Substrate; 2—lower confinement layer; 3—active layer, 4—upper confinement layer, 5—waveguide layer, 6—P^+ electrode layer, 7—N^+ electrode layer, 8—isolation trench, 9—distributed feedback Bragg grating layer, 10—random feedback grating layer; A—distributed feedback semiconductor laser region, B—random feedback region

2.3.2 *Simulation Study of a Three-Section Monolithic Integrated DFB Laser*

Based on the optical cross-injection coupling method, we propose a three-stage monolithic integrated DFB laser structure, which can be directly fabricated based on the current monolithic integration process for monolithic growth and stable performance for large-scale applications. The feasibility of this structure to produce high quality laser is verified by simulation before fabrication. Based on the transmission line model, the system of traveling wave rate equations of the integrated laser is derived, and the static and dynamic characteristics of the three-stage monolithic integrated DFB laser are simulated to provide theoretical support for the fabrication and application of the monolithic integrated photonic device. We use MATLAB simulation software to simulate and calculate the dynamic characteristics of the integrated laser to generate microwave signals, and analyze the effects of parameters such as laser grating coupling constants, bias currents of each segment and detuning on the microwave signals of the integrated laser, which provides theoretical support for the generation of microwave signals from integrated semiconductor lasers.

1. **Simulation principle and model of the semiconductor laser**

The structure of the three-section monolithic integrated DFB laser is shown in Fig. 2.45. The designed laser consists of two DFB lasers and a passive optical waveguide (WG). The laser emitted from the two DFB lasers is coupled by the passive waveguide, and the electrical and optical characteristics of the designed integrated laser are analyzed by using MATLAB and other software. The three-stage monolithic integrated laser adopts a symmetric structure with the same DFB laser structure at both ends. The specific growth structure is shown in Fig. 2.46, from bottom to top, N-type metal contact layer, substrate layer, lower confinement layer, quantum well (consisting of alternating potential well barrier layers), upper confinement layer, waveguide layer, and P-type metal contact layer, The alternately arranged potential well barriers (9 layers in total) constitute the active region of the DFB laser, the Bragg grating is etched in the upper confinement layer of the DFB laser, and the cavity length is 300 μm.

The above DFB laser uses a ridge waveguide structure for simulation calculations. In the laser, the ridge waveguide has the advantages of long primary mode cutoff wavelength, reduced waveguide size, wider single-mode operating band, and low equivalent impedance, so the ridge waveguide structure can be used in the monolithically integrated structure to optimize the integrated laser performance. The WG segment in the middle is 300 μm long and also adopts the ridge waveguide structure. The growth structure is shown in Fig. 2.47, from bottom to top, the N-type metal contact layer, substrate layer, waveguide transmission core layer, waveguide layer, and P-type metal contact layer. The thickness of the WG transmission core layer is

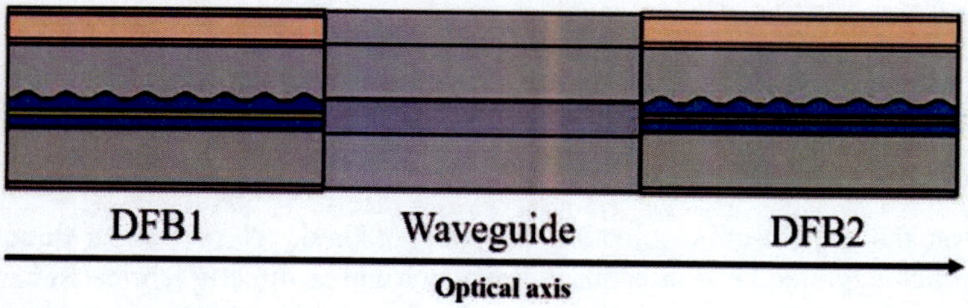

Fig. 2.45 Schematic diagram of the three-section monolithic integrated DFB-SL structure

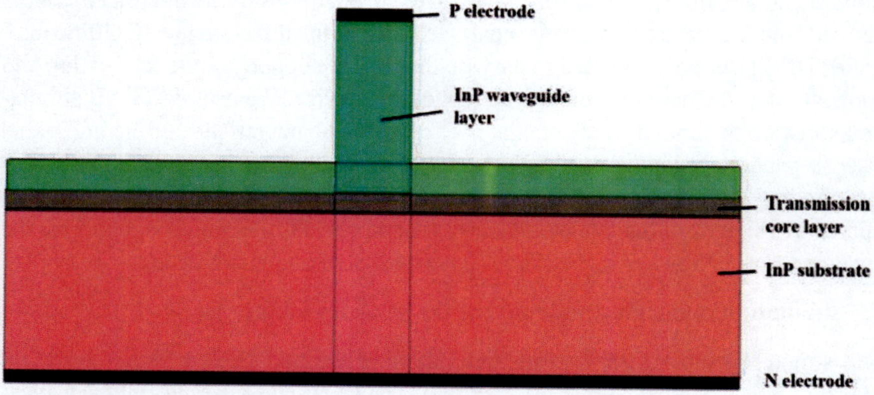

Fig. 2.46 Schematic diagram of the DFB segment epitaxial layer structure

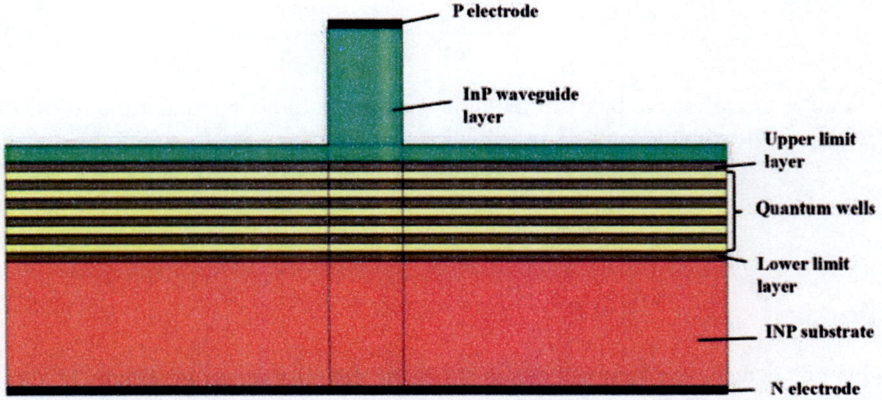

Fig. 2.47 Schematic diagram of the WG segment epitaxial layer structure

comparable to the total thickness of the upper and lower confinement layers and the active region in the DFB laser.

The Maxwell-Bloch equations for the electric field E, polarization P and particle number reversal W are derived using the two-energy atomic model, and based on this, we construct a set of traveling wave rate equations for a semiconductor laser and simulate its related characteristics. The two-energy atomic model is shown in Fig. 2.48, which uses quantum mechanical theory to quantize the atomic energy, considering only the upper and lower energy levels of the laser leap.

In the two-energy atomic model, the electric field Maxwell's equations are

$$\nabla\varepsilon(z,t)-\frac{1}{c^2}\frac{\partial^2\varepsilon(z,t)}{\partial t^2}=\mu_0\frac{\partial^2 P(z,t)}{\partial t^2} \tag{2.58}$$

where ε(z, t) is the electric field vector function, c = 3 × 108 m/s is the speed of light in a vacuum, ε is the dielectric constant, μ0 is the vacuum permeability, and P(z, t) is the polarization vector function. Assuming that the refractive index of the laser medium is uniform and the spatial pattern of linear polarization in the x and y directions varies with the propagation of the z-axis, the polarization of the field and matter is reduced to a scalar that propagates only in the z-direction, and Eq. (2.58) can be simplified to the following equation:

Fig. 2.48 Diagram of a two-level atomic model

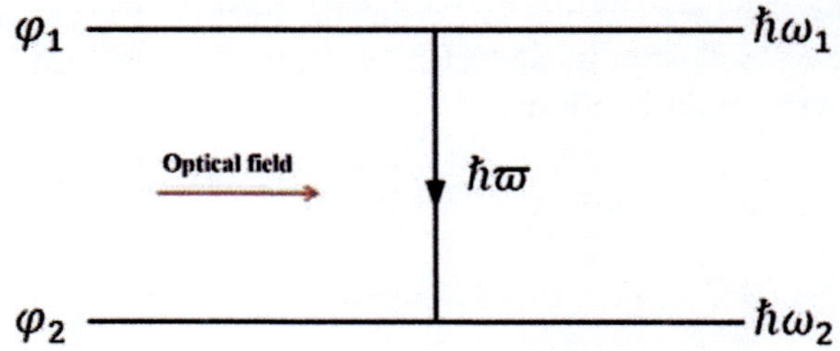

$$\frac{\partial^2 \varepsilon(z,t)}{\partial z^2} - \frac{\eta^2}{c^2}\frac{\partial^2 \varepsilon(z,t)}{\partial t^2} = \mu_0 \frac{\partial^2 P(z,t)}{\partial t^2} \tag{2.59}$$

η denotes the refractive index of the laser medium. z-direction, under the retarded envelope approximation, the above equation can be written as:

$$(z,t) = \frac{1}{2}E(z,t)\exp[i(kz - w_0 t)] + c^* \tag{2.60}$$

$$P(z,t) = \frac{1}{2}P(z,t)\exp[i(kz - \omega_0 t)] + c^* \tag{2.61}$$

where c* denotes the complex conjugate of the previous term, wave number k = $\eta\omega_0/c$, ω_0 is the angular oscillation frequency, E(z, t) is the electric field amplitude function, and P(z, t) is the polarization amplitude function. Ignoring the second-order small infinity, the following amplitude equations for the electric field and polarization can be obtained by substituting Eqs. (2.60) and (2.61) into Eq. (2.59):

$$\frac{\partial E(z,t)}{\partial z} + \frac{\eta}{c}\frac{\partial E(z,t)}{\partial t} = i\frac{k}{2\varepsilon_0\eta^2}P(z,t) \tag{2.62}$$

In the two-energy atomic model, there are absorption and emission of light, and the Bloch equation can be derived by considering the atomic quantum states. In the two-energy atomic model, the Hamiltonian quantity

$$\mathrm{H} = \mathrm{H}_0 - \mu \cdot \varepsilon \tag{2.63}$$

where H_0 is the Hamiltonian quantity unperturbed by the electric field vector, μ = e · r is the transition moment between two energy levels (r denotes the position vector, and the fundamental charge $e = 1.6 \times 10^{-19}$ °C). For the two-level eigenstates φ_1 and φ_2, the eigenenergy of each of the energy levels is $\hbar\omega_j$ ($j = 1, 2$; Planck's constant $\hbar = 6.626 \times 10^{-34}$ J s), and the interaction between the two energy levels is

$$\langle\varphi_j|H_0|\varphi_k\rangle = \hbar\omega_j\delta_{jk} \tag{2.64}$$

In the formula, δ_{jk} represents the Kronecker function (Kronecker function: if the input two independent variables are equal, the value of δ is 1, otherwise the value of δ is 0), and the angular frequency of the emission or absorption of the two energy levels is $\omega_A = \omega_2 - \omega_1$. When the light field exists, the quantum state $|\psi\rangle$ of the two-level atom is calculated by the linear superposition of the respective quantum states of the two-level:

$$|\psi\rangle = c_1(t)\exp(-i\omega_1 t)|\varphi_1\rangle + c_2(t)\exp(-i\omega_2 t)|\varphi_2\rangle \tag{2.65}$$

The above equation is replaced by Schrödinger's equation, and the coefficients c_1 and c_2 of the respective quantum states of the two energy levels are solved by the following Bloch coupling equation:

$$\frac{dc_1(t)}{dt} = \frac{c_2(t)}{i\hbar} \exp(-i\omega_A t)\langle\varphi_1|\boldsymbol{\mu}\cdot\boldsymbol{\varepsilon}|\varphi_2\rangle \tag{2.66}$$

$$\frac{dc_2(t)}{dt} = \frac{c_1(t)}{i\hbar} \exp(-i\omega_A t)\langle\varphi_2|\boldsymbol{\mu}\cdot\boldsymbol{\varepsilon}|\varphi_1\rangle \tag{2.67}$$

Using Avogadro's constant NA (NA $= 6.022 \times 10^{23}$), the medium macroscopic polarization function is defined by the following equation:

$$P = N_A\langle\psi|\mu|\psi\rangle \tag{2.68}$$

The following equations can be derived from Eqs. (2.62) and (2.65):

$$P = N_A\{p(t)\mu_{12} + p^*(t)\mu_{21}\} \tag{2.69}$$

where $\mu_{ij}(i, j = 1, 2)$ is the moment of transition from the lower to the upper energy level state and vice versa. The microscopic polarization function p(t) of each atom is given by the following equation:

$$p(t) = c_1^*(t)c_2(t)\exp(-i\omega_A t) \tag{2.70}$$

$$\mu_{ij} = \langle\varphi_j|\mu|\varphi_i\rangle \tag{2.71}$$

Finally, by substituting the above equations into Eqs. (2.63) and (2.64), the atomic polarization equation is obtained:

$$\frac{dp(t)}{dt} = -i\omega_A p(t) + \frac{i}{\hbar}E(t)\mu_{21}w(t) \tag{2.72}$$

The population inversion distribution of the two-level atomic model is $\omega(t) = |c2(t)|^2 - |c2(t)|^2$, and the population inversion equation is as follows:

$$\frac{dw(t)}{dt} = \frac{2}{i\hbar}E(t)\{p^*(t)\mu_{21} - p(t)\mu_{12}\} \tag{2.73}$$

The field and polarization equations are rearranged, and the complete laser rate equation, which is the same as the Lorentz chaos equation, taking the time variation of the particle number reversal in the laser medium into account. By differentiating Eq. (2.58) and combining Eqs. (2.66) and (2.69), the following macroscopic polarization function equation can be obtained:

$$\frac{d\boldsymbol{P}(z,t)}{dt} = -i(\omega_A - \omega_0)\boldsymbol{P}(z,t) + \frac{i\mu^2}{2\hbar}W(z,t)\big[E(z,t) + E^*(z,t)\exp\{-2i(kz - \omega_0 t)\}\big] \quad (2.74)$$

where ω_A is the angular frequency of light emitted or absorbed in two-level atoms, $\mu = |\mu 12|$, Planck's constant $\hbar = 6.626 \times 10^{-34}$ J s), $W(z, t) = N_A\omega(z, t)$ is the macroscopic particle number inversion function, and E(z, t) is the electric field function. From formula (2.70), the particle number inversion function equation can be obtained by the following formula:

$$\frac{dW(z,t)}{dt} = \frac{1}{i\hbar}\big[E(z,t)P^*(z,t) - E(z,t)P(z,t)\exp\{2i(kz - \omega_0 t)\} - c^*\big] \quad (2.75)$$

Only the variables that change slowly relative to the optical frequency are considered in the equation, so the fast vibration term related to the angular frequency $2\omega_0$ in Eqs. (2.71) and (2.72) can be neglected. The semiconductor laser requires an externally pumped laser, so an additional term is added to Eq. (2.72) to represent the actual pumped laser term. In addition, considering the existence of damped oscillations, the same term is added to Eqs. (2.59), (2.71) and (2.72) to finally obtain Maxwell-Bloch equations representing the electric field E, polarization P and particle number inversion W, respectively, under the slow-variance envelope approximation:

$$\frac{\partial E(z,t)}{\partial z} + \frac{\eta}{c}\frac{\partial E(z,t)}{\partial t} = i\frac{k}{2\varepsilon_0\eta^2}P(z,t) - \frac{\eta}{2T_{ph}c}E(z,t) \quad (2.76)$$

$$\frac{\partial P(z,t)}{\partial t} = -i(\omega_A - \omega_0)P(z,t) + \frac{i\mu^2}{2\hbar}E(z,t)W(z,t) - \frac{P(z,t)}{T_2} \quad (2.77)$$

$$\frac{dW(z,t)}{dt} = \frac{1}{i\hbar}\big\{E(z,t)P^*(\mathrm{z,t}) - E^*(z,t)P(z,t)\big\} + \frac{W_0 - W(z,t)}{T_1} \quad (2.78)$$

where W_0 is the particle number reversal value at the laser threshold, and Tph, T1 and T_2 denote the photon lifetime, polarization (transverse relaxation) time, and particle number reversal (longitudinal relaxation) time, respectively.

Assuming that the total length of the m-segment multi-segment laser Sr is L, and the length of each segment is $l_r(r = 1, 2, \ldots, m)$, the field is expressed as a superposition of left- and right-going waves, considering that the light emitted by the laser is quasi-monochromatic near the carrier frequency. The optical field function $E(z, t) = (E^+, E^-)T$ and the polarization function $P(z, t) = (P^+, P^-)T$ along the direction of the optical axis of the multi-segment laser lead to the following equation:

$$-i\partial_t E^{\pm} = v_g\left[(\pm i\partial_z - \beta(n))E^{\pm} - \kappa E^{\mp} + i\frac{\overline{g}}{2}\left(E^{\pm} - P^{\pm}\right)\right] \quad (2.79)$$

$$-i\partial_t P^{\pm} = -i\overline{\gamma}\left(E^{\pm} - P^{\pm}\right) + \overline{\omega}P^{\pm} \quad (2.80)$$

$$\beta(n) \stackrel{\text{def}}{=} \delta - i\frac{\alpha}{2} + \tilde{n}(n) + \frac{i}{2}g(n)$$
$$g(n) \stackrel{\text{def}}{=} g'(\tilde{n} - n_{tr})$$
$$\tilde{n}(n) = \frac{a_H g(n)}{2} \tag{2.81}$$

where v_g is the group velocity, κ is the grating coupling factor, β is the propagation factor, δ is the static detuning, α is the internal optical loss of the laser, g′ denotes the effective secondary gain value including the lateral constraint factor, n_{tr} is the laser transparent carrier density, α^H denotes the linewidth enhancement factor. The Lorentz function is used in the polarization equation to determine the gain dispersion of each laser segment in the frequency domain. $\overline{g}$ defines the Lorentz curve height ($\overline{g} > 0$), $\overline{\gamma}$ defines the full width of the Lorentz half height, and $\overline{\omega}$ defines the peak frequency.

The boundary condition of the optical field at the end face of the multi-segment laser is

$$E^+(0,t) = r_0 E^-(o,t) \quad E^-(L,t) = r_L E^+(L,t) \tag{2.82}$$

The field function norm ($|E(z, t)|^2 = |E^+|^2 + |E^-|^2$) represents the local photon density (that is, the z-direction power divided by the global constant $v_{g\sigma} \cdot hc/\lambda 0$), and the variable n(t) linearly enters the gain function g(n) to represent the average carrier density of the multi-segment laser, and the carrier nr(t) rate equation is

$$\partial_t n_r = I_r(z,t) - U_N - \frac{v_{gr}}{l_r}\Im m\left(\sum_{v=\pm} E^{v*}(2\beta^v E^v - i\overline{g}(E^v - P^v))\right)$$
$$U_N = A_r n_r + B_r n_r^2 + C_r n_r^3, r = 1, \ldots, m \tag{2.83}$$

where $I_r(z, t)$ is the injection current at the corresponding position, A^r indicates the SL non-radiation composite coefficient, which is the internal light loss caused by non-radiation capture (reflection, absorption), Br indicates the SL diatomic radiation composite coefficient, and Cr indicates the SL intermittent composite coefficient.

The values of the optical field, polarization and carrier density at the initial moment t0 are

$$E(z,t_0) = E_{in}(z), P(z,t_0) = P_{in}(z), n(t_0) = n_{in} \tag{2.84}$$

The final equation for the traveling wave rate of the multiband laser is obtained as follows:

$$\frac{d}{dt}n(z,t)|_{z\in S_r} = Ir(z,t) - \left(A_r n + B_r n^2 + C_r n^3\right) - v_{gr}\Im m\left(\sum_{v=\pm} E^{v*}(2\beta^v E^v - i\overline{g}(E^v - P^v))\right)_r \tag{2.85}$$

$$-i\frac{\partial}{\partial t}\begin{pmatrix} E(z,t) \\ P(z,t) \end{pmatrix} = \begin{pmatrix} H_0(\beta^\pm) + iv_{gr}\frac{\overline{g}}{2} & -iv_{gr}\frac{\overline{g}}{2} \\ -i\overline{\gamma} & \omega_P + i\overline{\gamma} \end{pmatrix}\begin{pmatrix} E \\ P \end{pmatrix} \tag{2.86}$$

$$\beta^\pm(n, |E|^2)|_{z\in S_r} = \sigma - i\frac{\alpha}{2} + \frac{i + \alpha H}{2}g'(n - n_{tr}) \tag{2.87}$$

2. **Simulation analysis of dynamic characteristics of the three-section monolithic integrated DFB laser**

In this section, the dynamic characteristics of a three-stage monolithic integrated DFB laser are simulated and analyzed using the traveling wave rate equation of a semiconductor laser.

(1) Effect of grating on the dynamic characteristics of a three-section monolithic integrated DFB laser.

The grating settings of the two-end lasers affect the dynamic characteristics of the three-stage monolithic integrated DFB laser. For the simulation of multi-segment integrated lasers, the laser size and related parameters are given priority. Three-section monolithic integrated laser excitation wavelength $\lambda = 1550$ nm, non-radiative complex coefficient A $= 1 \times 10^8$ s^{-1}, two-atom radiative complex coefficient B $=$ 7×10^{-10} cm^3 s^{-1}, Other complex coefficient C $= 1 \times 10^{-29}$ cm^6 s^{-1}, transparent carrier density $N_0 = 1 \times 10^{18}$ cm^{-3}, group velocity $v_g = 8.4 \times 10^7$ ms^{-1}, bipolar diffusion factor $a = 12$ cm^{-2} s^{-1}, maximum saturation gain $\chi = 72$ cm^{-1}, internal intensity loss coefficient $\alpha = 14.4$ cm^{-1}, and in-band relaxation time $\tau_k = 125$ fs. Considering the device size in the integration process, the total length L of the three-section integrated laser is set to 1050 μm, with the length of the DFB laser at both ends $L_{DFB} = 350$ μm and the length of the middle WG section $L_{WG} = 350$ μm.

The bias currents of both DFB lasers were set to 1.5 I_{th}, while the WG section bias current values and bandgap detuning were set to 0. The grating constant parameters of DFB laser 2 were changed (the DFB laser 2 grating coupling factor κ was set to 12.5, 25 and 37.5, respectively), and the time series, optical spectrum and frequency spectrum of the integrated laser were analyzed separately. Figures 2.49, 2.50, and 2.51 show the results of dynamic characteristics of κ values at 12.5, 25 and 37.5, respectively.

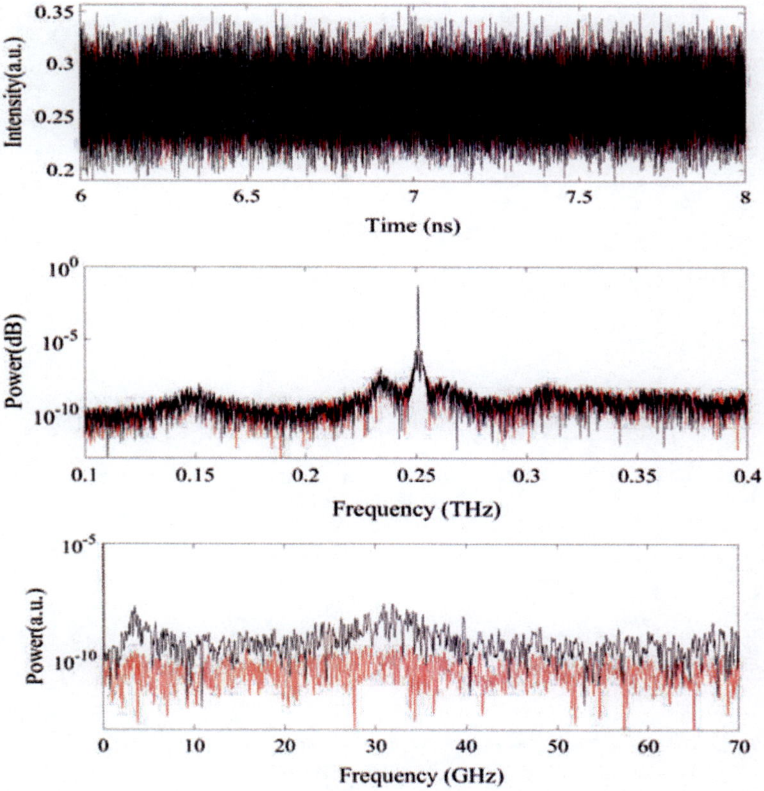

Fig. 2.49 Plot of the dynamic characteristics of the DFB laser 2 with grating constant $\kappa = 12.5$. **a** Time series, **b** optical spectra, **c** RF spectra

Figure 2.49 shows the dynamic signal results of the generated signal at the DFB laser 2 grating coupling factor $\kappa = 12.5$ in the three-stage monolithic integrated laser. The red curve indicates the characteristics related to the laser signal excited by DFB laser 1, and the black curve indicates the characteristics of the laser signal excited by DFB laser 2. The time series diagram (a) shows the irregular output of the signal; the two laser curves in the spectral diagram (b) completely coincide without frequency detuning; the peak frequency in the spectral diagram (c) is 1.342 GHz.

Figure 2.50 shows the dynamic signal results of the signal generated when the grating coupling factor $\kappa = 25$ of DFB laser 2 in the three-section monolithic integrated laser, the red curve is the characteristics related to the laser signal excited by DFB laser 1, and the black curve is the characteristics of the laser signal excited by DFB laser 2. The time series diagram (a) has no period and the time series has no regular output, but there is a clear difference between the time series intensity of the two lasers, and the time series intensity of DFB laser 2 is much larger than that of DFB laser 1; although the time series intensity of the two lasers is obviously different, the two laser curves in the spectral diagram (b) completely coincide, and

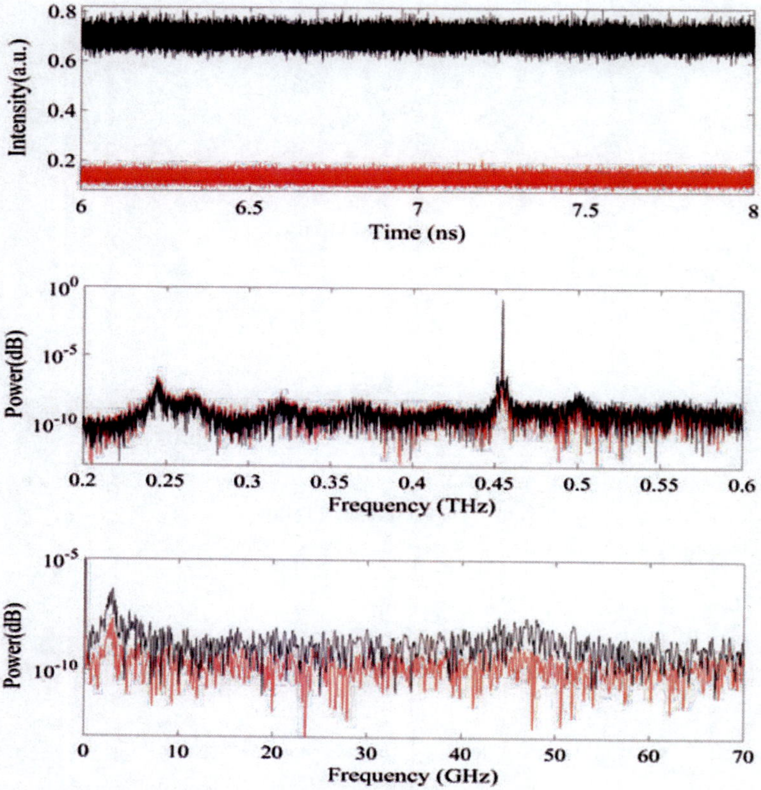

Fig. 2.50 Plot of the dynamic characteristics of the DFB laser 2 with grating constant κ = 25. **a** Time series, **b** optical spectra, **c** RF spectra

there is still no frequency detuning; the peak frequency of the spectral diagram (c) is 3 GHz, and the intensity is slightly increased compared with that of Fig. 2.49. The peak frequency of the spectrum (c) is 3 GHz and the intensity is slightly increased compared with Fig. 2.49.

Figure 2.51 shows the dynamic signal results of the generated signal at the grating coupling factor κ = 37.5 for the DFB laser 2 in the three-section monolithic integrated laser, the red curve and the black curve are the characteristics related to the laser signal excited by DFB laser 1 and DFB laser 2, respectively. The time series diagram (a) still has no period, the time series is irregularly output, the time series intensity of DFB laser 2 is larger than that of DFB laser 1, but the difference is smaller than the time series intensity of the two lasers in Fig. 2.50; the two laser curves in the spectral diagram (b) still have no frequency detuning, but the difference between the two peaks of the spectrum is significantly larger than that of the spectrum in Fig. 2.50; the peak frequency of the spectral diagram (c) is 3.6 GHz, and the intensity is comparable to that of Fig. 2.49.

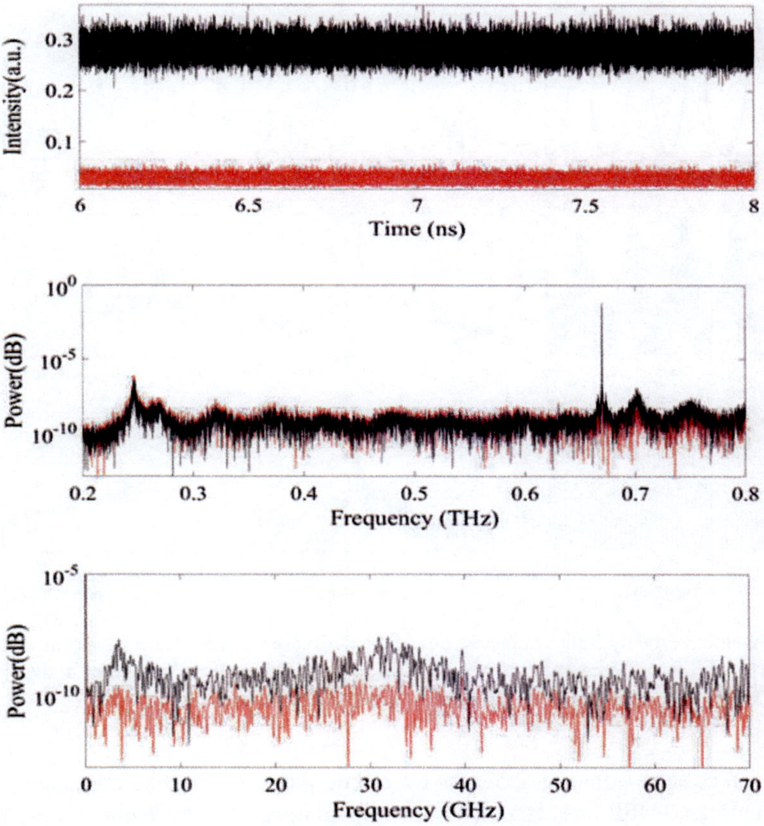

Fig. 2.51 Plot of the dynamic characteristics of the DFB laser 2 with grating constant κ = 37.5. **a** Time series, **b** optical spectra, **c** RF spectra

(2) Effect of current on the dynamic characteristics of a three-stage monolithic integrated DFB laser.

This subsection simulates and analyzes the effect of different section currents on the dynamic characteristics of the three-section monolithic integrated DFB laser, and the dimensional parameters and related basic parameters of the multi-section integrated laser are set as in subsection (1).

By using the control variable method, the bias current of DFB laser 2 in the three-stage DFB laser was changed (the bias current of DFB laser 2 was set to 2 Ith, 4 Ith, and 6 Ith, respectively), and the dynamic characteristics of the overall laser output signal were observed and analyzed. In this subsection, the grating coupling factor of the two DFB lasers (left DFB laser 1 and right DFB laser 2) is ensured to be the same, and the detuning of the two DFB lasers is set to 1.5 GHz, the bias current of DFB laser 1 is set to 1.5 I_{th}, and the bias current of the WG section is set to 0. The time series, spectrum and frequency spectrum of the integrated laser are analyzed separately, and the results are shown in Fig. 2.52. The results are shown in Fig. 2.52.

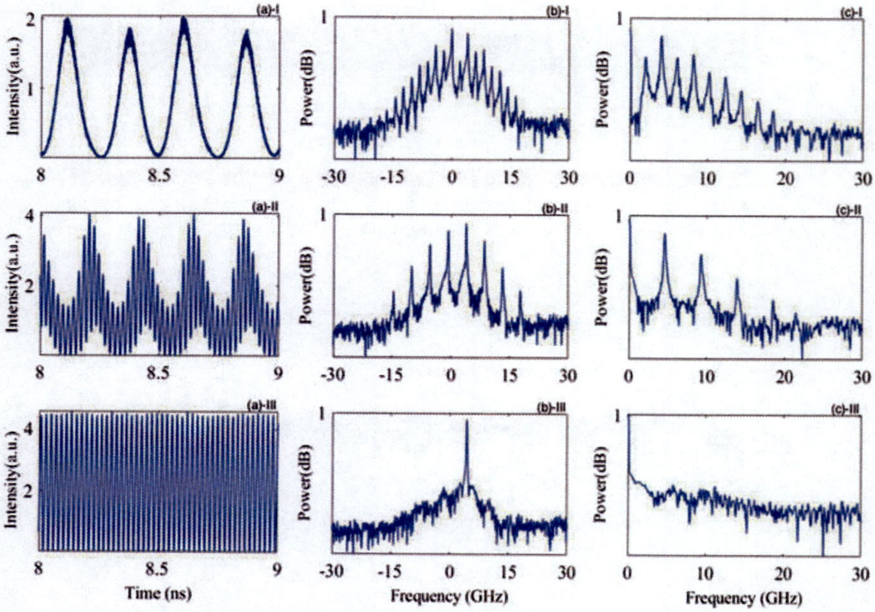

Fig. 2.52 Plot of the dynamic characteristics of the DFB laser 2 with different bias currents. I, II, III denote DFB laser 2 bias currents of 2 I_{th}, 4 I_{th}, 6 I_{th} respectively; **a**–**c** are time series, optical spectra and RF spectra of the integrated laser respectively

As can be seen from Fig. 2.52, the overall output signal of the three-band monolithic integrated DFB laser is a microwave signal under the above conditions. When the DFB laser 2 bias current is 2 Ith, the frequency locking of the multi-segment laser used in this section occurs, and there are two main modes in the spectrum, 458 GHz and 462 GHz, respectively, and the highest energy frequency component in the spectrum is the difference frequency of the two modes, 4.105 GHz. When the bias current of DFB laser 2 is 4 I_{th}, the frequency component caused by the beat frequency is reduced, and when the bias current of DFB laser 2 is 6 I_{th}, a strong injection lock state is formed, and the beat frequency component in the spectrum no longer exists.

Using the same control variable method, we changed the bias current of the WG section in the three-section monolithic integrated DFB laser (the WG bias currents were set to 2 I_{th}, 3 I_{th}, 4 I_{th}, and 5 I_{th}, respectively), and the dynamic characteristics of the overall laser output signal were observed and analyzed. The internal Bragg gratings of the two laser segments DFB laser 1 and DFB laser 2 were set as gratings with the same period, the bandgap detuning between DFB laser 2 and DFB laser 1 was set to 800 GHz, the bias current of DFB laser 1 was set to 1.5I_{th}, the bias current of DFB laser 2 was set to 5 I_{th}, and the detuning of the WG segment was zero, and the analysis was obtained as shown in Fig. 2.53. The results are shown in Fig. 2.53. It can be seen from Fig. 2.53 that under the above conditions, the overall output signal of the three-band monolithic integrated DFB laser is still a microwave signal, and a microwave signal with a frequency of 11.76 GHz is obtained when the bias current of the WG is 2 I_{th}, which coincides with the difference frequency between the main modes in the spectrum. Moreover, it can be seen that the detuning between DFB laser 2 and DFB laser 1 is not the same as the obtained difference frequency, which is the result of the combined effect of the mode and grating in the integrated laser cavity. And with the gradual increase of the waveguide bias current, the frequency of the integrated laser output microwave remains basically unchanged, indicating that when there is no detuning in the waveguide region, its bias current has little effect on the microwave generation, which may be caused by the injection locking of DFB laser 2.

(3) Effect of detuning on the dynamic characteristics of a three-stage monolithic integrated DFB laser.

First, the basic parameters related to the three-band monolithic integrated DFB laser are set according to subsection (1), which is identical to it. In this subsection, the bandgap detuning of the WG segment is varied and its effect on the overall laser output signal is analyzed. The internal Bragg gratings of the two laser segments in the three-band monolithic integrated DFB laser are set to gratings with the same period, the bandgap detuning of DFB laser 1 and DFB laser 2 are set to 5 GHz, the bias currents of both DFB laser 1 and DFB laser 2 are set to 1.5Ith, the injection current of WG segment is set to 2 Ith, and the bandgap detuning of WG segment is

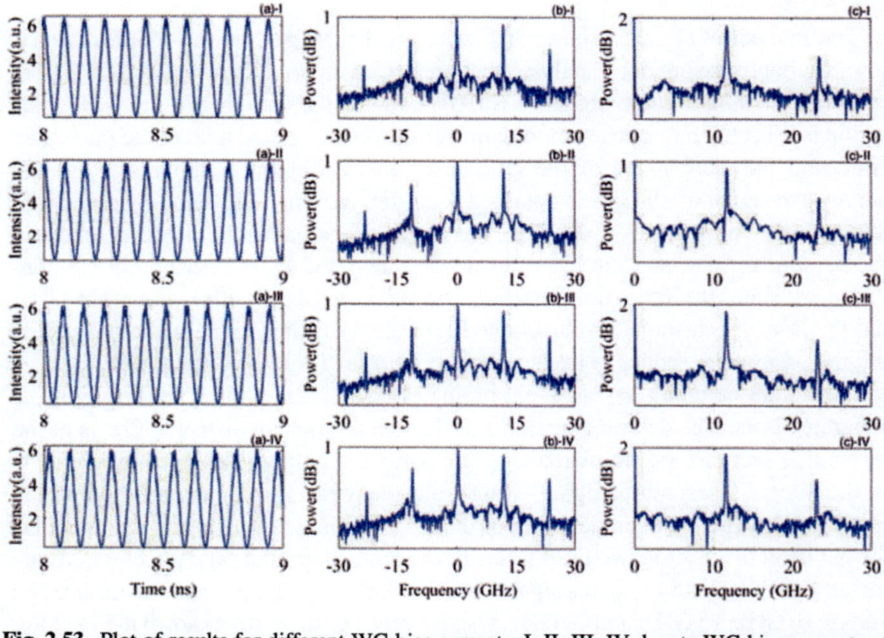

Fig. 2.53 Plot of results for different WG bias currents. I, II, III, IV denote WG bias currents of 2 Ith, 3 Ith, 4 Ith, 5 Ith; **a–c** are time series, optical spectra and RF spectra of the integrated laser respectively

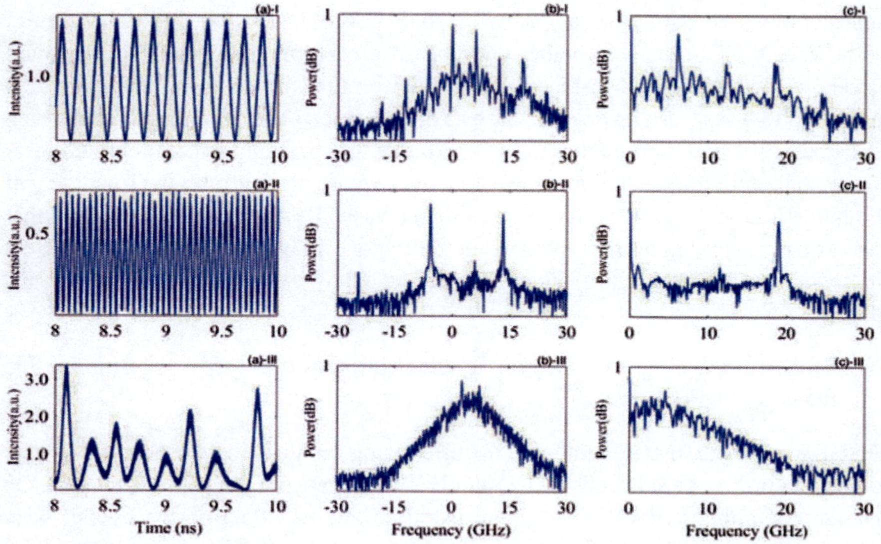

Fig. 2.54 Plot of results for different WG bandgap detuning. I, II, III denote WG bandgap detuning of 0 THz, 5 THz, 24 THz; **a–c** are time series, optical spectra and RF spectra of the integrated laser respectively

set to 0 THz, 5 THz, and 24 THz, respectively. 5, 24 THz, and the results are shown in Fig. 2.54.

The analysis of Fig. 2.54 shows that, when the detuning of the waveguide region is zero, similar to the previous analysis, the three-band monolithic integrated DFB laser can generate a microwave signal of 6.1 GHz. When the detuning of the waveguide region is 5 THz, it has a strong modulation effect on the optical field of the integrated laser, and the main mode of the DFB laser 1 laser region is suppressed, so the microwave signal of 19 GHz is obtained accordingly. When the detuning amount of waveguide area increases to 24 THz, the waveguide area has no modulation effect on the optical field, and the laser out of the integrated DFB laser enters the non-stationary state, the spectrum is raised, and the spectrum width is about 23 GHz, and the laser spectrum is also broadened, and the 3 dB line width is about 25 GHz, and the time series sequence has no obvious pattern. Therefore, the waveguide area with different detuning has different tuning effects on the optical field state of the integrated laser. The detuning amounts of the two lasers with different effects of the waveguide sections are simulated and calculated separately to analyze their effects on the overall laser output signal. The results shown in Fig. 2.55 are based on the condition that the bandgap detuning of the WG segment is 0 THz and the WG bias current is set to 2 I_{th}; the two DFB lasers have identical gratings and the bias currents are set to 1.5 I_{th}, and the detuning of the DFB laser 1 and DFB laser 2 are 0 GHz, 5 GHz, 10 GHz, 15 GHz, and 20 GHz, respectively. For the results shown in Fig. 2.56, the bandgap detuning of the WG segment is 5 THz and the WG bias current is 2 I_{th}; for the two DFB lasers with identical gratings and bias currents set to 1.5 I_{th},

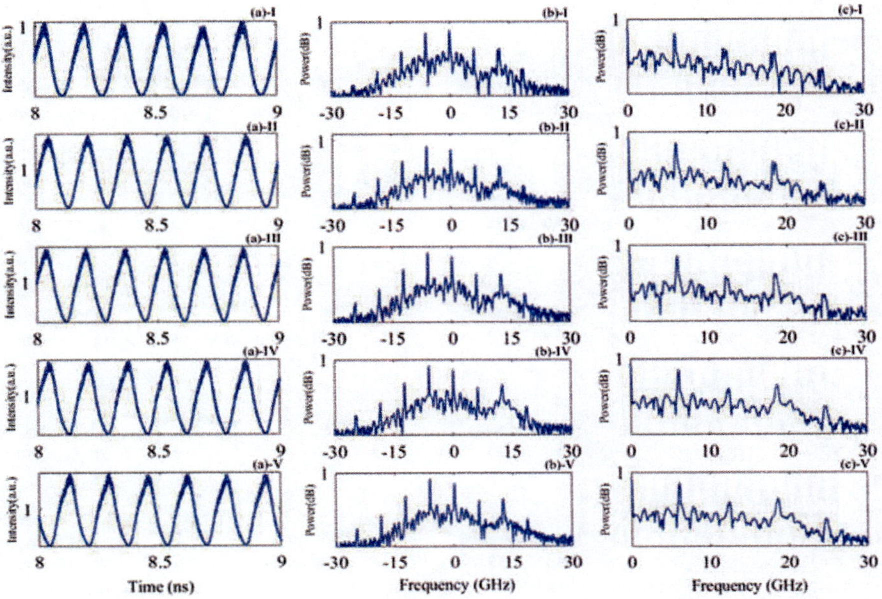

Fig. 2.55 Results of different detuning of DFB lasers for WG bandgap detuning of 0 THz. I, II, III, IV, V indicate two lasers with detuning of 0 GHz, 5 GHz, 10 GHz, 15 GHz, 20 GHz; **a–c** are time series, optical spectra and RF spectra of the integrated laser respectively

the detuning of DFB laser 1 and DFB laser 2 are 0 GHz, 5 GHz, 10 GHz, 15 GHz, 20 GHz, respectively. The conditions for the results are shown in Fig. 2.57 are a bandgap detuning of 24 THz in the WG section and a WG bias current of 0; the two DFB lasers with identical gratings and bias currents are both set to 1.5 I_{th}, the detuning amounts of DFB laser 1 and DFB laser 2 are 0 GHz, 5 GHz, 10 GHz, 15 GHz, 20 GHz, respectively.

For the analysis of Fig. 2.55, the three-section monolithic integrated DFB laser can produce 6.1 GHz microwave signals at a bandgap detuning of 0 THz in the WG band. The microwave signal of 6.1 GHz coincides with the difference frequency between the main modes in the spectrum, and the multi-section laser shows frequency locking, with the spectrum showing two main modes, which are 448.9 and 455.0 GHz. The frequency component with the highest energy in the spectrum is 6.1 GHz, and the time series diagram can show that the output time series of the multiband laser is pulsed with a period of 0.164 ns. With the change of the detuning of the two DFB lasers, the frequency of the integrated laser output microwave remains basically the same, indicating that the effect of the detuning of the two lasers on the microwave generation is small when there is no detuning in the WG section.

In the analysis of Fig. 2.56, when the WG segment mainly plays the role of modulation in the integrated laser, the main mode of DFB laser 1 is suppressed, and the microwave signal of 19.0 GHz is generated by the three-segment monolithic integrated DFB laser, and the output time series of the multi-segment laser is still

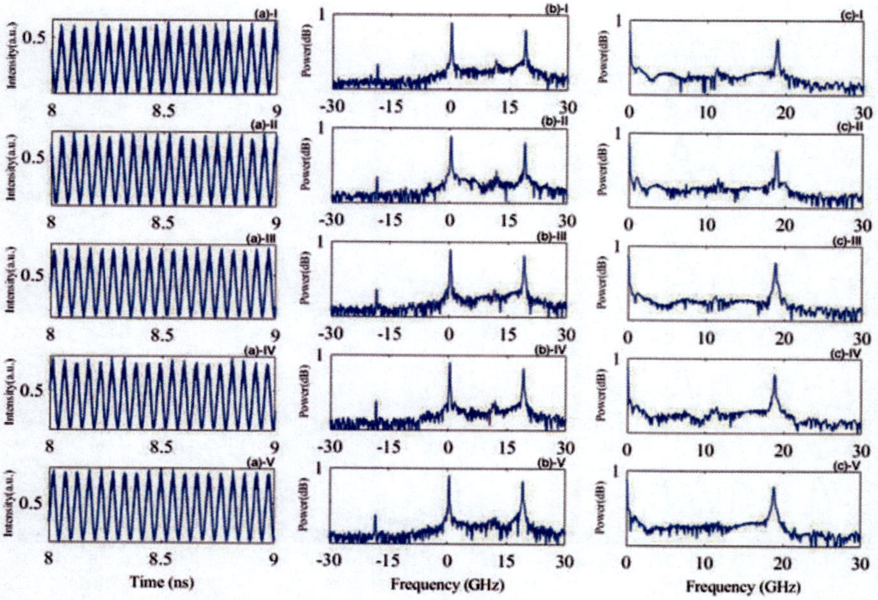

Fig. 2.56 Results of different detuning of DFB lasers for WG bandgap detuning of 5 THz. I, II, III, IV, V indicate two lasers with detuning of 0 GHz, 5 GHz, 10 GHz, 15 GHz, 20 GHz; **a–c** are time series, optical spectra and RF spectra of the integrated laser respectively

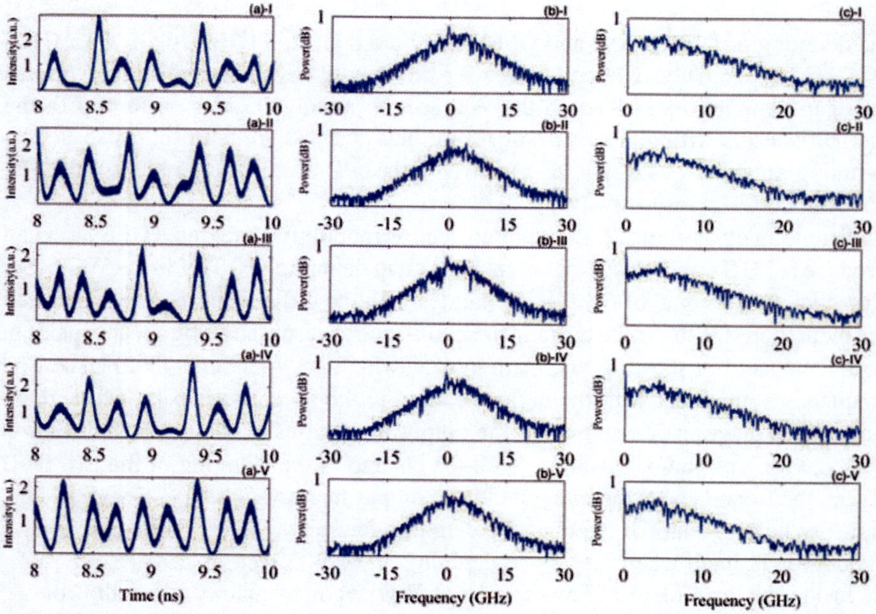

Fig. 2.57 Results of different detuning of DFB lasers for WG bandgap detuning of 24 THz. I, II, III, IV, V indicate two lasers with detuning of 0 GHz, 5 GHz, 10 GHz, 15 GHz, 20 GHz; **a–c** are time series, optical spectra and RF spectra of the integrated laser respectively

pulsed with a period of 0.053 ns. With the change of the detuning amount of the two DFB lasers, the frequency of the integrated laser output microwave remains basically unchanged, indicating that the detuning amount of the two lasers has a small effect on the microwave generation when the WG section mainly plays a modulating role.

For the analysis of Fig. 2.57, the WG segment detuning is at 24 THz, the WG segment is only used for transmitting waveguide, and the multi-segment monolithic integrated DFB laser generates the signal with spectral broadening, spectral elevation, and disorderly change in time series. With the change of the detuning of the two DFB lasers, the dynamic characteristics of the overall laser output signal are stable and unchanged, and the detuning of the two lasers still does not affect the integrated laser output signal. The analysis of Figs. 2.55, 2.56 and 2.57 shows that the detuning between the two lasers has little effect, regardless of whether the WG section mainly plays the role of gain, modulation or transmission, and the specific reason for this phenomenon still needs to be analyzed later.

2.4 Design of Drive and Temperature Control System for the Integrated Chaotic Laser

2.4.1 Hardware Circuit Design of the Drive and Temperature Control

At present, commercial temperature control sources and driving sources are widely used in photonic integrated chaotic lasers, but the commercial temperature control sources on the market have the problem of low control accuracy. The wavelength and threshold current of photonic integrated chaotic lasers are very sensitive to temperature changes. The change of wavelength and threshold current will lead to the instability of the chaotic state of its output, and the 0.1 mA adjustment accuracy provided by the commercial driving source can no longer meet the needs. In addition, the volume of the instrument of commercial temperature control source and driving source are large, and the separate existence of the two is not conducive to the integration of the entire system.

To solve these problems, our team designed a circuit system of high-precision temperature controller and direct current driver for photonic integrated chaotic lasers. The system can achieve the following functions: the temperature adjustment accuracy of photonic integrated chaotic laser is 0.01 °C, the driving current control accuracy is 0.01 mA and the stability of driving current is 0.02%. Experiments show that the fluctuation of the output wavelength of the laser can be stabilized within 9 p.m. by controlling the temperature within 120 min. Compared with the Newport driver, this design reduces the normalized mean square deviation of the output wavelength of the transmitter optical subassembly (TOSA) laser from 0.0044 to 0.0020, which improves the control stability.

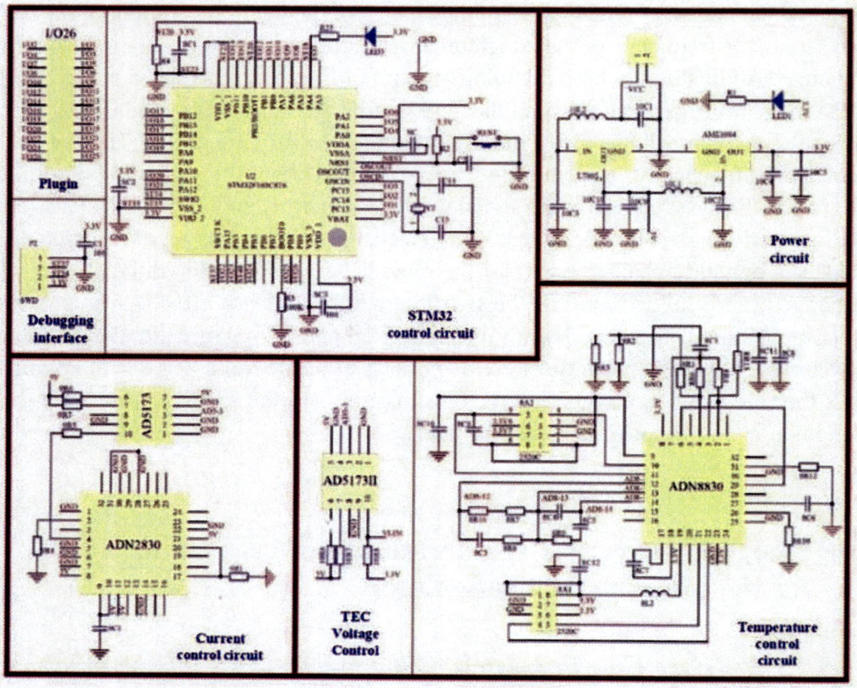

Fig. 2.58 Design schematic of high-precision temperature control and DC drive circuit system for photonic integrated chaotic lasers

1. **Overall design of temperature control and DC drive circuit system**

Figure 2.58 is a schematic diagram of the high-precision temperature control and direct current drive circuit system design for photonic integrated chaotic lasers. It consists of six parts: power supply circuit, STM32 control circuit, laser current control circuit, temperature control circuit, plugin interface and debugging interface.

The power supply circuit provides the required voltage for the entire circuit. The power supply voltage required by the STM32 and ADN8830 chips in the circuit is 3.3 V, and the supply voltage of the ADN2830 chip is 5 V. Therefore, the power supply circuit needs to generate two power supply voltages of 5 and 3.3 V. External 12 V DC (Direct current) power supply to the circuit. After filtering, it is connected to the selected l7805 chip to generate 5 V voltage. Part of the 5 V voltage generated by L7805 is transmitted to the circuit for power supply, and the other part is supplied to AME1084 to generate 3.3 V voltage.

The STM32 control circuit is responsible for signal processing, this part is connected with the plug-in interface circuit, through the plug interface circuit can be connected to the display screen and the key, and the key can be manipulated to set the temperature and current state of the current driving laser and displayed through the display screen to realize human–computer interaction. The debugging interface circuit can be programmed.

ADN2830 as the core of the laser current control circuit, according to the STM32 control signal to adjust, it is then converted into a current signal to be output to provide a constant current drive for the laser.

ADN8830 as the core of the laser temperature control circuit, also according to the master STM32 chip control Signal to adjust the temperature of laser. The resistance of the thermistor in the laser is measured by the Wheatstone bridge circuit to obtain the real-time laser temperature. According to this, the semiconductor cooler (TEC) refrigeration or heating is adjusted to achieve the set temperature value. Through the control of the TEC voltage control circuit, the temperature control and DC drive circuit system can adjust the required maximum voltage value at both ends of TEC for lasers with different types of TEC. The temperature adjustment of lasers with different TEC can be realized, which expands the application range of this circuit system.

2. **Design principle of the temperature control system**

According to the principle of TEC refrigeration and heating, the resistance value of the internal thermistor of laser is used to sense the temperature of the laser chip. Then the current flowing through the TEC is controlled according to the temperature change in the laser, and finally the desired temperature is achieved. The TEC terminal voltage regulation circuit is designed to realize the temperature control of lasers with different TEC.

The thermistor is close to the laser chip, which can detect the temperature of the laser chip and change its resistance value according to the temperature; The internal thermistor of the laser is usually a negative temperature coefficient (NTC) thermistor. The resistance of this type of thermistor tends to increase as the temperature decreases.

The formula for calculating the temperature of the NTC thermistor is:

$$R_T = \mathrm{R}e^{(B(1/T_1 - 1/T_2))} \tag{2.88}$$

$$B = \frac{\ln R_1 - \ln R_2}{1/T_1 - 1/T_2} \tag{2.89}$$

In the formula: T_1 and T_2 refer to the Kelvin temperature, 1 K = 273.15 (absolute temperature) + Celsius. Where $T_2 = (273.15 + 25)$ °C; RT is the resistance value of the thermistor at the temperature of T_1. R is the nominal resistance value of the thermistor at T_2 (normal temperature, usually 25 °C); R1 is the zero-power resistance value at temperature T_1; R_2 is the zero-power resistance value at temperature T2; B value is an important parameter of the thermistor.

The heat sink under the various components has the function of conducting heat, and the lower part is the TEC used to cool and heat the laser. The TEC is a semiconductor cooling device based on the Peltier effect, as shown in Fig. 2.59. Its working principle is that the energy required for the electron flow is provided by the

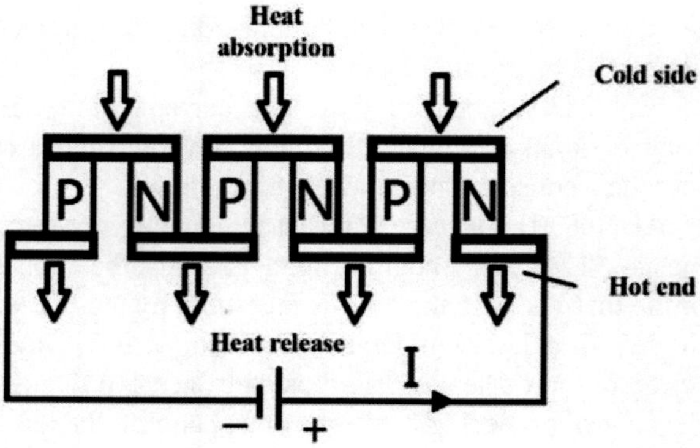

Fig. 2.59 Working principle diagram of semiconductor cooler

direct current power supply. After the power is turned on, the electrons start from the negative electrode (−), first pass through the P-type semiconductor, absorb heat here, and reach the N-type semiconductor, and then release the heat. Every time the electrons pass through an NP module, heat is sent from one side to the other side, causing a temperature difference to form the hot and cold ends. The hot and cold ends are respectively composed of two ceramic sheets, and the cold end should be connected to the heat source, that is, the location intended to be cooled. When the current passing through the TEC is a forward current, the upper surface exhibits the function of cooling, and the heat flows to the lower surface of the TEC. When the current flowing through the TEC is a reverse current, the heat will flow to the TEC, presenting the function of heating.

The semiconductor refrigeration system consists of a thermopile, a cold end heat exchanger, a hot end heat exchanger and a controller, and the thermopile is a refrigeration device. Since the thermopile is composed of multiple pairs of galvanic couples, as shown in the PN structure shown in Fig. 2.59, and each galvanic couple is connected in series for current. The galvanic couples are connected in parallel for heat flow. Therefore, when analyzing the performance of the thermopile, it is only necessary to analyze the cooling performance of the galvanic couple. The cooling capacity, voltage, output power and cooling coefficient of a pair of galvanic couples are respectively:

$$Q = AIT_C - K\Delta T - I^2R/2 \tag{2.90}$$

$$V = IR + A\Delta T \tag{2.91}$$

$$N = VI = I^2R + AI\Delta T \tag{2.92}$$

$$C = \frac{AIT_C - K\Delta T - I^2R/2}{I^2R + AI\Delta T} \tag{2.93}$$

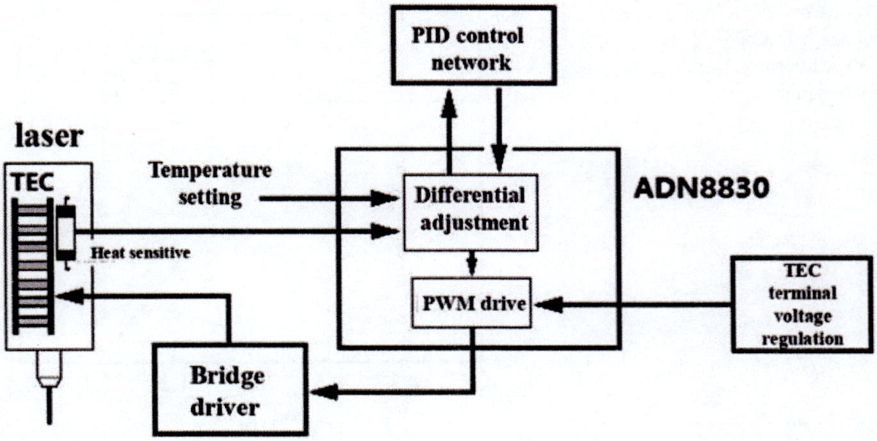

Fig. 2.60 Laser temperature control module schematic diagram

In the formula: Q is the Refrigerating capacity of the galvanic couple (W). I is the working current (A). K is the thermal conductivity of the galvanic couple (W/K). T is the temperature difference between the cold and hot ends (K). R is the resistance of the galvanic couple (Ω). A is the thermoelectric power of the galvanic couple (V/K). Tc is the temperature of the cold end of the couple (K).

As shown in Fig. 2.60, the temperature control circuit unit of the laser adopts ADN8830 temperature control chip as the core temperature control circuit. The control circuit is mainly composed of a temperature setting feedback unit, a PID (proportion integration differentiation) control network, an ADN8830 temperature control chip, a bridge drive circuit and a TEC terminal voltage adjustment unit.

3. **TEC terminal voltage regulation circuit**

In the process of researching the chip, the voltage applied to both ends of the laser TEC can be adjusted by controlling the output voltage of the VLIM of the temperature control chip ADN8830 [34]. The relationship between the maximum voltage output to the TEC terminal and the VLIM pin of ADN8830 in this design is:

$$V_{TEC,MAX} = (1.5 - V_{VLIM}) \times 4 \tag{2.94}$$

In the formula: $V_{TEC,MAX}$ is the equivalent DC voltage of the TEC terminal. VVLIM is the input voltage of the VLIM pin. V_{VLIM} cannot exceed 1.5 V, otherwise the circuit output will be disordered. The maximum voltage of $V_{TEC,MAX}$ is the power supply voltage of the chip, which is 3.3 V. Design the corresponding TEC voltage control circuit according to formula (2.94). When there is fault in the circuit, $0V_{VLIM}$ can be set to 1.5 V without cutting off the power supply to make the $V_{TEC,MAX}$ is 0, so as to stop TEC and reduce the impact on other control circuits. The corresponding relationship of formula (2.94) is shown in Fig. 2.61. When V_{VLIM} is less than 0.675 V,

Fig. 2.61 TEC maximum voltage VTEC, MAX and VVLIM corresponding relationship

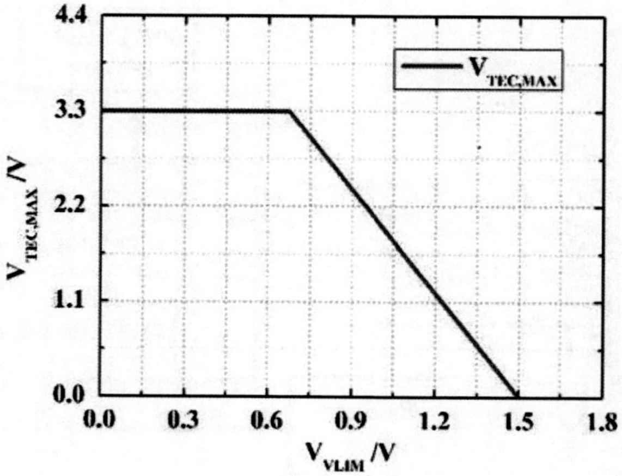

$V_{TEC,MAX}$ is a constant 3.3 V. When V_{VLIM} is greater than 0.675 V, $V_{TEC,MAX}$ has a linear relationship with V_{VLIM}, and decreases with the increase of V_{VLIM}.

Since the temperature of the laser needs to be controlled by controlling the TEC cooling chip, the circuit needs to meet the requirements of controlling both the TEC cooling and the TEC heating. The TEC refrigerator is made of the Peltier effect of semiconductor materials. Whether it is cooling or heating, and the rate of cooling and heating are determined by the magnitude and direction of its current. The size of the current can be controlled by the deviation voltage output by the PID network, and the control of the current flow direction requires the output control circuit to solve. The control current can be up to 2 A.

The output control unit of the circuit is shown in Fig. 2.62. The circuit uses a full-bridge drive circuit, and the transistor is FDW2520C, which has an N-type field effect transistor (metal oxide semiconductor, MOS) inside and a P-type MOS [35]. Two FDW2520Cs can be selected to form a full-bridge drive circuit, which can control the flow direction of the current by controlling the conduction of the MOS in the circuit. The current of the MOS is controlled by the pulse width modulation (PWM) wave [36]. The P1 and N1 pins output the PWM wave to the MOS P_1 and N_1, and the high and low levels in the PWM wave control the conduction of the MOS. P_1 and N_2 are a pair of control combinations. When the P_1 pin is low, the combination is turned on, and the current flows from TEC+ to TEC−. P_2 and N_1 form another control combination. When the N1 pin is low, the combination is turned on, and the current flows from TEC− to TEC+.

The difference between the input voltage of the temperature setting resistor and the input voltage of the laser thermistor is the source signal that causes the duty cycle of the PWM wave. The difference affects the flow direction of the current, and the difference affects the voltage amplitude output to the TEC by means of affecting the output of the PID network, so as to control the current size flowing through TEC. The control of the TEC is achieved through the above process. The equation corresponding to the output of the TEC terminal is as follows:

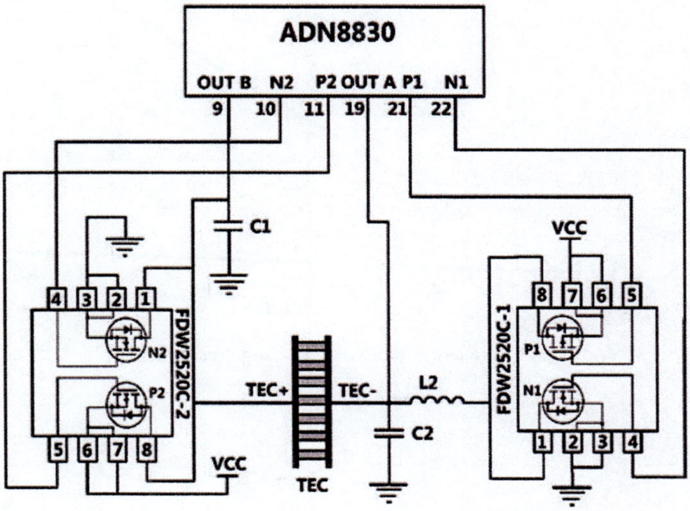

Fig. 2.62 Bridge drive circuit

$$V_{OUTA} = 4(V_{COMPOUT} - 1.5) + V_{OUTB} \tag{2.95}$$

$$V_{OUTB} = -14(V_{COMPOUT} - 1.5) + 1.5 \tag{2.96}$$

In the above two formulas, V_{OUT} A is the voltage of the OUT A pin, $V_{OUT\,B}$ is the voltage of the OUT B pin, and $V_{COMPOUT}$ is the voltage of the COMPOUT pin.

As can be seen from Fig. 2.63, with the increase of $V_{COMPOUT}$, both $V_{OUT\,A}$ and $V_{OUT\,B}$ decrease accordingly. It can be seen from Fig. 2.64 that with the increase of $V_{OUT\,A}$ and $V_{OUT\,B}$. Although the value of the voltage $V_{OUTA} - V_{OUTB}$ at both ends of the TEC is decreasing, its amplitude shows a trend of decreasing first and then increasing. $V_{OUT\,A} = V_{OUT\,B} = 1.5$ V is the critical point, the amplitude of the voltage difference is reversed at this point, the direction of the corresponding current also changes, and the cooling and heating of the TEC is also changed accordingly. From the formula (2.94) and formula (2.95), we can know that the size of $V_{OUT\,A}$ and $V_{OUT\,B}$ is affected by $V_{COMPOUT}$, and the size of the absolute value of the difference reflects the size of the adjustment voltage.

Since the adjustment of the temperature of the laser is a dynamic process. Therefore, during the adjustment process, the input voltage of the laser thermistor is constantly changing, and the corresponding $V_{COMPOUT}$ is also constantly changing. When the temperature is about to reach the set value, the voltage difference between the input terminal voltage and the output terminal voltage is adjusted by oscillation at 0 V. causing VTEMPCTL oscillates around 1.5 V. Then through the COMPOUT terminal input, VOUTA and VOUTB are caused in oscillate around 1.5 V, causing their difference to oscillate around 0 V and the current flow of the TEC is constantly

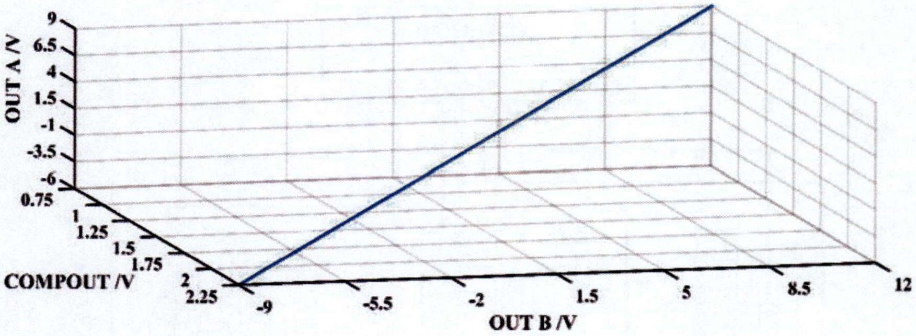

Fig. 2.63 V_{OUTA}, V_{OUTB} and $V_{COMPOUT}$ foot voltage change relationship

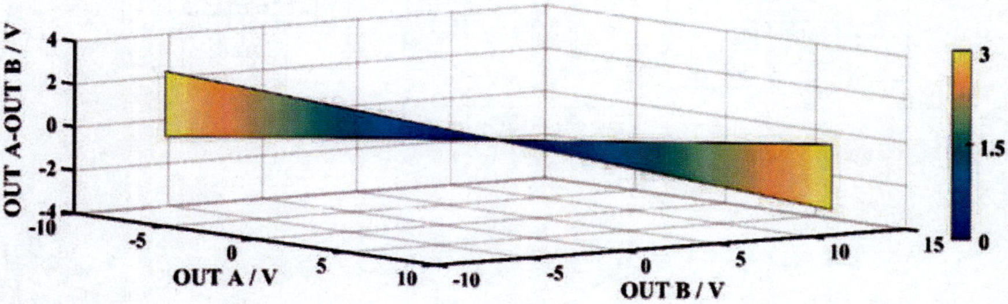

Fig. 2.64 TEC terminal voltage $V_{OUTA}-V_{OUTB}$ and V_{OUTA} and V_{OUTB} relationship

adjusted, which promotes the constant adjustment of the TEC cooling and heating, so as to stabilize the temperature and reach the set value.

The principle block diagram of the voltage control circuit at the TEC terminal of the laser is shown in Fig. 2.65. The AD5173 chip is used to form a voltage control circuit to control the voltage at the VLIM terminal of the ADN8830. AD5173 is a high-precision digital potentiometer. Due to its high stability, the temperature coefficient is 35 ppm/°C (1 ppm = 0.0001%, that is, one millionth), and it can be programmed to control it to improve the accuracy and automation of the control. The control of the digital potentiometer is realized by the STM32 chip. By adjusting the resistance value of the digital potentiometer, the regulation of the voltage loaded to the VLIM terminal is realized. The VLIM control voltage further control the amplitude of the PWM wave generated inside the ADN8830, and finally realizes the adjustment of the voltage amplitude loaded to the TEC terminal of the laser [37].

4. **Temperature setting and feedback setting circuit**

The temperature setting and feedback unit is shown in Fig. 2.66. The temperature setting pin of the laser in the figure is the 4-pin TEMPSET pin. The laser temperature can be set by setting the size of resistance value connected to this pin. In the circuit design, this pin is connected with a digital potentiometer to realize the setting of the required temperature value. Pin 2 THERMIN is connected to one pin of the

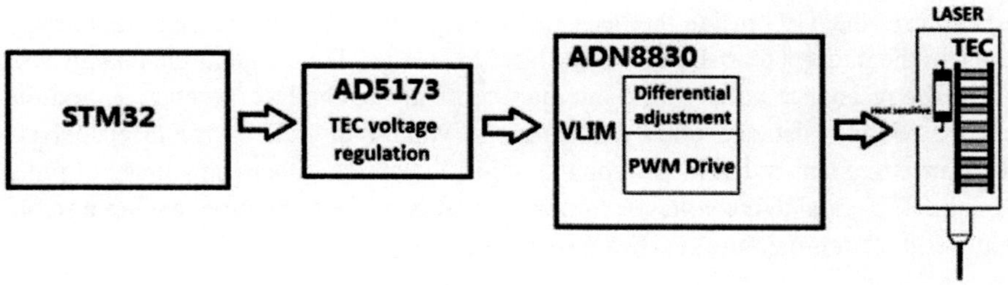

Fig. 2.65 Laser TEC terminal voltage control diagram

thermistor of the laser, the other pin of the thermistor is connected to the ground, and pin 7 VREF provides a reference voltage of 2.5 V.

Pins 2 and 4 are connected to pin 7 through resistors R_2 and R_4 respectively. The resistance value and characteristics of these two resistors must be consistent. Because the reference voltage provided by pin 7 is equivalent to providing a voltage source: the voltage of pin 4 is the voltage divided by digital potentiometer R_3 in the voltage divider circuit composed of R4 and digital potentiometer R3. The voltage of pin 2 is the voltage divided by the thermistor in the voltage divider circuit composed of R2 and the thermistor. The temperature of the laser is adjusted according to the voltage of pins 2 and 4. If the temperature of the laser needs to be changed, the resistance of the digital potentiometer connected to the 4-pin need to be changed. The change of the resistance value of the digital potentiometer will cause the voltage divided by the digital potentiometer to change, and then change the voltage value input to pin 4. The change caused by this setting prompts the temperature control circuit to adjust,

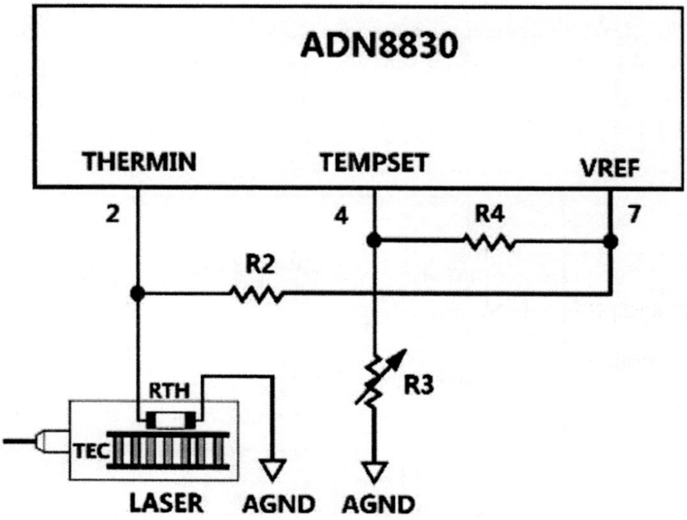

Fig. 2.66 Temperature setting adjustment principle diagram

and control the TEC inside the laser for corresponding cooling or heating, thereby causing the internal temperature of the laser to change. The temperature change will cause the resistance value of the internal thermistor of the laser to change, and the change of the resistance value will cause the voltage divided by the laser thermal resistance to change, that is, the voltage of pin 2 changes. When the voltage of pin 2 is stable and equal to the voltage of pin 4, the temperature regulation reaches a stable state, and the temperature reaches the set value.

5. **Temperature regulation control circuit**

As shown in Fig. 2.67, in the process of processing the temperature of the laser, the temperature setting terminal voltage (4-pin voltage) and the temperature feedback terminal voltage (2-pin voltage) are input to the input amplifier. The amplifier amplifies the signal differential voltage of the two pins, and its gain magnification is 20. After processing, the amplifying circuit will output the signal into the subsequent PID control network, and the corresponding relationship between the output signal and the two input signals is:

$$V_{TEMPCTL} = 20(V_{TEMPSET} - V_{THERMIN}) + 1.5 \tag{2.97}$$

Among them: $V_{TEMPCTL}$ is the output voltage of the input amplifier, which is output to pin 12 (TEMPCTL). VTEMPSET is the input voltage of the temperature setting terminal (pin 2). THERMIN is the input voltage of the laser thermistor terminal. It can be known from the equation that when the temperature of the laser reaches the set value, V_{TEMPSE} is equal to $V_{THERMIN}$, and the value of $V_{TEMPCTL}$ becomes 1.5 V.

The error signal is input to the compensation amplifier. The peripheral circuit of the compensation amplifier is a compensation network composed of resistors and capacitors. In this section, the PID temperature compensation network is used. As shown in Fig. 2.67, the temperature compensation control network is connected to pins 12, 13 and 14, and the transfer function of pins 12 and 13 is Z_1, and the transfer function of pins 13 and 14 is Z_2, then the gain of the compensation amplifier is Z_2/Z_1. The output signal of the compensation network is input to the linear amplifier and

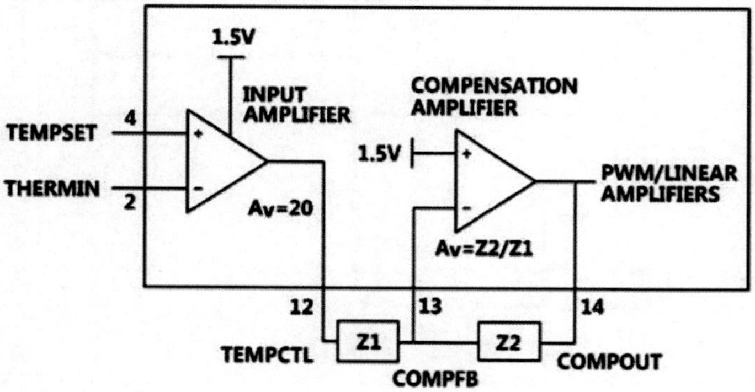

Fig. 2.67 Temperature setting and feedback voltage processing

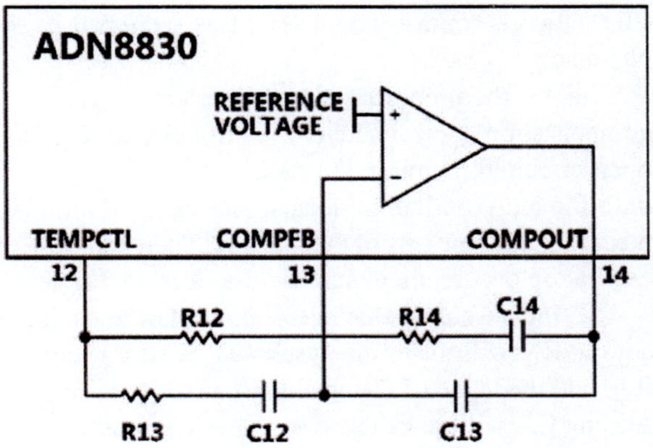

Fig. 2.68 PID control network

the PWM amplifier. The outputs of the above two amplifiers are directly connected to the TEC cooler of the laser to control the cooling and heating of the TEC.

The connection diagram of the PID compensation network is shown in Fig. 2.68. The s-domain transfer function of the PID network is as follows:

$$G = \frac{(1 + sR_{14}C_{14})[1 + s(R_{12} + R_{13})C_{12}]}{sR_{12}(C_{14} + C_{13})(1 + sR_{13}C_{12})[1 + sR_{14}(C_{14}C_{13})/(C_{14} + C_{14})]} \tag{2.98}$$

In the formula: R_{12} and R_{14} form a proportional amplifier circuit. R_{12} and C_{14} form an integration circuit. C_{12} and R_{14} form a differentiation circuit. R_{13} plays the role of current limiting. Moreover, it can flexibly adjust the distribution of circuit zeros and poles by adjusting the size of R_{13}. C_{13} is used to compensate for the zero caused by the equivalent series resistance of the output capacitor.

The larger the proportional coefficient realized by the proportional amplifying circuit, the faster the adjustment speed, but the amplitude of the oscillation will also increase. The smaller the integral control coefficient is, the stronger the integral effect is the faster the control is. However, if the integral coefficient is too large, the adjustment effect is not obvious; Differential control enables system control to predict changes in parameters and control errors in advance. The corresponding relationship between the normalized gain crossover frequency in the circuit and the compensation network parameters is as follows:

$$f_{0\,\mathrm{dB}} \propto \frac{1}{2\pi R_{12}C_{14}} \tag{2.99}$$

The above formula shows that the normalized gain junction frequency is inversely proportional to R_{12} and C_{14}. In order to maintain the stability of temperature control, the normalized gain junction frequency f0 dB should be less than the thermal time constant of the TEC cooler and thermistor in the laser. However, this time constant is not specific and very difficult to characterize. In order to be able to set appropriate

PID network control parameters, this section uses the following steps to set each element.

First set the proportional amplifier circuit part, only connect R_{12} and R_{14} in the compensation control network control circuit. C_{14} is short-circuited, other parts are open-circuited. Increase the ratio of R_{14}/R_{12}, that is, increase the proportional gain until the loop oscillation occurs, and then the ratio is reduced by half, the gain can meet the requirements of most lasers. Then add capacitor C_{14}, reduce the value until oscillation occurs, then increase the value to 2 times then short-circuit.

R_{13}, increase the value of C_{12} until Start to oscillate. At this time, reducing C_2 or adding R_{13} will make the system more stable, but the value of R_{13} is greater than R_{12}, and the value of C_{12} is at least an order of magnitude smaller than that of C14. Adding C_{13} will make the system more stable.

6. **The overall design of the drive system**

The current control adopts the current control chip ADN2830, the maximum current that the chip can output is 200 mA. The output current of the laser is related to the resistance of the PSET pin, and the corresponding relationship is:

$$I_{LD} \propto 1.23V/R_{PSET} \tag{2.100}$$

In the formula, 1.23 V is the constant voltage of the PSET pin to the ground; RPSET is the resistance connected between the ADN2830 and the ground. ILD is the injection current flowing through the laser.

The LD current drive circuit is shown in Fig. 2.69. The PSET of the current control chip ADN2830 is connected with IBMON to form a closed-loop feedback control circuit. The laser current can be adjusted by adjusting the resistance of the PSET terminal. In order to improve the stability and automation of the current control, the system uses a digital potentiometer to control the resistance of the PSET end of the ADN2830. Due to the TEC voltage control and temperature setting of the temperature control unit are also controlled by digital potentiometer, so the digital potentiometer should meet the requirements that multiple digital potentiometer chips can be independently controlled by the I^2C port of STM32 chip.

The AD1 and AD0 pins of the digital potentiometer AD5173 chip correspond to the AD1 and AD0 control bits of its address control word. Set the corresponding control bit according to the level of the corresponding pin in the hardware circuit, which can realize the independent control of the target potentiometer. Since AD5173 has two address control bits, AD0 and AD1, the I^2C port of STM32 can realize independent control of 22 digital potentiometers, that is 4 digital potentiometers. Write the corresponding control program of AD5173, realize the adjustment of the resistance value of the digital potentiometer through the SDA and SDL of the I^2C port of the STM32, and then realize the control of the LD current.

7. **Design of printed circuit boards**

Figure 2.70 is the four-layer wiring diagram of the printed circuit board (PCB) of the circuit, figure (a) and (d) are the top layer and the bottom layer respectively, which

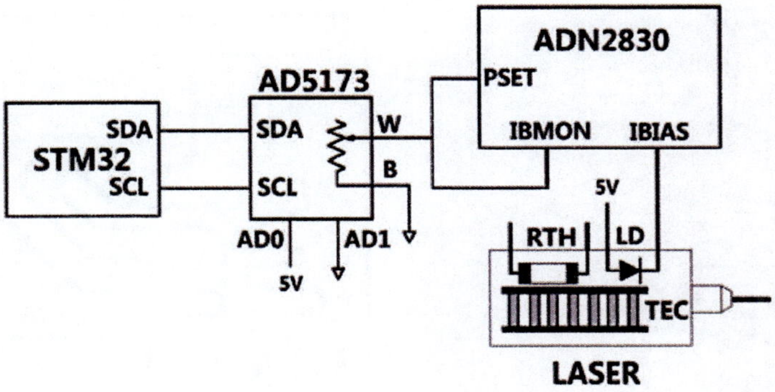

Fig. 2.69 LD current drive circuit diagram

are the wiring layers of the data signal. This layer of welding electronic components and interface devices, using a large area of copper can reduce the electromagnetic interference between electronic components. Figure (b) and (c) shows the power supply layer and the ground layer respectively, the power supply layer is divided into 3.3 and 5 V two parts. Different voltage areas are further divided according to the input/output terminals of the components, and the star and bus layout are used to prevent crosstalk between loops and input/output signals. Similarly, the ground layer is also divided according to this principle.

Figure 2.71 is the 3D-effect drawing of the PCB board of the core control circuit. Among them, 1 is the power module of the circuit, which provides 3.3 and 5 V power supply voltage for the circuit. At the input end of the power supply, there is a diode that controls the current flow direction to prevent the voltage from being to prevent voltage inversion, causing damage to the circuit and causing damage to the circuit. 2 is the STM32 control circuit, which can control the chip and peripheral devices through programs. 3 is External interface circuit, through which other components are connected. 4 is the STM32 debugging interface circuit. 5 is the laser current control interface, through which the on/off of the driving current of the laser chip can be controlled. And the signal of laser attenuation and damage can be fed back through this port to realize the protection of the driving circuit. 6 is the laser drive current control module. 7 is the TEC voltage control module. 8 is the laser temperature control circuit. 9 is the laser temperature control interface, which controls the on/off of the laser temperature regulation. This interface also feeds back laser temperature information and TEC terminal voltage information, which can be used for laser temperature protection.

8. **Operation panel design**

In order to improve the convenience of human–computer interaction of the design system, an operation control panel is designed. There are buttons on the operation panel, which can set the temperature and current of the system. At the same time, a display screen is arranged on the panel to display the current setting state of the laser and the adjustable content.

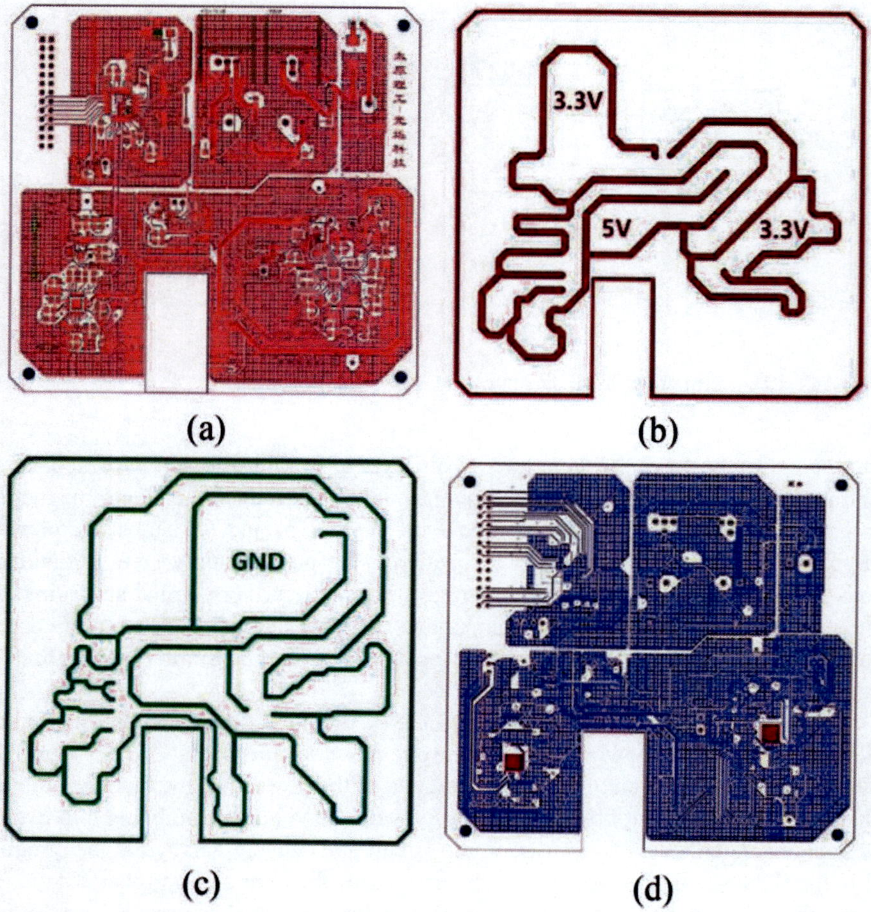

Fig. 2.70 PCB four-layer wiring diagram. **a** Top floor; **b** power supply layer; **c** stratum; **d** bottom

Figure 2.72 shows the display screen selected for the design. The screen is a 1.3-in. OLED display screen with 128 × 64 dot matrix. It has four connection pins, which are respectively responsible for the analog signal ground (GND) and the analog signal power (VCC) pin of the power supply, and SCL and SDA pins responsible for data communication. Among them, SCL is the reference clock input terminal of the display screen, and SDA is the control data input terminal.

Figure 2.73 shows the keying and keying circuit used in the system. The key is a single pole double throw switch with a light-emitting diode indicator in the middle. When the key is pressed, the key is on. The three keys SC, SU and SD in the figure are all low voltage level active. SC is used to switch the current regulation content, whether to adjust the temperature or the current; SU is used to adjust the increase of the set value. SD is used to adjust the reduction of the set value.

Fig. 2.71 Core control board 3D renderings

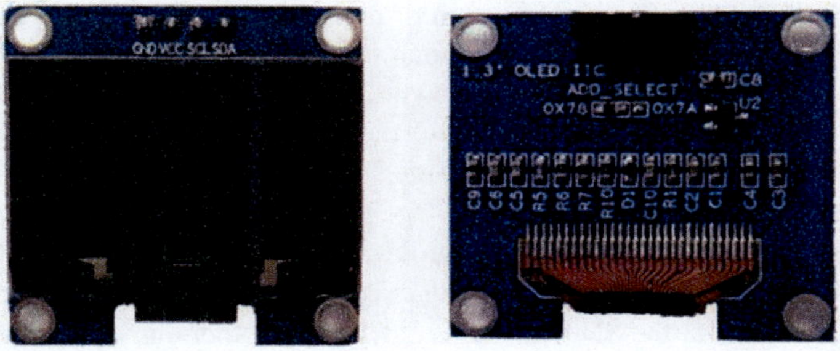

Fig. 2.72 1.3-in. OLED display

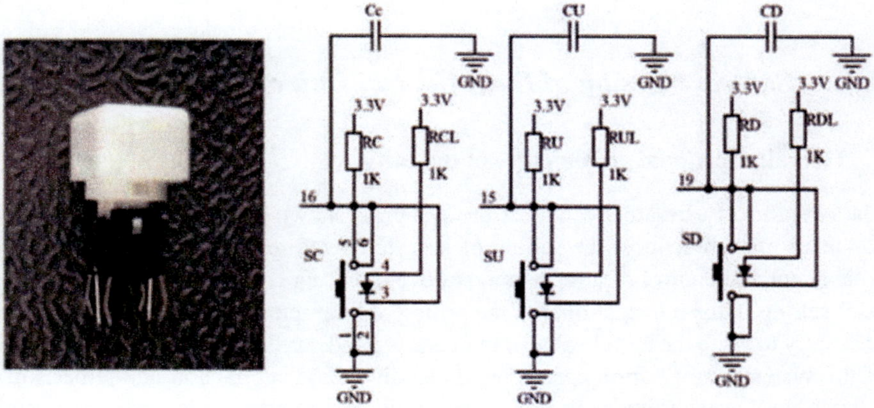

Fig. 2.73 Control buttons and connection diagram

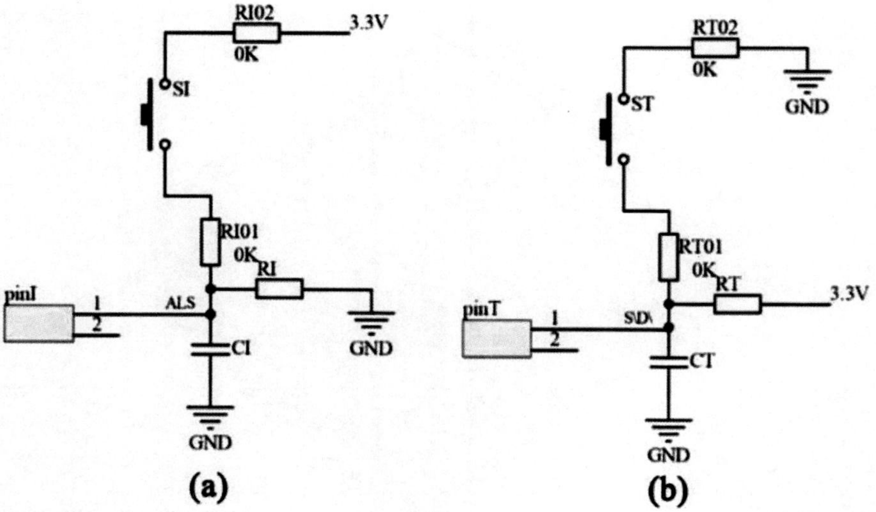

Fig. 2.74 Current and temperature switch circuit connection diagram

Figure 2.74 is the connection diagram of the switch button circuit connected to ADN2830 and ADN8830. Since the switch input terminal of ADN2830 is active to high level, its schematic design is shown in Fig. 2.74a. When the switch is not pressed, the output to the ALS terminal of the ADN2830 is grounded through the resistor RI, and the voltage is zero at this time. When the button is pressed, the remote terminal of the resistor RI divides the voltage from the 3.3 V power supply section, and the voltage output to the ALS is high level. Although the switch control level of ADN8830 is active at low level, the principle is the same. When the button is not pressed, the voltage at the SD terminal is 3.3 V at the other end of the resistor RT, which is a high level. When the button is pressed, the SD terminal is grounded to a low level.

2.4.2 *Software Design of Each Control Unit of the System*

1. **Overall functional architecture of the software**

The overall software architecture of the system is shown in Fig. 2.75. The system has three main functions: the setting of laser temperature and driving current, the adjustment and control of laser temperature and driving current, and the display of laser setting status information. For the setting of laser temperature and current, it is necessary to write the initial value of temperature and current in the internal memory of the system, and perform corresponding addition and subtraction on the basis of the initial value according to the key information, and convert it into the temperature or current value through data processing operation. In the process of temperature and

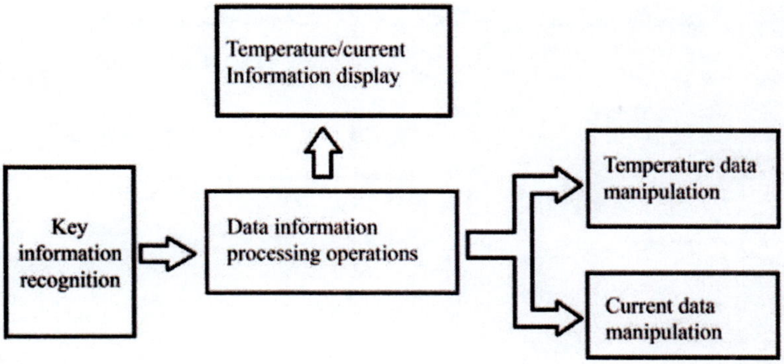

Fig. 2.75 System software architecture

current data operation, it is necessary to write the corresponding comparison formula of temperature and current according to the characteristics of the hardware circuit, and convert it into the control information to be output according to the formula, and the interactive communication to the controlled chip can be realized. In the process of temperature display, the current temperature and driving current information should be displayed according to the results of data processing, and the current adjustable content should be displayed.

The software development of the design system is completed based on the MDK-ARM software development environment. On the basis of the designed hardware circuit containing the ARM chip STM32F103, the development environment required by the system is built on the PC side. The above hardware environment is mainly composed of the underlying hardware of the chip, memory address allocation and the drive of the clock crystal oscillator. The program of the system mainly realizes the processing and transformation of data, the reading and identification of the hardware interface, the control of the communication control word between the chips, and the communication control between the chip and external devices, etc. Figure 2.76 shows the interface of the software development engineering framework built.

2. **Button program design**

The button program enables the adjustment of the temperature and driving current controlled by the laser to be realized by the button. The program flow is shown in Fig. 2.77, which mainly completes the reading, recognition and storage conversion of the button information by the main control chip STM32.

First, the program completes the initialization of the STM32 pins used, the interrupt program used to process the keys, and the timers that control the scanning frequency and response time of the keys; Then apply for three registers BUFS, BUFT and BUFI, use to store the parameter values after key adjustment. Among them, BUFS is used to store the control result of the temperature/current switch key (SWITCH). Because the amount of adjustment that this register needs to deal with

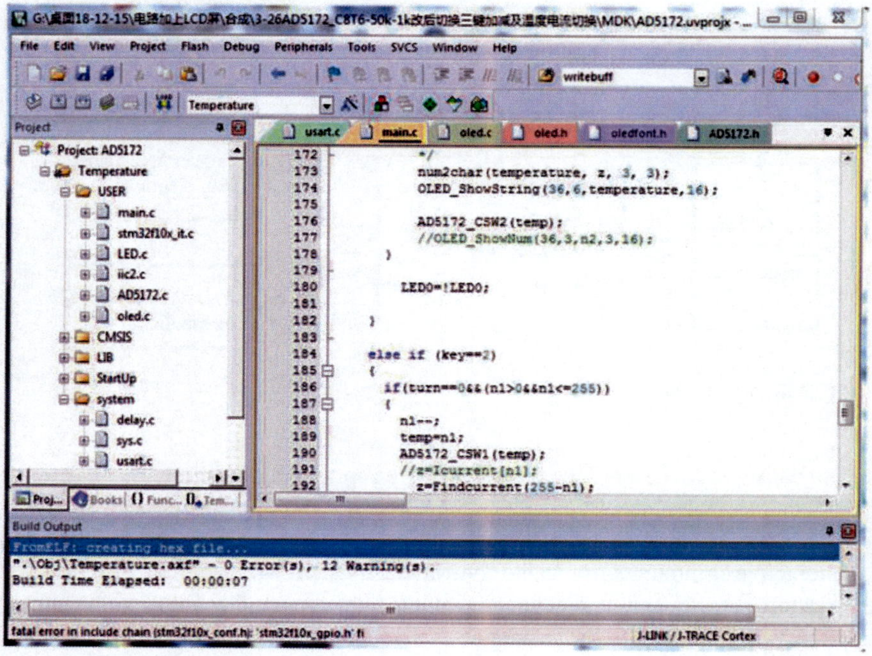

Fig. 2.76 Software editing interface

is only two—temperature and current, the value stored in BUFS is only set to two kinds—0 and 1. Its default initial value is 0. When the value in BUFS is 0, the laser drive current can be set. When the SWITCH button is pressed, the operation of BUFS = ! BUFS is performed, that is, the value in BUFS is reversed. At this point, the value in BUFS becomes 1. When the value in BUFS is 1, it means that the temperature of the laser can be set at present. In this way, the conversion of laser temperature and current regulation is realized.

The key scanning control process is as follows: when the prototype is turned on, the system is initialized, and the initialization value in BUFS is 0. The initialization value in BUTI is 255, which corresponds to the minimum value of the output laser control current. The initialization value in BUFT is 128, and the corresponding laser temperature is 25 °C. After the initialization is completed, the program starts to scan the keys. First determine whether to press the button, if not, the program continues to scan the keys. If it is detected that a key is pressed, it is further judged whether the key is a SWITCH key. If it is, the value in BUFS is reversed to 1, which means that the temperature of the laser can be adjusted currently. Otherwise, if the SWITCH key is not pressed, the default is to adjust the current. Then determine whether there is an UP or DOWN button pressed. If it is pressed, the value in BUFS is read and judged. If the value in BUFS is 0, the corresponding addition and subtraction operations will be performed on the value in the electric current register BUTI. If the value in BUFS is 1, the corresponding addition and subtraction operations are performed on

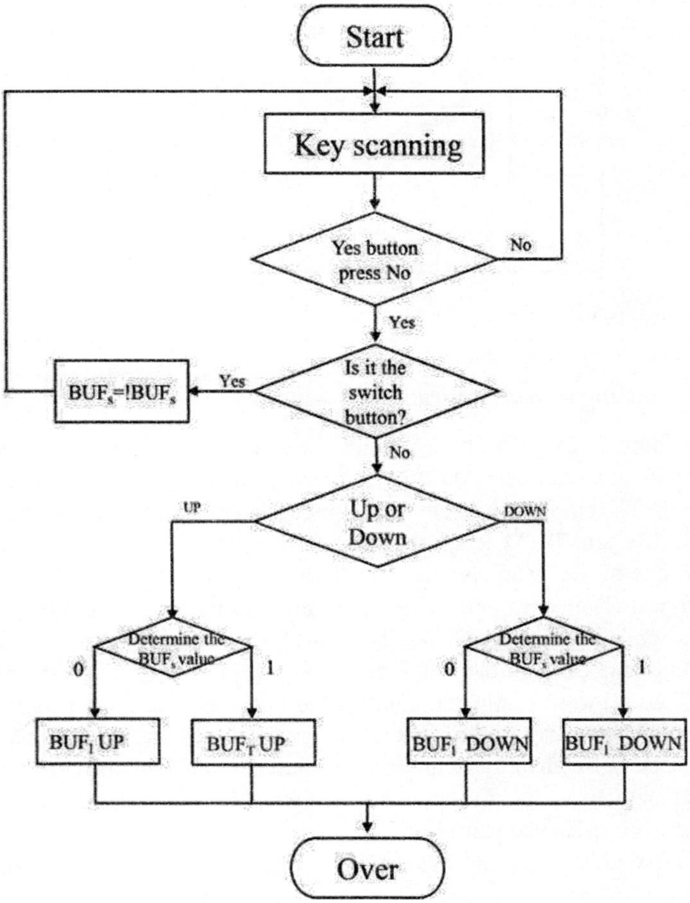

Fig. 2.77 Key control program flow chart

the value in the temperature register BUTT. The setting of the laser drive current and temperature is realized through the above adjustment process. When the value in BUFT and the value in BUTI reach the maximum value, press the UP key again, and the program will no longer process the key; Similarly, after reaching the minimum value, press the DOWN key again, and the program will no longer process.

3. **Temperature and current control program design**

The temperature control is essentially realized depend on the resistance of the thermistor in the laser. Therefore, in order to realize the program design of laser temperature control, it is necessary to obtain the characteristic curve equation of the laser thermistor, and write the corresponding control program according to the equation of the characteristic curve. Similarly, the setting of the laser driving current also needs to design the corresponding control program by solving the corresponding current-resistance characteristic curve equation.

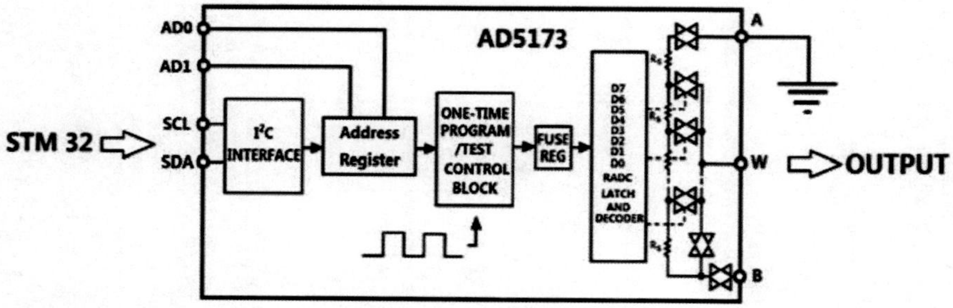

Fig. 2.78 Digital potentiometer control diagram

4. **Control setting circuit program**

In order to improve the stability and automation of the designed circuit, the system uses the digital potentiometer AD5173 to control the resistance of the PSET terminal of ADN2830. The chip is a dual-channel, 8-bit digital potentiometer, so the variation range of BUFT and BUTI is set to 0–255. Figure 2.78 is the control schematic diagram of one of the channels, the STM32 chip controls the AD5173 through the I2C output port. The control clock signal is output through SCL, and the writing of the word is controlled through SDA. The control word is transferred to the one-time programming area through the I2C interactive port of AD5173, and the processed data is stored in the fuse register. Control the output of the 8-bit parallel control port of D0–D7 according to the value of this register, so as to set the on/off of the internal resistance brush of the chip. From the number of bits in the register, it can be concluded that there are 28 kinds of control outputs, that is, the digital potentiometer has 256 kinds of resistance values.

The corresponding relationship between the resistance value of the digital potentiometer and the 8-bit resistance value control signal is:

$$R_{WA} = \frac{D - 255}{256} R_{AB} + 2R_W \tag{2.101}$$

In the formula: D is the decimal value corresponding to the 8-bit resistance control binary word; R_{AB} is the total resistance value corresponding to the digital potentiometer. R_W is the resistance value of the contact brush of the potentiometer.

In this design system, not only the temperature control circuit uses the AD5173 digital potentiometer, but also the laser drive current control circuit and the TEC voltage control circuit also use the digital potentiometer, that is, the design system needs to control three AD5173 digital potentiometers in total.

Since there are only two I2C output ports of the STM32 chip. One of them is used to control the display, so the other needs to control multiple digital potentiometers. This requires the identity setting of different digital potentiometers to facilitate identification and control. The digital potentiometer provides Identity address setting ports AD0 and AD1, and each setting port can set two values: 0 and 1. The setting of its identity address is realized through the connection of the hardware circuit, and it is

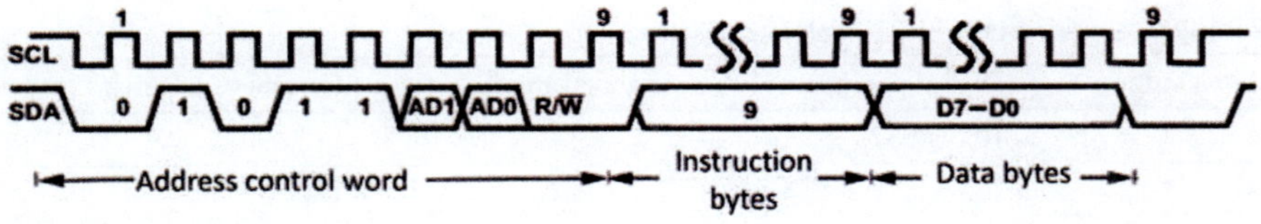

Fig. 2.79 Digital potentiometer control timing

controlled by high and low levels, that is, grounded and connected to 5 V voltage, and then the corresponding software address control word is set according to the hardware address. As a result, a single I^2C port of the STM32 chip can independently control four AD5173 digital potentiometers.

Figure 2.79 shows the control sequence of the digital potentiometer. The beginning of the SDA control sequence is the address control word of AD5173. If the address control word is written, the control word of the $R/\overline{W}$ bit should be set to 0. For example, the control word is 01011010. the last bit is 0 means to write the control word to the chip, the second bit is 1, and the third bit is 0, indicating that the I2C signal controls the digital potentiometer with AD1 foot grounded and AD0 connected to 5 V in the hardware circuit. As shown in the time series in Fig. 2.79, in order to realize the control of the chip resistance, not only the data transmission signal SDA but also the timing signal SCL needs to be provided. AD5173 identifies the control signal according to the high and low levels of the SCL signal and the state of the SDL signal at the rising and falling edges.

Whether the digital potentiometer control program is right or wrong can be analyzed by a logic analyzer. The logic analyzer can collect the timing of timing sequence output. This design system uses the Saleae Logic16 logic analyzer, which supports 17 kinds of communication protocols, including I^2C, JTAG and SPI, etc. It is very convenient to use and can transmit data to the computer through USB. It has 16 acquisition channels, and each channel supports 4 trigger modes: rising edge, falling edge, high level and low level.

5. **Key value operation output program**

To convert the value controlled by the key into the final set temperature or the output value of the laser drive current. It is necessary to determine the relationship equation between the value in the register and the final output value. First of all, it is necessary to determine the relationship between the laser temperature and the laser thermistor and the relationship between the laser driving current and the current setting resistance.

Table 2.7 shows the corresponding relationship between the current setting resistance and the laser drive current. The input value column is the set value of the button, and the set value is also the corresponding data value written into the digital potentiometer. The resistance value column of the digital potentiometer is the resistance value output by the digital potentiometer under this input value. The current value column is the lasers driving current value output by the control circuit under the set resistance value. The drive current values shown in the table are the actual measured currents at the corresponding input values.

Table 2.7 Relation between digital potentiometer and LD current

Input value	Digital potentiometers resistance value/kΩ	Current of LD/mA
0	10.0609375	116.698
1	10.021875	112.387
2	9.9828125	108.385
…	…	…
122	5.2953125	20.695
123	5.25625	20.555
124	5.2171875	20.417
…	…	…
253	0.178125	11.231
254	0.1390625	11.192
255	0.1	11.151

Table 2.8 is the corresponding chart between the resistance of the laser thermistor and the temperature of the laser. Since the resistance value of the temperature setting resistor of the laser is consistent with the resistance value of the thermistor in the laser. Therefore, the table reflects the corresponding relationship between the laser temperature setting resistance and the laser temperature. According to the above-mentioned comparison table, the relationship equation between the corresponding parameters is solved, and then the fitting curve is obtained. In this design, MATLAB is used to solve the fitting curve equation. First, it is necessary to read the values in the Excel table, and then write the corresponding solving program, and then obtain the equation.

Figure 2.80 is the interface for solving the fitting curve equation using MATLAB, and the underlined part of the red line in the figure is the fitting curve equation

Table 2.8 Relationship between laser temperature and thermistor

Temperature/°C	Thermistor/kΩ
12	18.296
13	17.423
14	16.600
…	…
24	10.449
25	10.000
26	9.5757
…	…
50	3.5880
51	3.4541
52	3.3254

obtained by the solution. Figure 2.81 shows the corresponding relationship between the temperature of the laser, the driving current of the laser and the set resistance. The data points with asterisks in the figure (a) are the corresponding temperature values of the thermistor in the laser at a given resistance value. The green curve is the fitted curve for the data. In figure (b), the asterisk data is the actual measured laser drive current value under the set resistance. And the green curve is the fitting curve. As shown in Fig. 2.81, the coincidence degree between the fitting curve and the data points is pretty good.

The digital potentiometer control process is shown in Fig. 2.82. First, read the data value in the BUF, and then compare the data value read this time with the last value

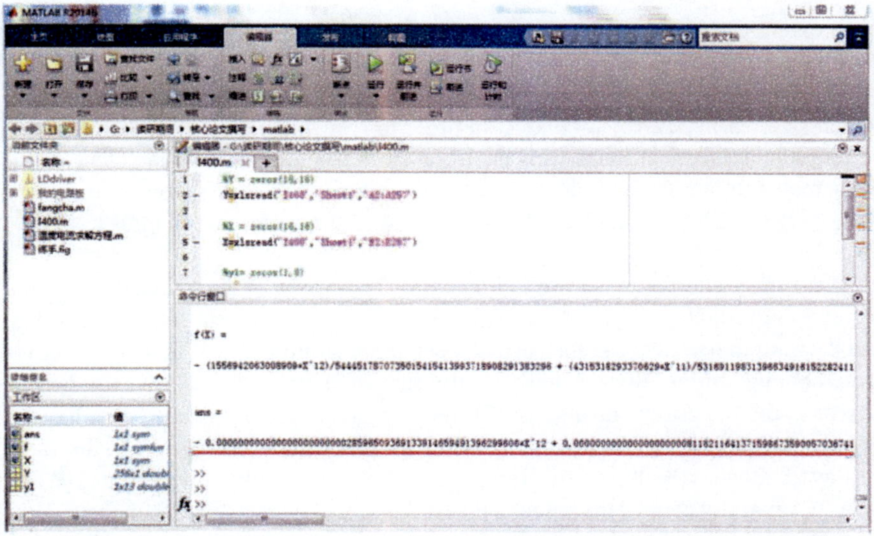

Fig. 2.80 MATLAB Solve the fitting equation

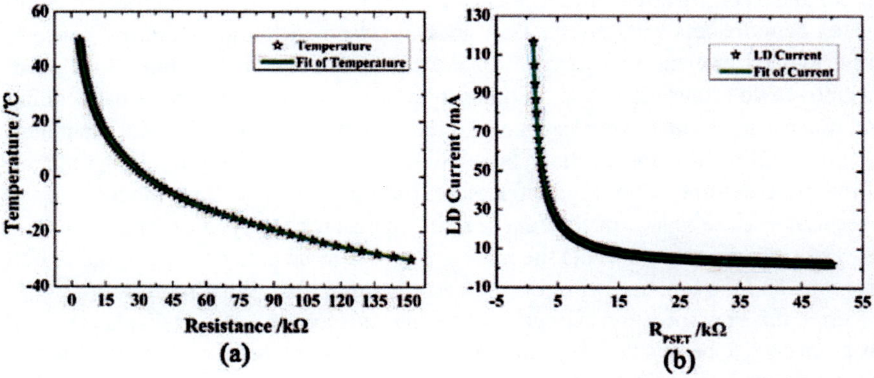

Fig. 2.81 The relationship between the temperature and current of the laser and its set resistance

to judge. If the value does not change, continue to read the BUF value. If the value is different from the last read, read the value of BUFS to determine whether to adjust the current or adjust the temperature. Then generate the corresponding control address according to the adjusted object, which is used to control the corresponding digital potentiometer. The program generates the corresponding control data according to the value in the BUF, and the data is binary code. Then the binary control address and control data are converted into corresponding high and low levels, and output to the digital potentiometer through the I2C port to realize the control of the digital potentiometer. And then complete the adjustment of laser temperature or laser drive current.

6. **Screen display control program design**

The content displayed on the screen is the content currently being adjusted: lasers drive current or laser temperature, laser drive current value and laser temperature value. The physical parameters of the OLED display screen used in this system are: screen size 1.3 in., 128 × 64 pixels. In order to realize the display of temperature, current and adjustment items, it is necessary to set three lines of characters. Since the screen column has only 64 pixels, the dot matrix size of the word in each row cannot exceed 21 × 21, and the dot matrix size of the words used in this system is 16 × 16.

The display interface is shown in Fig. 2.83. The top row of the screen displays the currently adjustable content, the middle row displays the current set current value, and the bottom row displays the currently set temperature value. When the system is turned on, the initial setting is to adjust the current value of the laser. In the display interface, the uppermost display is "Current", indicating that the current value can be adjusted at present. When pressing the toggle key (SWITCH) to adjust, the value displayed above will change to "Temperature", indicating that the current UP and DOWN keys only control the value in the temperature key register. When the current or temperature is adjusted by pressing the button, the program will refresh and display the corresponding display block.

As shown in Fig. 2.84, the data of the multiple averages of the laser drive current output in the control circuit are stored in the Icurrent [ ] data array in this design. Store the temperature data values of the laser generated from the temperature-resistance fit curve in the Ttemperature [ ] array. When the key control signal changes, the data in the above two arrays are called. In order to improve the convenience of data calling and processing, write the corresponding Findcurrent (u8 num) and Findtemperature (u8 num) calling functions. Among them: Findcurrent (u8 num) is the calling function of the drive current value. Findtemperature (u8 num) is the calling function of the laser temperature value. In the function, u8 represents the type of data: unsigned character type, num represents the num + 1th data to be called, execute the search program, and the function will give the corresponding data value found.

Since the key operation output control program and the screen display control program need to be processed synchronously, that is, when the key setting is adjusted, the corresponding digital potentiometer is controlled and written, and the corresponding display content on the screen should also be changed. Therefore, their

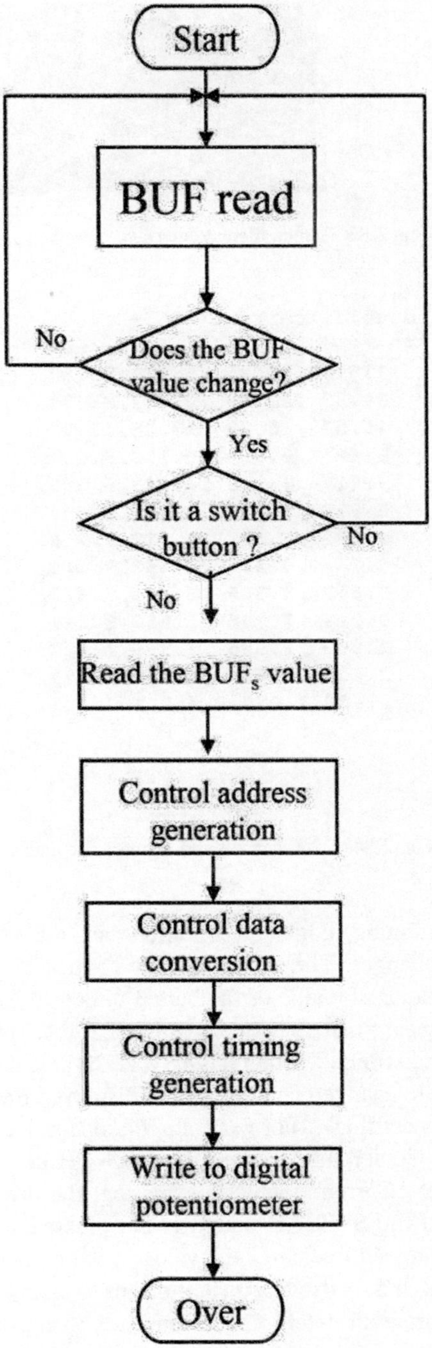

Fig. 2.82 Digital potentiometer control flow chart

Fig. 2.83 Screen display interface

```
double Icurrent[256]=
|{
   115.00,96.95 ,83.72,73.67,
   24.33 ,23.39 ,22.52,21.71,
   13.52 ,13.23 ,13.08,12.85,
   9.44  ,9.3   ,9.157,9.017,
   7.21  ,7.126 ,7.043,6.964,
   5.807 ,5.751 ,5.697,5.644,
   4.889 ,4.85  ,4.812,4.775,
   4.2   ,4.171 ,4.142,4.114,
   3.687 ,3.665 ,3.643,3.622,
   3.294 ,3.276 ,3.259,3.241,
   2.969 ,2.955 ,2.941,2.927,
   2.709 ,2.697 ,2.685,2.673,
   2.489,2.478,2.469,2.46 ,2.


};
```

```
double Ttemperature[256]=
{

35.973 ,35.313 ,34.726 ,34.197 ,
30.081 ,29.887 ,29.704 ,29.531 ,
27.912 ,27.823 ,27.736 ,27.653 ,
26.808 ,26.757 ,26.708 ,26.660 ,
26.143 ,26.110 ,26.078 ,26.047 ,
25.699 ,25.676 ,25.654 ,25.632 ,
25.382 ,25.365 ,25.348 ,25.332 ,
25.143 ,25.130 ,25.118 ,25.105 ,
24.958 ,24.947 ,24.937 ,24.927 ,
24.809 ,24.801 ,24.793 ,24.784 ,
24.687 ,24.681 ,24.674 ,24.667 ,
24.586 ,24.580 ,24.575 ,24.569 ,
24.500 ,24.495 ,24.490 ,24.485 ,
24.426 ,24.422 ,24.418 ,24.414 ,
24.363 ,24.359 ,24.355 ,24.352 ,
24.307 ,24.303 ,24.300 ,24.297 ,


};
```

Fig. 2.84 Laser current and temperature data storage

running judgment conditions are the same and they are all when the key value changes. The screen display control flow chart is shown in Fig. 2.85: First, read the BUF value of the button to determine whether its value has changed. If not, it means that no button is pressed. If there is a key to determine the type of press. First, determine whether it is a SWITCH key. If yes, it means that the options to be adjusted are changed, and the words displayed on the upper end of the display are switched accordingly. For example, the operation prompt displayed before the operation is "temperature". When the SWITCH key is pressed, the displayed content changes to "Current", which means that the driving current of the laser can be adjusted. If the SWITCH button is not pressed, it means that the UP or DOWN button is pressed to change the value in the corresponding register. Then judge the value of BUFS, if its value is 0, it means to change the driving current of the laser. The main program enters the current processing operation subroutine, and reads the corresponding current value according to the information in the BUFI. Then extract and convert the read value, and call the screen display program to display. When the value of BUFS is 1, the same method is used for the display of laser temperature.

Fig. 2.85 Screen display control flow chart

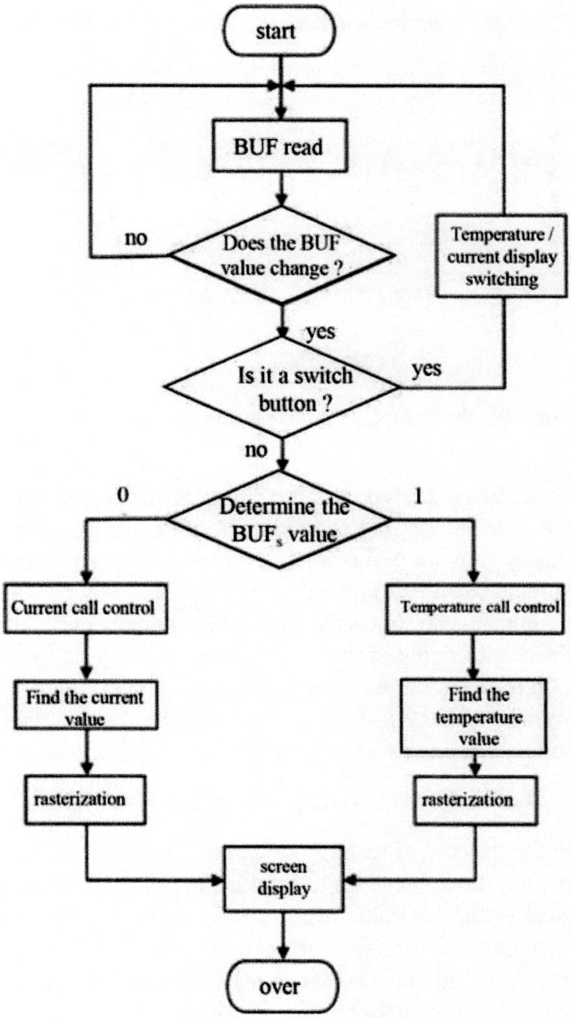

In this way, the human–computer interaction of the system is completed, and the visualization of the operation is realized.

2.4.3 *Output Characteristics and Analysis of the System*

1. **Supply voltage module test**

After the design of the laser temperature and LD drive current control system is completed, the output of the transformer circuit that provides the power supply for the system is tested first. Since the laser drive current is at the milliamp level, it

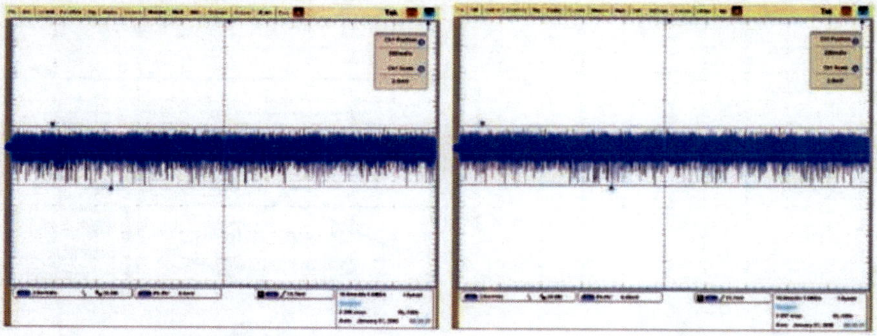

(a) 3.3V (b) 5V

Fig. 2.86 Power supply module output ripple. **a** 3.3 V; **b** 5 V

is necessary to suppress the ripple of the output voltage of the circuit. Therefore, the voltage control conversion circuit and filter output control circuit dedicated to this system are designed. Use an oscilloscope to test the output performance of the voltage conversion module.

Figure 2.86 shows the measured output ripple of the circuit, in which figure (a) is the output ripple of the 3.3 V supply voltage, and its peak-to-peak value is 6.6 mV. Figure (b) is the waveform diagram of the output ripple of the 5 V power supply voltage, its peak-to-peak size is 6.48 mV. The above test results show that the voltage supply circuit of this system meets the design requirements.

2. **Laser drive current performance test**

In the process of testing the output current of the LD, the output current is also at the mA level. In order to reduce the influence of human factors during the testing process, the test leads are also connected to the driving current output end of the circuit through wires. The test current used is Agilent's 34410A desktop multimeter, and the maximum output current can reach 116.698 mA. Test the stability of the LD control current output by the circuit, record every 10 s, and the recording time is 100 min. The measured data is shown in Fig. 2.87. Figure 2.87a–d are the waveforms of LD currents of four magnitudes of 20 mA, 40 mA, 60 mA, and 80 mA within 100 min, respectively. The corresponding current fluctuation amplitudes are 0.003 mA, 0.008 mA, 0.009 mA and 0.008 mA, respectively. The corresponding current stability is 0.015%, 0.02%, 0.015% and 0.01%, respectively, and the current stability is in the range of 0.02% [38].

Figure 2.88 shows that every 10 min of the output current is a time period, and the current fluctuation range in each section is counted. It can be seen from the figure that the current fluctuation of each section is within 0.004 mA.

3. **Laser temperature control performance test**

The temperature control of the laser is mainly to control the center wavelength of the output light. In this experiment, a spectrometer of APEX model AP2041B was used

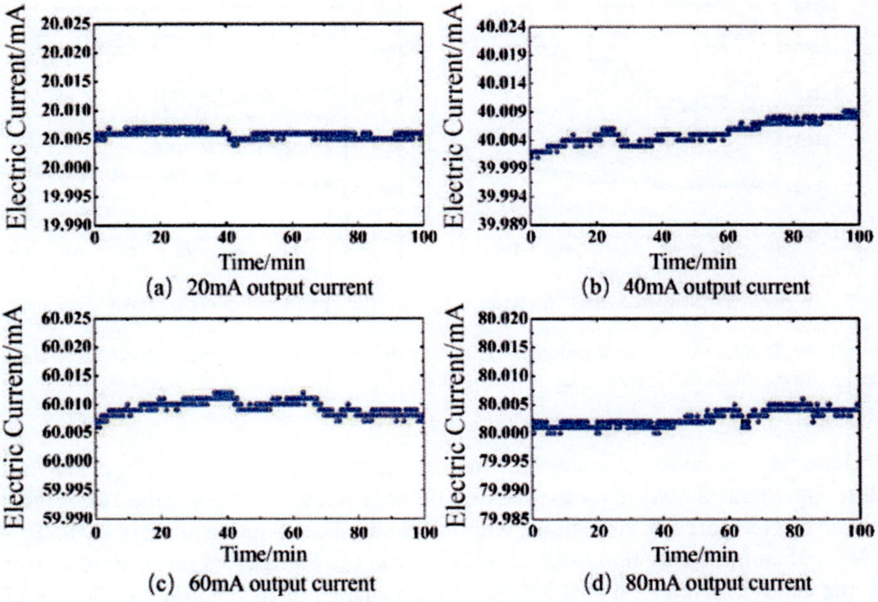

Fig. 2.87 LD drive current stability. **a** 20 mA output current; **b** 40 mA output current; **c** 60 mA output current; **d** 80 mA output current

Fig. 2.88 Variation of current output range per 10 min

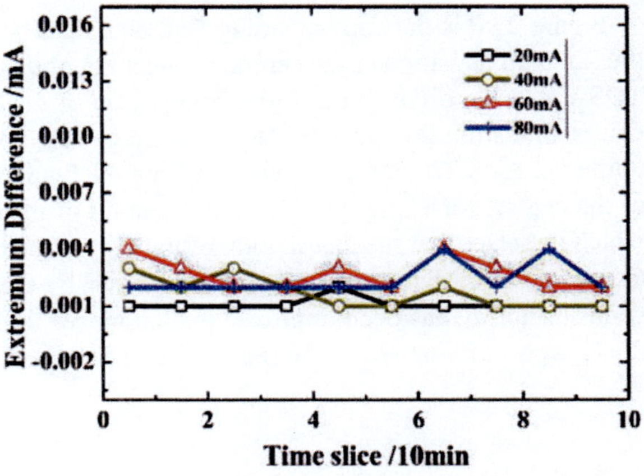

to test the central wavelength of the laser output light at a resolution of 1.12 p.m. The center wavelength is recorded every 12 s, and the recording time is 120 min.

The DFB laser is tested with Newport's driving source. Under the condition that the temperature is 25 °C and the LD operating current is twice the threshold current, the fluctuation range of the central wavelength of the laser output light is 8 p.m. within 120 min. The design circuit under the same conditions, the fluctuation amplitude of the central wavelength of the laser output light is 7 p.m.

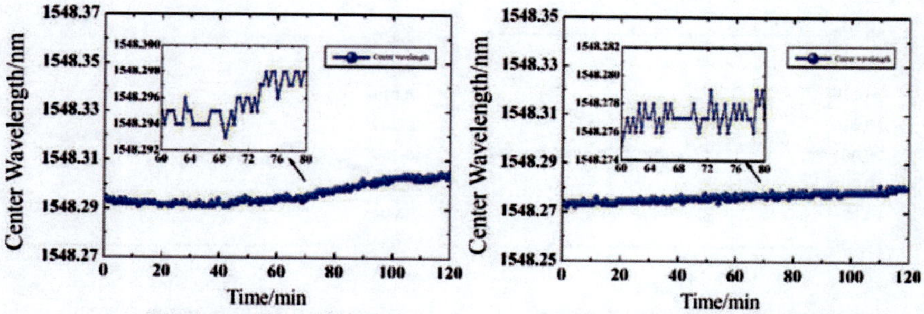

(a) Newport drives TOSA lasers (b) This design drives the TOSA laser output

Fig. 2.89 TOSA laser output wavelength variation with time. **a** Newport driven TOSA laser; **b** the design drives the TOSA laser output

However, the control effects of the two driving sources on the TOSA laser are very different. Figure 2.89a is the variation map of the center wavelength of the TOSA laser driven by Newport's driving source within 120 min. The variation range is 1548.291–1548.305 nm, the variation range is 14 p.m., and the normalized mean square error of the central wavelength is 0.0044 [39]. The change of the central wavelength of the TOSA laser control output by this design is shown in Fig. 2.89b. The normalized mean square error is 0.0020. The control system of this design improves the center wavelength control effect of TOSA laser, which makes up for the problem that the Newport drive source has a poor effect on the wavelength control of the TOSA laser.

Figure 2.90 is the corresponding Gaussian distribution map drawn according to the standard deviation of the central wavelength of the output light controlled by the TOSA laser. In order to make the comparison of the discrete data reflected by the Gaussian distribution map clearer, the average value of the Gaussian distribution is unified at 1548.279 nm. It can be seen from the figure that the Gaussian distribution of the central wavelength of the control output of the designed system is narrower, which indicates that the distribution of the output wavelength is more concentrated. Compared with the control of the TOSA laser by the Newport driving source, its control stability has been improved, which makes up for the poor control effect of the Newport driving source on the TOSA laser.

2.5 Broadband Chaotic Signal Source

In order to solve the problems of the uneven spectrum and narrow bandwidth of chaotic lasers and further promote the industrial application of chaotic lasers, the authors designed and developed a high-bandwidth, high-flatness broadband chaotic signal generator based on the amplified spontaneous emission (ASE) noise perturbation and semiconductor laser inter-injection method. The signal generator consists of two DFB lasers and a semiconductor optical amplifier. The ASE noise generated by the semiconductor optical amplifier is used to perturb the DFB lasers to generate

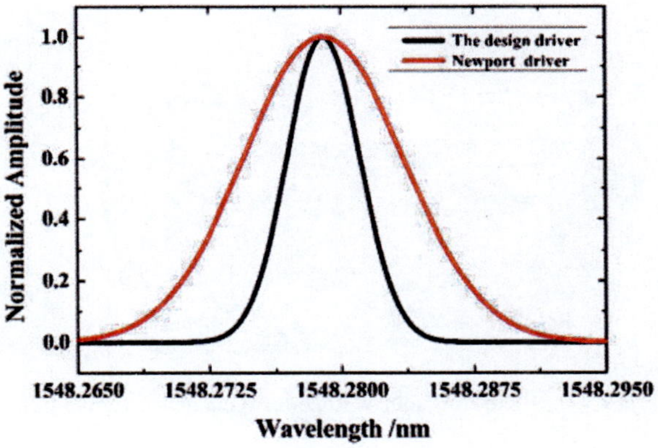

Fig. 2.90 Gaussian distribution of TOSA output wavelength controlled by different driving sources

chaotic lasers, and the beat frequency effect of the DFB laser inter-injection is used to further achieve spectrum shaping and bandwidth enhancement. With the feedback strength of 9.096%, frequency detuning of − 32.75 GHz and coupling strength of 1.635, the generator can generate a chaotic laser with a spectral bandwidth of more than 50 GHz, a flatness of ± 2.5 dB and a spectral linewidth of 0.56 nm. Further, the output spectra and spectral states of the chaotic laser were studied at different frequency detuning, and the control of the chaotic laser bandwidth was realized.

2.5.1 Structure and Principle

1. **Appearance and driver module design**

The physical diagram of the 50 GHz broadband chaotic signal generator is shown in Fig. 2.91, figure (a) shows the appearance of the prototype, and figure (b) shows the internal structure of the prototype. The prototype is designed and developed by the Key Laboratory of New Sensors and Intelligent Control of the Ministry of Education of Taiyuan University of Technology and Accelink Co. The broadband chaotic laser source includes two DFB lasers, two 90:10 fiber couplers, one semiconductor optical amplifier, one polarization controller, one optical attenuator and four optical isolators. The simple structure and low cost will help to make the chaotic laser source practical and marketable. The prototype has four output ports: monitoring signal 1, output signal 1, output signal 2, and monitoring signal 2, and the broadband chaotic signal is output from the output signal 1 port. The monitoring signal port can monitor the injected light in real time to prevent the laser from being damaged by excessive injection power. The output signal port can monitor the change of the output light in real time to obtain a suitable bandwidth of the chaotic laser.

The DFB laser and SOA inside the broadband chaotic laser source are soldered on a self-designed PCB board, as shown in Fig. 2.92. The wires are soldered to the pins of each device and connected to the serial port of the pin to realize the connection to

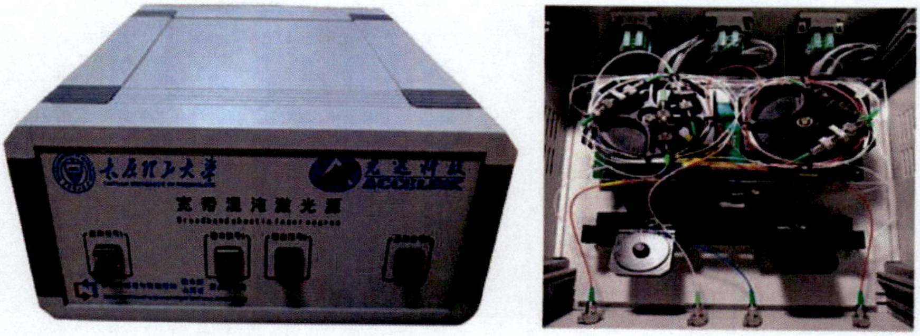

Fig. 2.91 50 GHz broadband chaotic signal generator diagram

the driver circuit of the device. In order to make the DFB laser and SOA dissipate heat effectively, a piece of copper is connected to the bottom of the device. After the test of the driver circuit, each device works stably. The design of the driver circuit board replaces the traditional commercial laser fixture, reduces the development cost and size, and facilitates the development of the integrated prototype.

The prototype drive source was designed by the authors' group, and the structure of the drive control module is shown in Fig. 2.93 [34]. The module is designed to meet the wide range of current regulations of SOA, and it can realize stable and accurate control of bias current and TEC temperature of DFB laser 1, DFB laser 2 and SOA, which means the adjustment of output power and central wavelength of DFB laser, and the effective control of ASE noise and laser inter-injection power of SOA. The DFB laser output is a broadband chaotic laser output when the set bias current and TEC temperature are input via the key operation of the driver source.

2. **Experimental setup diagram and working principle**

The experimental setup for generating broadband chaotic lasers based on ASE noise perturbation joint cross-injection is shown in Fig. 2.94, and the blue area shows

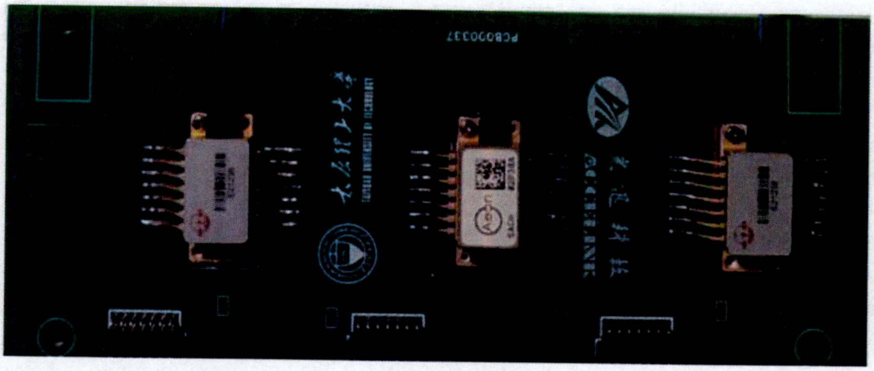

Fig. 2.92 Broadband chaos laser source drives circuit board

Fig. 2.93 Drive control module structure diagram

the internal structure of the broadband chaotic signal generator. The system uses two common commercial DFB lasers with similar wavelengths and performance to generate broadband chaotic lasers with suitable frequency detuning and coupling intensity. In the system, the ASE noise generated by the SOA disturbs the DFB laser1 to generate the chaotic laser, and the SOA achieves bi-directional amplification of the mutual injection power to achieve strong light injection, which is conducive to enhancing the beat frequency effect of the DFB laser. The PC controls the polarization state, the VOA controls the coupling strength, and the ISO (isolation $\geq$ 48 dB) prevents the external light from affecting the laser.

The output of the DFB laser 1 is divided into two paths through the left 90:10 fiber coupler, and the 90% path is connected to the SOA, which perturbs the chaotic laser generated by the DFB laser 1 and outputs it through the 10% path. The output

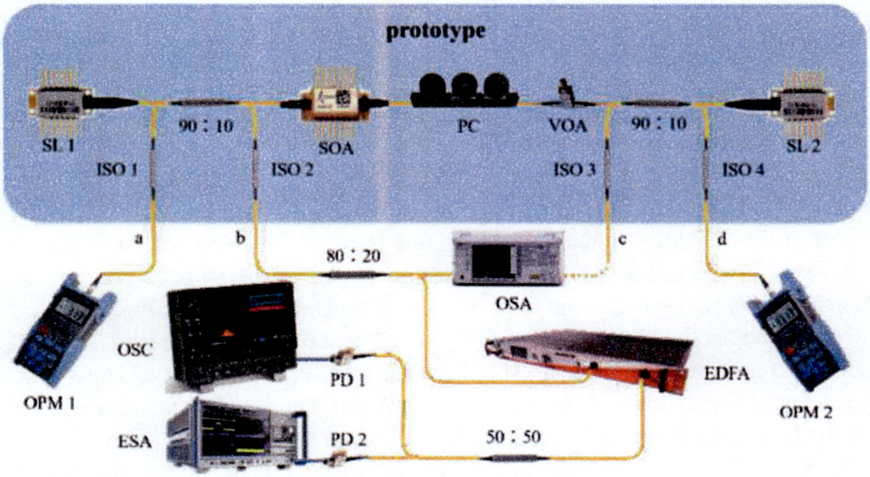

Fig. 2.94 Bandwidth chaotic signal generator experimental device diagram

of DFB laser 1 is amplified and connected to DFB laser 2 through the 90% path of the PC, VOA, and right 90:10 fiber coupler.

The prototype has four ports, a and b are the monitoring end and output end of DFB laser 1, c and d are the output end and monitoring end of DFB laser 2, respectively. The monitoring end monitors the change of feedback light through the optical power meter, so as to calculate the change of feedback intensity and coupling intensity, and to prevent the laser from being damaged by too much feedback light intensity. The feedback intensity is the ratio of the feedback optical power to the output power of DFB laser 1, and the coupling intensity is the ratio of the injected power to the output power of DFB laser 1. The output of DFB laser 1 is divided into two paths through an 80:20 fiber coupler, and the 20% path is connected to a 0.03 nm resolution spectrum analyzer, which monitors the output power of DFB laser 1 and the change of the central wavelength. The 80% output signal is amplified by the erbium-doped fiber amplifier, and then divided into two channels by a 50:50 fiber coupler. One way is connected to PD1 with a bandwidth of 50 GHz, which converts the optical signal into the electrical signal, and a spectrum analyzer with a bandwidth of 50 GHz monitors the spectral state change of the chaotic electrical signal; the other way is connected to PD2 and a high-speed real-time oscilloscope with a bandwidth of 36 GHz, which monitors the time series change of the output signal in real time. The output of DFB laser 2 is connected to the OSA to monitor the change of its output power and central wavelength, so as to obtain the change of frequency detuning $\Delta\nu$, that is, $\Delta\nu = \Delta\nu_2 - \Delta\nu_1$, where $\Delta\nu_2$ is the central frequency of DFB laser 2 and $\Delta\nu_1$ is the central frequency of DFB laser 1.

2.5.2 *Output Characterization*

1. **Typical output state**

The bias current of the DFB laser1 was set to 1.3 times the threshold current, at which the output power was 0.403 mW, its temperature was controlled at 25 °C, and the central wavelength of the output spectrum was 1548.177 nm. Adjust the driver module of SOA, set the bias current of SOA to 1. The ASE noise output is 0.04 mW. Therefore, the feedback intensity of the DFB laser 1 is 9.096%. At this time, the DFB laser 1 is perturbed by the ASE noise generated by the SOA and outputs the chaotic laser with a − 5 dB linewidth of 2.09 GHz. This is shown in Fig. 2.95.

Further adjust the bias current of DFB laser 2 to 2.6 times the threshold current, and set its temperature at 25.1 °C. The chip output power is 1.87 mW and the central wavelength of the output spectrum was 1548.439 nm. The coupling strength is 1.635 and the frequency detuning is − 32.75 GHz.

In this case, the output characteristics of the broadband chaotic signal generator are shown in Fig. 2.96. Figure 2.96a shows the spectrum of the chaotic signal, which is limited by the bandwidth range of the spectrometer, the bandwidth coverage should be more than 50 GHz, and the flatness is ± 2.5 dB. Figure 2.96b shows the optical

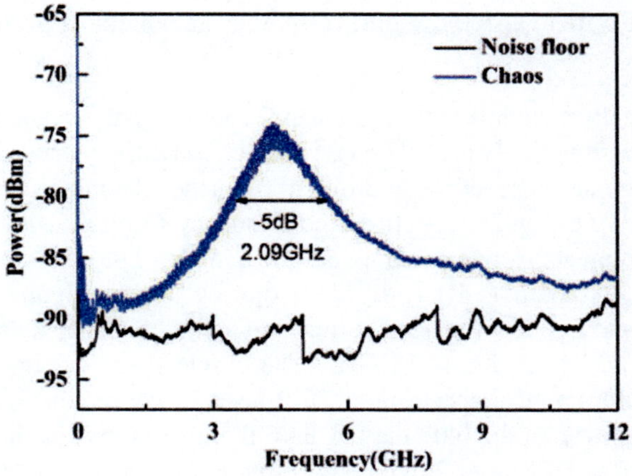

Fig. 2.95 ASE noise perturbation produces chaotic laser power

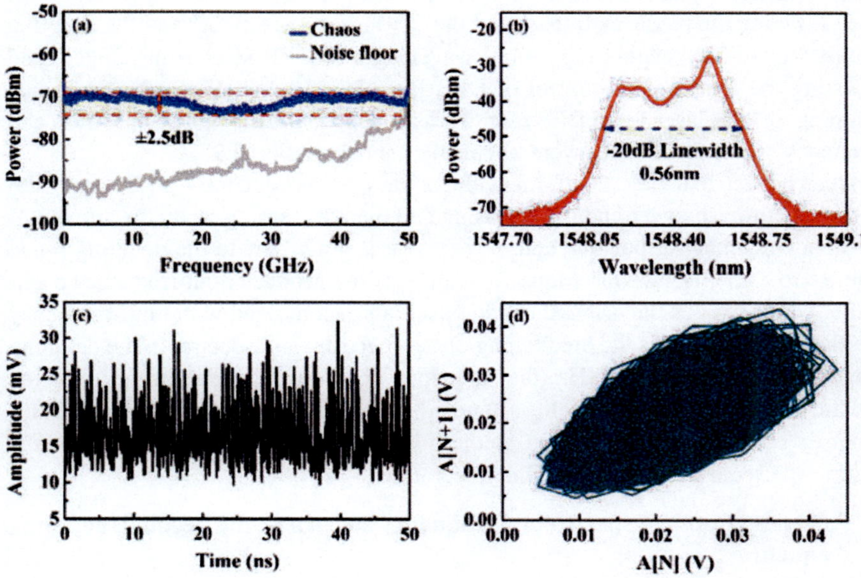

Fig. 2.96 Output characteristics of wideband chaotic signal generator. **a** Frequency spectrum; **b** spectrum; **c** time series; **d** phase diagram

spectrum of the chaotic signal, with a line width of 0.56 nm at − 20 dB, and the linewidth is substantially broadened. Figure 2.96c shows the time series of the chaotic signal, which shows a random distribution of large values with a peak-to-peak voltage of 22.7 mV. Figure 2.96d shows the phase diagram of the chaotic signal, which exhibits the typical chaotic attractor characteristics.

2. **Analysis of optical spectrum and frequency spectrum at positive detuning frequency**

By adjusting the drive module of the broadband chaotic signal generator, it is possible to control the bandwidth of the chaotic laser, which has different states of the optical spectrum and frequency spectrum at different frequency detuning.

The variation of the optical spectrum and frequency spectrum with the amount of frequency detuning at positive detuning is shown in Fig. 2.97. Figure (a-I)–(e-I) are optical spectra, and figure (a-II)–(e-II) are frequency spectra. Figure (a)–(e) are the optical spectra and frequency spectra correspond to the frequency detuning amounts at 51.75, 44.12, 37.5, 29.5, and 18 GHz. The purple curve in Fig. 2.97a-I is the free-running output optical spectrum of DFB laser 1, whose corresponding center wavelength is shown as the blue dashed line; the green curve is the free-running optical spectrum of DFB laser 2, whose corresponding center wavelength is shown as the black dashed line; the red curve is the spectrum of the chaotic laser output from DFB laser 1. In the frequency spectra shown in Fig. 2.97a-II, the gray curve is the noise floor, and the blue curve is the spectrum of the chaotic laser output from DFB laser 1. Fixing the feedback intensity of the DFB laser 1 at 9.096% and the coupling intensity at 1.635, keeping the central wavelength of DFB laser 1 unchanged, and adjusting the temperature control of DFB laser 2 at 18.2–20.9 °C, the frequency detuning of DFB laser 1 and DFB laser 2 was changed, and the optical spectrum and frequency spectrum changes were obtained as shown in Fig. 2.97.

It can be seen that the – 20 dB linewidth of the spectrum and the – 5 dB bandwidth of the spectrum show a trend of increasing and then decreasing when the amount of frequency detuning at positive detuning increases gradually. In the detuning range from 18 to 29.5 GHz, the laser taps the frequency to introduce high-frequency oscillation, which excites the original chaotic oscillation combined with high-frequency oscillation and achieves the broadening of the spectrum and spectra. In the detuning range from 37.5 to 51.75 GHz, the frequency detuning increases further, and the chaotic oscillations are gradually separated from the high-frequency oscillations, which is caused by the reduction of the beat frequency effect, and therefore the chaotic spectrum bandwidth gradually decreases.

3. **Analysis of optical spectra and frequency spectra at the negative detuning frequency**

The variation of optical spectra and frequency spectra with the amount of frequency detuning at negative detuning is shown in Fig. 2.98. Fixing the feedback intensity of the DFB laser 1 at 9.096%, the coupling intensity at 1.635, keeping the center wavelength of DFB laser 1 unchanged, adjust the temperature control of DFB laser 2 at 23–28.7 °C, then the frequency detuning of DFB laser 1 and DFB laser 2 varied as shown in figure (a)–(e), which were – 6.87 GHz, – 28.87 GHz, – 32.75 GHz, – 51.37 GHz, and – 77.25 GHz, respectively.

It can be seen that the – 20 dB linewidth of the optical spectrum and the – 5 dB bandwidth of the frequency spectrum show an increasing and then decreasing trend when the frequency detuning at negative detuning increases gradually. In the

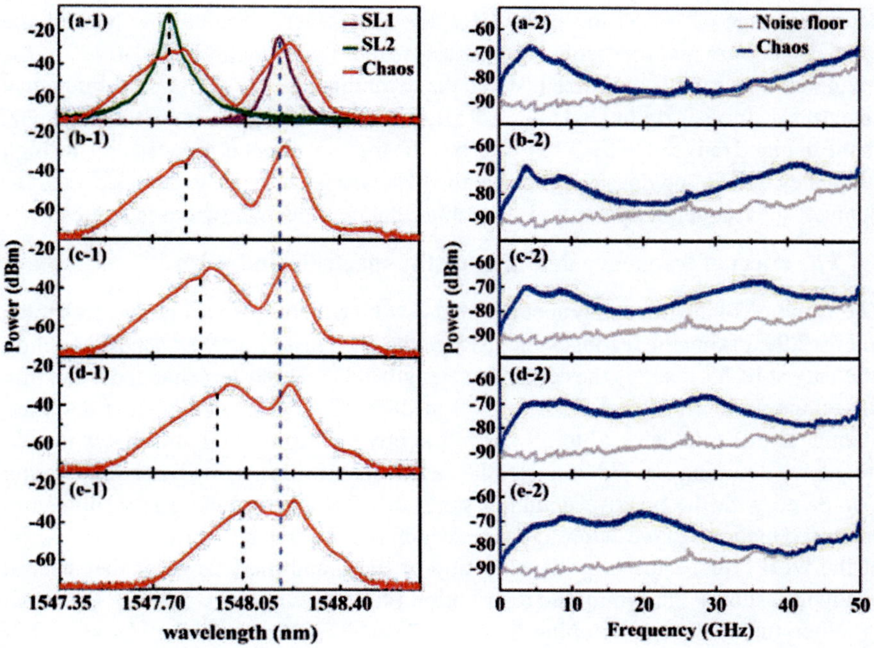

Fig. 2.97 Change of spectrum (I) and spectrum (II) with frequency detuning in positive detuning

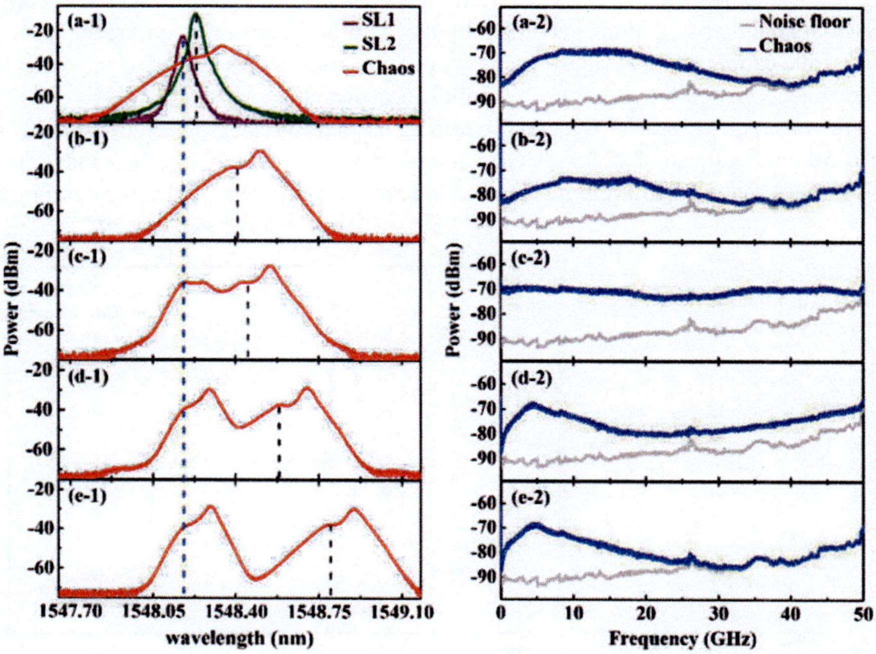

Fig. 2.98 The change of spectrum (I) and frequency spectrum (II) detuning in negative detuning

detuning range of − 6.87 to − 32.5 GHz, the beat effect is gradually enhanced, the optical spectrum and the frequency spectrum are broadened, and the flatness of the frequency spectrum is optimized. When the detuning is − 32.75 GHz, the frequency spectrum flatness is optimized to ± 2.5 dB, with the coverage of more than 50 GHz. In the detuned range of − 51.37 to − 77.5 GHz, the beat effect is reduced and the high frequency oscillation does not enhance the bandwidth, so the chaotic spectra start to depress, the frequency spectrum is no longer flat, and the bandwidth is reduced.

4. **The effect of frequency detuning on the spectral bandwidth**

The variation of the frequency spectrum bandwidth with frequency detuning is shown in Fig. 2.99. Fixing the feedback intensity of the DFB laser 1 at 9.096%, the coupling intensity at 1.635, keeping the center wavelength of DFB laser 1 unchanged, adjusting the temperature control of DFB laser 2 at 28.7–17.6 °C, the frequency detuning amount at this time is − 77 to 59 GHz. It is obvious from the figure that when the frequency detuning is − 32.75 GHz, the − 5 dB bandwidth is 50 GHz at maximum. Compared with the positive detuning state, the chaotic bandwidth is significantly enhanced in the negative detuning range, which is due to the red-shift of the spectrum of the DFB laser1 caused by the injection of light, and the change of the spectral bandwidth shows an asymmetric distribution [40].

When the amount of detuning in − 28.87 to 4.37 GHz range, the chaotic bandwidth increases less due to the injection locking of the DFB laser 2, which increases the relaxation oscillation frequency of the DFB laser 1. While in the range of 4.37–22.75 GHz and − 28.87 to − 35.87 GHz, when the detuning amount increases, the original chaotic oscillations generated by the ASE noise perturbation are superimposed and coupled with the high-frequency periodic oscillations caused by the optical inter-injection, and the chaotic bandwidth is broadened.

However, the amount of frequency detuning is not the greater the better, the better. In the detuning range of 22.75–59 GHz and − 35.87 to − 77.25 GHz, the bandwidth decreases gradually with the increase of the detuning amount. When the detuning

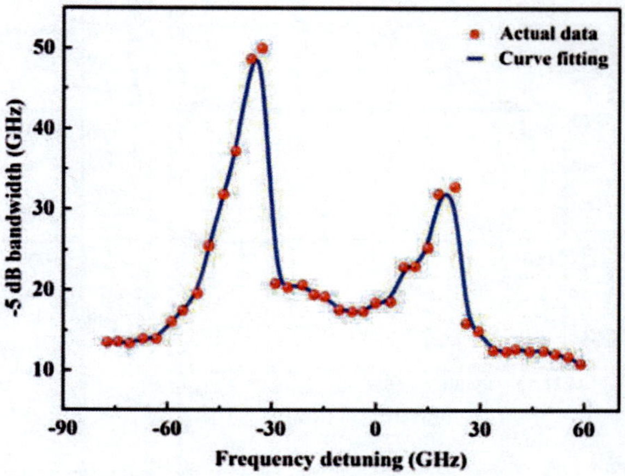

Fig. 2.99 Spectrum bandwidth variation with frequency detuning

amount is too large, the outgoing laser of DFB laser 2 is at the edge of the chaotic laser gain spectrum of DFB laser 1, and the effect of frequency slapping is reduced, and the influence of mutual injection disturbance on the original chaotic oscillation is smaller.

5. **Discussion and analysis**

The 50 GHz broadband chaotic signal source is firstly based on the introduction of external optical feedback by ASE scattering of SOA, then the dynamic balance of carrier number and photon number inside the DFB laser1 is disrupted, while the relaxation oscillations are excited. The nonlinear interaction between the relaxation oscillations and the external cavity mode increases the freedom of the laser, thus generating a chaotic laser.

Further, the optical field of the chaotic laser in the cavity of DFB laser 1 is beat-frequency coupled with the optical field of DFB laser 2. That is the mutual the beat-frequency between the relaxation oscillation mode of DFB laser 1 and the main mode of DFB laser 2. As the laser produces beat-frequency effect in a certain optical frequency detuning range, the beat-frequency coupling between different optical frequency components makes the energy transfer from the relaxation oscillation peak to the high frequency position to produce high-frequency periodic oscillation, so the broadband high-frequency periodic oscillation is superimposed and coupled with the original chaotic oscillation to realize the generation of spectrum flat and broadband chaotic laser. The non-linear interaction between lasers of similar wavelengths during mutual injection excites a large number of new optical frequency components, and the spectrum is further broadened compared to the optical feedback introduced by ASE scattering.

At the same time, the laser generates high-frequency chaotic oscillations during the non-locked injection, while the locked injection makes it difficult to generate high-frequency oscillations because the energy of the power spectrum is mainly concentrated around the relaxation oscillation frequency, so that the bandwidth spreading of the chaotic laser with different amplitudes is achieved. The laser is asymmetrically distributed in the positive and negative detuning range, and the nonlinearity of the laser output is more abundant in the negative detuning range. This is due to the fact that the injection intensity and frequency detuning of the semiconductor laser affect the change of the chaotic state, and the DFB laser1 will show different oscillations when the injection intensity and frequency detuning of the laser are changed.

References

1. Argyris A, Hamacher M, Chlouverakis KE, et al. Photonic integrated device for chaos applications in communications. Phys Rev Lett. 2008;100(19): 194101.
2. Chlouverakis KE, Argyris A, Bogris A, et al. Hurst exponents and cyclic scenarios in a photonic integrated circuit. Phys Rev E. 2008;78(6):0662151–5.
3. Syvridis D, Argyris A, Bogris A, et al. Integrated devices for optical chaos generation and communication applications. IEEE J Quant Electron. 2009;45(11):1421–8.
4. Argyris A, Grivas E, Hamacher M, et al. Chaos-on-a-chip secures data transmission in optical fiber links. Opt Express. 2010;18(5):5188–98.
5. Toomey JP, Kane DM, McMahon C, et al. Integrated semiconductor laser with optical feedback: transition from short to long cavity regime. Opt Express. 2015;23(14):18754–62.
6. Toomey JP, Argyris A, McMahon C, et al. Time-scale independent permutation entropy of a photonic integrated device. J Lightwave Technol. 2017;35(1):88–95.
7. Tronciu VZ, Mirasso C, Colet P, et al. Chaos generation and synchronization using an integrated source with an air gap. IEEE J Quant Electron. 2010;46(12):1840–6.
8. Sunada S, Harayama T, Arai K, et al. Chaos laser chips with delayed optical feedback using a passive ring waveguide. Opt Express. 2011;19(7):5713–24.
9. Sunada S, Harayama T, Arai K, et al. Random optical pulse generation with bistable semiconductor ring lasers. Opt Express. 2011;19(8):7439–50.
10. Wu JG, Zhao LJ, Wu ZM, et al. Direct generation of broadband chaos by a monolithic integrated semiconductor laser chip. Opt Express. 2013;21(20):23358–64.
11. Yu LQ, Lu D, Pan BW, et al. Monolithically integrated amplified feedback lasers for high-quality microwave and broadband chaos generation. J Lightwave Technol. 2014;32(20):3595–601.
12. Sunada S, Fukushima T, Shinohara S, et al. A compact chaotic laser device with a two-dimensional external cavity structure. Appl Phys Lett. 2014;104(24): 241105.
13. Liu D, Sun CZ, Xiong B, et al. Nonlinear dynamics in integrated coupled DFB lasers with ultra-short delay. Opt Express. 2014;22(5):5614–22.
14. Mingjiang Z, Yuhang X, et al. A hybrid integrated short-external-cavity chaotic semiconductor laser. IEEE Photonics Technol Lett. 2017;29(21):1911–4.
15. Soriano MC, Garcíaojalvo J, Mirasso CR, et al. Complex photonics: dynamics and applications of delay-coupled semiconductors lasers. Rev Mod Phys. 2013;85(1):421–70.
16. Park JK, Woo TG, Kim M, et al. Hadamard matrix design for a low-cost indoor positioning system in visible light communication. IEEE Photonics J. 2017;9(2):1–10.
17. Xiang SY, Wen AJ, Zhang H, et al. Effect of gain nonlinearity on time delay signature of chaos in external-cavity semiconductor lasers. IEEE J Quant Electron. 2016;52(4):1–7.
18. Wang A, Wang D, Gao H, et al. Time delay signature elimination of chaos in a semiconductor laser by dispersive feedback from a chirped FBG. Opt Express. 2017;25(10):10911.
19. Wang AB, Wang YC, Wang JF. Route to broadband chaos in a chaotic laser diode subject to optical injection. Opt Lett. 2009;34(8):1144–6.
20. Lang R, Kobayashi K. External optical feedback effects on semiconductor injection laser properties. IEEE J Quant Electron. 1980;16(3):347–55.
21. Rontani D, Locquet A, Sciamanna M, et al. Time-delay identification in a chaotic semiconductor laser with optical feedback: a dynamical point of view. IEEE J Quant Electron. 2009;45(7):879–1891.
22. Bjerkan L, Royset A, Hafskjaer L, et al. Measurement of laser parameters for simulation of high-speed fiberoptic systems. J Lightwave Technol. 2002;14(5):839–50.
23. Olshansky R, Hill P, Lanzisera V, et al. Frequency response of 1.3 μm InGaAsP high speed semiconductor lasers. IEEE J Quant Electron. 1987;23(9):1410–8.
24. Cartledge JC, Srinivasan RC. Extraction of DFB laser rate equation parameters for system simulation purposes. J Lightwave Technol. 1997;15(5):852–60.

25. Andre PS, Pinto AN. Extraction of DFB laser rate equations parameters for optical simulation purposes. In: Conference on telecommunications, Conftele'9 conference on telecommunications, Conftele; 1999.
26. Wang AB. Broadband chaos generation and chaos OTDR. Taiyuan: Taiyuan University of Technology; 2014.
27. Ye J, Li H, Mcinerney JG. Period-doubling route to chaos in a semiconductor laser with weak optical feedback. Phys Rev A. 1993;47(3):2249.
28. Ahmed M, Yamada M, Abdulrhmann S. Numerical modeling of the route-to-chaos of semiconductor lasers under optical feedback and its dependence on the external-cavity length. Int J Numer Model Electron Networks Devices Fields. 2009;22(6):434–45.
29. Lei YM, Zhang HX. Homoclinic and heteroclinic chaos in nonlinear systems driven by trichotomous noise. Chin Phys B. 2017;26(3):242–50.
30. Kuang JH, Sheen MT, Wang SC, et al. Post-weld-shift in dual-in-line laser package. IEEE Trans Adv Packag. 2001;24(1):81–5.
31. Lin YM, Liu WN, Shi FG. Laser welding induced alignment distortion in butterfly laser module packages: effect of welding sequence. IEEE Trans Adv Packag. 2002;25(1):73–8.
32. Hsu YC, Huang WK, Sheen MT, et al. Fiber-alignment shifts in butterfly laser packaging by laser-welding technique: measurement and finite-element method analysis. J Electron Mater. 2004;33(1):40–7.
33. Lin YM, Eichele C, Shi FG. Effect of welding sequence on welding-induced-alignment-distortion in packaging of butterfly laser diode modules: simulation and experiment. J Lightwave Technol. 2005;23(2):615–23.
34. Yang SJ, Zhang JZ, Liu Y, et al. Design of precise temperature control and driving circuit for chaotic laser. J Shenzhen Univ Sci Eng. 2018;35(5):516–22.
35. Jin XL, Su J, Jin PX, et al. Study on a high-precision digital temperature-control system for all-solid-state single-frequency green laser. Chin J Lasers. 2015;42(9):80–6.
36. Wang D, Lv Y. Design of modulation constant current driving circuit for laser diode. Mod Electron Tech. 2010;7:92–4.
37. Wang C, Wang C, Jiang XB, et al. A novel cascaded diode H-bridge multi-level rectifier. Power Syst Technol. 2015;39(3):829–36.
38. Chen YY, Li HY, Wang RF, et al. Waveform scheme of deep brain stimulation based on principle of PWM. Comput Eng Appl. 2015;16:259–64.
39. Guo FL, Xu GP, Huang BK. Constant temperature control systems for semiconductor laser based on DRV595. Laser Technol. 2017;41(5):734–7.
40. Takiguchi Y, Ohyagi K, Ohtsubo J. Bandwidth-enhanced chaos synchronization in strongly injection-locked semiconductor lasers with optical feedback. Opt Lett. 2003;28(5):319–21.

Chapter 3
Chaos Brillouin Distributed Optical Fiber Sensing

In conventional Brillouin distributed optical time-domain sensing technology, the spatial resolution is difficult to be reduced to less than 1 m due to the limitation of the pulse width which must be larger than the Brillouin phonon lifetime (10 ns). In the optical correlation-domain system, the sensing distance is only tens of meters owing to the periodic correlation peaks. In other words, the conventional Brillouin distributed optical sensing technology suffers from a trade-off problem between the measurement range and the spatial resolution, and it is difficult to realize the accurate positioning of the measured parameters, which seriously limits its practical applications. In this chapter, a novel chaotic Brillouin distributed optical sensing system is proposed, where the broadband chaotic laser instead of the pulse light is used as the sensing signal. The contradiction between the measuring range and spatial resolution has been overcome because of the chaos auto-correlation characteristics. The δ function characteristics of the chaotic laser eliminate the restriction of the periodic correlation peaks on the sensing distance. Therefore, chaos sensing technology has broken the bottleneck of the trade-off problem between the large measurement range and the high spatial resolution in the traditional technology. Ultimately, chaotic Brillouin distributed optical sensing technology has experimentally achieved a 3.5 mm spatial resolution and a 10.2 km measuring range.

3.1 Research Status of Distributed Optical Fiber Sensing

3.1.1 Introduction

Optical fiber sensing is a novel sensing technology that emerged with the development of the optical fiber and optical fiber communication technology in the 1980s. In the optical fiber sensing system, the light wave is the probe signal, and the optical fiber

M. Zhang et al., *Novel Optical Fiber Sensing Technology and Systems*, Progress in Optical Science and Photonics 28, https://doi.org/10.1007/978-981-99-7149-7_3

is both the transmission and the sensing medium, which can sense and detect environmental variables such as temperature, pressure, strain, electric field, and displacement. The fiber itself is uncharged, small size, lightweight, easy bending, excellent anti-electromagnetic interference and anti-radiation performance, especially suitable for flammable and explosive, strictly limited space and strong electromagnetic interference and other harsh environments. In modern information technology, sensing technology has been regarded as the third high-tech technology, and as a sign of a country's information strength. Therefore, as an important part of the sensor network, optical fiber sensing technology has received great attention and extensive research since it came out, and achieved considerable development [1–4].

Fiber sensing can be divided into three categories according to its sensing range: point [5], quasi-distributed [6], and full-distributed [7]. The point sensor is also known as the discrete optical sensor, which uses a single sensing unit to sense and measure the parameter changes in a small range near a predetermined point. Commonly used point-sensing units include fiber Bragg grating (FBG), Mach–Zehnder interferometer, and other sensors specially designed to measure a certain quantity, which can accurately measure the quantity change at a certain position.

The quasi-distributed sensors can realize multi-point sensing by arranging multiple sensing units to form a sensing unit array. This type of sensing system connects multiple points sensing units in a certain order and uses wavelength-division multiplexing, time-division multiplexing, and frequency-division multiplexing technologies to share one or more channels to form a distributed system. The system can be regarded as a point sensor or a distributed sensor, so it is called quasi-distributed optical fiber sensing technology.

In the full-distributed optical fiber sensing system, the fiber is both a signal transmission medium and a sensing unit, i.e., the sensing/transmission element can realize the monitoring of any position along the fiber. Therefore, the distributed fiber sensing system is highly valued by domestic and foreign counterparts and has become a research hotspot nowadays. The distributed optical fiber sensors are mainly based on optical scattering in the fiber. According to the difference of the scattering light signal, it is mainly divided into three types: Rayleigh, Brillouin, and Raman. The distributed fiber sensing technology based on Rayleigh scattering is mainly optical time domain reflectometry (OTDR) technology. It realizes the detection of fiber loss, bending, breakpoint, vibration, etc., by monitoring the power, polarization state, phase change, and other information of the Rayleigh scattering signal. The distributed optical fiber sensing technology based on Raman scattering can monitor the object change because of the sensitivity of the light intensity of Raman-anti-Stokes light to the temperature change. The Raman-based sensing technology has been very mature and has achieved commercialization.

Distributed optical fiber sensing technology based on Brillouin scattering has become one of the most representative sensing technologies because of its ability to monitor external parameters such as temperature, vibration, and strain. After decades of development, a series of results have been achieved, and the sensing performance has been greatly improved. It is not only small size, lightweight, and strong anti-electromagnetic interference but also can realize the continuous measurement of

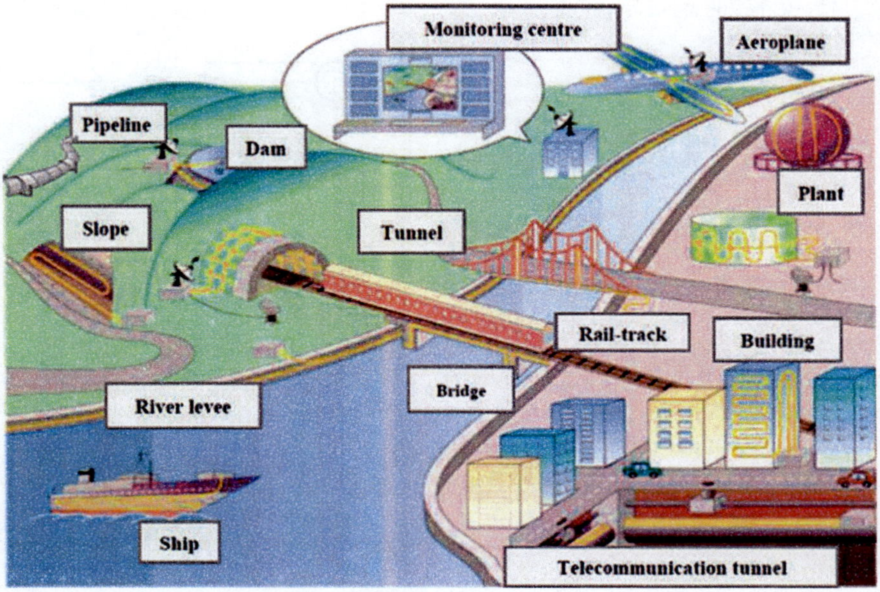

Fig. 3.1 Various application of Brillouin distributed fiber sensors

temperature and strain at any position along the fiber. It has the advantages of high measurement accuracy, long sensing distance, and high spatial resolution, which is very significant and has broad application prospects in the fields such as oil and gas transportation pipeline, power network, submarine optical cable, urban infrastructure, highway, railway bridge tunnel road network, large structure health monitoring and geological disaster detection (Fig. 3.1).

3.1.2 Brillouin Distributed Optical Fiber Sensing

In 1972, E. P. Ippen and others at Bell Labs in the United States first observed the phenomenon of stimulated Brillouin scattering (SBS) [8]. Initially, Brillouin scattering was only used for studying the properties of optical fibers [9–11]. In 1989, D. Culverhouse at the University of Kent discovered the linear relationship between the Brillouin frequency shift (BFS) and temperature for the first time [12], T. Horiguchi and others in Nippon Telegraph and Telephone Company (NTT) found the positive linear relationship between the BFS and strain [13]. The discovery of the linear relationship between the BFS and temperature or strain has laid the cornerstone of fiber Brillouin sensing and unveiled the curtain of its research.

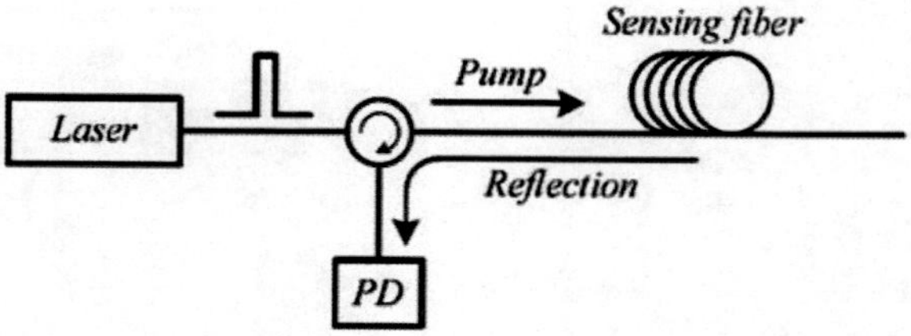

Fig. 3.2 Schematic diagram of the BOTDR

At present, according to the difference between the probe signal and its sensing mechanism, Brillouin distributed optical fiber sensing technology can be mainly classified into five types: Brillouin optical time domain reflectometry (BOTDR), Brillouin optical time domain analysis (BOTDA), Brillouin optical frequency domain analysis (BOFDA), Brillouin optical correlation domain reflectometry (BOCDR), and Brillouin optical correlation domain analysis (BOCDA).

1. **Brillouin optical time domain reflectometry**

The BOTDR technique realizes the fully distributed sensing by using the linear relationship between the variation of power or frequency shift of the spontaneous Brillouin scattering (SpBS) light in the fiber and the temperature or strain. T. Kurashima first proposed the BOTDR system in 1993, and the basic principle is shown in Fig. 3.2. The pulse pump light, angular frequency of ω_0, is injected into the sensing fiber from one end and then the backward SpBS with the frequency of $\omega_0 \pm \Omega_B$ is generated. The BFSs along the fiber can be obtained via Lorentz-fitting the Brillouin gain signals and the temperature or strain information is further got.

Meanwhile, the position information can be obtained by measuring the time delay between the pulse and the scattering light according to the location principle of the OTDR technology. Combined with the temperature (strain) information and position information above, the distributed temperature (strain) sensing can be finally realized. BOTDR has the advantage of single-end measurement and simpler working principles, which makes it more practical. However, the intensity of SpBS is very faint (about two orders of magnitude lower than Rayleigh scattering) and the frequency shift relative to the incident light is small (about 11 GHz at the incident wavelength of 1550 nm), so the signal detection of BOTDR system is rather hard.

In the SpBS-based sensing system, faint Brillouin signals can be obtained by direct detection or coherent detection. In the direct detection BOTDR, the backscattering light is filtered by using the optical filter with high precision and stability, such as Mach–Zehnder interferometer [14], Fabry-Pérot interferometer [15] or fiber grating [16], where the Rayleigh scattering light is filtered and then SpBS light is obtained. Direct detection has very strict requirements on the optical filter, whose bandwidth must be less than 11 GHz and the central wavelength be stable. Owing to the high BFS of 11 GHz, it is difficult to obtain the beat signal between the SpBS light and the reference light. Therefore, in the actual measuring system, the beat signal is often detected by down-conversion to the low-frequency domain [17].

At present, a single light source is used in the common coherent detection methods. The output light of the light source is divided into the pulse pump wave and the continuous wave (CW) by the coupler, and then various methods, such as the reference light source [18], reference fiber [19], acoustic-optic cyclic frequency shift [20] and microwave electro-optic modulation frequency shift [21], are used to generate local reference light with BFS, which puts the beat signal at the low-frequency domain and reduces the requirements for photoelectric detectors and signal processing system.

Nowadays, the BOTDR technology can realize 100 km distributed sensing [22] thanks to the high signal-to-noise (SNR) of microwave coherent heterodyne detection. If combined with distributed Raman/Brillouin amplification configuration, the sensing distance can be extended to 150 km [23]. The traditional spatial resolution is restricted to more than 1 m limited by the 10 ns phonon lifetime in fiber. BOTDR can break through the limitation of phonon lifetime and achieve the spatial resolution of decimeter level by using double pulse [24] and equivalent pulse optical fitting [25]. In addition, the biggest advantage of the BOTDR technology is single-ended measurement and easy to use. It is especially suitable for some occasions that can only be measured at one end of the fiber. Many companies in China and abroad have developed commercial BOTDR products and put them into the market. The representative productions of distributed BOTDR sensing mainly include the AQ8603 of YOKOGAWA Electric Corporation, DTSS of Sensornet Co., Ltd., AV6419 of 41st Institute of CETC, and so on.

2. **Brillouin optical time domain analysis**

SpBS-based BOTDR technology has the advantage of single-ended measurement. However, it is difficult to detect and the sensor performance is greatly restricted due to the faint SpBS signal. The SBS-based BOTDA sensing method was first proposed by Horiguchi et al. in 1989. A schematic representation of the typical BOTDA system is shown in Fig. 3.3 [26]. The CW generated by the LD1 is modulated as the pulse pump light, and the probe light (Stokes light) is generated by the LD2, whose frequency is about one BFS lower than the pump light. The pulse pump and the probe light are injected into both ends of the fiber simultaneously and propagate counter. When the two beams meet in the fiber, part of the energy of the pulse pump light is transferred to the Stokes light through the acoustic field due to the SBS amplification. The probe light powers varying with time are measured at the pump end by changing the probe frequency, and the Brillouin gains at different positions of the fiber are obtained, so as to demodulate the distribution of the BFS along the fiber. The distributed sensing of temperature or strain is realized by using the linear relationship between BFS and ambient temperature or strain. This sensing system is called gain-type BOTDA.

In the gain BOTDA, because the energy of the pump pulse wave is continuously transferred to the continuous wave (Stokes light), the energy of the pump pulse decreases continuously along the fiber, which is not conducive to long-distance transmission. Therefore, the research group of Xiaoyi Bao from the University of Ottawa proposed the loss-type BOTDA in 1993 [27]. As shown in Fig. 3.3, the frequency of laser1 output is one BFS lower than that of laser2 in the loss-type BOTDA, i.e., the pulse wave is the Stokes light and the continuous wave is the pump light. Pulse

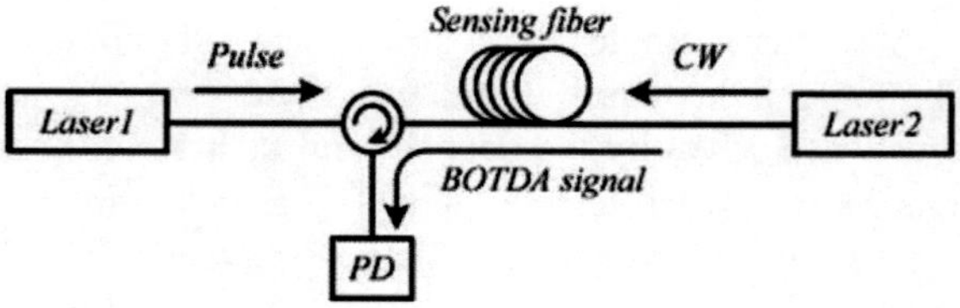

Fig. 3.3 Schematic diagram of the BOTDA

energy is promoted with the increasing of sensing distance which can realize a larger measurement range. In fact, both gain-type and loss-type configurations encounter a problem of non-local effect caused by the pump consumption [28], which will affect the measurement accuracy under a long distance.

There is a trade-off between the spatial resolution and the measurement accuracy in the Brillouin time-domain distributed optical sensing technology. The sensing system must use a narrow pulse in order to obtain a high spatial resolution. However, the probe light with narrow pulse width causes the BGS to be broadened, resulting in the decrease of the measurement accuracy of BFS. In addition, the probe wave with narrow pulse width means the shorter spatial length of the interaction between the pump wave, probe wave, and the phonon, which weakens the Brillouin signal and increases the detection error, leading to the resolution reduction of strain or temperature. In order to overcome the above difficulties, researchers have proposed a variety of methods. In 2008, Bao's proposed BOTDA based on differential pulse pair (DPP-BOTDA), which improves the spatial resolution to 0.2 m [29], while ensuring the accuracy of BFS. In addition, the team continuously improved the DPP-BOTDA technology and realized the sensing of 0.5 m spatial resolution on a 50 km fiber [30] by combining with pulse coding and optical differential parametric amplification. Yongkang Dong et al. extended the sensing distance of the BOTDA to 100 km [31] and 75 km [32] respectively by using time/frequency division multiplexing technology in 2011. Online erbium-doped fiber amplifier (EDFA) was introduced into the frequency division multiplexing scheme to further expand the sensing distance to 150 km [33] in 2012.

At the same time, L. Thévenaz used electro-optic microwave frequency shift to achieve Brillouin sideband modulation. Based on this, a single-source BOTDA was proposed to improve the stability of the system [34], and the BGS characteristics of the single-mode fiber (SMF) were studied [35]. L. Thévenaz proposed Brillouin echoes technology to improve the spatial resolution of BOTDA and realized distributed sensing with 5 cm spatial resolution and 3 MHz frequency shift accuracy on the 5 km fiber in 2008 [36]. In addition, the team has also conducted a lot of research on the long-distance BOTDA. For example, the Simplex code was used to encode the pump pulse, which greatly improved the signal-to-noise ratio (SNR) and realized the 1 m spatial resolution over the 50 km measurement range [37, 38]. The distributed Raman remote amplification was used for long distance Brillouin distributed sensing. The temperature sensing experiment was carried out with a 2 m spatial resolution on the 100 km optical fiber [39]. A detailed theoretical analysis about the influence of pump consumption on long distance sensing of BOTDA was demonstrated by the team in 2013. The detection limit of optical power that the adverse effects of pump consumption can be ignored was obtained, which provided an important guide for the power budget of the BOTDA [40]. In the same year, the SNR of the sensing system was improved by using the Brillouin gain and loss and the bipolar Golay complementary code for pulse coding. The temperature sensing was realized with a 2 m spatial resolution on the 100 km optical fiber [41]. In addition, Yunjiang Rao et al. proposed the hybrid distributed Raman amplification BOTDA, which realized the long-distance sensing of 154.4 km and the spatial resolution of 5 m [42] in 2014. Chao Zhang et al. applied pulse coding technology to bidirectional Raman pump BOTDA, analyzed the influence of pulse coding on the system, and obtained 2.5 m spatial resolution and 1 °C temperature measurement accuracy over about 50 km [43]. Zinan Wang et al. used the wavelet denoising algorithm based on Bayesian shrinkage to improve the SNR and reduced the measurement uncertainty from $\pm$ 1.52 MHz to $\pm$ 1.03 MHz on the 155 km sensing fiber [44] in 2016. In the same year, they combined the pulse coding technology with the non-local mean algorithm and realized 8 m spatial resolution on the 158 km sensing fiber [45]. Pulse pre-pumping [46], dark pulse [47], and other technologies have also been proposed and realized the distributed temperature or strain sensing with long distance and high spatial resolution.

BOTDA technology has been quite mature and basically met the practical requirements in sensing accuracy, sensing distance, and spatial resolution after 30 years of development.

3. **Brillouin optical frequency domain analysis**

Spatial resolution is an important index in the distributed optical fiber sensing system. The spatial resolution of the Brillouin optical time-domain system is restricted to 1 m magnitude due to the phonon lifetime (10 ns). To further improve the spatial resolution, except for the differential pulse, dark pulse, and others, BOFDA has also been proposed.

BOFDA, proposed by D. Garus et al. in 1996, is a distributed fiber sensing method combining the SBS effect and optical frequency domain reflection technology [48]. Compared with the BOTDA, BOFDA has the advantages of high spatial resolution and low detection light power. The essence of BOFDA is to realize spatial positioning based on the composite baseband transmission function of the measured fiber. This function relates the complex amplitude of the probe and the pump wave transmitted through the fiber to the geometric length of the fiber and realizes measuring point positioning by calculating the impact response function of the fiber.

The working principle of BOFDA is shown in Fig. 3.4. Continuous pump wave and probe wave after amplitude modulation by electro-optic intensity modulator (EOM) are injected into both ends of the sensing fiber, respectively. The frequency difference between the pump and the probe is approximately equal to the BFS. Changing the modulation frequency of the EOM, the intensity of the probe and pump light is detected by the photoelectric detector (PD), whose output signal is sent to the vector network analyzer to calculate the baseband transmission function of the sensing fiber. Inverse fast Fourier transform (IFFT) is applied to the baseband transmission

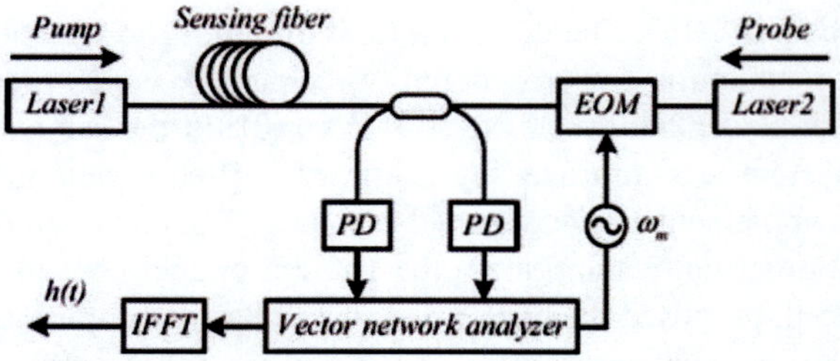

Fig. 3.4 Schematic diagram of the BOFDA

function of the sensing fiber. For the linear system, this result can be similar to the impulse response function of the sensing system, which contains the temperature or strain information along the fiber.

Garus et al. conducted distributed temperature and strain sensing experiments using BOFDA, where 5 °C temperature resolution, 0.01% strain resolution, and 3 m spatial resolution were obtained on a 1 km sensing fiber [49]. The BOFDA, proposed by Gallus, used the IFFT to convert the data to the time domain for processing. This method assumed the linear approximation relationship between the Brillouin signal and the Brillouin gain, not considering the nonlinear effect of the Brillouin amplification. In 2002, the research group of A. Minardo in Italy improved the signal processing method of BOFDA. Taking into account the influence of the Brillouin amplification nonlinear effect, the expression of the BOFDA signal was strictly deduced. Without using the IFFT, a cost function was introduced, and the data were completely analyzed and processed in the frequency domain [50]. Based on this, the research group conducted a lot of research on BOFDA. In 2004, the improved BOFDA distributed sensing experiment was carried out to verify the feasibility of this method [51]. In 2007, the intensity modulation was exerted on the pump wave in BOFDA, and the lock-in amplifier was used for heterodyne detection, without vector network analyzer and high-speed detector, which effectively reduced the system cost [52]. In 2008, the BOFDA was used to study the Brillouin characteristics of the high birefringence micro-structure fiber [53]. In 2009, the single-ended BOFDA was realized by using the reflection of the fiber end [54]. In 2012, an iterative algorithm was adopted to eliminate the pseudo-signal effect of sound field modulation and achieved the high spatial resolution of 3 cm [55]. In 2014, BOFDA was used to study the Brillouin characteristics of the polymer fiber [56].

BOFDA, which has low hardware requirements and cost, has high spatial resolution (about 1 mm) and measurement accuracy in theory. It can be measured in parallel without high-speed data sampling and grabbing. But the spatial resolution is limited by the pseudo-signal generated by sound field modulation [55]. At the same time, BOFDA sensing distance is limited and the measurement time is long, and there is no practical instrument reported.

4. **Brillouin optical correlation domain reflectometry**

In view of the limitation of phonon lifetime on the spatial resolution of Brillouin optical time domain system, Y. Mizuno group of Tokyo University proposed an optical correlation domain reflection sensing method based on SpBS in 2008.

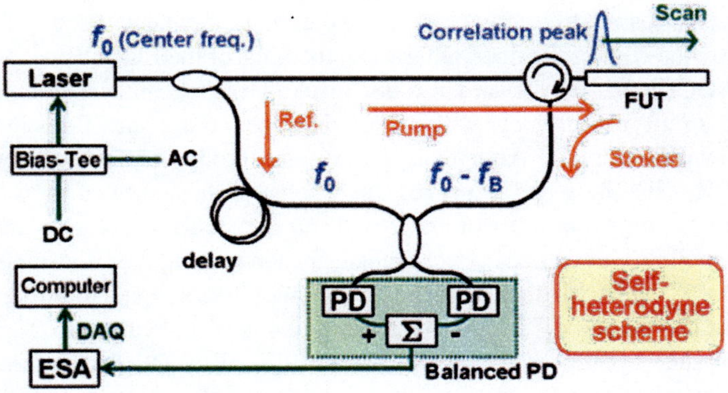

Fig. 3.5 Schematic diagram of the BOCDR

This technology combined the correlation peak localization method and OTDR technology, and its basic principle is shown in Fig. 3.5.

The laser is intensity modulated by the sinusoidal signal and the output continuous light is split into two beams through the coupler. One beam is injected into the delay fiber as the reference light, and the other beam passed through the optical circulator injects into the sensing fiber as the pump light. The pump generates SpBS in the sensing fiber, and the generated Brillouin backscattering light returns to the detection end through the circulator. A series of correlation peaks appear in different positions of the sensing fiber at intervals, and the frequency difference between the Brillouin scattering light and the local reference light at the correlation peak position is exactly equal to the BFS of the fiber at this position. However, the frequency difference is always changing owing to the non-correlation of the two beams at other positions [57]. The sensing fiber has only one correlation peak by selecting the appropriate modulation frequency of the light source, and then the modulation frequency is scanned to change the position of the correlation peak in the fiber continuously, and then realize the distributed sensing of the whole fiber.

$$d_m = \frac{\mathrm{v_g}}{2f_m}, \tag{3.1}$$

$$\Delta z = \frac{\mathrm{v}_g \Delta \upsilon_B}{2\pi f_m \Delta f}, \tag{3.2}$$

Spatial resolution and the sensing distance of BOCDR can be expressed by the Eqs. (3.1) and (3.2). Here, f_m is the modulation frequency of the light source, v_g is the group velocity in the fiber, Δf is the frequency modulation depth, and Δv_B is the Brillouin gain spectrum width. It can be seen from the above two equations that there is a trade-off between the spatial resolution Δz and the sensing length $d_{m,}$ and the spatial resolution is closely related to the modulation frequency f_m and the frequency modulation depth Δf. When the f_m is lower, the effective sensing distance will be longer, and the Δz will increase. However, the Δf cannot be unlimitedly increased to maintain the spatial resolution because the maximum frequency modulation depth is limited by the BFS of the fiber. Otherwise, the SNR of the BOCDR will decrease.

Since BOCDR was put forward in 2008, Yosuke Mizuno realized the single-ended distributed strain measurement with 100 m distance, 40 cm spatial resolution and 50 Hz dynamic sampling. After that, the research group continuously optimized the BOCDR technology and improved the sensing performance of the system. In 2009, a distributed strain measurement with 66 cm spatial resolution and 50 Hz dynamic sampling was realized at 1 km sensing distance using time division selection technology [58]. In 2010, dual-frequency modulation was adopted to further expand the sensing distance to 1.5 km and improve the spatial resolution to 27 cm [59]. In 2016, Yosuke Mizuno et al. proposed an ultra-fast BOCDR device to achieve a high-speed acquisition of up to 100 kHz with poor spatial resolution, small dynamic range and deteriorated measurement accuracy. The experiment verified the real-time measurement of 1 kHz dynamic strain [60]. In addition, the research group also verified the feasibility of the slope-assisted BOCDR device, and proved that this method could be used to realize the measurement of strain or temperature whose scale was less than spatial resolution, and successfully expanded the sensing distance to 10 km [61].

The advantages of BOCDR are single-ended measurements and high spatial resolution. However, the structure of BOCDR is complex and the signal demodulation is difficult. Compared with BOTDR and BOTDA, the sensing distance of BOCDR is shorter and it is still in the experimental research stage.

5. **Brillouin optical correlation domain analysis**

Optical correlation domain analysis based on SBS is a high resolution distributed sensing scheme earlier than BOCDR, proposed by K. Hotate of Tokyo University in 2000 [62]. BOCDA also uses the method of correlation peak localization, and its basic structure is shown in Fig. 3.6. The laser is divided into two beams after sinusoidal modulation, and one is used as the probe wave by shifting the frequency of BFS (ν_B). The other one is used as the pump wave after intensity modulation of the electro-optic modulator (EOM) for lock-in amplification detection. Similar to the BOCDR, a series of correlation peaks appear at different positions of the sensing fiber equidistantly, and the frequency difference between the pump and the probe at the position of the correlation peak is exactly equal to the BFS of the fiber, and the SBS continues to occur. In other positions, the frequency difference between the pump and the probe is always changed which makes the Brillouin interaction between them very weak. Continuously changing the modulation frequency of the light source, positions of the correlation peak can be moved to realize the distributed sensing of the whole fiber. The spatial resolution Δz and the sensing length d_m of BOCDA can also be expressed by the Eqs. (3.1) and (3.2), which are also mutually restricted.

In order to solve the trade-off between the sensing distance and spatial resolution in the traditional BOCDA system [63], Kazuo Hotate et al. adopted the time-gating technology [64] and the dual lock-in detection technology [65], respectively. The effective sensing distance of the system is increased, and the distributed sensing with

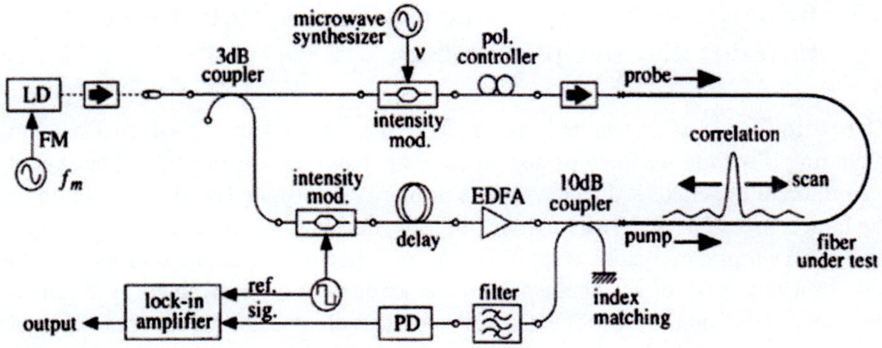

Fig. 3.6 Schematic diagram of a typical experiment structure of BOCDA [62]

spatial resolution of 7 cm is finally realized at the distance of 1 km. In addition, the light source intensity modulation is used to suppress the systematic noise, and the dual modulation combined with differential measurement is proposed to improve the SNR and optimize the sensing performance of the system [66]. At the same time, high spatial resolution (millimeter level) can be obtained in the case of short sensing distance. At present, the distributed sensing with about 1.6 mm spatial resolution at 5 m sensing distance, and 200 Hz dynamic strain have been realized [67, 68]. In addition, the research group of Zadok used the correlation function characteristics of amplified spontaneous emission noise (ASE) signal with the single correlation peak to construct the BOCDA with 4 mm spatial resolution and 5 cm sensing distance [69]. At the same time, the research group used continuous light with phase modulation by pseudo-random bit sequence (PRBS) [70–73] and Gray code [74] as the light source to achieve a more flexible BOCDA for sensing position scanning. However, in order to obtain higher spatial resolution, higher modulation rate and modulation devices were needed which would undoubtedly increase the cost of the system. The prominent advantage of BOCDA is higher spatial resolution. But this protocol is complex and affected by the periodic correlation peak, which limits the sensing distance to several or hundreds of meters.

Based on the above analysis, distributed Brillouin sensing technology has been extensively studied, and its wide application in the engineering field will inevitably become the future development trend. However, the light source of BOTDR and BOTDA is pulse light, whose width needs to be increased in enlarging the sensing distance, which will seriously reduce the spatial resolution. Although the spatial resolution of the distributed fiber sensing based on BOCDR and BOCDA is high (up to millimeter level or even sub-millimeter level), it cannot guarantee a long sensing distance. Therefore, the trade-off between spatial resolution and sensing distance in distributed Brillouin fiber sensing technology is still to be studied.

3.2 Brillouin Scattering Characteristics of Chaotic Laser Injecting into an Optical Fiber

The performance of distributed optical fiber sensing systems based on Brillouin scattering is mainly weighed by spatial resolution and sensing distance. The spatial resolution of the chaos sensing system is primarily determined by the bandwidth of the light source. The wider bandwidth will bring the shorter coherence length, which will obtain a higher spatial resolution. In the reflective Brillouin sensing system based on the SpBS effect, the sensing distance is mainly determined by the SBS threshold of the light source. The higher threshold will make the higher optical power injected into the optical fiber, and obtain the higher SNR and the longer transmission distance. If SBS occurs and the Brillouin gain reaches saturation, there is almost no light passing through the end of the fiber, which limits the transmission distance. Moreover, the nonlinear effect caused by SBS will introduce more noise and reduce the SNR of the system. In addition, the spontaneous SBS effect will also accelerate the depletion of the pump power, which severely restricts the sensing distance of the analytical Brillouin system with double-ended incidence. Combining the above factors, considering that the chaotic laser has the characteristics of wide bandwidth and high SBS threshold, it is introduced into distributed optical fiber sensing system based on Brillouin scattering. The analysis of Brillouin scattering characteristics of the chaotic laser in optical fiber has important guiding significance for improving the spatial resolution and sensing distance.

3.2.1 Theory of Stimulated Brillouin Scattering

Fiber is a low-loss and high-speed optical transmission medium. However, due to the certain inhomogeneity of the medium itself, part of the light waves in the optical fiber will occur the various backscattering as shown in Fig. 3.7 [75].

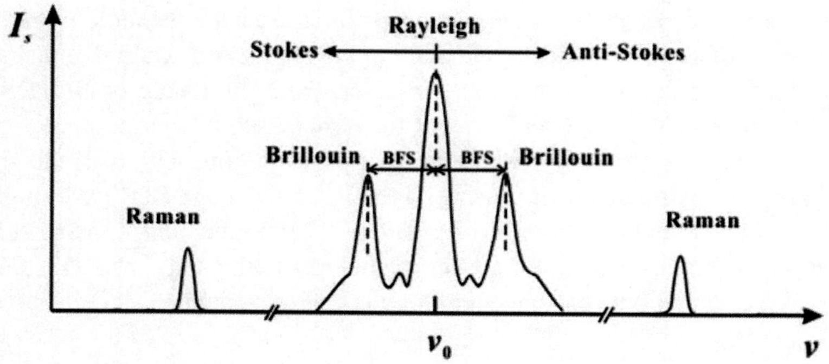

Fig. 3.7 Typical backscattering diagram in fiber

Rayleigh scattering is an elastic scattering caused by the inhomogeneity of fiber local density and composition, and its center frequency is the same as the incident light. The Brillouin scattering and Raman scattering are inelastic scattering, whose frequency changes relative to the incident light. When the scattering frequency is higher than the incident frequency, it is called anti-Stokes light. Conversely, it is called Stokes light. In Brillouin scattering, this frequency change is called Brillouin frequency shift (BFS).

Brillouin scattering is essentially caused by the interaction between incident light and acoustic phonons [75–79]. As mentioned in Sect. 1.2, it can also be divided into SpBS and SBS according to the different generation modes of acoustic phonons, which are introduced respectively in the following.

The particles (atoms, molecules, or ions) that make up the medium form the continuous elastic mechanical vibration in the medium because of the self-heating motion, which causes the medium density to change periodically with time and space, resulting in a spontaneous acoustic field inside the medium. The acoustic field makes the refractive index of the medium periodically modulated and propagates at the sound velocity V_a in the medium. The variation is like a grating (called acoustic grating). The scattering will be generated via acoustic field interaction when the pump light injecting into the medium, and then the scattering light, whose frequency is related to the sound velocity due to the Doppler shift, is stimulated and named as SpBS [76].

The physical model of SpBS in fibers is shown in Fig. 3.8 [77]. The angular frequency of the incident light is set to be ω, and the moving grating reflects the incident light through the Bragg grating without considering the dispersion effect of the fiber on the incident light. When the acoustic grating and the incident light are in the same direction, the frequency of the scattering light moves down relative to the incident light due to the Doppler effect. At this time, the scattering light is called the Brillouin Stokes light, whose angular frequency is ω_S, as shown in Fig. 3.8 (a). When the grating and the incident light move in the opposite direction, the scattering light moves up relative to the frequency of the incident light. At this time, the scattering light is called Brillouin anti-Stokes light, whose angular frequency is ω_{AS}, as shown in Fig. 3.8b.

The SBS process can be described as the nonlinear interaction between the pump light and the Brillouin Stokes light through the acoustic wave. The basic principle is shown in Fig. 3.9. The counter-propagated pump and Stokes light interfere in the fiber, and the interference electric field makes the fiber produce periodic deformation or elastic vibration through the electrostrictive effect, leading to the stimulation of the acoustic field. The propagation direction of the acoustic field in the fiber is consistent with that of the pump light, and the refractive index of the fiber changes periodically to form the refractive index grating moving along the fiber at the sound velocity. The refractive index grating is produced by the pump and the Stokes light, which scatters the pump light through Bragg diffraction. Similar to SpBS, the scattering light of the pump light makes frequency shift down due to the Doppler effect of the

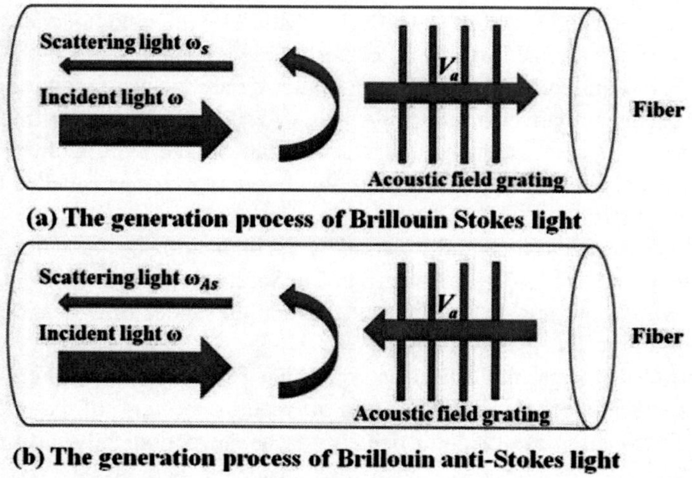

Fig. 3.8 Physical model of SpBS in optical fiber [77]

moving grating, forming the light with a new frequency, called Stokes light. In this way, the pump light transfers energy to the Stokes light, which makes the Stokes light amplified. The amplified Stokes light interacts with the pump light to excite a stronger acoustic field, which in turn acts on the Stokes light. Therefore, pump light, Stokes light, and acoustic wave fields continue to interact with each other, continuously enhancing the Brillouin scattering effect and finally reaching a stable state.

It can be concluded that the frequency of the pump light ν_P and that of probe light ν_S must meet the prerequisite of ν_P -$\nu_S = \nu_B$ in the process of SBS, where ν_B is the BFS. In this way, when the pump light power reaches a certain degree, the counter-propagating pump and probe light in the fiber interfere in some positions, and the acoustic field is excited by the electrostrictive effect. The acoustic field then induces the refractive index grating, which will couple the two beams. In quantum mechanics, the pump photon disappears during the scattering process. At the same time, a Stokes photon and a phonon are generated, and the process also follows the

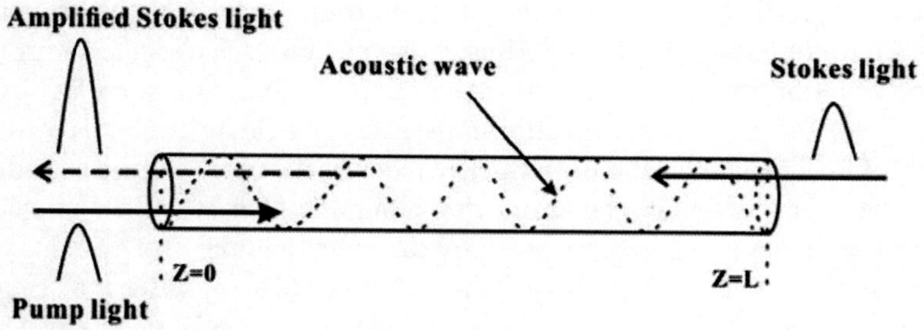

Fig. 3.9 Physical model of the stimulated Brillouin scattering in optical fiber

law of conservation of energy and momentum in physics. Therefore, the final result is that the power of the pump is transferred to the probe, and the acoustic field is enhanced. Therefore, this process can also be vividly described as the stimulated Brillouin amplification of the probe [75–79].

3.2.2 *Brillouin Backscattering Properties of the Chaotic Laser*

Figure 3.10 illustrates the experimental setup for measuring the characteristics of the chaos Brillouin backscattering light. The chaotic laser is obtained from a distributed feedback laser diode (DFB-LD) disturbed by the feedback light, as shown in the dashed box in Fig. 3.10. The isolator (ISO) prevents the undesired light from injecting into the DFB-LD, which may affect the generation of the chaotic laser. The 20% output of the 20:80 optical coupler is used to monitor the state of the chaotic laser in real time, and the other output is employed as the input signal which is injected into the single-mode fiber (SMF, G.655). Since the polarization state is easily changed when the chaotic laser is propagating in the SMF, we introduce a polarization scrambler (PS) in our experiment, which makes the chaotic laser lose polarization characteristics to avoid the influence of polarization state. The input power is controlled by an EDFA, whose maximum output power is 2 W. The refractive index matching liquid is used to suppress the strong Fresnel reflection at the end of the fiber. An optical spectrum analyzer (OSA) with an ultra-high resolution of 5 MHz is used to measure and analyze the optical spectra of the chaotic laser and Brillouin backscattering light. An electrical spectrum analyzer (ESA) with a 26.5 GHz bandwidth and an oscilloscope (OSC) with a 36 GHz bandwidth, an 80 GS/s sampling rate are utilized to measure the power spectrum and the time series of the chaotic laser, respectively.

In order to explore the Brillouin scattering mechanism of the chaotic laser in the fiber and provide theoretical guidance for the chaotic Brillouin distributed optical fiber sensing, the characteristics of chaos Brillouin backscattering spectra, injection power, and SBS threshold are theoretically analyzed and experimentally studied.

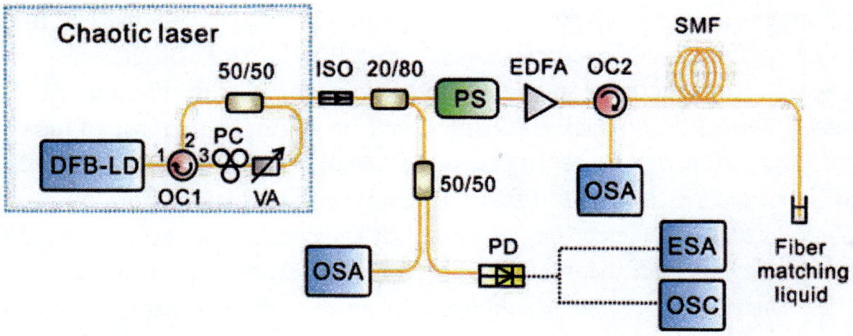

Fig. 3.10 Experimental device for Brillouin backscattering of the chaotic light in the fiber

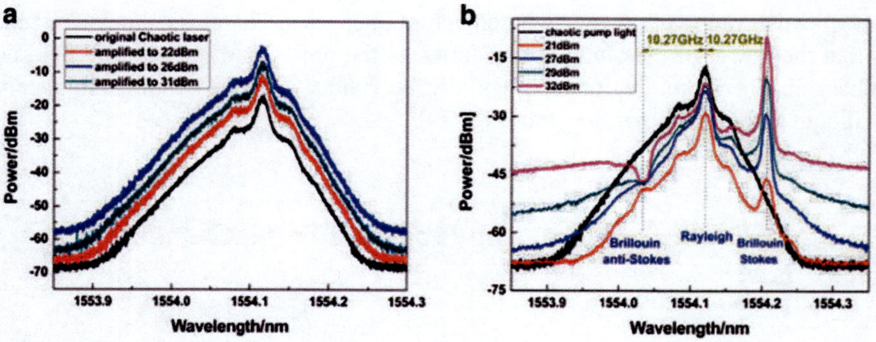

Fig. 3.11 Chaotic laser and Brillouin backscattering spectrum. **a** Original chaotic spectrum and chaotic spectra amplified by EDFA, **b** original chaotic pump spectrum and Brillouin backscattering spectra of chaotic light with different powers injected into 5.05 km fiber

1. **Brillouin backscattering spectra of the chaotic laser**

After the output power of the chaotic laser is amplified to different power values by EDFA and then injected into the SMF of different lengths, different Brillouin scattering spectra can be obtained. It is particularly important to ensure that the characteristics of the chaotic laser will not change after EDFA amplification. As shown in Fig. 3.11a, the black line represents the original chaotic laser spectrum with a linewidth of 1.7 GHz, and the others are the chaotic laser spectra amplified to 22 dBm, 26 dBm, and 31 dBm respectively. It can be seen that the chaotic laser spectra with different magnifications are all displayed as overall amplification. Measured by the 5 MHz high-resolution spectrometer, the linewidth and center wavelength of the amplified chaos are the same as the original one, and the properties of the 1.7 GHz linewidth chaotic laser have not been changed after being amplified by EDFA.

Based on the above analysis, as shown in Fig. 3.11b, the spectra of Brillouin backscattering light with different input powers injecting into the 5.05 km SMF have been experimentally studied. The black curve is the chaotic pump light with linewidth of 1.7 GHz, and the other curves are the Brillouin backscattering spectra obtained via different injection powers. The BFS is 10.27 GHz. When the input power is low, the SpBS is stimulated and the power of anti-Stokes light is too weak to be recognized. With the increase of the input power, there appears SBS, and the intensity of Brillouin backscattering light grows much stronger, especially the Stokes light.

It is noteworthy that there appears a dip in the position of the chaotic anti-Stokes frequency of the Brillouin backscattering spectrum. With the increase of the chaotic injection power, the dip degree increases gradually, which is not observed when the traditional laser is as a pump light. To analyze this phenomenon, we study the Brillouin backscattering spectrum generated by the continuous laser from the DFB-LD, and compare it with the chaos Brillouin backscattering spectrum, as shown in Fig. 3.12. When the input power is low, the location of the anti-Stokes frequency in the Brillouin backscattering light spectrum shows a peak in the case of a chaotic pump light of 21 dBm as shown in Fig. 3.11b, and a DFB-LD pump light of 4.3 dBm is as

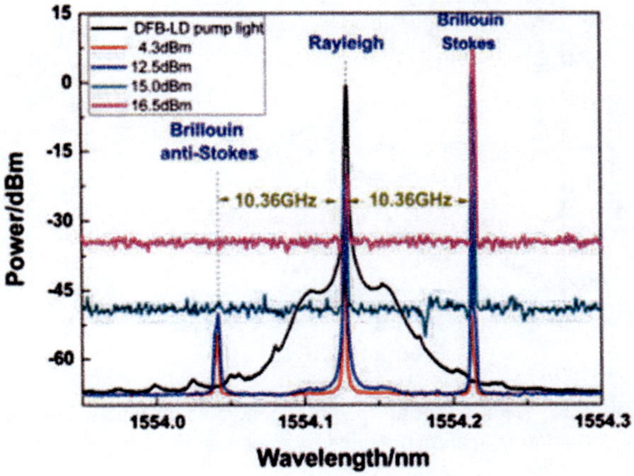

Fig. 3.12 Brillouin backscattering spectra of original DFB-LD pump light and different input powers injected into 5.05 km fiber

shown in Fig. 3.12. The reason is that in SpBS process, the power of the generated anti-Stokes increases gradually with the increase of the input power, without any energy conversion. When the injection power is high enough, there appears a dip at anti-Stokes frequency of the chaos Brillouin backscattering spectrum (as the pink curve shown in 32 dBm in Fig. 3.11b), but the anti-Stokes light of the DFB-LD Brillouin scattering spectrum disappears (as the pink curve shown in 16.5 dBm in Fig. 3.12).

The physical mechanism of the above phenomena is as follows. The SBS can be described as a nonlinear interaction between the pump and the Stokes light through an acoustic wave in the fiber. When the pump power increases to a certain value (the SBS threshold), an interference effect occurs between the pump and the Stokes light and then most of the pump light converts into the Stokes light. Chaotic laser has the characteristic of self-similarity. This means that the chaotic laser in the field of the Brillouin anti-Stokes frequency has similar properties to that in the field of the central wavelength. Furthermore, the chaotic laser covers the field of the anti-Stokes frequency as shown in Fig. 3.11. Thus, the chaotic laser at the anti-Stokes frequency transforms into Stokes light in the SBS process, which causes the Stokes light power increases and there appears a dip at the location of the anti-Stokes frequency. As for the spectrum of the DFB-LD backscattering light, the amplified spontaneous emission (ASE) noise caused by the EDFA enlarges while the input optical power increases. However, the ASE noise does not participate in the SBS process because it does not satisfy the SBS generation conditions. Moreover, the power of the ASE noise is so high that the anti-Stokes light signal is submerged by the noise. Therefore, there only exists Brillouin Stokes light in SBS process caused by DFB-LD.

The power spectrum of the Brillouin backscattering signal generated by the chaotic laser in the fiber is shown in Fig. 3.13. The black curve is the intrinsic noise base of the electrical spectrometer, and the red curve is the power spectrum of the Brillouin backscattering signal of the chaotic laser in the fiber in Fig. 3.13a. It can be seen that the chaos Brillouin spectrum still has the characteristic of high bandwidth, covering

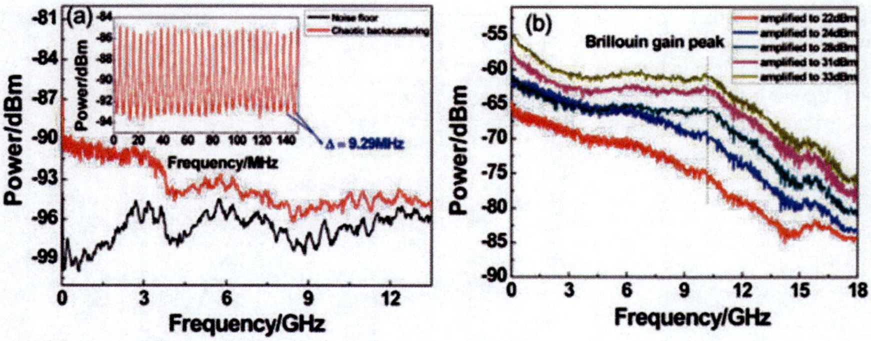

Fig. 3.13 Backscattering spectrum generated by chaotic light. **a** The backscattering spectrum **b** the backscattering spectra obtained by injecting different optical powers into the fiber

the frequency range of 15 GHz. To analyze the power spectrum details of the signal, the frequency range of the spectrometer is set to 150 MHz. It can be seen that the spectrum is periodic, as shown in the small and medium box in the figure, and the period is 9.29 MHz, which is equal to the cavity length of the chaotic laser source, indicating that the chaos Brillouin backscattering signal still carries the time delay signature of the chaotic laser. The backscattering power spectra with different injection powers are further analyzed, as shown in Fig. 3.13b. When the magnification of EDFA is small, the spectrum shows overall amplification, such as the curves of 22 and 24 dBm in the figure. As the injection power gradually increases, the Brillouin gain peak caused by the self-beat of the backscattering is visible and performs broadband. The frequency of gain peak is equal to the BFS, which is about 10.27 GHz.

In the experiment, the time series of backscattering signals generated by the chaotic laser in the fiber is measured by high-speed real-time OSC, as shown in Fig. 3.14a. Its time series oscillates randomly and performs the noise-like variation. As shown in Fig. 3.14b, the chaotic autocorrelation curve maintains the property of δ-like function.

To further demonstrate the interpretation that the power dip at chaos Brillouin anti-Stokes frequency is not caused by the absorption of the anti-Stokes light, the chaotic laser and the tunable laser are injected into the 10 km SMF in the opposite or the same direction, respectively.

(1) The interaction between the chaotic laser and the continuous light (at anti-Stokes frequency) when they are injected in the opposite direction.

As shown in Fig. 3.15, the chaotic laser is amplified by EDFA and injected into one end of the SMF through an optical circulator (OC2). The continuous laser output by TLS is injected into the other end of the SMF, and the output power is controlled by a variable optical attenuator (VA2).

The optical spectrum is analyzed by a 5 MHz high resolution spectrometer and the results are shown in Fig. 3.16. The red curve is the backscattering spectrum of the chaotic laser directly measured when closing the TLS output, and the spectra of

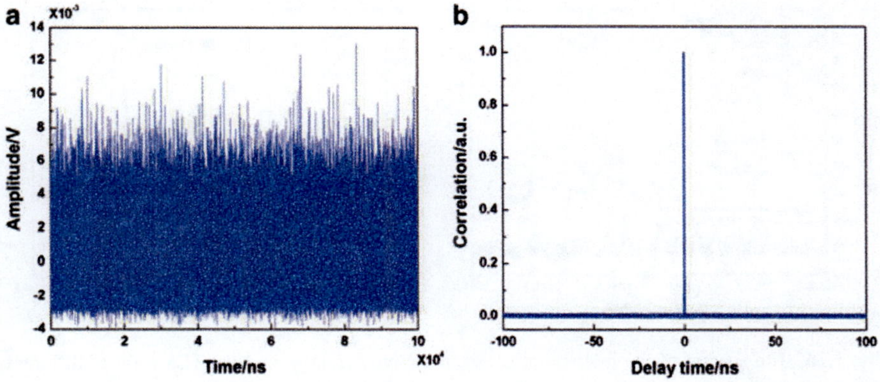

Fig. 3.14 The time series of chaotic backscattering light. **a** Backscattering time series of chaotic light, **b** autocorrelation curve of backscattering time series of chaotic light

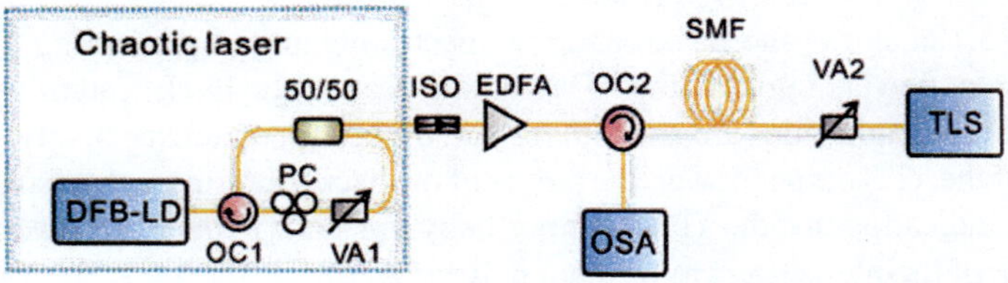

Fig. 3.15 The experimental device of the interaction of the light output from the chaotic laser and the tunable laser (f = anti-Stokes light frequency) in the opposite direction

anti-Stokes, Stokes, and Rayleigh light are as smooth as before in Fig. 3.16a. The black curve is the original spectrum of TLS when closing the chaotic laser, directly measured by OC2. It can be seen that there are multi-order sidebands on both sides of the central wavelength, but the side mode suppression ratio is relatively high. The other curves are the interaction spectra when the output of the chaotic laser is constant and the output power of TLS is set to 0.3 mW, 1 mW, 5 mW, 8 mW, 10 mW, 12 mW, and 15 mW, respectively. With the increase of the TLS output power, the interaction spectra are almost completely coincident. Only the power at the anti-Stokes frequency ascends with the increase of TLS output power, indicating that the anti-Stokes light is not absorbed in the transmission process.

(2) The interaction between the chaotic laser and the continuous light (at anti-Stokes frequency) when they are injected in the same direction.

Figure 3.17 shows the experimental setup for the interaction between the chaotic laser and the continuous light when they are injected in the same direction. The chaotic laser is amplified by EDFA and injected into one end of the 50:50 fiber coupler. The TLS lights with different output powers are injected into the other end of the coupler, and then injected into the SMF by OC2. The fiber end is placed in the refractive index

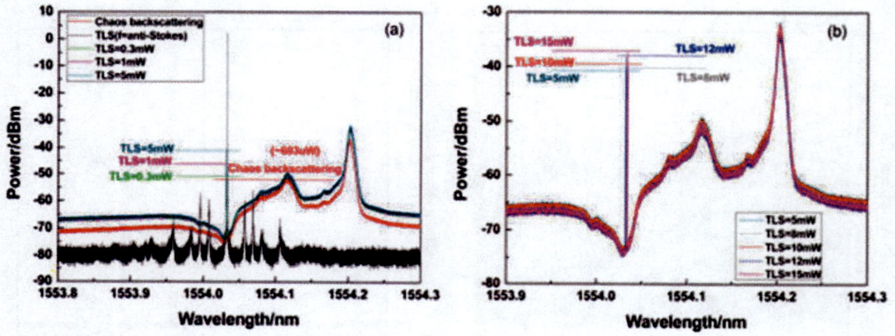

Fig. 3.16 The spectra of the interaction between the chaotic laser and the TLS light of different optical powers (f = anti-Stokes optical frequency). **a** TLS < 5 mW, **b** TLS > 5 mW

matching liquid to absorb its strong Fresnel reflection. Their interaction spectra are also analyzed by a 5 MHz high resolution spectrometer.

Figure 3.18a shows the backscattering spectra obtained by injecting TLS light with different powers into the fiber. The middle peak is Rayleigh scattering spectra, and the left and right sides are anti-Stokes and Stokes spectra respectively. With the increase of the TLS output power, the power of the backscattering light also increases gradually, indicating that the TLS laser normally transmits in the fiber. Based on this, the spectra of its interaction are further analyzed.

As shown in Fig. 3.18b, the red curve is the backscattering spectrum of chaotic laser when closing the output of the TLS, and the black curve is that of the TLS laser when the output of chaotic laser is closed. The other curves are the interaction spectra under different TLS output powers. With the ascend of the TLS power, the interaction power slowly increases and that at anti-Stokes frequency position increases gradually, indicating that the anti-Stokes light is not absorbed.

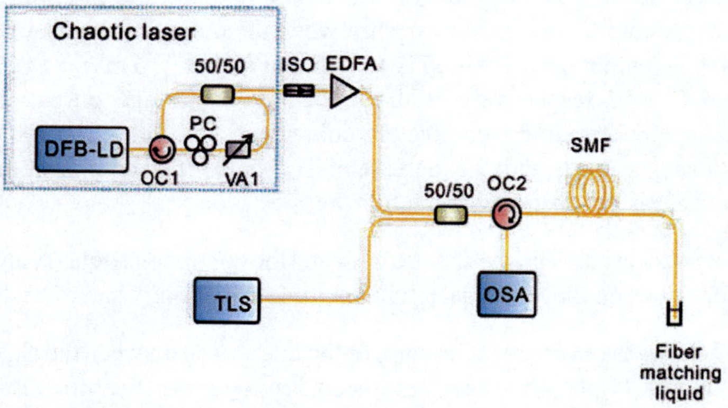

Fig. 3.17 The experimental device of the interaction between the chaotic laser and the light output from the tunable laser (f = anti-Stokes light frequency) in the same direction

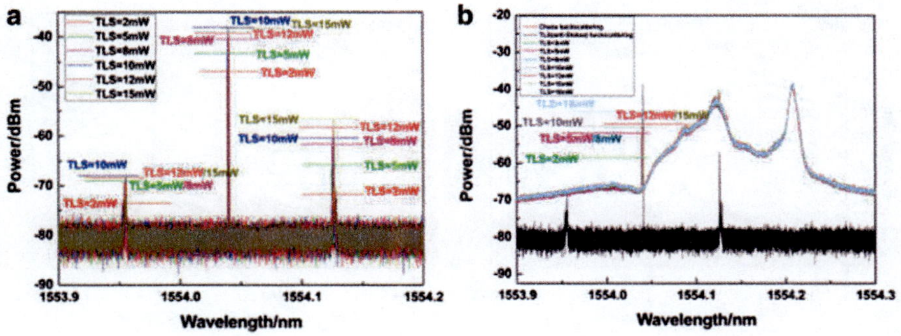

Fig. 3.18 The results of the interaction between TLS output light of different powers (f = anti-Stokes light frequency) and chaotic laser in the same direction. **a** The backscattering spectra obtained by injecting the TLS output light of different powers into the fiber, **b** the spectra of the interaction between chaotic laser and TLS output light acting in the same direction

(3) The interaction between the chaotic laser and the continuous light (at Stokes frequency) when they are injected in the opposite direction.

The experimental setup is the same as Fig. 3.15, and their interaction spectra are analyzed by 5 MHz high resolution spectrometer. In Fig. 3.19a, the red curve is the backscattering spectrum of the chaotic laser when the output of the TLS is closed. At this time, the power measured by the optical power meter is 900 μW, and the spectra at the anti-Stokes, Stokes, and Rayleigh frequencies are all smooth and burr-free, which is consistent with the above experimental phenomena, indicating that the output of the chaotic laser is stable. The black curve is the TLS laser when the chaos output is closed. The other curves are the interaction spectra when the chaos output is unchanged and the output power of TLS is set to 0.3 mW, 1 mW, 5 mW, 8 mW, 10 mW, 12 mW, and 15 mW, respectively. In order to observe each curve more clearly, the spectrum of TLS = 8 mW is separately marked in Fig. 3.19b and the other curves are amplified and narrowed on this basis. With the increase of the TLS output power, the interaction power ascends gradually. When the output power is more than 10 mW, the interaction spectrum begins to descend and the power of that decreases. After the interaction, there are bifurcations and burrs in the spectra of anti-Stokes, Stokes, and Rayleigh light, which might be caused by the strong interaction and the large power gain.

(4) The interaction between the chaotic laser and the continuous light (at Stokes frequency) when they are injected in the same direction.

The experimental setup is shown in Fig. 3.17. As depicted in Fig. 3.20a, the backscattering spectra are obtained by injecting TLS laser with different powers into the fiber. The middle peak is the Rayleigh scattering spectrum, and the left and right sides are anti-Stokes and Stokes spectra respectively. With the increase of TLS power, the backscattering power ascends gradually, indicating a normal transmission process.

The red curve in Fig. 3.20b is the backscattering spectrum of the chaotic laser measured when the TLS output is closed. The black curve is the backscattering

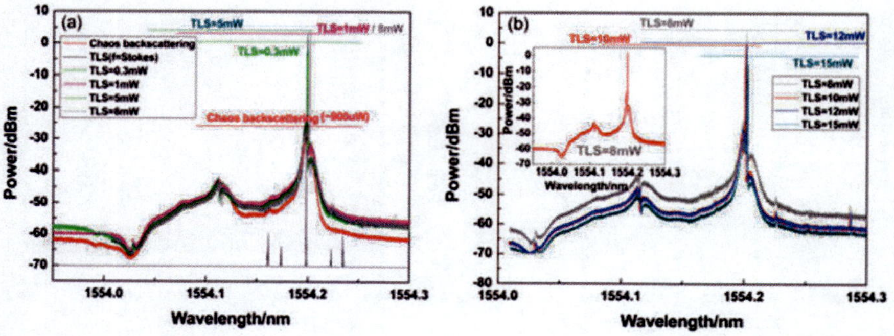

Fig. 3.19 The spectra of the interaction between the chaotic laser and TLS (f = Stokes optical frequency) of different optical powers in the opposite direction. **a** TLS < 8 mW, **b** TLS > 8 mW

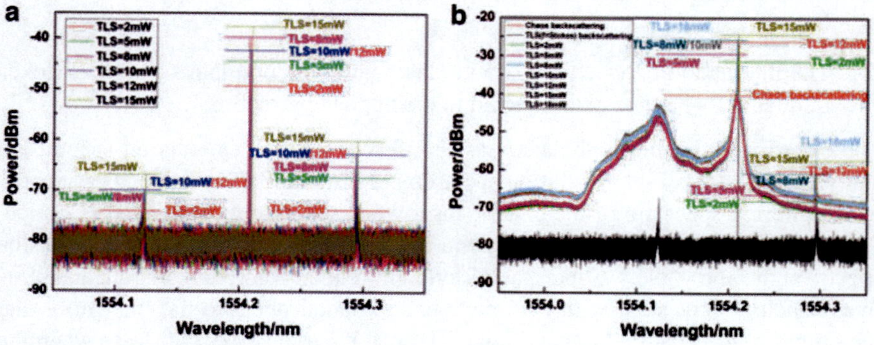

Fig. 3.20 The results of the interaction between TLS output light of different powers (f = Stokes light frequency) and chaotic laser in the same direction. **a** Backscattering spectra obtained by TLS injection of different powers into the fiber, **b** spectra of the same direction of the chaotic laser and TLS output light

spectrum of the TLS laser measured when the chaotic laser is closed. The other curves are the interaction spectra between backscattering spectra generated by the chaotic laser and those generated by the TLS laser under different injection powers. With the gradual increase of TLS power, the interaction power slowly ascends, and the power at Stokes frequency also gradually increases.

(5) The interaction between the chaotic laser and the continuous light (at chaotic laser frequency) when they are injected in the opposite direction.

The experimental setup is also analogous with that in Fig. 3.15. As shown in Fig. 3.21a, the red curve is the chaos backscattering spectrum directly measured by closing the TLS output. The black curve is the measured spectrum of TLS laser by OC2 when the chaotic laser output is closed. The other curves are the interaction spectra when the chaotic laser output is constant and the output power of TLS is set to 0.3 mW, 1 mW, 5 mW, 8 mW, 10 mW, 12 mW, and 15 mW, respectively. With

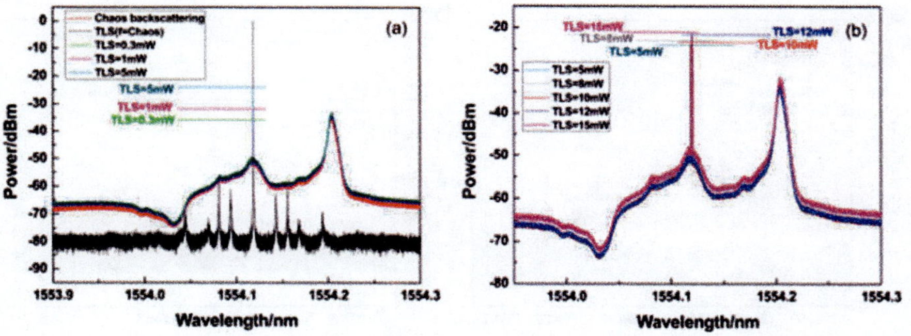

Fig. 3.21 The spectra of the interaction between the chaotic laser and TLS ($f =$ chaotic optical frequency) of different powers in the opposite direction. **a** TLS $<$ 5mW, **b** TLS $>$ 5mW

the increase of TLS output power, the interaction power ascends gradually and the spectra move up as a whole, which is still no burr or distortion in the spectrum.

(6) The interaction between the chaotic laser and the continuous light (at chaotic laser frequency) when they are injected in the same direction.

The experimental setup is also analogous with that in Fig. 3.17. As shown in Fig. 3.22a, the backscattering spectra are obtained when the TLS laser with different injection powers. The middle peak is the Rayleigh scattering spectrum, and the left and right sides are anti-Stokes and Stokes spectra respectively. With the increase of the TLS output power, the backscattering power also ascends gradually, indicating a normal TLS transmission process. As depicted in Fig. 3.22b, the red curve is the backscattering spectrum of the chaotic laser when the TLS output is closed, and the black curve is the backscattering spectrum of the TLS laser measured by closing the chaotic laser. The others are the interaction spectra under different TLS powers. As the output power of TLS increases, the interaction power ascends slowly.

Based on the above experiments and analysis, it is proved that the depression at anti-Stokes frequency of the chaos Brillouin backscattering light is due to the spectral correlation of the chaotic laser. In other words, the chaos at the central wavelength has similar properties to that at anti-Stokes frequency. Therefore, the chaotic laser at the anti-Stokes frequency is transformed into the Stokes components, and the depression occurs in the process of the SBS.

2. **The influence factors of linewidth of chaotic Brillouin Stokes light**

Firstly, the relationship between the linewidth of Brillouin scattering Stokes light and the injection power is discussed. The data of the Brillouin backscattering spectrum obtained by injecting chaotic laser into 5.05 km fiber is processed, and the relationship between the Stokes linewidth and the chaos injection power is obtained, as the red line shown in Fig. 3.23a. With the increase of injection power, the Stokes linewidth is almost constant. Continuously increasing the injection power, the Stokes linewidth of the light gradually decreases. When the power reaches a certain value, the Brillouin

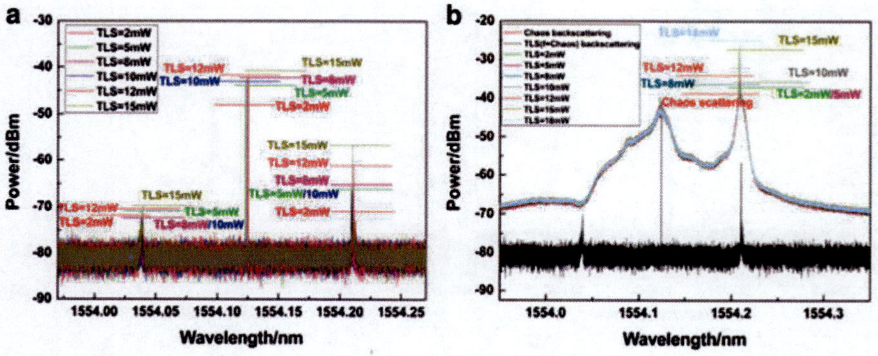

Fig. 3.22 Different powers of the TLS output light (f = chaotic optical frequency) and the chaotic laser in the same direction. **a** Backscattering spectra obtained by injecting TLS light with different powers into the fiber, **b** spectra of interaction of the chaotic laser and the TLS output light in the same direction

scattering Stokes linewidth tends to remain constant. The theoretical reason can be explained from the perspective of energy conservation. The linewidth of Stokes light changes little with the increase of the chaos injection power in the SpBS process. With the increase of the chaos injection power, the SBS occurs, and the pump power is almost transformed into the Stokes light. At this time, the Stokes energy is more and more concentrated, so its linewidth decreases gradually. When the chaos injection power is much larger than the SBS threshold, the Stokes power tends to be saturated and its energy is almost constant, so is its linewidth. The effects of the injection power on Stokes linewidth at different fiber lengths of 3.18, 10.35, 15.41, 21.64, and 24.82 km are further studied, and the same variation relationship is obtained. In view of its similar trends, this section selects three different lengths of fiber, as shown in Fig. 3.23. However, the larger injection power is required to obtain obvious chaos Brillouin scattering. If the injection power is relatively low, the Stokes component is very weak and its −3 dB linewidth cannot be measured, especially for the short fiber. Therefore, for a 3.2 GHz linewidth chaotic laser, there is no corresponding measurement point in the figure when the injection power is less than 23 dBm.

The relationship between Stokes linewidth and fiber length is analyzed and shown in Fig. 3.24. Stokes linewidth decreases with the increase of fiber length, when the injection power is constant, and tends to remain unchanged when the fiber length exceeds 10.35 km. The theoretical reason is that the Stokes power becomes larger along with the increase of fiber length, which would make the energy concentrated and the linewidth narrow. When the length continuously increases until it exceeds the effective length, the Stokes power tends to be saturated and its linewidth tends to be constant.

Compared the Figs. 3.23 and 3.24, the Stokes linewidth obtained by the chaotic laser with the linewidth of 1.7 GHz is narrower than that obtained by the chaotic laser of 3.2 GHz under the same fiber length and injection power, implicating that the change of Stokes linewidth is consistent with that of chaos linewidth. This is

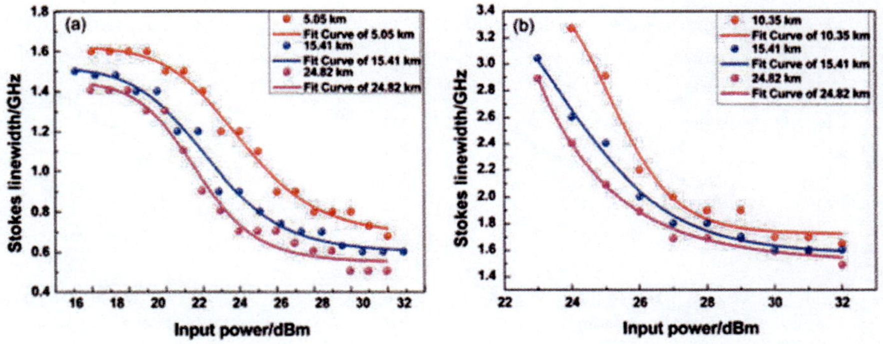

Fig. 3.23 Relationship between Brillouin Stokes linewidth and injection power under different fiber lengths. **a** 1.7 GHz linewidth chaotic laser, **b** 3.2 GHz linewidth chaotic laser

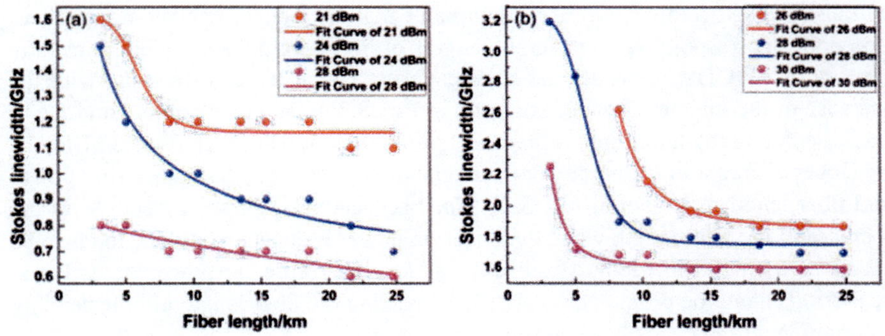

Fig. 3.24 Relationship between the Brillouin Stokes linewidth and the fiber length under different injection powers. **a** 1.7 GHz linewidth chaotic laser, **b** 3.2 GHz linewidth chaotic laser

because that the Brillouin gain spectrum $g(v)$ is equal to the convolution of the SBS intrinsic gain spectrum $g_B(v)$ and the pump power spectrum $I_P(v)$, i.e., $g(v) = g_B(v) * I_P(v)$ [49]. When the bandwidth of the pump power spectrum is much larger than that of the SBS intrinsic gain spectrum, the bandwidth of $g(v)$ is approximately equal to that of $I_P(v)$. Therefore, broadening the pump signal could result in a gain spectrum broadening.

According to the complex relationship between the Stokes linewidth and injection power or fiber length, the basis guidance for selecting appropriate injection power and fiber length has been provided in practical sensing applications.

3. **The influence factors of chaos Brillouin scattering power**

The relationship between the backscattering power and injection power under different fiber lengths is further analyzed. It can be seen from Fig. 3.25a that the backscattering power gradually ascends with the increase of the injection power. The backscattering power begins to increase sharply when the injection power reaches a

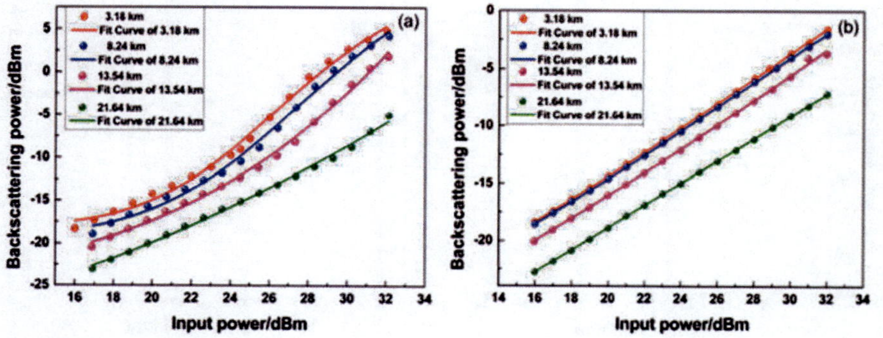

Fig. 3.25 Relationship between backscattering light power and injection power under different fiber lengths. **a** 1.7 GHz linewidth chaotic laser, **b** 3.2 GHz linewidth chaotic laser

certain value (SBS threshold). It is remarkable that the injection power at the inflection points corresponding to different lengths of fibers are different. For the chaotic laser with 3.2 GHz linewidth, the backscattering power keeps increasing with the increase of the injection power, as shown in Fig. 3.25b, and that cannot continue to increase due to the limitation of the maximum output power of EDFA (33 dBm).

Under different injection powers, the relationship between backscattering power and fiber length is shown in Fig. 3.26. The backscattering power ascends with the increase of the fiber length when the injection power is constant and that tends to be saturated when the fiber length exceeds 15.41 km. Therefore, the power the Brillouin scattering cannot be promoted by simply increasing the fiber length after exceeding a certain length in chaos Brillouin sensing system.

4. **The threshold characteristics of chaos stimulated Brillouin scattering**

There are many definitions of SBS threshold [80–84], and this section chooses one of the most common one. The SBS threshold is the injection power when the Brillouin

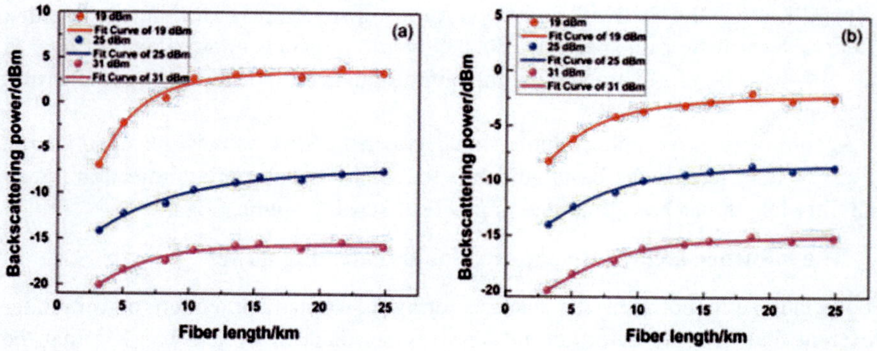

Fig. 3.26 The relationship between backscattering power and fiber length under different injection powers. **a** 1.7 GHz linewidth chaotic laser, **b** 3.2 GHz linewidth chaotic laser

backscattering power is μ times as much as the injection power [1], where μ is 0.001. The SBS threshold can be approximately expressed as [84].

$$P_{th} \approx 21\frac{A_{eff}b}{g_B L_{eff}}\left(1+\frac{\Delta v_{source}}{\Delta v_B}\right) \tag{3.3}$$

where P_{th} is the SBS threshold, A_{eff} is the effective area of the fiber, Δv_{source} is the linewidth of source, and the correction factor b is between 1 and 2, which depends on the relative polarization direction of the pump wave and Stokes wave. L_{eff} is the effective length of the optical fiber, $L_{eff} = [1-exp(-\alpha L)]/\alpha$. Here, L is the fiber length, α is the fiber attenuation coefficient, and g_B is the Brillouin gain coefficient whose value is close to $4*10^{-11}$ m/W and it's independent of the wavelength. It can be seen from Eq. (3.3) that the SBS threshold power increases with the linewidth broadening of the source.

Figure 3.27 shows the relationship between the SBS threshold and the fiber length. The SBS threshold decreases with the increase of the fiber length. When the fiber length is same, the SBS threshold of broadband chaotic laser is higher than that of the narrow chaos. In order to further compare the SBS threshold of the chaotic laser and normal laser, the SBS threshold power of DFB-LD is experimentally measured and shown in Fig. 3.27. Under the same conditions, the SBS threshold powers of the chaotic laser with 1.7 GHz linewidth or 3.2 GHz linewidth are about 15 dB or 19 dB higher than that of DFB-LD laser, respectively. Therefore, the chaotic laser with a higher SBS threshold power has a significant application in distributed optical fiber sensing systems based on SpBS.

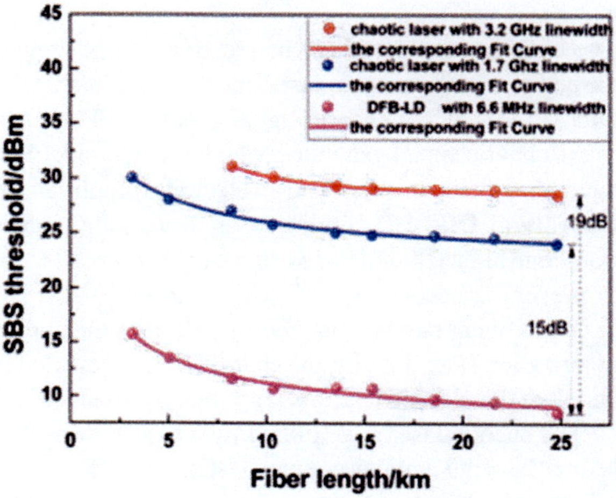

Fig. 3.27 Relationship between the SBS threshold and the fiber length

3.3 Chaotic Brillouin Optical Correlation Domain Reflectometry

Distributed Brillouin sensing technology has been extensively employed in many fields and performs enormous development prospect [85–89]. Remarkably, the use of pulse signal in the BOTDR [90] and BOTDA [91] could not only improve the sensing distance but also seriously limit the spatial resolution. Although the spatial resolution of BOCDR [92] and BOCDA [62] could reach a millimeter level, a long sensing distance cannot be guaranteed. Therefore, the trade-off problem between the spatial resolution and sensing distance restricts the practical application of distributed Brillouin sensing method. Compared with the traditional laser source, the chaotic laser of broadband, noise-like and the strong anti-interference could be used to realize the measurement of high-precision, and spatial resolution, which are all irrelative with the distance.

In this section, a distributed fiber sensing system based on chaotic Brillouin optical correlation domain reflection (chaotic BOCDR) technology is proposed. When the optical path of the reference light is equal to that of the probe light, the chaos Stokes light and the chaos reference light are highly correlated. The BGSs at different positions could be obtained by continuously adjusting the length of the reference path. Finally, a distributed temperature sensing has been achieved with a spatial resolution of 96.25 cm along 155 m sensing fiber.

3.3.1 Sensing Mechanism of Chaotic BOCDR

The experimental setup of chaotic BOCDR is shown in Fig. 3.28. As depicted in dashed box 1, the tunable laser output is modulated through an electro-optic modulator (EOM) by the chaotic signal generator. The dashed box 2 shows the structure of the chaotic signal generator, which includes a DFB-LD, a feedback loop, a broadband photodetector (PD), and an electrical amplifier. In the experiment, the feedback intensity of DFB-LD is adjusted by a variable optical attenuator (VA1). Polarization controller (PC1) is used to match the polarization state. The ISO1 is applied to prevent unnecessary feedback to the DFB-LD. The chaotic signal required for the experiment can be obtained by selecting the appropriate feedback intensity and polarization state. Finally, the chaos laser is converted into the electric chaos by PD1, and then the electric chaos is amplified by an electrical amplifier.

The chaotic laser is amplified by a high-power EDFA and then divided into two beams by 1:99 optical coupler (OC). The lower path (1%) is directly used as a reference chaos whose length is represented by L_{Ref}. The detection position is selected or located by using variable optical delay line with maximal range of 0~20 km and highest resolution of 0.3 μm. The other path (99%) is directly injected into the fiber under test (FUT) as pump chaos, and L_X is the fiber length from the OC2 to the detection position.

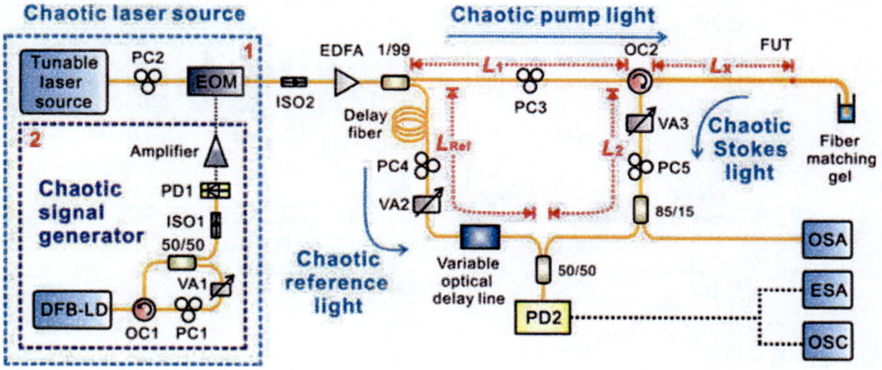

Fig. 3.28 Experimental setup of chaotic BOCDR

When the pump chaos is injected into the FUT, the SpBS is generated and the spectrum is defined as BGS, which has the shape of Lorentz function. Here, the center frequency of the BGS is about 10.8 GHz lower than that of the incident laser. The beat signal between the reference chaos and backscattering wave is detected through a 3 dB coupler. The locating principle of proposed system is realized by adjusting the optical path of the reference wave, as shown in Eq. (3.4).

$$L_{Ref} = L_1 + L_2 + 2L_X \tag{3.4}$$

Here, L_{Ref} is the length of reference path. L_1 is the length of the fiber from the 1:99 coupler to the OC2 red end, and L_2 is the length of the fiber from the VA3 to the end of 50:50 coupler.

When the optical path of the reference wave is equal to that of the probe wave, the backscattering wave and the reference chaos have the same chaotic state and a strong interference effect would be operated, where the Stokes component could be extracted. The BGS at different positions of the fiber can be obtained by continuously adjusting the length of the reference path. The real-time polarization state of two optical paths is controlled by manually adjusting the polarization controller in this experiment. The spectral change of the beat frequency is monitored by the OSA. After the beat wave is converted into the electrical signal by PD2 with 45 GHz bandwidth, it is observed by 26.5 GHz ESA and 6 GHz real-time OSC.

3.3.2 Analysis of Light Source Characteristics

In the experiment, the bias current of 1550 nm DFB-LD is 33 mA (1.5 times the threshold current), and the feedback intensity is locked at −10 dB. The laser emitted by the tunable laser source (1550 nm) is modulated by the single-feedback chaotic signal through the EOM. Figure 3.29a shows the optical spectrum of the chaotic laser,

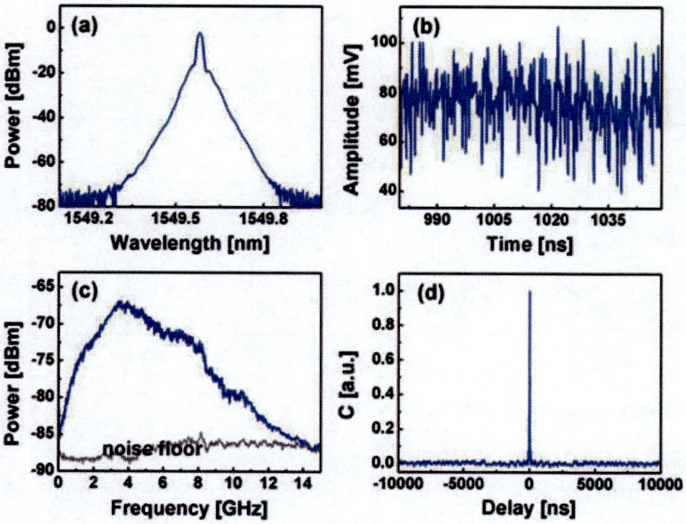

Fig. 3.29 Characteristics of the chaotic laser. **a** Spectrum, **b** time series, **c** power spectrum, **d** autocorrelation

and the linewidth is approximately 71.91 MHz measured by delayed self-heterodyne [93]. According to Eq. (3.5), the calculated coherence length is 88.53 cm.

$$L_C = \frac{c}{\pi n \Delta f} \tag{3.5}$$

Here, L_C is the coherent length of the chaotic laser, c is the propagation velocity of light in vacuum, n, and Δf represent the refractive index of fiber and the linewidth of the laser source respectively. The coherence length is equivalent to the spatial resolution of the sensing system. Figure 3.29b is the time series of the chaotic signal. It can be seen that it has fast and irregular oscillation changes without the obvious period. The power spectrum, covering range of 0~14 GHz, is shown by the blue curve in Fig. 3.29c. Figure 3.29d is the autocorrelation curve of the δ-like function with the sequence length of 10,000 ns, whose noise structure is very low.

When the chaotic laser is amplified to 1.25 W by EDFA and injected into the G.655 single-mode fiber of 155 m, the beat signals of reference wave and the Stokes wave are shown in Fig. 3.30. In Fig. 3.30a, the blue curve represents the optical spectrum of the reference chaos, and the green curve represents that of the chaotic Stokes wave. The component enclosed by the gray curve is Rayleigh scattering wave, which is much lower than Stokes wave. The red curve represents the beat optical spectrum after the interference. Figure 3.30b is the BGS measured by ESA, whose -3 dB bandwidth is around 19.2 MHz.

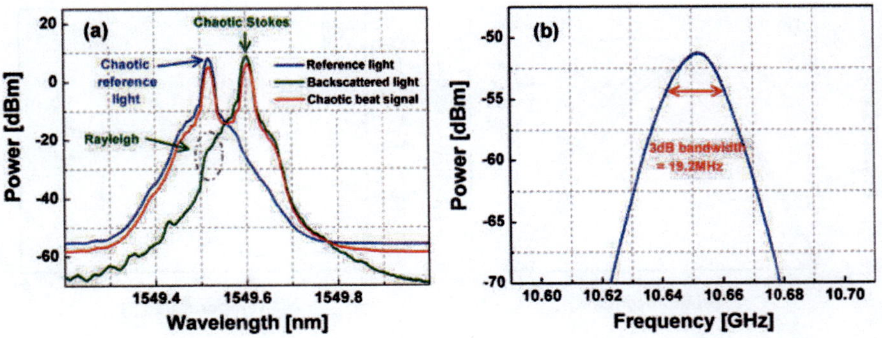

Fig. 3.30 Beat signals of the chaotic reference wave and the chaotic Stokes wave. **a** Optical spectrum, **b** BGS

3.3.3 Results of Temperature Measurement

Figure 3.31 shows the structural schematic diagram of FUT in the experiment. The last 50 m of the SMF with a total length of 155 m is placed inside the thermostat. The variable optical delay line of the reference path is adjusted to make the optical path of the reference path equal to that of the scattering wave, that is, the correlation position is located in the middle of the fiber in the thermostat. Then adjusting the temperature setting of the thermostat and the temperature measurement experiment was carried out at an interval of 5 °C from 25 to 45 °C.

Figure 3.32a is the measured BGS at different temperatures. It can be seen that the BGS gradually moves to the direction of high frequency with the increase of the temperature. The BFSs at different temperatures are obtained from the BGS shown in Fig. 3.32b. It can be found that the BFS linearly ascends with the temperature increase, and the slope of the fitting curve is approximately 1.07 MHz/°C, which means the measured temperature coefficient is 1.07 MHz/°C, consistent with the theoretical value. At the same time, the correlation coefficient of the fitting curve is 0.99784, which indicates a fine linear relationship between BFS and temperature.

In order to obtain the temperature resolution of this scheme, the temperature of the thermostat is adjusted to explore the minimum temperature range that the system

Fig. 3.31 Structure diagram of sensing fiber in experiment

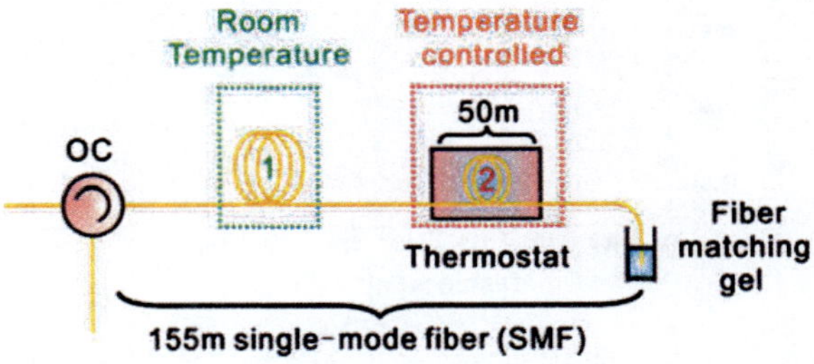

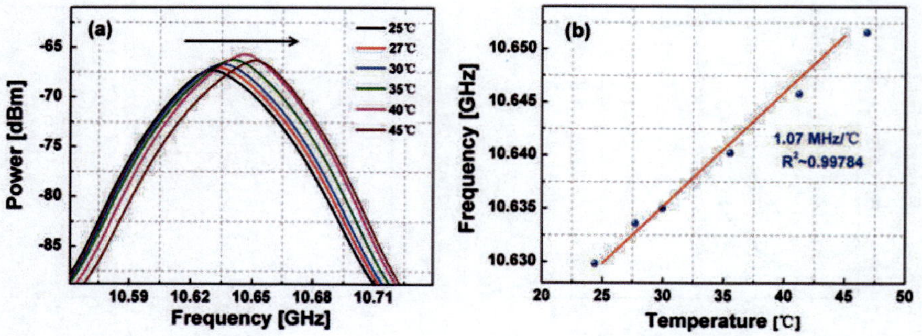

Fig. 3.32 Relationship between the BFS and temperature. **a** BGS at different temperatures, **b** fitting curve of temperature coefficient

can resolve. From 25 to 27 °C, the change of the BGS can be obviously observed. Therefore, the resolution of the system for temperature is ± 1 °C.

In order to confirm the spatial resolution, fibers of 50, 10 and 3 m are placed in the thermostat. Firstly, the last 50 m FUT is placed in the thermostat. The temperature is set at 10 and 45 °C respectively, and the room temperature is maintained at 24 °C. As mentioned above, the different lengths of the delay fiber is adjusted to match the optical path so that the positions can be continuously scanned from the beginning to the end, and the temperature distribution measurement of the whole 155 m FUT is realized.

Figure 3.33 shows the BFS distribution along the FUT, and the temperature change region can be obviously observed. In Fig. 3.33a, the variation of the BFS is about 15 MHz, which matches the temperature change at 14 °C. Figure 3.33b shows that the frequency shift is about 20 MHz and the corresponding temperature difference is 21 °C.

Next, the fiber is changed into 10 m or 3 m, as shown in Fig. 3.34. The FUT of 110~120 m is placed in the thermostat, and the temperature is set to 13 °C and 45 °C, respectively.

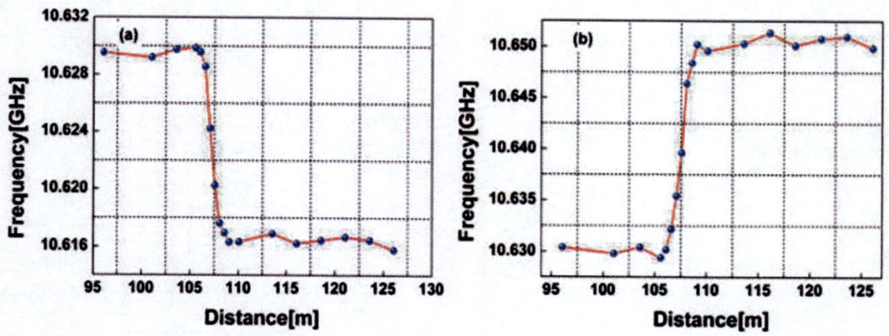

Fig. 3.33 BFS distribution along the FUT. **a** 50 m, 10 °C, **b** 50 m, 45 °C

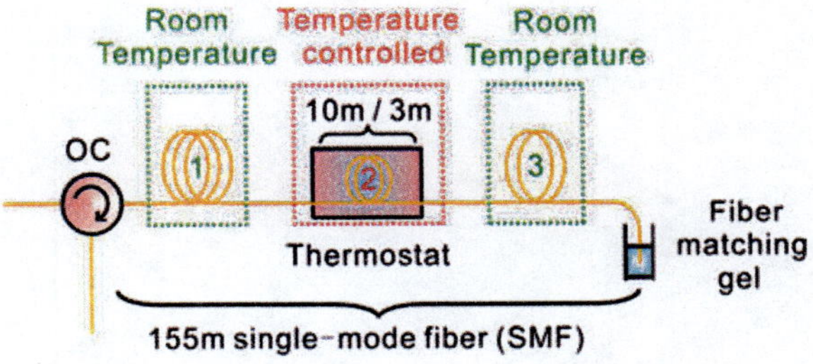

Fig. 3.34 Structure diagram of sensing fiber in experiment

Figure 3.35 shows the experimental measurement results at the low temperature. Figure 3.35a is the three-dimensional distribution of the BGS and Fig. 3.35b is the BFS distribution along the FUT. When the room temperature is maintained at 23 °C, the BFS is about 11 MHz, which matches the temperature change at 10 °C. Figure 3.36 is the three-dimensional distribution of BGS and BFS distribution when the temperature of the thermostat is set at 45 °C. Under the condition of 21 °C temperature difference, the BFS is about 20 MHz.

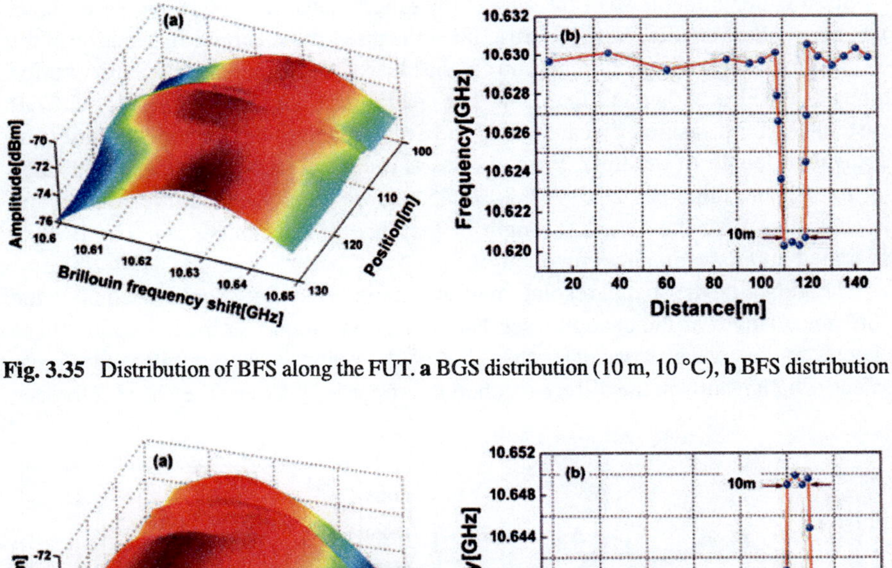

Fig. 3.35 Distribution of BFS along the FUT. **a** BGS distribution (10 m, 10 °C), **b** BFS distribution

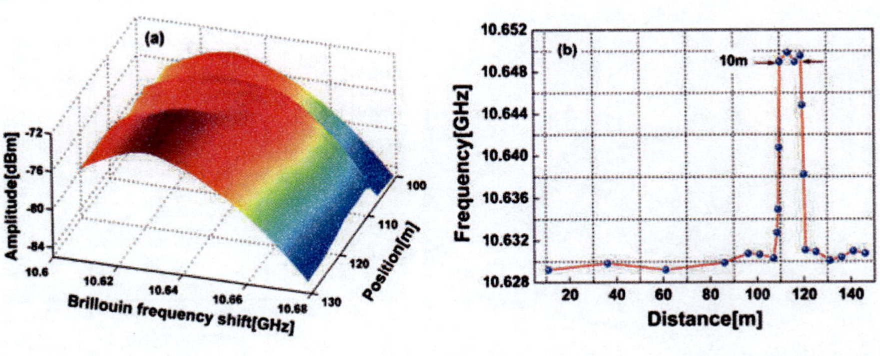

Fig. 3.36 Distribution of BFS along the FUT. **a** BGS distribution (10 m, 45 °C), **b** BFS distribution

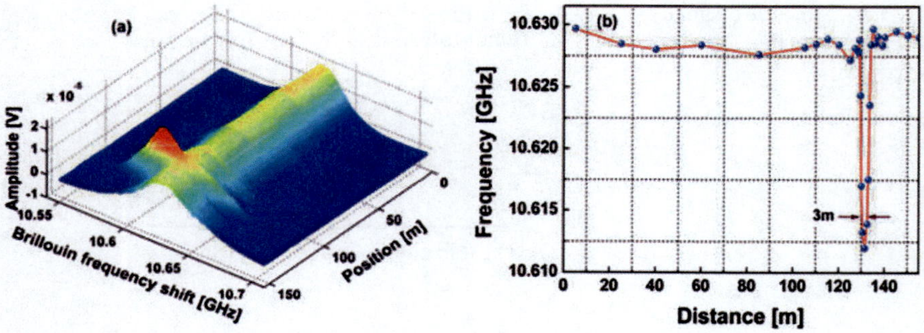

Fig. 3.37 Distribution of BFS along the FUT. **a** BGS distribution (3 m, 50 °C), **b** BFS distribution

Finally, the FUT of 130~133 m is placed in the thermostat. The temperature of the thermostat is set to 8 °C. Figure 3.37a is a three-dimensional distribution of the BGS along the FUT, and the temperature change region can be clearly observed. Figure 3.37b is the BFS distribution along the FUT. It can be seen that the BFS is about 17 MHz, corresponding to the temperature change of 15 °C.

Setting the temperature of the optical fiber thermostat to 50 °C, the measurement results are shown in Fig. 3.38. Figure 3.38a is the three-dimensional distribution of the BGS and the high temperature region can be clearly identified. The BFS distribution along the FUT is shown in Fig. 3.38b. The spatial resolution of the chaotic BOCDR system can be measured by calculating the average length of the rise- and fall-time equivalent length of the fiber. From the inset, it can be seen that the average value of the rise and fall area of 10~90% is 96.25 cm, which is the spatial resolution and is consistent with the coherent length of the chaotic laser of 89 cm. The BGS is averaged 300 times in data processing.

Theoretically, the spatial resolution of the chaotic BOCDR should be equal to the coherence length of the chaotic laser. However, it is found that the linewidth of the chaotic Stokes wave is narrower than that of the pump wave due to the nonlinear effect, which results in the different coherent properties between the chaos reference

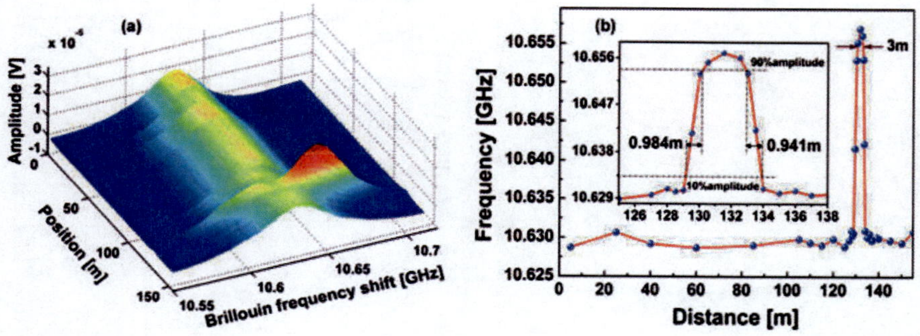

Fig. 3.38 Distribution of BFS along the FUT. **a** BGS distribution (3 m, 50 °C), **b** BFS distribution

and Stokes wave. At the same time, the nonlinear amplification causes decoherence, leading to the increase of the coherence length. In addition, if a broader chaotic laser is used as the light source, a higher spatial resolution of centimeter level can be obtained.

3.4 Chaotic Brillouin Optical Correlation Domain Analysis

To fundamentally solve the trade-off problem between the measurement range and the spatial resolution, several novel laser sources have been inspired and employed. For example, phase-coding using PRBS [70, 71], ASE source [69]. However, the periodicity and code length of the sequence would still restrict the sensing performance. ASE source will introduce unsolvable spectral overlap problems in the system, which seriously deteriorate the SNR and limit the sensing distance to several meters. In Sect. 3.3, the chaotic BOCDR has been theoretically proposed and experimentally verified with a centimeter-level spatial resolution. However, the sensing length was only 155 m due to the limitation of SpBS in the sensing fiber [94].

In this section, we propose a distributed fiber sensing system based on chaotic Brillouin optical correlation domain analysis (chaotic BOCDA) technology, whose sensing distance is enlarged due to the SBS effect with maintaining a high spatial resolution.

3.4.1 Sensing Mechanism of Chaotic BOCDA

1. **Physical mechanism**

BOCDA based on SBS mainly depends on the coupling among the pump wave, probe wave, and the complex amplitude of Brillouin acoustic wave. This process can be described mathematically by the SBS coupling equation.

The process of the SBS can be described by the following coupled equations without considering the loss [95].

$$\left[\frac{\partial}{\partial t} + v_g \frac{\partial}{\partial z}\right] E_1 = i\kappa_1 E_2 \rho \tag{3.6}$$

$$\left[\frac{\partial}{\partial t} - v_g \frac{\partial}{\partial z}\right] E_2 = i\kappa_1 E_1 \rho * \tag{3.7}$$

$$\left[\frac{\partial}{\partial t} + \frac{\Gamma}{2} + i\delta\omega_B\right] \rho = i\kappa_2 E_1 E_2^* \tag{3.8}$$

Here, v_g is the group velocity, and $\Delta\omega_B(z)$ is the deviation of the BFS from its mean value. The acoustic damping rate Γ related to the linewidth of the BGS is $2\pi\Delta\nu_B$.

$\kappa_1 = \pi v g \gamma \rho / 2n\lambda\rho_0$, $\kappa_2 = \pi n \varepsilon_0 \gamma \rho / 4\lambda v_a$ are the coupling coefficient. Here, ε_0 is the dielectric constant in vacuum. E_1, E_2, and ρ are the slowly varying amplitudes of the pump, the probe, and the Brillouin acoustic wave, respectively. They are functions of t which represents time and z which represents position along the entire fiber.

The perturbation theory is used to solve the coupled equation, and the gain function that the probe wave propagates through the FUT offers the mathematical support for the demodulation of the BGS. The specific equation is as follows [62]:

$$g = \frac{v_g \overline{P}_1}{A_{eff}} \int_{-\infty}^{\infty} d\zeta \int_{-\infty}^{\infty} \frac{d\omega}{2\pi} g_B(\zeta, \omega) S_b(\zeta, \omega) \tag{3.9}$$

where $\zeta = z/v_g$, $\bar{P}_1$ is the average pump power over time. A_{eff} is the effective core area, and $g_B(\zeta,w)v_g d\zeta$ is the BGS of the fiber segment of the length $v_g d\zeta$ at position ζ. Notably, $S_b(\zeta,w)$ is defined as the beat spectrum of the pump-probe at position ζ, implying that the probe's BGS at position ζ is determined by the overlap integral of the g_B and the pump-probe beat spectrum. Consequently, the total Brillouin gain is obtained from its integral.

The equation above reasonably expounds general acquisition of the BGS. At the correlation peak, the beat spectrum S_b is approximately δ of frequency. When the beat spectrum is shifted along ω direction by varying the pump-probe mean-frequency difference, the gain that the probe undergoes at the correlation peak varies according to the BGS there. On the other hand, the beat spectrum S_b is broadened at low correlation positions. When the beat spectrum is shifted, the gain for the probe remains small and almost constant. Thus, the BGS at the correlation peak is reflected in the variation of the probe power at the fiber output end.

In addition, the complex amplitude function of the acoustic field density distribution is very important in BOCDA system. Its waveform determines not only the width and intensity of the correlation peak, but also the localization method of the system. The complex amplitude function of the acoustic density variations is given by Cohen et al. [69]:

$$\rho(z,t) = jg_1 \int_0^t \exp[-\Gamma_A(t-t')] E_1\left(t' - \frac{z}{v_g}\right) E_2^*\left[t' - \frac{z}{v_g} + \theta(z)\right] dt' \tag{3.10}$$

Here, g_1 is the electrostrictive effect coefficient. Suppose that the FUT length is L and the pump and probe light are injected into both ends of the fiber. The position-dependent temporal offset $\theta(z)$ is defined as $\theta(z) = (2z–L)/v_g$. When $\nu = \nu_B$, the effective acoustic field will be confined into a narrow range due to $\Gamma_A = 1/(2\tau)$, whose scope corresponds to the coherence length of the source and locates at the middle position $\theta(z/2) = 0$.

When the sinusoidal modulation signal is used as the laser source in the conventional scheme, the periodic correlation peak is obtained. In this case, it is necessary to adjust the delay fiber so that there is a single correlation peak in the FUT, and then change the frequency of the sinusoidal modulation signal to completely scan the FUT. In recent years, the broadband laser with a partially coherent state has been applied to generate a single correlation peak along the arbitrary long FUT.

2. **Distributed sensing system**

The schematic diagram of the experimental setup for chaotic BOCDA is illustrated in Fig. 3.39. The chaotic laser, as plotted in the dashed box, is generated by the optical feedback structure, which consists of a DFB-LD without internal isolator and a feedback loop. The output of the DFB-LD is injected into the external feedback loop. The appropriate feedback state is adjusted to drive the DFB-LD into the chaotic oscillation by adjusting the VA and the PC1. The output of the chaotic laser passes through the ISO and then is split into two beams by an optical coupler (20:80). One of the beams (20%), serving as the chaotic pump wave, is amplified by EDFA1 after passing through the PC2 and then is launched into one end of the FUT via the OC2. The other beam (80%), as the chaotic probe wave, is modulated in a suppress-carrier, double-sideband format by the EOM, driven by the microwave generator (MWG). The bias of EOM is automatically adjusted by the bias control circuit board to select different operating modes. The chaotic probe signal is transmitted through the variable optical delay line, amplified by the EDFA2, and then through the ISO launched into the opposite end of the FUT. Moreover, a polarization scrambler (PS) is inserted in the probe path to suppress the polarization dependence of the Brillouin signal. Finally, the chaotic probe wave is amplified by the chaotic pump wave along the FUT and filtered by the optical bandpass filter via the OC2. Moreover, 20% of the filtered output is used to monitor the optical spectrum by the OSA, and 80% is connected to a digital optical power meter (OPM) with an integrating sphere sensor for power acquisition. The structure of the FUT, consisting of a 906 m single-mode fiber (G.655), in which a 1.03 m fiber near 883.8 m is placed in a fiber thermostat and the remaining fiber is placed in room temperature, is illustrated in the solid line box.

In the experiment, the bias current of the DFB-LD with center wavelength of 1550 nm is set to 33 mA, which is 1.5 times threshold current. Then the feedback intensity is changed by adjusting the tunable attenuator and polarization controller to obtain chaotic laser sources with different linewidths. Figure 3.40a displays the optical spectrum of the chaotic laser, whose linewidths of −3 dB, −10 dB and −20 dB are measured by a high-resolution OSA, minimum of 5 MHz, are 2.02 GHz, 6.63 GHz and 19.04 GHz, respectively. As shown in Fig. 3.40b, compared with the traditional continuous wave, the chaotic laser has a wide power spectrum, almost covering the range of 0~15 GHz. Figure 3.40c is the time series of the chaotic laser measured by a real-time OSC with 36 GHz bandwidth whose maximum sampling rate is 80 GS/s. It can be seen that the time series performs fast, irregular oscillation and noise-like characteristics. The central amplitude is non-zero value because the used PD is a negative gain type. Figure 3.40d further displays the autocorrelation curves of the chaotic laser with an extremely narrow correlation peak.

Figure 3.41 shows the optical spectra and power spectra of different branches in the chaotic BOCDA system. In the pump branch, the optical spectrum of the chaotic laser with a center frequency v_0 is illustrated as the black curve. In the probe

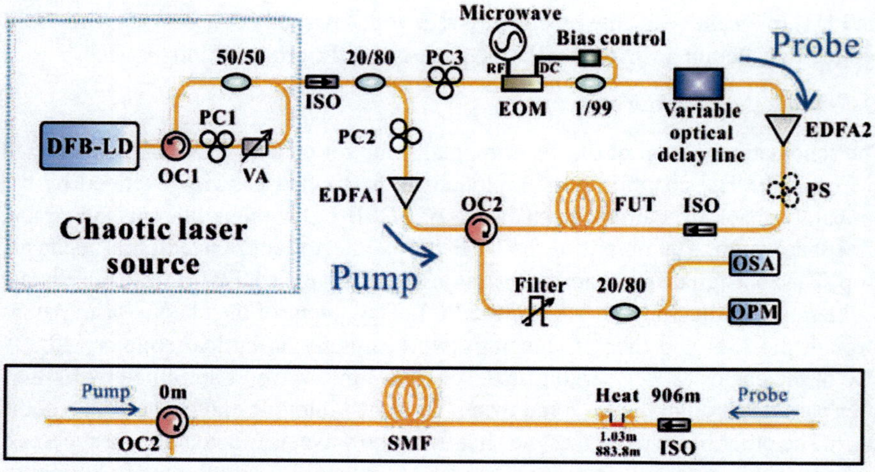

Fig. 3.39 Schematic diagram of the experimental setup for chaotic BOCDA

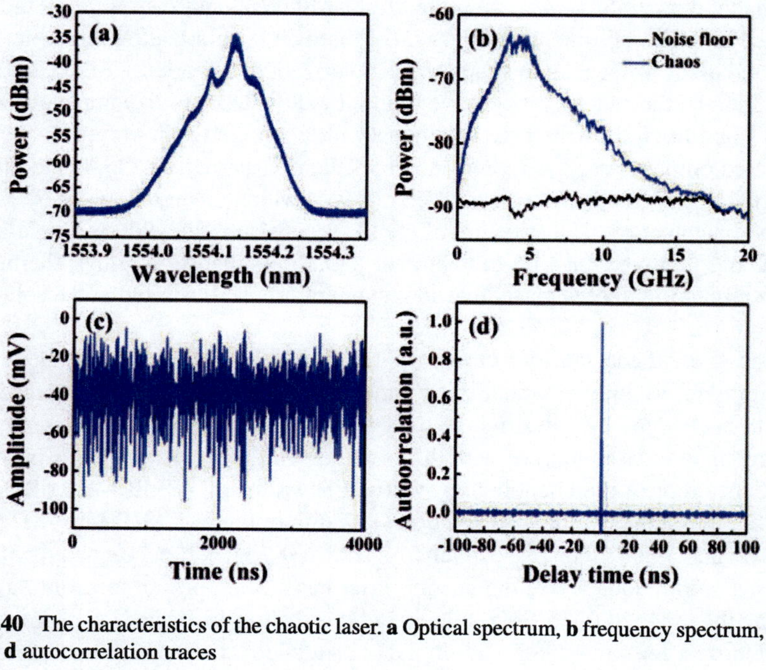

Fig. 3.40 The characteristics of the chaotic laser. **a** Optical spectrum, **b** frequency spectrum, **c** time series, **d** autocorrelation traces

branch, the optical spectrum of the chaotic laser with the double-sideband sinusoidal modulation at $\nu_0 \pm \nu$ is depicted as the red curve. The modulation frequency ν is commonly selected in the vicinity of the Brillouin frequency ν_B. When the chaotic pump and probe wave meet at a certain location of the FUT, the first-order low-frequency sideband $\nu_0 - \nu$ is subject to SBS amplification. The optical spectrum of

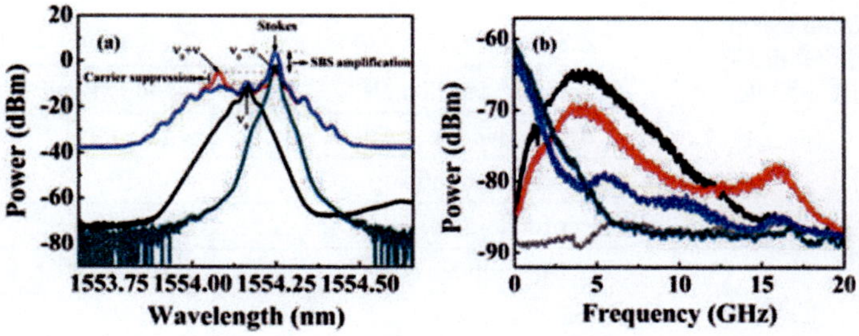

Fig. 3.41 Optical spectra and frequency spectra of different branches in the system

the amplified chaotic probe wave is shown as the blue curve. It is seen that there is an effective amplification when the modulation frequency ν matches the ν_B. The dark cyan line gives the optical spectrum of the filtered Stokes wave with a bandwidth of 6 GHz.

The corresponding power spectra are shown in Fig. 3.41b which are different from that of the chaotic BOCDR system. It is difficult to demodulate the beat information of the chaotic pump and probe wave from the power spectrum, so the BGS of the system cannot be obtained. The essence of the BOCDA technology is the power transfers from the pump wave to the probe wave and the transmission efficiency is highest when the frequency difference matches the BFS of the fiber. When the frequency is swept around the BFS, the chaotic BGS can be equivalently obtained by collecting the average power of the chaotic Stokes wave. In the experiment, the power is simultaneously obtained by utilizing a digital OPM with an integrating sphere sensor and the modulation frequency of signal the generator is swept from 10.5 to 10.7 GHz with the step of 1 MHz.

According to Sect. 1.2, the BOCDA system usually realizes the scanning of the whole fiber by changing the position of the correlation peak along the FUT. There is no exception in the chaotic BOCDA. More importantly, the correlation peak is unique and non-periodic in this system as shown in Fig. 3.40d. Only within this correlation peak, the SBS amplification between the chaotic pump and probe wave could be effectively generated. Therefore, the position of the correlation peak can be scanned over a range of the FUT by a variable optical delay line, which is composed of the cascading fibers of different lengths and two programmable optical delay lines, one of which has a large delay range of 0~20 km with the delay resolution of 30 cm, and the other has the more precise delay range of 0~168 mm with the delay resolution of 0.3 μm. Firstly, the initial position of the correlation peak can be located at the end of FUT by adjusting the length of the cascading fibers. Then the scanning of the correlation peak is accomplished by cooperatively adjusting these two programmable optical delay lines. In order to obtain optimal system performance, this section preliminarily explores the setting of key devices and their parameters in the system.

Table 3.1 Main technical indicators of EOSPACE electro-optic modulator

Parameter	Coefficient
Operating wavelength	C-and L-Band
3 dB bandwidth	> 12.5 GHz
RF extinction ratio	> 12 dB
Insertion loss	< 3 dB
DC Bias port V_π	< 10 V
Modulation port V_π	< 5.5 V

3. **Parameter selection of system key devices**

According to the aforementioned contents, only when the frequency difference of the pump and the probe wave is around ν_B, the SBS amplification between them can occur effectively. Therefore, the EOM must be used to shift the frequency of the chaotic laser. The main technical parameters of the used EOM are shown in Table 3.1. At present, most EOM adopted $LiNbO_3$ crystal with high bandwidth and electro-optic coefficient, low transmission loss and dispersion. In actual applications, the external factors, such as ambient temperature, mechanical vibration and polarization state variation, would also indirectly change the refractive index of the crystal through the photoelectric and thermoelectric effects, resulting in the drift of the working point of the EOM, which can be expressed by the change of the EOM transmission curve in Fig. 3.42.

It can be seen from Fig. 3.42 that when the bias voltage of the EOM is constant, the real transmission characteristic curve moves. The actual operating point deviates from the initial value, and the modulation efficiency is seriously affected, which directly leads to the deterioration of the measurement results of the distributed optical fiber sensing system. For example, in the pulse-based system, the shift of operating

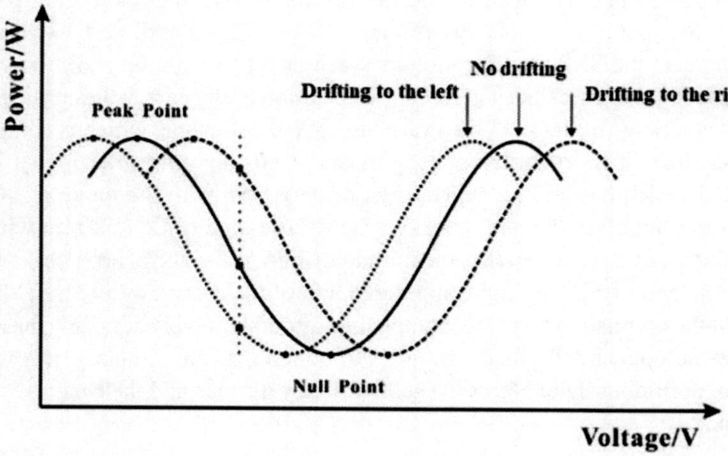

Fig. 3.42 The schematic diagram of the drifting phenomenon of the operating point of the EOM

point can limit the extinction ratio and introduce a large amount of direct current substrates, and thus the SNR severely deteriorates. In this system, the input RF signal of the EOM is ahigh-frequency sinusoidal signal, but the actual V_{pp} will change accordingly when its operating point drifts, which will introduce the unstable factors into the interaction between the pump and probe waves. Therefore, it is necessary to introduce the bias control system in order to automatically adjust the bias current of the EOM without manual compensation.

Since the phenomenon of the EOM operating point drift was found, researchers around the world have proposed many solutions, and the most common method is the introduction of EOM bias control system. At present, there are many methods for EOM bias control, which can be roughly divided into two categories: direct detection method [96] and disturbance method [97]. The direct detection method uses the PD to monitor the optical power of the EOM output signal and refers to the closed-loop feedback control method in modern automatic control technology. The hardware circuit combined with the software algorithm is used to realize the automatic control of EOM bias voltage. The disturbance method is to introduce a weak disturbance signal at the input of the EOM, and its actual working state is obtained by detecting the parameter change of the disturbance signal at the output. The former has a fast response, high stability, simple structure, good practicability, and there are a large number of mature products on the market. In the research, we use the Mini-MBC3 bias control circuit board produced by a domestic company.

The performance of the bias circuit board can be reflected by the static extinction ratio of the EOM, and the maximum and minimum stable optical intensity of the EOM can be obtained. The minimum optical intensity of the EOM with and without the bias control circuit is tested for 120 min when there is no input RF signal, as shown in Fig. 3.43. When the EOM input wave is relatively stable, it can be seen that when the bias voltage control is not used, the minimum light intensity increases slowly with the measurement time. When the bias control is used, the minimum optical intensity is very stable, and the power is floating within 0.2 μW. Thus, the problem of EOM operating point drifting could be well solved by using the bias control circuit, and its working stability has been greatly improved.

The BOCDA is based on the SBS process in the fiber, which requires that the frequency difference between the counter-propagating pump and the probe wave is exactly ν_B. Considering the typical BFS range in the fiber, the MWG used in this system is the N5173B EXG Analog Signal Generator produced by Keysight. The frequency of the sinusoidal electrical signal is covering 9 kHz~13 GHz, the frequency resolution is 0.001 Hz and the amplitude range is −20~19 dBm. More importantly, the MWG has rich scan modes, which supports step-scan (frequency and amplitude at the same interval or logarithmic interval frequency step), list-scan (any list of frequency step and amplitude step), and the dwell time of the frequency point ranges from 100 μs to 100 s. In addition, it also has multiple trigger modes such as free-running, trigger key, external, timer, bus (GPIB, LAN, USB), and its rich functions can make the system successfully realize the automatic frequency sweep function.

The selection of the frequency sweep range and the step of the MWG directly determines the acquisition time of the signal and also affects the obtaining of BGS.

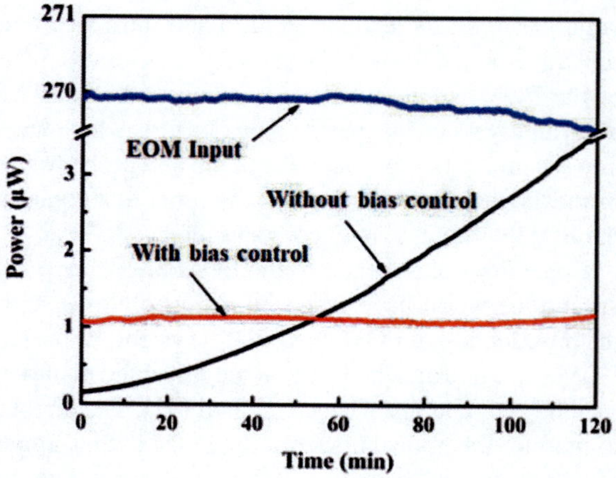

Fig. 3.43 Comparison of minimum output light intensity of EOM with and without bias control circuit

At the beginning, considering the typical BFS in the fiber, the initial frequency range is set to 10~11 GHz and the frequency sweep step is set to 500 kHz. The measured point is located in the thermostat, but the thermostat is not working (that is, at room temperature). This is only to verify the correctness of the gain spectrum from the perspective of temperature sensing. At this time, the black curve represents the obtained BGS as shown in Fig. 3.44. It can be seen that the envelope of the entire spectrum is very wide, which obviously does not fit the Lorentz type. However, there is an obvious peak at the top that corresponds to a frequency of 10.612 GHz, which is also consistent with the typical value of the BFS in the fiber at room temperature. For further verification, the temperature of the thermostat is set to 53 °C, and the BGS of is measured again. The measurement result is shown by the red curve in Fig. 3.44, which is still a broad spectrum, but the peak at the top appears frequency shift. It can be seen from the partially enlarged view that the central frequency corresponding to the peak of the BGS moves from 10.612 to 10.645 GHz when the temperature changes from 26 to 53 °C

It is easy to find that the noise structure is largely resulting from the excessive frequency sweep range. In order to test this conjecture, the positioning point is located at a certain point outside the FUT (this point should not exist in theory because the optical circulator and the optical isolator connected at both ends of the FUT have the isolation effect) by further adjusting the long-distance delay line. The actually measured gain spectrum is shown as the black curve in Fig. 3.45a. It is still a broad spectrum, but the preliminarily verified gain peak that can sense temperature changes disappears. The correlation peak is located in the FUT at room temperature again and the gain spectrum obtained under the same measurement condition is shown in the red curve in Fig. 3.45a. The sharp peak of the perceived temperature change can be obviously seen, which is almost submerged due to the random uncertainty of the noise. In addition, Fig. 3.45b shows the difference result of two curves. The SNR is improved to 4.37 dB and the −3 dB linewidth is 43 MHz. In fact, the scanning range could be set at 10.5–10.7 GHz to obtain the BGS efficiently.

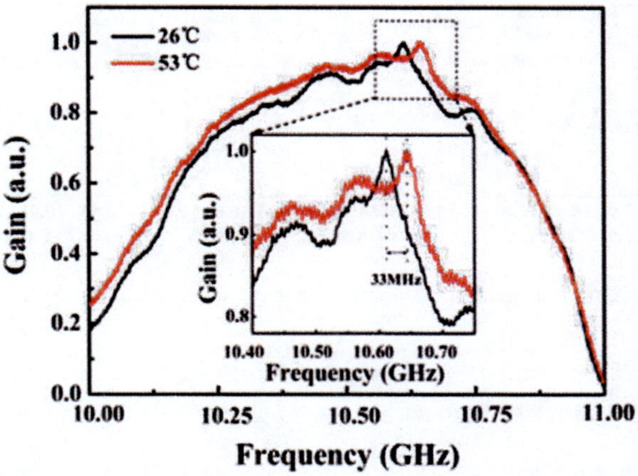

Fig. 3.44 Chaotic brillouin gain spectrum

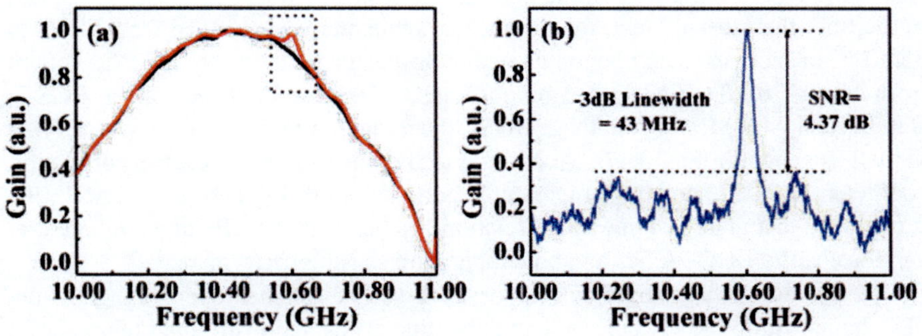

Fig. 3.45 Chaotic Brillouin gain spectrum

However, the different scanning steps would also affect the quality of the BGS and then deteriorating measurement accuracy of temperature. Therefore, the gain spectra are respectively measured in the frequency range of 10.5~10.7 GHz when the scanning step is 100 kHz, 200 kHz, and 1 MHz, as shown in Fig. 3.46. The three curves are fitting the Lorentz type, and the linewidths are 33.4 MHz, 29.3 MHz and 44.9 MHz respectively. Theoretically, taking not considering the measurement time, the frequency accuracy of BGS will be higher with the decrease of the frequency sweep step. However, it can be intuitively that the BGS is deteriorated by the frequency burrs obviously and the error of center frequency is increased when the frequency step is too small. Thus, some postprocessing methods, such as denoising, fitting, and etc., should be adopted to optimize the BGSs and BFSs. When the sweep frequency step is 1 MHz, the original BGS with a relatively stable and smooth curve, high BFS accuracy and guaranteed measurement time can be obtained.

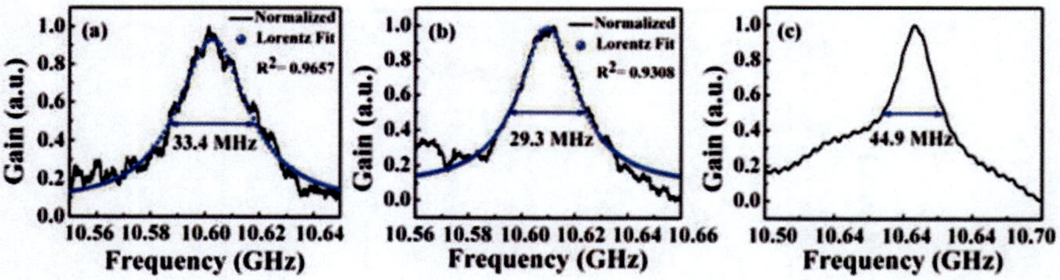

Fig. 3.46 Chaotic Brillouin gain spectra corresponding to different sweep frequency steps of microwave source

4. **Distributed temperature sensing**

Here, the temperature sensing ability of the chaotic BOCDA system is systematically studied and analyzed. The relationship of the chaotic BGS with temperature is illustrated in Fig. 3.47. Figure 3.47a shows the temperature dependence of the BGS in the FUT. First, the single correlation peak of the chaotic laser is located in the fiber thermostat. Next, the thermostat temperature is changed from 23.97 to 55.30 °C in succession. A phenomenon that the center frequency of the BGS moves from 10.608 to 10.646 GHz can be observed, where the linewidth of the BGS is stabilized at 44.9 MHz. From these BGSs, the temperature dependence with BFS is plotted as shown in Fig. 3.47b. According to the linear fitted curve, the correlation coefficient is 0.9975 and the temperature coefficient of this system is 1.24 MHz/°C. It is proved that there is a fine linear relationship between the BFS and temperature, and the chaotic BOCDA has a good temperature sensing performance.

The position of the single correlation peak can be scanned over a range of the FUT by adjusting the variable optical delay line. Thus, the distributed temperature sensing measurement results of the system shown in Fig. 3.48 can be obtained. Figure 3.48a shows a three-dimensional plot of the BGS measured along the FUT. The 1 m long heated fiber near 884 m can be distinguished. In the experiment, the optical fiber thermostat is set to 55 °C, which is closed to the maximum temperature (60 °C) of this instrument, and the room temperature is maintained constant at 25 °C. Figure 3.48b plots the measured distribution of the BFS along the FUT. The inset

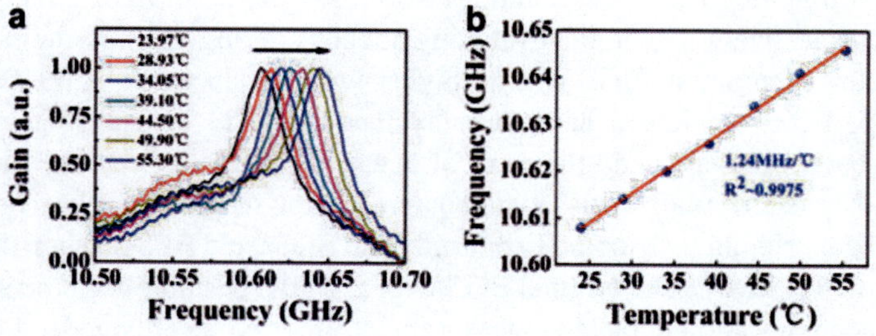

Fig. 3.47 Relationship of the chaotic BGS with temperature. **a** Temperature dependence of BGS in the FUT and **b** temperature dependence of the BFS

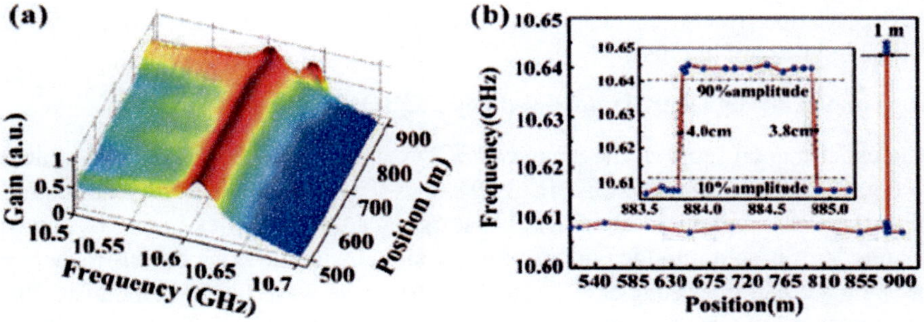

Fig. 3.48 Measurement results of distributed temperature sensing. **a** Distribution of the BGS along the FUT and **b** distribution of the BFS along the FUT

view shows the enlargement of the BFS distribution along the 1 m long heated fiber. The change of the BFS is approximately 37 MHz, which corresponds to a 30 °C temperature variation. The spatial resolution of the chaotic BOCDA system reaches 4 cm, which can be calculated by the average value of 10~90% of the rise and fall time equivalent length in meter for the heated fiber section [29]. According to the SBS amplification mechanism between the chaotic pump light and probe light, the spatial resolution of the system is theoretically determined only by the coherence length of the chaotic laser. According to the equation $L_C = c/\pi n\Delta f$, the coherence length L_C is related to the spectral width Δf, where $c = 3 \times 10^8$ m/s is the light speed and $n = 1.5$ is the refractive index of the fiber. According to Fig. 3.48a, the spectral linewidth of the chaotic laser is 2.020 GHz, so the theoretical spatial resolution of the chaotic BOCDA system is 3.15 cm, which is almost consistent with the experimental results. Moreover, the spatial resolution can be further improved to a millimeter or sub-millimeter level by increasing the bandwidth of the chaotic laser.

3.4.2 Long Distance Chaotic BOCDA

A chaotic BOCDA has been experimentally demonstrated with a 4 cm spatial resolution over a 906 m measurement range [98]. However, the weak amplitude autocorrelation of the chaotic signal occurring at the delay time of the external cavity, i.e., time delay signature (TDS), could increase the background noise of BGS and decrease the performance of the chaotic BOCDA system. The effect of the injection current and feedback strength on the TDS suppression is theoretically analyzed. By optimizing the two free parameters, the chaotic laser with a good TDS suppression has been applied in the distributed optical fiber sensing system [99]. However, the sensing distance is seriously limited due to the accumulation of system noise with the length of the fiber. To overcome this problem, we propose a time-gated optimization scheme to suppress background noise such as residual TDS and non-zero autocorrelation structure, which further extends the sensing distance with no loss of the spatial resolution [100].

1. **Time delay signature suppressed chaotic BOCDA**

The experimental setup of the chaotic BOCDA with suppressed TDS is illustrated in Fig. 3.49. The dashed box in Fig. 3.49 shows the chaotic laser source consisting of the DFB-LD without internal light isolators and a fiber feedback loop. The output chaotic laser is split into the pump and probe signal with the same center frequency of ν_0 by a 20:80 fiber coupler. The upper as the probe wave (80%) is modulated in a suppress-carrier, double-sideband format by the EOM, which is driven by MWG. The modulation frequency is in the vicinity of the Brillouin shift ν_B of the FUT. The probe wave through a programmable optical delay generator (PODG), a PS, and an ISO2 is launched into the FUT. The PODG and EDFA1 are utilized to control the position of the correlation peak and boost up the probe wave to 7 dBm, respectively. The PS is inserted after EDFA1 for suppressing the polarization dependence of the SBS interaction. And the pump wave (20%) through the PC3 and EDFA2 is injected into the opposite end of the FUT. The EDFA2 is employed to boost up the pump power to 27 dBm. After undergoing the SBS amplification by the chaotic pump wave along the FUT, the chaotic probe wave is filtered by the optical band-pass filter via the OC2. The optical power of the filtered Stokes wave is recorded by an OPM.

The structure of the FUT is consisted of a 3.2 km SMF (G.655), in which a 60 m fiber near 3 km is placed in a fiber thermostat. In addition, the temperature of the fiber thermostat is set to 55 °C and the room temperature is maintained constant at 25 °C. The measured point is located in the thermostat by adjusting PODG and then the relevant temperature information can be collected. The location of a correlation peak is scanned by a variable optical delay line to map the temperature along the FUT.

Figure 3.50 displays the δ-like autocorrelation trace of the chaotic laser, which indicates its strong anti-interference ability. As depicted in the autocorrelation curve, a dominant correlation peak at zero delay position could be observed, and the SBS amplification interaction between the chaotic pump and probe wave with the same chaotic state only occurs in the dominant correlation peak. Thus, the spatial resolution

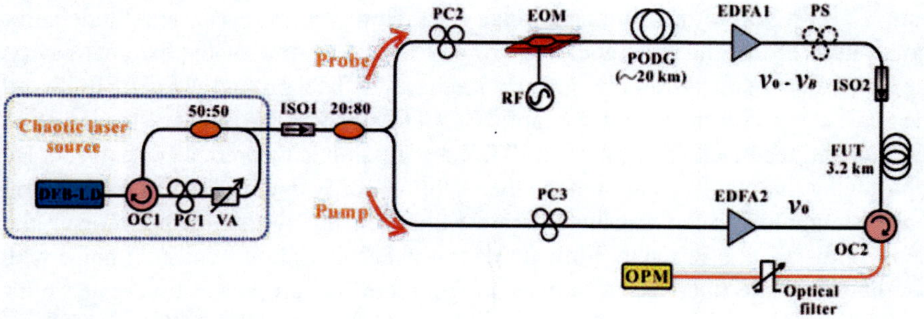

Fig. 3.49 Experimental setup of the chaotic BOCDA with suppressed time delay signature system

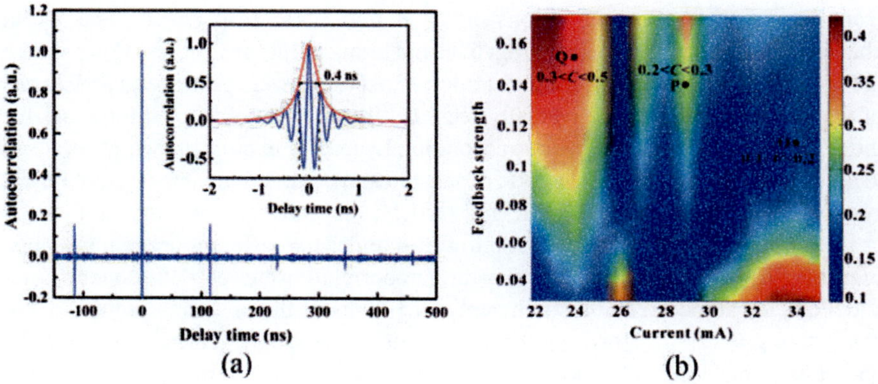

Fig. 3.50 Debugging result of chaotic laser time delay characteristics. **a** Autocorrelation trace of the chaotic laser, **b** the distribution map of the correlation. coefficient at the external feedback delay $\tau = 115$ ns under different injection currents and feedback strengths. Three representative operating points, i.e., O (34, 0.112), P (29, 0.139) and Q (24, 0.156) are arbitrarily chosen from three TDS distribution regions, respectively

is theoretically determined by the full width at half maximum (FWHM) of the peak. A magnified view of the peak is illustrated in the inset. Around this peak, there is a slight fluctuation with a period of the laser relaxation oscillation. According to the Gaussian-fitted peak width, the FWHM of the correlation peak is 0.4 ns and the theoretical spatial resolution is 4 cm. There are also residual side peaks near the main peak, which are the result of a weak amplitude autocorrelation of the chaotic signal occurring at the delay time of the external cavity $\tau = 115$ ns. Such residual off-peak SBS amplification contributes an additional noise mechanism, which can largely limit the SNR.

Figure 3.50b illustrates the distribution map of the correlation coefficient at the external feedback delay $\tau = 115$ ns under different bias currents and feedback strengths. Firstly, we experimentally analyze the effect of the injection current and feedback strength on the TDS suppression of the chaotic semiconductor laser. The feedback strength is scaled with the feedback ratio k mentioned before and defined as the power ratio of the feedback wave to the laser output. The chosen external cavity length is approximately 11.5 m and the corresponding external feedback delay τ is 115 ns. It can be seen that the evolution of the TDS could be roughly divided into three regions, i.e., $0.1 < C < 0.2$, $0.2 < C < 0.3$ and $0.3 < C < 0.5$. Especially, the TDS suppression region with $0.1 < C < 0.2$ accounts for a large proportion of the whole area. Moreover, it is experimentally found that the TDS suppression becomes easier along with the increase of the injection current, which is in accordance with the previous theoretical analysis. Therefore, the TDS can be effectively suppressed by optimizing the two parameters of the injection current and feedback strength.

Figure 3.51 illustrates the output characteristics of the chaotic laser source under the different operating points of Fig. 3.50b. Here, three representative operating points, i.e., O (34, 0.112), P (29, 0.139), and Q (24, 0.156) are arbitrarily chosen from three TDS distribution regions shown in Fig. 3.50b, respectively. The output chaotic characteristics corresponding to the operating points of O, P, and Q are shown in Fig. 3.51a–c, respectively. From the optical spectra, power spectra, and time series of the chaotic laser, it can be visually seen that there are no obvious differences for these three cases. However, the corresponding Lyapunov exponents, which are used to quantitatively measure asymptotic expansion and contraction rates in a dynamical system [101], are 0.0819, 0.0511, and 0.0123, respectively. This means that the chaos output performs different chaotic states under the different operating points. Figure 3.51a-IV~c4-IV further display the autocorrelation curves of the chaotic laser. Videlicet, the autocorrelation coefficient is 0.179 under the operating point O at the delay time $\tau = 115$ ns. However, the coefficient is 0.280, 0.374 corresponding to the operation points Q, P at the delay time, respectively. As shown in Fig. 3.51c-IV, the autocorrelation trace under the operating point Q has nine side-lobes separately at $\tau, 2\tau, \ldots, 9\tau$ with gradually reduced height. At the same time, there are several high sidelobes in the autocorrelation curve at the operating point P. These side lobes cause additional noise when stimulated Brillouin scattering amplification occurs between the chaotic pump and probe wave, which increases the background noise of the obtained BGS and limits the improvement of the sensing distance. In comparison, the number of the side peaks under the operating point O is significantly less than that at the working points Q and P.

The outputs from the chaotic laser under the above three operating points O, P, and Q are served as the pump and probe waves respectively. They are then injected into the FUT, with the chaotic pump entering in one end and the probe light entering in the other end. When the chaotic pump and probe waves meet at one location of the FUT, the chaotic probe wave is subject to SBS amplification. The BGS is achieved by recording the average power of the filtered chaotic Stokes probe versus the modulation frequency. The BGS obtained by averaging 20 repeating measurements are shown in Fig. 3.52. Figure 3.52a~c correspond to the operation points O, P, and Q, respectively. The blue line represents the correlation peak located at 1.6 km of the FUT with an ambient temperature of 25 °C. The red line represents the correlation peak located in the hot spot section (i.e., at the middle of the 60 m heated fiber) with the temperature of 55 °C.

The blue line represents the correlation peak located at 1.6 km of the FUT with an ambient temperature of 25 °C. The red line represents the correlation peak located at the middle of the 60 m heated fiber with a temperature of 55 °C.

It can be seen that if the TDS is suppressed better, the background of the obtained BGS will be lower. Based on the BGS, the signal-to-background ratio (SBR), defined as the amplitude ratio of the signal peak to the background peak, is experimentally measured to quantitatively evaluate the quality of the BGS. With the optimization of TDS suppression, the SBR increases from 1.12 to 2.44 dB when the temperature is 25 °C. Similarly, for the temperature of 55 °C, the BGS under the operation point O significantly improves compared with that under the operation point Q, i.e., the lower background of the obtained BGS. At the same time, we find that as to the operation point Q, the BGS with the correlation peak in the heated section is clearly worse than the one with the correlation peak outside the heated section. This is because there

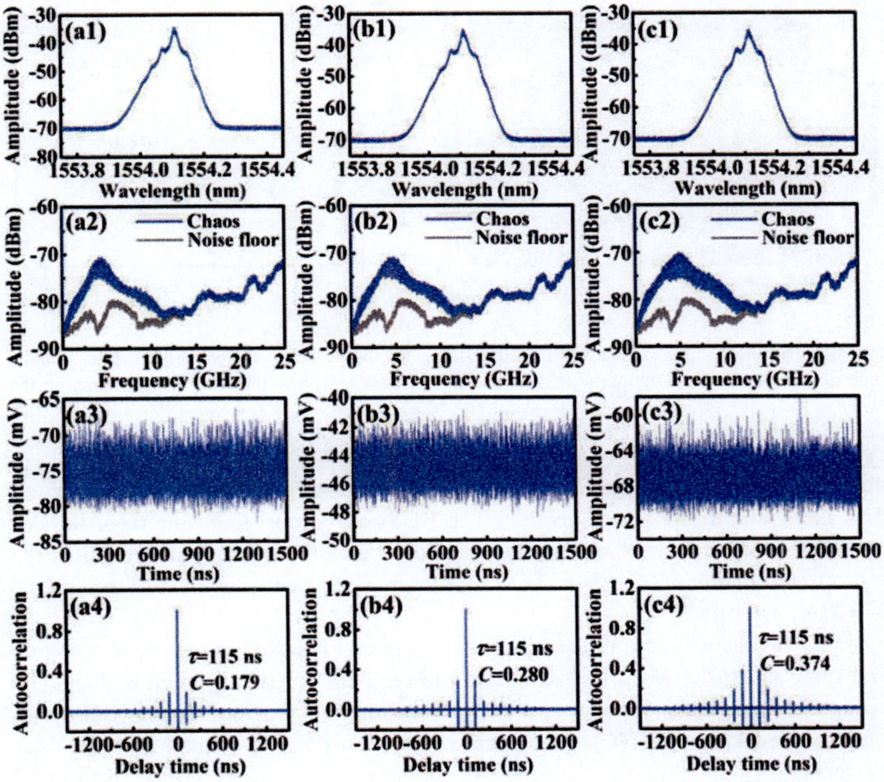

Fig. 3.51 The characteristics of chaotic laser at different operating points. **I** Optical spectra, **II** power spectra, **III** time series, **IV** autocorrelation traces, **a**, **b** and **c** correspond to the three operating points respectively, i.e., O (34, 0.112), P (29, 0.139) and Q (24, 0.156) shown in Fig. 3.53

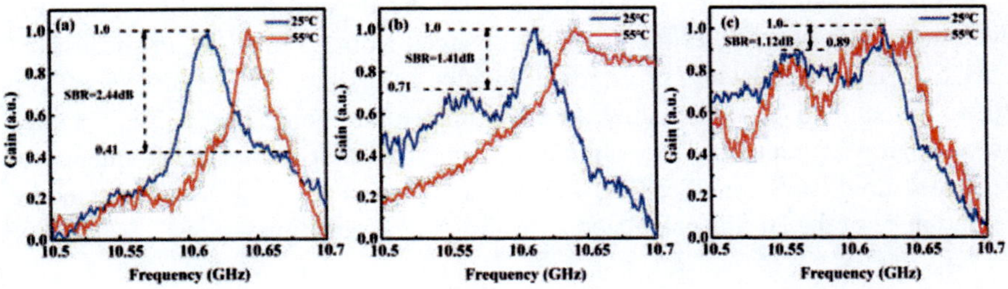

Fig. 3.52 The Brillouin gain spectra corresponding to the above operation points. **a** O, **b** P and **c** Q

are some TDS-induced side-lobes (1~3τ) in the heated fiber and other side-lobes (> 3τ) outside this fiber when the correlation peak is located at the middle of the 60 m heated fiber. The side-lobes are located in the fiber sections of different temperatures, which gives rise to the inhomogeneous off-peak amplifications. This eventually leads to the deterioration of the BGS. However, with regard to the operation point O, all

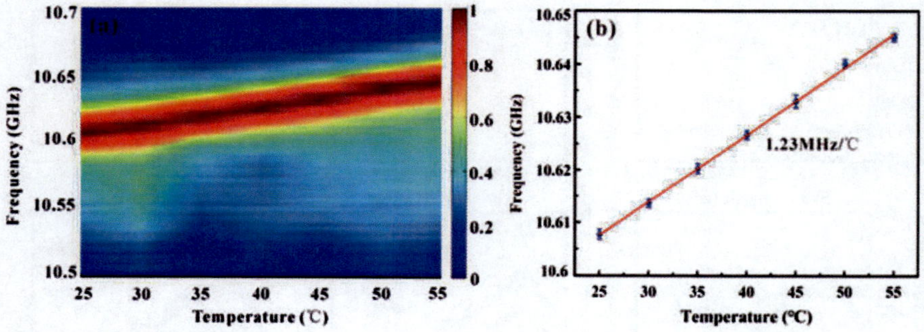

Fig. 3.53 Relationship between the chaotic BGS with temperature. **a** Two-dimensional map of chaotic BGS at different temperatures in the FUT, **b** chaotic BFS at different temperatures in the FUT

side-lobes are in the heated fiber due to the TDS suppression when the correlation peak is located in the heated section. The BGSs are almost the same regardless of the correlation peak in or outside the heated section. Thus, it is vital to select the appropriate chaotic state with the TDS suppression to improve the BGS of the chaotic BOCDA system.

Next, the chaos output under the operation point O is used as the sensing signal. The relationship between the BGS and the temperature of the fiber is displayed in Fig. 3.53. Figure 3.53a shows a two-dimensional plot of the chaotic BGS with temperature along the FUT. The single correlation peak is located into the fiber thermostat by adjusting the PODG and the temperature is changed from 25 to 55 °C in succession. A phenomenon that the center frequency of the BGS moves from 10.616 to 10.658 GHz can be obviously observed, where the linewidth of the BGS is stabilized at 45 MHz. And there is about a 37-MHz BFS at the hot spot section, which matches well with the real temperature of 30 °C.

Besides, in our measurement of the temperature-dependence of BGS and BFS, the temperature of the fiber thermostat is changed from 25 to 55 °C in succession. Each 5 °C is an interval and seven representative temperature points like 25, 30, 35, 40, 45, 50, and 55 °C are selected. At each temperature point, the BGS is obtained by averaging 20 repeating measurements. Supposed that the center frequency of each measured BGS, i.e., the BFS, is denoted as f_i ($I = 1, 2, \cdots, 20$). The standard deviation δ_f of the BFS is calculated according to the equation below:

$$\delta_f = \sqrt{\frac{\sum_{i=1}^{n}\left(f_i - \overline{f}\right)}{n(n-1)}} \tag{3.11}$$

where $\overline{f} = \sum_{i=1}^{n} \frac{f_i}{n}$ is the average value of the BFS, and $n = 20$ is the measurement times. Among these representative temperature points, the maximum standard deviation of the BFSs is taken as the measurement uncertainty of the BFS. Therefore,

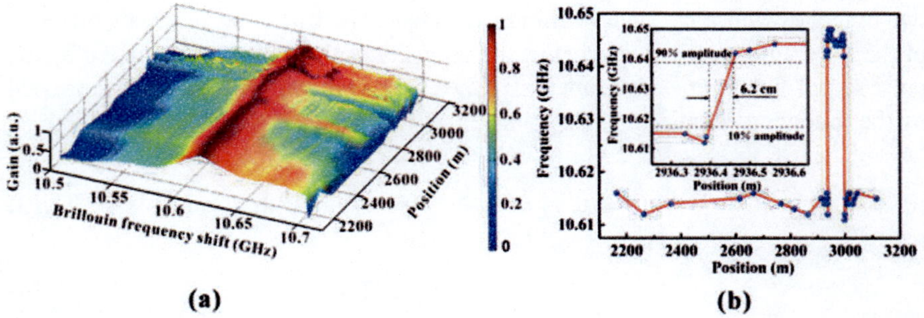

Fig. 3.54 Measured distribution of the BFS along the FUT. **a** Measured distribution of the BGS along the FUT, **b** that of the BFS along the FUT

the largest value of uncertainty of the local BFS is ± 1.2 MHz with the temperature coefficient of 1.23 MHz/°C, as shown in Fig. 3.53b.

Next, Fig. 3.54a shows a three-dimensional plot of the BGS measured along the FUT. Obviously, the BFS at the hot spot section is distinguished. In the experiment, the temperature of the fiber thermostat is set to 55 °C and the room temperature is maintained constant at 25 °C. Figure 3.54b plots the measured distribution of the BFS along the FUT. And the spatial resolution can be measured by the average value of 10~90% of the rise and fall time equivalent length in the meter for the hot spot section, where the rise and fall time equivalent length are 6.2 and 8.5 cm, respectively. Therefore, the spatial resolution can reach approximately 7.4 cm when a 60 m long section is heated in the thermostat. The experimentally obtained spatial resolution is nearly consistent with the above theoretically expected one. In fact, the approximate 100 resolution points are addressed in practice. In Fig. 3.54a, we can see that there is a big frequency interval at 2700, 2900, and 3200 m, because the fiber may be interfered by other external factors.

2. **Time-gated chaotic BOCDA**

As is known above, in the chaotic BOCDA system, the chaotic probe and pump wave transmitted in opposite directions are excited at the position z in the fiber to generate Brillouin acoustic field. The complex amplitude function of the density distribution of the acoustic field is as follows [69]:

$$Q(z,t) = \frac{1}{2T_B}\int_0^t \exp\left(\frac{t\prime - t}{2T_B}\right) A_P\left(t\prime - \frac{z}{\mathrm{v}_g}\right) A_S^*\left[t\prime - \frac{z}{\mathrm{v}_g} + \theta(z)\right] dt\prime, \quad (3.12)$$

where v_g is the group velocity of light in fiber, τ is the acoustic phonon lifetime, $A_P(t)$ denotes the instantaneous complex envelope of the pump wave entering the fiber at $z = 0$ and propagating in the positive z direction. $A_S(t)$ denotes the probe wave propagates from $z = L$ in the negative z direction. The position-dependent temporal offset $\theta(z)$ is defined as $\theta(z) = (2z–L)/v_g$, where L is the fiber length. The pump and the probe wave come from the same chaotic source, and then they are divided into two branches whose frequency difference matches the Brillouin frequency offset ν_B of the fiber. In the proposed scheme, the intensity of the pump is modulated by the pulse signal and the envelope function of the chaotic pump/probe wave are expressed by the following equations:

$$A_P(z=0,t) = A_{P0}u(t)rect\left(\frac{t}{\tau_{\text{Pulse}}}\right), \tag{3.13}$$

$$A_s(z=L,t) = A_{s0}u(t), \tag{3.14}$$

Here *u(t)* is a common, normalized envelope function with an average magnitude of unity with regard to chaotic signals, and A_{p0} and A_{s0} are constant magnitudes of the pump and probe, respectively, rect (ξ) equals 1(|ξ| < 0.5) or zero and τ_{pulse} is the duration of the pump amplitude pulse that is greater than phonon lifetime τ. The expectation value of the acoustic field magnitude at position z (for $t \ll \tau_{\text{pulse}}$) is

$$\overline{Q(t,z)} = \frac{A_{P0}A_{S0}^*}{2T_B}\int_0^t rect\left(\frac{t\prime}{\tau_{Pulse}}\right)\exp\left(\frac{t\prime - t}{2T_B}\right)$$
$$\overline{u\left(t\prime - \frac{z}{\mathrm{v}_g}\right)u^*\left[t\prime - \frac{z}{\mathrm{v}_g} + \theta(z)\right]}dt\prime = C\langle\theta(z)\rangle, \tag{3.15}$$

where $C\langle\theta(z)\rangle$ is the cross-correlation function between the chaotic pump and probe wave. The chaotic signal has fast time sequence, irregular oscillation, noise-like characteristic, and δ-like autocorrelation curve whose correlation peak is extremely narrow. In Eq. (3.15), the acoustic field is confined to a single correlation peak width, and the SBS amplification interaction at the center of the fiber where $\theta(L/2) = 0$ only occur in it. Thus, the spatial resolution is theoretically determined by the FWHM of the correlation peak [102]. Besides, pulse modulation theoretically provides 1/0 coefficient distribution for the acoustic field. There is acoustic field distribution only in the correlation peak of the chaotic signal at rect (ξ) = 1, and the acoustic field at other positions is zero. The SBS amplification in correlation peak is stimulated when the pump pulse reaches position *z* and grows up to the steady-state condition. After the pulse duration time τ_{pulse}, the correlation peak disappears and the SBS acoustic field quickly decays to zero.

The schematic diagram of the conventional chaotic BOCDA is illustrated in Fig. 3.55a. The chaotic signal is divided into the probe and pump wave, which are launched into opposite ends of the fiber. After the counter-propagating pump and probe waves with the same coherent state have gone through the same optical path, a unique non-period correlation peak, in which occurs the stable SBS, is generated at this location according to the synthesis of optical coherence function (SOCF). However, there are also residual side peaks near the main peak resulting from the TDS of the chaotic laser itself. At the same time, the natural oscillation of the chaotic signal causes the non-zero average value of the correlation function outside the peaks. These factors will stimulate a weak SBS acoustic field, so the intensity of incoherent

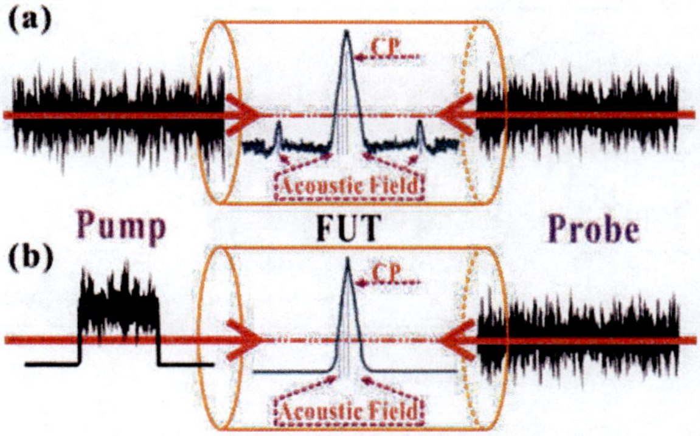

Fig. 3.55 The model diagram of the principle of the chaotic BOCDA system sensing. **a** The conventional chaotic BOCDA system, **b** the time-gated chaotic BOCDA system

the acoustic field persistently accumulates along the fiber, and the system noise level continues to increase.

For the suppression of the background noise structure, we propose a new time-gated chaotic BOCDA scheme, as depicted in Fig. 3.55b. After pump pulse modulation, the cross correlation between the chaotic pump and probe waves, i.e. the correlation peak, is confined into the pulse duration time due to the 0/1 switched modulation, which is set to zero in the rest time. The side-lobe peak at the chaotic delay position can be zero by selecting the appropriate pulse width and the time of the integral function of the cross-correlation between the pump light and the probe light is shortened. The mean value of the inherent oscillation of chaos is also infinitely close to zero, and the intensity of the incoherent acoustic field in the system is suppressed, that is, the noise base of the system is suppressed.

However, it is impossible in the experimental process that the amplitude of the pump path reaches zero as described in Eq. (3.13). As shown in Fig. 3.56a, the blue line represents the time series signal of the conventional chaotic BOCDA and the red line represents the backscattering sequence of the chaotic laser after pulse modulation. It can be seen that the scattering intensity around 280~400 ns is obviously enhanced. From the chaotic autocorrelation curve in Fig. 3.56b, it can be seen that the time delay cavity length has not been changed in both cases, which remains at $\tau_d = 123.8$ ns, but the TDS decreases from $C_1 = 0.205$ to $C_2 = 0.053$. At the same time, the autocorrelation curve noise base is partially suppressed after pulse modulation, then the weak acoustic grating in the fiber and the system noise base are all greatly suppressed. Based on this, the above theory is experimentally verified in a long-distance chaotic BOCDA system.

The experimental setup of the time-gated chaotic BOCDA is depicted in Fig. 3.57. The chaotic output laser is split into the pump and probe branches by a 90:10 fiber coupler. The upper branch is used as the probe wave and the lower branch is the pump light. The probe wave (90%) is modulated in a double-sideband format by an EOM, driven by a sine-wave whose central frequency ν_B matches the BFS of the sensing fiber. Thus, the probe wave generates double-sideband of $\nu_0 \pm \nu_B$ and the carrier of central frequency ν_0 is suppressed. Then the modulated probe light enters a PODG and the optical length of the probe wave is manually adjusted to realize the location of

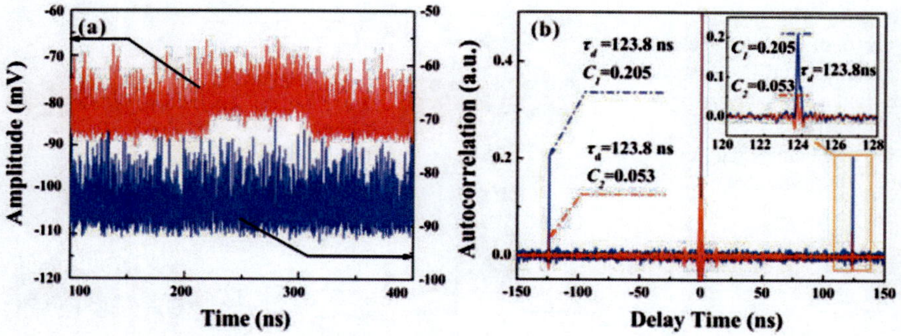

Fig. 3.56 **a** Chaotic pump time series and **b** autocorrelation characteristic with and without the pulse modulation

the correlation peak. The probe wave is amplified by the EDFA and injected into the PS to avoid the influence of Brillouin gain polarization sensitivity. Finally, the probe wave through the isolator is launched into the end of the FUT at $z = 10.2$ km. The pump wave is modulated by another EOM, driven by an optimized square pulse. The optimal pulse duration is 120 ns, and the repeating period is 110 μs corresponding to the time-of-flight through the FUT. The modulated signal, maintaining the center frequency, is amplified by EDFA and launched into the front of the FUT by the incident end of the circulator at $z = 0$. As shown in Fig. 3.58a, there is a slight periodic oscillation near the center peak of the autocorrelation curve of the chaotic signal. According to the Gaussian-fitted full width at half maximum, the theoretical spatial resolution is 2.7 cm. The distribution of the FUT is shown in Fig. 3.58b. The structure of the FUT is made up of a 10 km SMF (G.655), in which a hot zone about 150 m is placed near 10 km and connected to the band-pass filter through the reflector end of the circulator. The low-frequency wave (Stokes wave) of the probe signal amplified is filtered out by a band-pass filter (BPF). The optical power of the filtered Stokes wave is recorded by an OPM and the BGS is obtained by power information demodulation.

First, we further compare the BGS in the chaotic BOCDA systems with and without the pump pulse modulation. Figure 3.58a–c show the BGS at different positions of 5.0, 8.5, and 10.0 km sensing fiber in different devices, respectively. Obviously, in the chaotic BOCDA system without pulse modulation, the background structure of the BGS increases with the increase of the sensing fiber length and conversely keeps low background structure in the time-gated chaotic BOCDA system. Moreover, the BGS linewidth in the time-gated chaotic BOCDA system is stabilized at 49.6 MHz. The SBR is experimentally measured to quantitatively evaluate the equality of the extracted BGS [102]. As shown in Fig. 3.58d, the obvious increment of SBR is obtained along the entire FUT with the time-gated scheme. When the sensing fiber is longer than 8.0 km, the SBR of the chaotic BOCDA system without the pulse modulation is almost constant at 1.00 dB, which means the signals of BFS are submerged in the background structure induced by the off-peak amplification.

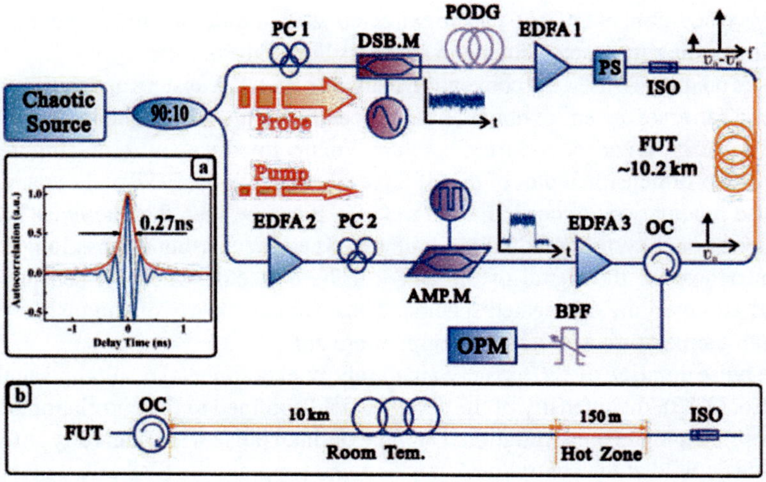

Fig. 3.57 Experimental setup of the time-gated chaotic BOCDA system

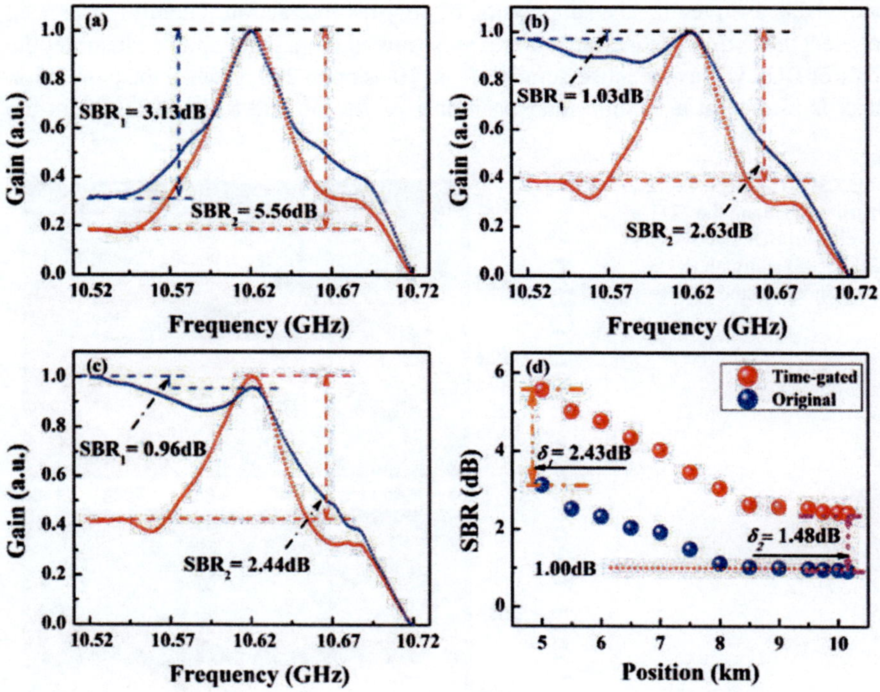

Fig. 3.58 Comparison of BGS between time-gated and conventional chaotic-BOCDA systems at the different fiber positions

However, the signal of the BFS can be extracted when the sensing fiber reaches 10 km by utilizing the time-gated scheme to suppress the off-peak interaction.

Based on this, we used the conventional chaotic BOCDA system and the improved device to calibrate the entire optical fiber at specific points. The variable temperature zone is set to 55 °C and the room temperature is maintained at about 21 °C. Figure 3.59 shows a two-dimensional plot of the BFS measured along the FUT without and with the pulse modulation. According to Fig. 3.59a, it can be seen that the noise base of the conventional system gradually dominates when the sensing fiber is longer than 8.0 km. However, the signal of the BFS can be extracted when the sensing fiber reaches 10 km in the time-gated scheme. Thus, we can clearly distinguish between the room temperature and variable temperature zone.

The pulse duration of 120 ns is used in the above experimental results and analysis. As in Eq. (3.15), the intensity of the acoustic field confined to the correlation peak is closely related to the pulse duration. Only the optimal pulse width can fully stimulate the acoustic field at the central peak position and suppress the SBS effect at the off-peak. According to Fig. 3.56a, the backscattering wave will reflect obvious pulse modulation information when the extinction ratio of pump pulse modulation is large enough, so as to realize the suppression of off-peak interaction. Therefore, we need to select the optimal extinction ratio for pulse modulation. Figure 3.60a illustrates the SBR of BGS versus the pulse duration from 10 ns up to 200 ns when the correlation peak is located at a specific fiber position $z = 7.5$ km and SBR ≈ 0.0 dB in the

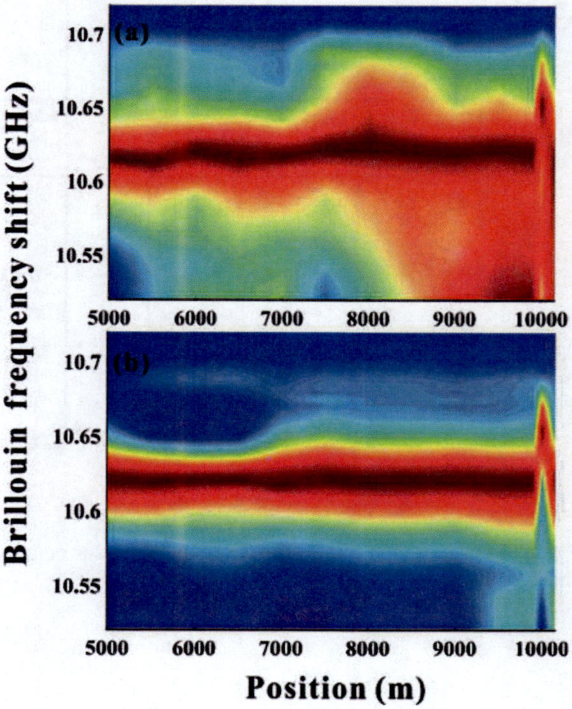

Fig. 3.59 The BFS distribution along the FUT in the chaotic BOCDA systems without (**a**) and with (**b**) the time-gated scheme

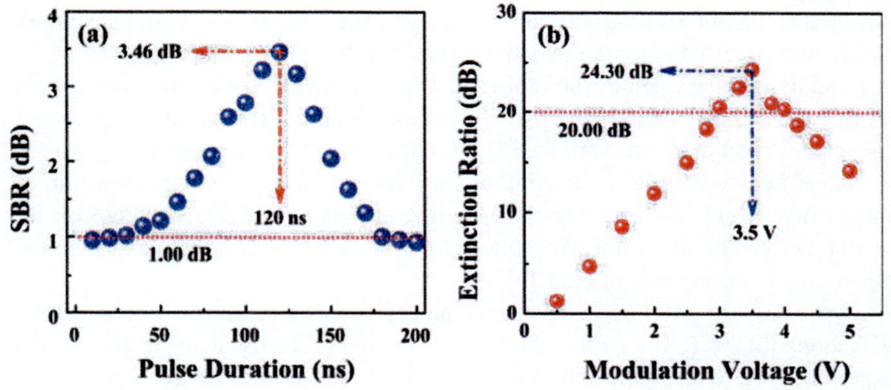

Fig. 3.60 Time-gated chaotic-BOCDA system pump. **a** Selection of pulse duration, **b** selection of modulation voltage

conventional system. After 60~160 ns pulse modulation, the central peak could be effectively stimulated while the incoherent acoustic field being suppressed, and SBR reaches the maximum of 3.46 dB when the pulse duration is 120 ns. At this time, the central peak acoustic field is fully excited by the pulse signal, and the SBS effect at non-central peak positions such as chaotic sidelobes is suppressed, which makes noise accumulation reach the minimum level. The acoustic field does not have enough time to reach the steady-state condition when the pulse width is shorter than 50 ns. When the pulse width is longer than delay cavity length $\tau_d = 123.8$ ns, more noise is accumulated and the SNR of the system deteriorates from the appearance of SBS acoustic field outside the correlation peaks such as the chaotic side-lobes and natural oscillation.

Figure 3.60b shows the variation trend of extinction ratio with pulse modulation under different modulation voltages. When the RF voltage is 3.5 V, the extinction ratio approaches the maximum 24.3 dB and the incoherent acoustic field is largely inhibited. At the same time, in this section, the EOM is used as the modulation device of pump pulse and the maximum extinction ratio is still low due to the characteristics of the device itself. Therefore, the modulation devices with higher extinction ratio can be used to optimize the system.

In order to ensure the accuracy of temperature measurement, the temperature coefficient is measured and calibrated again in time-gated chaotic BOCDA, as shown in Fig. 3.61. The single correlation peak is located into the hot-zone by adjusting PODG. The temperature is changed from 21 to 55 °C in succession. A phenomenon that the center frequency of the BGS moves from 10.616 to 10.658 GHz can be obviously observed. From these BGS, we can plot the temperature-dependence of BFS and the uncertainty of the estimated BFS as shown in Fig. 3.61b. According to the fitting curve, the temperature coefficient is 1.22 MHz/°C and the maximum standard deviation in the reconstruction of the local BFS is $\pm$ 1.8 MHz with 25 times of the repeating measurements. This measurement uncertainty of the BFS is a little higher than that in the previous chaotic BOCDA system, which is probably related to the modulation of the pump pulse on the chaotic light. The measurement uncertainty is achieved by calculating the maximum standard deviation of the BFSs. Thus, the uncertainty is determined by the BGS, and it will be high with the broadening of the BGS. As we know, the measured BGS is a convolution of the pump spectrum and the actual Brillouin spectrum of the FUT. In the time-gated chaotic BOCDA system, the pulse modulation on the chaotic pump light broadens its optical spectrum to some extent, which ultimately leads to the linewidth of the BGS increasing from 45 to 49.6 MHz. The linewidth widening of the BGS gives rise to the increase in the measurement uncertainty of the BFS.

Figure 3.62 shows a three-dimensional plot of the BGS and the distribution of the BFS along the FUT. The BFS at the hot spot section is clearly distinguished. In the experiment, the temperature of the hot-zone is set to 55 °C and the room temperature is maintained at 21 °C. The spatial resolution of the time-gated chaotic BOCDA system can be measured by the mean value of 10~90% of the rise and fall time equivalent length in meter, where the rise and fall time equivalent length are 8.86 cm and 9.24 cm, respectively. Therefore, the spatial resolution can approximately approach 9 cm.

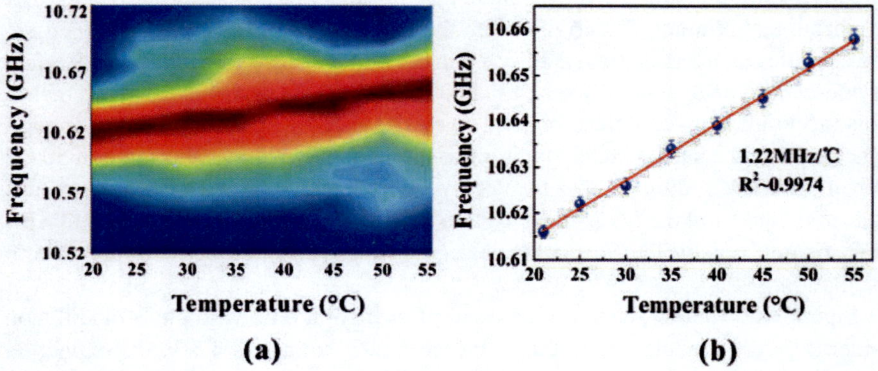

Fig. 3.61 The relationship of the BGS with temperature. **a** Temperature-dependence of the BGS along the FUT, **b** that of the BFS along the FUT

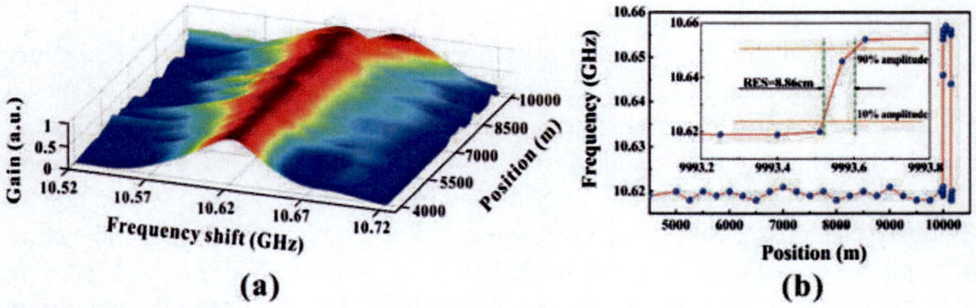

Fig. 3.62 Measured distributions of the **a** BGS and **b** BFS along the FUT

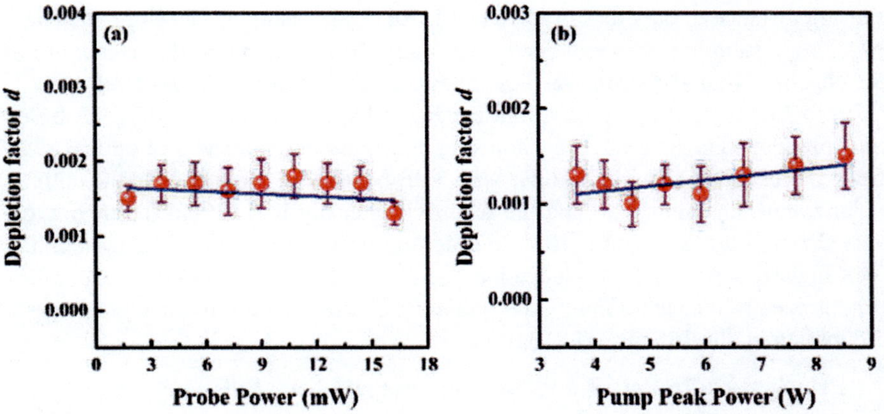

Fig. 3.63 Influence analysis of pump depletion in time-gated chaotic BOCDA system. **a** Depletion factor d as a function of the probe power for the lower frequency sideband, where a fixed input pump peak power is 6 W, **b** depletion factor d as a function of the pump peak power for a fixed input probe power of 3.6 mW

Besides, we further analyze the effect of pump pulse depletion in the time-gated chaotic BOCDA system as shown in Fig. 3.63. The chosen fiber length of 1 km is comparatively short. This is because that a short fiber has a good uniformity to ignore the effect of the BFS fluctuation, and on the other hand, it can avoid the decrease of SNR induced by fiber attenuation. The duration of the pump pulse is 120 ns. The chaotic probe wave operates in the double-sideband suppress-carrier mode. Here, a depletion factor d is introduced to characterize the amount of depletion. Figure 3.63a shows the depletion factor *d* as a function of the probe power for the lower frequency sideband, where a fixed input pump peak power is 6 W. Figure 3.63b represents the depletion factor *d* versus the pump peak power while keeping the probe power constant (3.6 mW). We can see that the pump depletion is checked to be below 0.2%, having a negligible effect on the measurements. One probable reason is that the chaotic pump wave only plays an important role in suppressing the off-peak amplification in the SBS interaction process between the chaotic pump and probe light, and is not involved in the energy transfer during the whole process of transmission. Another possible reason is that the double-sideband modulation using a symmetric probe wave is much more tolerant to pump depletion. Therefore, the pump pulse plays an important role in the suppression of the off-peak amplification, which is not accompanied by the pump depletion.

3.4.3 Millimeter-Level-Spatial-Resolution BOCDA

Recently, the BOCDA system can achieve millimeter-level spatial resolution, but the periodic sinusoidal signal and ASE signal with low power spectral density lead to the sensing distance of only tens of meters [62, 70]. The sensing distance of the chaotic BOCDA system has been extended to 10.2 km. However, due to the bandwidth of the chaotic signal, the spatial resolution remains centimeter level [98–100].

In this section, we propose a millimeter-level-spatial-resolution BOCDA based on a broadband chaotic laser. By adjusting the external parameters of optical feedback, a broadband chaos of 10 GHz with 3 dB bandwidth is obtained at the optimal polarization matching state and the feedback intensity is 0.12 and the theoretical spatial resolution is 3.0 mm [103]. In order to improve the SNR of the system, the lock-in detection is first introduced into the chaotic BOCDA, and the measurement time is greatly reduced. Finally, the spatial resolution can approximately approach 3.5 mm along the 165 m FUT [102].

1. **Principal realization of millimeter-lever-spatial-resolution**

In this experiment, the single feedback loop structure is used to generate the chaotic laser. Its dynamic characteristics meet the L-K rate equation, and the theoretical expression is Eqs. (2.1)~(2.5). It can be seen from the above equation that different chaotic states can be obtained by changing the polarization matching state and the feedback intensity. The feedback strength (FBS) is defined as the power ratio of feedback light to the DFB laser's output. In general, a stable chaotic laser can be generated by selecting the appropriate feedback strength and matching the feedback light with the polarization state of free output DFB.

We know that the TDS, autocorrelation non-zero background, and bandwidth of chaotic laser can directly affect the capabilities of the chaotic BOCDA system. The first two features have been exhaustively analyzed and optimized [99, 100]. This section will focus on the influence of chaotic bandwidth on the system. In the chaotic BOCDA system, only the Brillouin acoustic field is excited within the chaotic autocorrelation center peak, and the spatial resolution depends on the full-width at half-maximum of the center peak. In addition, the width of the central peak is inversely proportional to the chaotic bandwidth, and when the bandwidth is wider, the center peak is narrower. Therefore, the theoretical spatial resolution will be higher along with the widening of the chaotic bandwidth in the chaotic BOCDA system. In the BOCDA system based on ASE noise source [95], the spatial resolution depends on the secondary calculation of the spectral width. Due to the inaccuracy of the spectrum reading and the equation error, the theoretical spatial resolution of this method has a calculation error, so it is not used in the chaotic system. Figure 3.64 shows the typical characteristics of chaotic lasers in different states. By increasing the bandwidth from 3 to 10 GHz, the full-width at half-maximum (FWHM) of the CP of autocorrelation curve gradually decreases.

Finally, the center peak of 0.03 ns is obtained in the 10 GHz chaotic bandwidth. At this time, the theoretical spatial resolution of the system can reach 3.0 mm.

To study the route to broadband chaos, we measured the variations in bandwidth with those of FBS when matching different states of polarization, as depicted in Fig. 3.65. It's remarkable that the bandwidth presented an initially decreasing trend that subsequently ascended via increasing the FBS. The TDS is largely suppressed and the chaotic bandwidth reaches near maximum in the optimal region where the

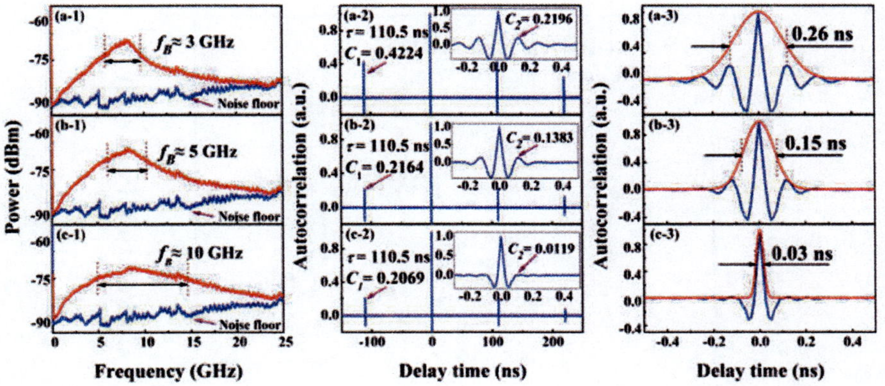

Fig. 3.64 Characteristic diagram of chaos in three typical status: **1** frequency spectra, **2** autocorrelation trace, **3** correlation peak with Gauss fitting

FBS varies from 0.10 to 0.15. Different from the feedback strength, the polarization matching states cannot be described quantitatively. In the experiment, the optimal matching state A and the general matching state B can be described qualitatively. Figure 3.65 shows that the chaotic bandwidth is significantly wider in the matching state A, and the maximum difference of bandwidth near the optimal operating point is 6 GHz. Therefore, in the experiment, a TDS of 0.2069 can be obtained by setting the FBS at 0.12 and the PMS at A, and the chaotic bandwidth was maximized to about 10 GHz.

In fact, the relaxation oscillation peaks around the central CP play the main role in restricting the FWHM of the CP. The restriction of relaxation oscillation on Gaussian fitting of center peak decreases with the increase of bandwidth, which can be observed from the Gaussian fitting curve in Fig. 3.64a-III. Serving as narrow-band chaos, the relaxation peaks, 1st order or even 2nd order, can be clearly revealed.

Fig. 3.65 Variations of bandwidth of chaos with feedback strength in different polarization matching states

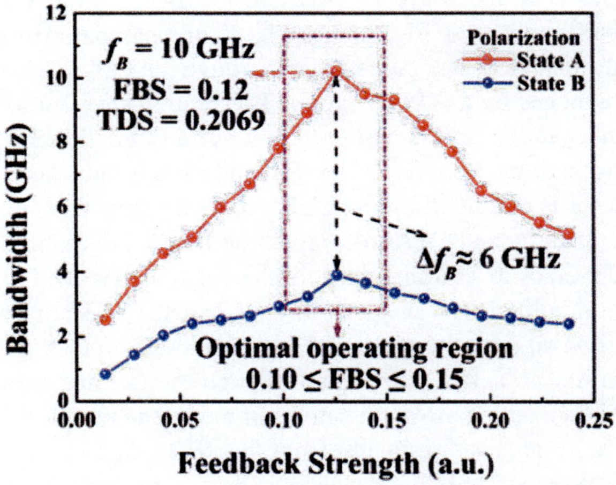

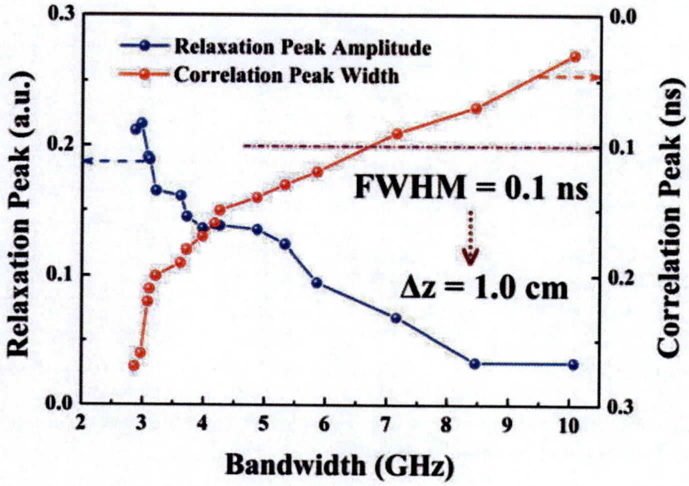

Fig. 3.66 Relationship between the amplitude of relaxation oscillation peaks (blue) and the FWHM of correlation peaks with Gaussian fitting (red)

By increasing the bandwidth, the 1st order peaks decrease from 0.2196 to 0.0119, while the 2nd order peaks ultimately vanish. Gaussian fitting, thus, can be actualized with optimal parameters. Figure 3.66 shows the relationship between the amplitude of the relaxation oscillation peak (blue) and the FWHM of the CP with Gaussian fitting (red). The FWHM is gradually compressed along with the weakened relaxation oscillation. Especially, when the bandwidth is around 7.5 GHz (or wider), the FWHM approximately equals to 0.1 ns (or narrower), implying that the spatial resolution can reach 1.0 cm correspondingly. When the chaotic bandwidth continues to increase, the theoretical spatial resolution of the system can reach millimeter level.

2. **Broadband chaos distributed sensing**

Based on the above 10 GHz chaotic laser, a broadband chaotic BOCDA system is built as shown in Fig. 3.67, and its theoretical spatial resolution is 3 mm. The output of the chaotic laser source passes through the PC1 and is split into the probe and pump branches by a 90:10 coupler. The probe wave (90%) is modulated in the double-sideband, suppress-carrier format by an EOM, driven by the MWG. The probe wave through the PODG, EDFA, PS, and ISO, is launched into a FUT. Here, the chaotic laser is amplified by the EDFA to 11.0 dBm. The pump wave (20%) is boosted to 5 dBm by an EDFA to compensate for the later amplitude-modulated signal that is driven by an arbitrary waveform generator (AWG). Then, the modulated pump wave is amplified to an average power of 33 dBm. The pump and probe waves are launched into two opposite ends of the FUT through an OC. After SBS amplification occurs in the FUT, two waves pass through the OC and inject into the tunable BPF. The filtered output Stokes signal is sampled by a lock-in amplifier (LIA), whose reference frequency also originates from the AWG. In addition, the LIA is controlled by the MWG to ensure that data acquisition is synchronous with frequency scanning.

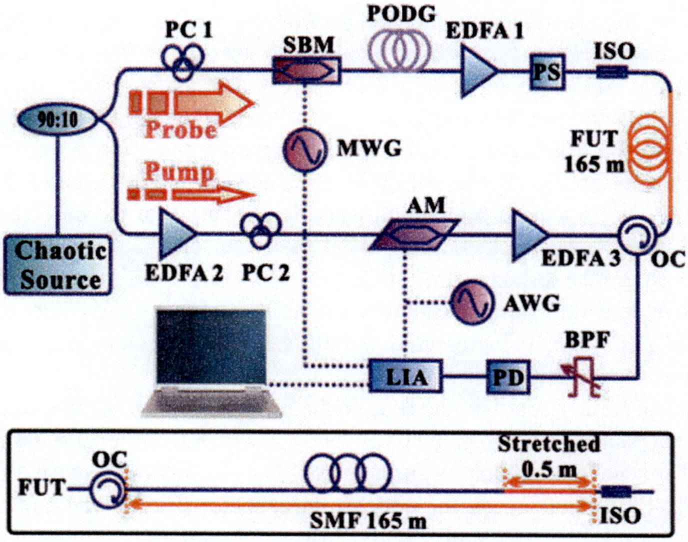

Fig. 3.67 Experiment setup of millimeter-scale spatial resolution BOCDA based-on a broadband chaotic laser

The chaotic correlation peak in the FUT should be narrow enough for achieving the millimeter-level ultra-high resolution. However, as the correlation peak generated by the interaction of the pump and probe wave in the fiber becomes narrower, the weak stimulated Brillouin gain is generated in the FUT. Moreover, the SBS amplification caused by the non-central peak of the chaotic signal introduces additional noise, which accumulates along the fiber. Thus, the SBS gain signal is easily submerged into the noise, which seriously affects system performance. The analysis of the SNR of the chaotic BOCDA system can be estimated as [98]:

$$SNR \approx \frac{1}{2} g_0 v_g \left| A_{p0} \right|^2 \sqrt{T / f_B}, \tag{3.16}$$

Here, $g_0 = 0.2\ (\mathrm{W} \bullet \mathrm{m})^{-1}$ is SBS gain coefficient of FUT, $f_B = 10$ GHz is the bandwidth of the chaotic laser, and T is the response time of the signal detection. Considering that the SBS threshold of chaos is much higher than that of the conventional continuous-wave laser, the mean optical power of chaotic pump light, $|A_{p0}|^2$, is approximately 2 W to inspire the Brillouin amplifications effectively. Note that the features of chaotic laser, e.g., the width of central CP and the TDS, and performance of chaotic sensors will not be affected by such high amplification.

In the narrowband chaotic BOCDA systems, an optical power meter (OPM) is used for recording the filtered output Stokes power and avoiding thermal noise of PD. However, in the 10 GHz bandwidth chaotic BOCDA system, the response time of the optical power meter is 0.3 μs, the calculated SNR is merely 0.2 dB, and the Brillouin gain signal is almost completely submerged into the noise. Thus, averaging

25 repetitions or more of measurements is inevitable for enhancing the SNR. Further, as a single measurement takes about one minute, this is a rather time-consuming process.

Consequently, lock-in detection is utilized for the first time in the chaotic BOCDA systems. An AWG is employed to provide the chopping and reference frequency of 200 kHz. In addition, the chaotic pump signal is intensity modulated by a sinusoidal signal and the noise caused by the non-peak amplification of the chaotic signal is effectively suppressed, which is similar to the time-gated chaotic BOCDA scheme. In signal acquisition, the response time of the PD is the same as the reference frequency of the lock-in detection, and the calculated SNR of the lock-in detection broadband chaotic BOCDA system is increased to 0.8 dB, which is about 4 times of the optical power acquisition system.

Besides, we further compare the BGS in the broadband chaotic BOCDA systems with different data acquisition methods as shown in Fig. 3.68. A tensile strain section of 2000 $\mu\varepsilon$ is set at the terminal position in the FUT. Owing to the sharp differences in SNR, the BGS obtained by the optical power meter (Fig. 3.68a) and the lock-in detection (Fig. 3.68b) is significantly different regardless of the proportion of background noise or Brillouin frequency offset. Obviously, in the lock-in detection system, because of the high SNR, the signal-to-background ratio (SBR) is up to 6.33 dB and the BFS is around 102 MHz, which is in agreement with the standard 100 MHz. However, due to the inferior SNR and slightly broadening of the BGS based time-gated, which increases the uncertainty of frequency shift, the BFS of the OPM scheme is 95 MHz, which is obviously far from the actual value. The SBR is just 1.94 dB even though 25-repetition averaged values per time are implemented.

Moreover, in the lock-in detection system, each BGS is obtained via a single measurement where MWG is scanned from 10.7 to 11.1 GHz with a 2 MHz step. It has no requirement of averaging, and it only consumes 0.2 s, which is only determined by the frequency-sweeping speed of MWG and the sampling rate of LIA with the sensitivity of 10 μV and maximum sampling rate of 1 kHz. In addition,

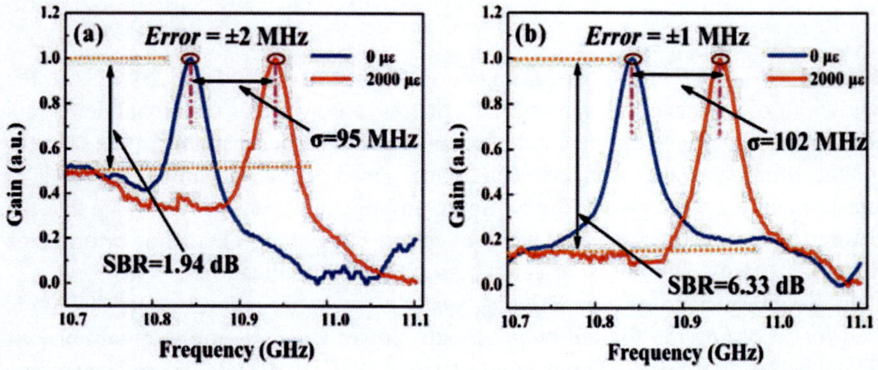

Fig. 3.68 Comparison of the BGS in the broadband chaotic BOCDA systems with different data acquisition methods: **a** optical-power-meter-based time-gated scheme, **b** lock-in detection scheme

the measurement error of peak frequency is ± 1 MHz at each resolution point. In the OPM-based system, the single measurement takes about 2 min and the center frequency measurement error of single resoluion point is around ± 2 MHz. The utilization of lock-in detection is a notable improvement in real-time monitoring for the chaotic BOCDA setups.

Figure 3.69 shows the map of BFS distribution along the FUT in the broadband chaotic BOCDA with different data acquisition methods. Owing to poor SNR, accurately identifying the stretched section in the OPM scheme is practically impossible and the measurement uncertainty of BFS is as high as ± 3.5 MHz. On the contrary, in the lock-in detection system, the strain tensile section can be clearly identified, and the measurement uncertainty of BFS is only ± 2 MHz, so the overall performance is significantly better than that of the former. Note that the uncertainty of BFS recording over the whole FUT is slightly greater than the error of peak frequency measuring at a single point, which could be attributed to the accumulation of the incompletely inhibited background noise.

Figure 3.70a shows the three-dimensional plot of the BGS measured along the FUT, in which the BFS in the strain zone changed significantly. In the experiment, the strain of the stretched section is set to 2000 με and the BFS is about 102 MHz. Figure 3.70b plots the measured distribution of the BFS along the FUT. The corresponding fiber lengths of the rising and fall edges are 3.4 mm and 3.6 mm respectively. Therefore, the average spatial resolution of the system is 3.5 mm, which is consistent with the theoretical spatial resolution of the system.

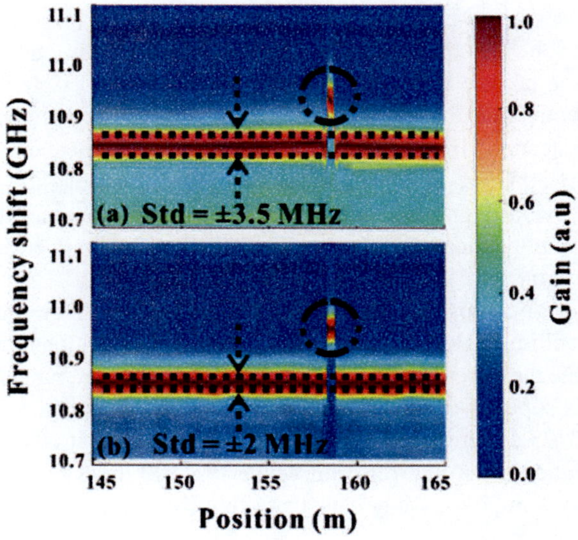

Fig. 3.69 The map of BFS distribution of broadband chaotic BOCDA system in different acquisition modes

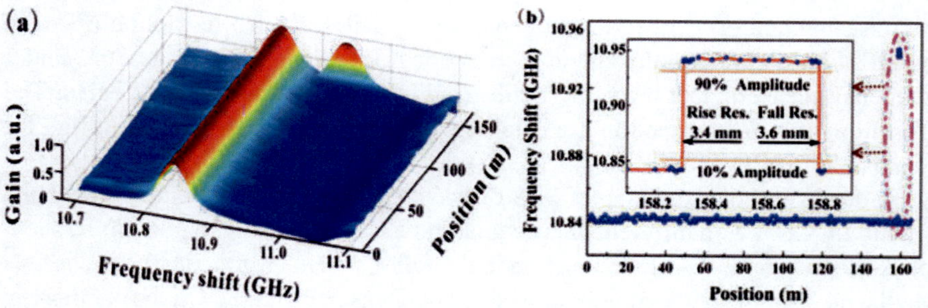

Fig. 3.70 Measured distributions of the BGS. **a** and BFS, **b** along the FUT

3. Advantages of broadband chaotic BOCDA

In order to obtain a millimeter-level ultra-high spatial resolution, a special three-electrode LD with a large push–pull-type current modulation and the frequency-modulation amplitude of 33 GHz is used in the conventional direct-modulation BOCDA system, where an intensity modulator is essential to compensate for the intensity chirp. However, the measurement range is limited to 5 m owing to the poor SNR resulting from the high modulation frequency of the LD [67]. Similarly, in the phase-modulated correlation-domain systems [70–71], the spatial resolution is given by the bit duration, e.g., 1 cm FWHM of CP for a bit length of 100 ps, i.e., 10 Gbit/s phase modulation rate. Consequently, a higher modulation rate is required to further promote the spatial resolution to sub-millimeter, but it is hard to get the MWG and EOM with bandwidth of 100 GHz. For comparison, the main advantage of the chaotic BOCDA is that the spatial resolution is theoretically determined via the chaotic bandwidth, where the chaos of 10 GHz is easily obtained in a single optical feedback loop setup. In addition, the experiment of dual-wavelength optical injection [101] has verified the generation of the bandwidth-enhanced chaotic laser of 30 GHz, which could achieve a resolution of sub-millimeter level theoretically. Several highly-nonlinear fiber sections will be employed to enhance the Brillouin amplification and obtain higher SBS power gain over millimeter-scale or shorter segments.

The proposed broadband chaotic BOCDA system is utilized to realize the millimeter-level spatial resolution and relatively longer measurement. In this system, the measurement of 47,000 resolving points (i.e., range/resolution = 165 m/3.5 mm) is superior to the ordinary direct-current-modulation BOCDA [67], in which only 3125 sensing points are addressed. Moreover, it takes only 0.2 s for a single resolution point interrogation which is a 100 times faster than the double modulation system [104]. We believe that if not considering the complexity of the system, even more than 2,500,000 (i.e., 10 km/4 mm) will be further confirmed via combining the lock-in detection technique and pulse-modulated time-gated scheme.

In addition, the main drawback of the current chaotic BOCDA is the location of the single CP scanned by a PODG, which results in an inconvenient and time-consuming system. Recently, a novel method of multiple CPs interrogated concurrently has been proposed to decrease the measurement time observably, taking about 1 s for the measurement of 20,000 resolution points [74] and about 4400 s to access overall 51,000 sensing points [105]. However, it brings about a deteriorated SNR and an inferior strain measurement accuracy. Thus, a great deal of average is required for higher performance. To avoid the limitation of PODG in utilization, the chaotic correlation optical time-domain reflectometry technology [106] could be used to address the sensing points along the FUT. Moreover, the method of resolving the CP in the time domain will be further explored to realize a point-by-point real-time chaotic BOCDA system.

References

1. Kurashima T, Horiguchi T, Tateda M. Distributed-temperature sensing using stimulated Brillouin scattering in optical silica fibers. Opt Lett. 1990;15(18):1038–40.
2. Bao XY, Chen L. Recent progress in Brillouin scattering based fiber sensors. Sensors. 2011;11:4152–87.
3. Barrias A, Casas JR, Villalba S. A review of distributed optical fiber sensors for civil engineering applications. Sensors. 2016;16(5):748.
4. Motil A, Bergman A, Tur M. State of the art of Brillouin fiber-optic distributed sensing. Opt Laser Technol. 2018;78:81–103.
5. Fernicola V, Crovini L. Digital optical fiber point sensor for high-temperature measurement. J Lightwave Technol. 1995;13(7):1331–4.
6. Wu M. Fiber fence alarm system based on long-distance quasi-distributed FBG sensor. Chongqing: Chongqing University; 2007.
7. Ukil A, Braendle H, Krippner P. Distributed temperature sensing: review of technology and applications. IEEE Sens J. 2015;12(5):885–92.
8. Ippen EP, Stolen RH. Stimulated Brillouin scattering in optical fibers. Appl Phys Lett. 1972;21(11):539–41.
9. Rich TC, Pinnow DA. Evaluation of fiber optical waveguides using Brillouin spectroscopy. Appl Opt. 1974;13(6):1376–8.
10. Rowell NL, Thomas PJ, Van Driel HM, et al. Brillouin spectrum of single-mode optical fibers. Appl Phys Lett. 1979;34(2):139–41.
11. Lagakos N, Bucaro JA, Hughes R. Acoustic sensitivity predictions of single-mode optical fibers using Brillouin scattering. Appl Opt. 1980;19(21):3668–70.
12. Culverhouse D, Farahi F, Pannell CN, et al. Potential of stimulated Brillouin scattering as sensing mechanism for distributed temperature sensors. Electron Lett. 1989;25(14):913–5.
13. Horiguchi T, Kurashima T, Tateda M. Tensile strain dependence of Brillouin frequency shift in silica optical fibers. IEEE Photonics Technol Lett. 1989;1(5):107–8.
14. De Souza KP, Lees GP, Wait PC, et al. Diode-pumped Landau-Placzek based distributed temperature sensor utilizing an all-fibre Mach-Zehnder interferometer. Electron Lett. 1996;32(23):2174–5.
15. Kee HH, Lees GP, Newson TP. All-fiber system for simultaneous interrogation of distributed strain and temperature sensing by spontaneous Brillouin scattering. Opt Lett. 2000;25(10):695–7.

16. Wait PC, Hartog AH. Spontaneous Brillouin-based distributed temperature sensor utilizing a fiber Bragg grating notch filter for the separation of the Brillouin signal. IEEE Photonics Technol Lett. 2001;13(5):508–10.
17. Alahbabi MN, Lawrence NP, Cho YT, et al. High spatial resolution microwave detection system for Brillouin-based distributed temperature and strain sensors. Meas Sci Technol. 2004;15(8):1539–43.
18. Kurashima T, Horiguchi T, Izumita H, et al. Brillouin optical-fiber time domain reflectometry. IEICE Trans Commun. 1993;76(4):382–90.
19. Iida D, Ito F. Low-bandwidth cost-effective Brillouin frequency sensing using reference Brillouin-scattered beam. IEEE Photonics Technol Lett. 2008;20(22):1845–7.
20. Shimizu K, Horiguchi T, Koyamada Y, et al. Coherent self-heterodyne Brillouin OTDR for measurement of Brillouin frequency shift distribution in optical fibers. J Lightwave Technol. 1994;12(5):730–6.
21. Song MP. The technique of Brillouin scattering-distributed optical fiber sensing based on microwave electrooptical modulation. Acta Opt Sinica. 2004;24(8):1111–4.
22. Alahbabi MN, Cho YT, Newson TP. 100 km distributed temperature sensor based on coherent detection of spontaneous Brillouin backscatter. Meas Sci Technol. 2004;15(8):1544–7.
23. Alahbabi MN, Cho YT, Newson TP. 150-km-range distributed temperature sensor based on coherent detection of spontaneous Brillouin backscatter and inline Raman amplification. J Opt Soc Am B. 2005;22(6):1321–4.
24. Koyamada Y, Sakairi Y, Takeuchi N, et al. Novel technique to improve spatial resolution in Brillouin optical time-domain reflectometry. IEEE Photonics Technol Lett. 2007;19(23):1910–2.
25. Wang F, Zhan W, Zhang X, et al. Improvement of spatial resolution for BOTDR by iterative subdivision method. J Lightwave Technol. 2013;31(23):3663–7.
26. Horiguchi T, Tateda M. BOTDA-nondestructive measurement of single-mode optical fiber attenuation characteristics using Brillouin interaction: theory. J Lightwave Technol. 1989;7(8):1170–6.
27. Bao XY, Webb DJ, Jackson DA. 32-km distributed temperature sensor based on Brillouin loss in an optical fiber. Opt Lett. 1993;18(18):1561–3.
28. Geinitz E, Jetschke S, Opke UR, et al. The influence of pulse amplification on distributed fibre-optic Brillouin sensing and a method to compensate for systematic errors. Meas Sci Technol. 1999;10(2):112–6.
29. Li W, Bao XY, Li Y, et al. Differential pulse-width pair BOTDA for high spatial resolution sensing. Opt Expr. 2008;16(26):21616–25.
30. Liang H, Li W, Linze N, et al. High-resolution DPP-BOTDA over 50 km LEAF using return-to-zero coded pulses. Opt Lett. 2010;35(10):1503–5.
31. Dong Y, Chen L, Bao XY. Time-division multiplexing-based BOTDA over 100 km sensing length. Opt Lett. 2011;36(2):277–9.
32. Dong Y, Bao XY, Chen L. High-performance Brillouin strain and temperature sensor based on frequency-division multiplexing using nonuniform fibers over 75-km fiber. In: 21st International conference on optical fiber sensors; 2011.
33. Dong Y, Chen L, Bao X. Extending the sensing range of Brillouin optical time-domain analysis combining frequency-division multiplexing and in-line EDFAs. J Lightwave Technol. 2012;30(8):1161–7.
34. Niklès M, Thévenaz L, Robert PA. Simple distributed fiber sensor based on Brillouin gain spectrum analysis. Opt Lett. 1996;21(10):758–60.
35. Niklès M, Thévenaz L, Robert PA. Brillouin gain spectrum characterization in single-mode optical fibers. J Lightwave Technol. 1997;15(10):1842–51.
36. Foaleng SM, Tur M, Beugnot J-C, et al. High spatial and spectral resolution long range sensing using Brillouin echoes. J Lightwave Technol. 2010;28(20):2993–3003.
37. Soto MA, Bolognini G, Di Pasquale F, et al. Simplex-coded BOTDA fiber sensor with 1 m spatial resolution over a 50 km range. Opt Lett. 2010;35(2):259–61.

38. Soto MA, Bolognini G, Pasquale FD, et al. Long-range Brillouin optical time domain analysis sensor employing pulse coding techniques. Meas Sci Technol. 2010;21(9): 094024.
39. Angulo-Vinuesa X, Martin-Lopez S, Nuno J, et al. Raman-assisted Brillouin distributed temperature sensor over 100 km featuring 2 m resolution and 1.2 °Σ uncertainty. J Lightwave Technol 2012;30(8):1060–65.
40. Thévenaz L, Mafang SF, Lin J. Effect of pulse depletion in a Brillouin optical time-domain analysis system. Opt Expr. 2013;21(12):14017–35.
41. Soto MA, Le Floch S, Thévenaz L. Bipolar optical pulse coding for performance enhancement in BOTDA sensors. Opt Expr. 2013;21(14):16390–7.
42. Rao YJ, Jia XH, Wang ZN, et al. 154.4km BOTDA based on hybrid distributed Raman amplifications. In: 23rd international conference on optical fibre sensors; 2014.
43. Zhang C, Rao YJ, Jia XH, et al. Brillouin optical time domain analyzer based on bi-directional Raman amplification. Acta Phys Sinica. 2010;59(8):5523–7.
44. Qian X, Wang Z, Sun W, et al. Long-rang BOTDA denoising with multi-threshold 2D discrete wavelet. In: Asia-Pacific optical conference. Optical Society of America; 2016.
45. Qian X, Wang Z, Wang S, et al. 157km BOTDA with pulse coding and image processing [C]. In: Six European workshop on optical fiber sensors. International Society for Optics and Photonics; 2016.
46. Kishida K, Li C-H. Pulse pre-pump-BOTDA technology for new generation of distributed strain measuring system. In: 2nd international conference on structural health monitoring of intelligent infrastructure; 2006.
47. Brown AW, Colpitts BG, Brown K. Dark-pulse Brillouin optical time-domain sensor with 20-mm spatial resolution. J Lightwave Technol. 2007;25(1):381–6.
48. Garus D, Gogolla T, Krebber K, et al. Distributed sensing technique based on Brillouin optical-fiber frequency-domain analysis. Opt Lett. 1996;21(17):1402–4.
49. Garus D, Gogolla T, Krebber K, et al. Brillouin optical-fiber frequency-domain analysis for distributed temperature and strain measurements. J Lightwave Technol. 1997;15(4):654–62.
50. Bernini R, Crocco L, Minardo A, et al. Frequency-domain approach to distributed fiber-optic Brillouin sensing. Opt Lett. 2002;27(5):288–90.
51. Bernini R, Minardo A, Zeni L. Stimulated Brillouin scattering frequency domain analysis in a single-mode optical fiber for distributed sensing. Opt Lett. 2004;29(17):1977–9.
52. Bernini R, Minardo A, Zeni L. Self-demodulated heterodyne frequency domain distributed Brillouin fiber sensor. IEEE Photonics Technol Lett. 2007;19(6):447–9.
53. Minardo A, Bernini R, Urbanczyk W, et al. Stimulated Brillouin scattering in highly birefringent microstructure fiber: experimental analysis. Opt Lett. 2008;33(20):2329–31.
54. Minardo A, Bernini R, Zeni L. Brillouin optical frequency-domain single ended distributed fiber sensor. IEEE Sens J. 2009;9(3):221–2.
55. Bernini R, Minardo A, Zeni L. Distributed sensing at centimeter-scale spatial resolution by BOFDA: measurements and signal processing. IEEE Photonics J. 2012;4(1):48–56.
56. Minardo A, Bernini R, Zeni L. Distributed temperature sensing in polymer optical fiber by BOFDA. IEEE Photonics Technol Lett. 2014;26(4):387–90.
57. Hotate K, He ZY. Synthesis of optical-coherence function and its applications in distributed and multiplexed optical sensing. J Lightwave Technol. 2006;24(7):2541–56.
58. Mizuno Y, He ZY, Hotate K. Measurement range enlargement in Brillouin optical correlation-domain reflectometry based on temporal gating scheme. Opt Expr. 2009;17(11):9040–6.
59. Mizuno Y, He ZY, Hotate K. Measurement range enlargement in Brillouin optical correlation-domain reflectometry based on double-modulation scheme. Opt Expr. 2010;18(6):5926–33.
60. Mizuno Y, Hayashi N, Fukuda H, et.al. Ultrahigh-speed distributed Brillouin reflectometry. Light Sci Appl 2016;5(12):e16184.
61. Lee H, Noda K, Mizuno Y, et al. Distributed temperature sensing based on slope-assisted Brillouin optical correlation-domain reflectometry with over 10 km measurement range. Eletron Lett. 2019;55(5):276–8.
62. Hotate K, Hasegawa T. Measurement of Brillouin gain spectrum distribution along an optical fiber using a correlation-based technique: proposal, experiment and simulation. IEICE Trans Electron. 2000;83(3):405–12.

63. Hotate K. Fiber distributed Brillouin sensing with optical correlation domain techniques. Opt Fiber Technol. 2013;19(6):700–19.
64. Hotate K, Arai H, Song KY. Range-enlargement of simplified Brillouin optical correlation domain analysis based on a temporal gating scheme. SICE J Control Measur Syst Integr. 2008;1(4):271–4.
65. Song KY, Hotate K. Enlargement of measurement range in a Brillouin optical correlation domain analysis system using double lock-in amplifiers and a single sideband modulator. IEEE Photonics Technol Lett. 2006;18(3):499–451.
66. Song KY, He Z, Hotate K. Effects of intensity modulation of light source on Brillouin optical correlation domain analysis. J Lightwave Technol. 2007;25(5):1238–46.
67. Song KY, He Z, Hotate K. Distributed strain measurement with millimeter-order spatial resolution based on Brillouin optical correlation domain analysis. Opt Lett. 2006;31(17):2526–8.
68. Song KY, Hotate K. Distributed fiber strain sensor with 1-kHz sampling rate based on Brillouin optical correlation domain analysis. IEEE Photonics Technol Lett. 2007;19(23):1928–30.
69. Cohen R, London Y, Antman Y, et al. Brillouin optical correlation domain analysis with 4 millimeter resolution based on amplified spontaneous emission. Opt Expr. 2014;22(10):12070–8.
70. Zadok A, Antman Y, Primerov N, et al. Random-access distributed fiber sensing. Laser Photonics Rev. 2012;6(5):L1–5.
71. Denisov A, Soto MA, Thévenaz L. Going beyond 1000000 resolved points in a Brillouin distributed fiber sensor: Theoretical analysis and experimental demonstration. Light Sci Appl. 2016;5: e16074.
72. London Y, Antman Y, Preter E, et al. Brillouin optical correlation domain analysis addressing 440 000 resolution points. J Lightwave Technol. 2016;34(19):4421–9.
73. López-Gil A, Martin-Lopez S, Gonzalez-Herraez M. Phase measuring time-gated BOCDA. Opt Lett. 2017;42(19):3924–7.
74. Elooz D, Antman Y, Levanon N, et al. High-resolution long-reach distributed Brillouin sensing based on combined time-domain and correlation-domain analysis. Opt Expr. 2014;22(6):6453–63.
75. Boyd RW. Nonlinear optics. 3rd ed. New York: Academic Press; 2007.
76. Agrawal GP. Nonlinear fiber optics. 4th ed. New York: Academic Press; 2007.
77. Zhang XP. Fully distributed optical fiber sensing technology. Beijing: Science Press; 2013.
78. Tu XB. Distributed fiber sensing technique employing vector Brillouin optical time domain analyzer. Changsha: National University of Defense Technology; 2015.
79. Sun Q. Long-range and high-resolution distributed fiber sensing system based on Brillouin optical time domain analyzer employing pulse coding technique. Changsha: National University of Defense Technology; 2016.
80. Hasi W, Liu ZW, Teng P, et al. Study on stimulated Brillouin scattering pulse waveform. Acta Phys Sin. 2007;56(2):878–82.
81. Williams KJ, Esman RD. Brillouin scattering: beyond threshold. Opt Fiber Commun IEEE 1996;227–228.
82. Wang RG. Study of Brillouin scattering mechanism in optical fiber and its application. Nanjing: Nanjing University; 2012.
83. Shen YC, Song MP, Zhang XM, et al. Analysis and measurement of stimulated Brillouin scattering threshold in single mode fiber. Chin J Lasers. 2005;32(4):497–500.
84. Zhang C, Yu WF, Li ZL, et al. Numerical study on scattering properties for the stimulated Brillouin scattering fiber. Acta Opt Sin. 2015;35(3):248–54.
85. Parker T, Farhadiroushan M, Handerek V, et al. The simultaneous measurement of strain and temperature distributions from Brillouin backscatter. Opt Tech Smart Struct Struct Monit. 1997;1:1–6.
86. Horiguchi T, Shimizu K, Kurashima T, et al. Development of a distributed sensing technique using Brillouin scattering. J Lightwave Technol. 1995;13(7):1296–302.
87. Zhou DP, Li WH, Chen L, et al. Distributed temperature and strain discrimination with stimulated Brillouin scattering and Rayleigh backscatter in an optical fiber. Sensors. 2013;13(2):1836–184.

88. Liu DM, Sun QZ. Distributed optical fiber sensing technology and its applications. Laser Optoelectron Prog. 2009;46(11):29–33.
89. Williams D, Bao XY, Chen L. Investigation of combined Brillouin gain and loss in a birefringent fiber with applications in sensing. Chin Opt Lett. 2014;12(12):1231011–7.
90. Kurashima T, Horiguchi T, Izumita H, et al. Brillouin optical fiber time domain reflectometry. IEICE Trans Commun 1993;E76-B(4):382–390.
91. Song MP, Li ZC, Qiu C. A 50 km distributed optical fiber sensor based on Brillouin optical time domain analyzer. Chin J Lasers. 2010;7(6):1426–9.
92. Mizuno Y, Zou WW, He ZY, et al. Proposal of Brillouin optical correlation-domain reflectometry (BOCDR). Opt Expr. 2008;16(16):12148–53.
93. Horak P, Loh WH. On the delayed self-heterodyne interferometric technique for determining the linewidth of fiber lasers. Opt Expr. 2006;14(9):3923–8.
94. Ma Z, Zhang MJ, Liu Y, et al. Incoherent Brillouin optical time-domain reflectometry with random state correlated Brillouin spectrum. IEEE Photonics J. 2015;7(4):6100407.
95. Govind P, Agrawal. Nonlinear Fiber Optics. Academic Press; 2019.
96. Sekine K, Hasegawa C, Kikuchi N, et al. A novel bias control technique for MZ modulator with monitoring power of backward light for advanced modulation formats. In: Optical fiber communication and the national fiber optic engineers conference; 2007.
97. Tang YF, Li HZ, Zhan WD, et al. Research of auto control about bias voltage of high speed and power electro-optic modulator. In: International conference on electronics and optoelectronics; 2011.
98. Zhang JZ, Zhang MT, Zhang MJ, et al. Chaotic Brillouin optical correlation domain analysis. Opt Lett. 2018;43(8):1722–5.
99. Zhang JZ, Feng CK, Zhang MJ, et al. Brillouin optical correlation domain analysis based on chaotic laser with suppressed time delay signature. Opt Expr. 2018;26(6):6962–72.
100. Zhang JZ, Wang YH, Zhang MJ, et al. Time-gated chaotic Brillouin optical correlation domain analysis. Opt Expr. 2018;26(13):17597–607.
101. Zhang MJ, Liu TG, Li P, et al. Generation of broadband chaotic laser using dual-wavelength optically injected Fabry-Pérot laser diode with optical feedback. IEEE Photonics Technol Lett. 2011;23(24):1872–4.
102. Wang YH, Zhang MJ, Zhang JZ, et al. Millimeter-level-spatial-resolution Brillouin optical correlation-domain analysis based on broadband chaotic laser. J Lightwave Technol. 2019;37(15):3706–12.
103. Zhang Q, Wang YH, Zhang MJ, et al. Distributed temperature measurement with millimeter-level high spatial resolution based on chaotic laser. Acta Phys Sin. 2019;68(10): 104208.
104. Kim YH, Lee K, Song KY. Brillouin optical correlation domain analysis with more than 1 million effective sensing points based on differential measurement. Opt Expr. 2015;23(26):33241–8.
105. Ryu G, Kim G, Song KY, et al. Brillouin optical correlation domain analysis enhanced by time-domain data processing for concurrent interrogation of multiple sensing points. J Lightwave Technol. 2017;35(24):5311–6.
106. Dong XY, Wang AB, Zhang JZ, et al. Combined attenuation and high-resolution fault measurements using chaos-OTDR. IEEE Photonics J. 2015;7(6):6804006.

Chapter 4
Brillouin Distributed Optical Fiber Sensing Based on Disordered Signals

Among distributed fiber optic sensors based on Brillouin scattering, Brillouin optical coherence domain systems are favored by domestic and international researchers because they provide high spatial resolution and relatively long measurement range. The spatial resolution and measurement range of conventional periodic sinusoidally modulated Brillouin optical coherence domain systems can be adjusted by varying the modulation amplitude and modulation frequency of the laser source to achieve random access sensing measurements by detecting the correlation peaks at specific locations in the fiber. However, since multiple periodic correlation peaks are generated, the precise measurement range of the system is limited to two adjacent correlation peaks.

In 2012, Zadok et al. used a high-speed pseudo-random bit sequence to phase encode the signal, extending the sensing distance to several kilometers [1]. In addition, they proposed to combine time domain and coherence domain analysis to provide high spatial resolution and long-range measurements by coherence domain Brillouin analysis, reduce acquisition time by interrogating multiple correlation peaks simultaneously, and phase modulate the source using Gray code to reduce noise caused by residual non-peak Brillouin interactions to improve the signal-to-noise ratio and increase the sensing distance [2]. In addition, to further improve the spatial resolution of the system, the Brillouin pump and probe light were modulated by broadband spontaneous radiation amplification noise in a fiber amplifier, resulting in a spatial resolution of 4 mm in a 5 cm long quartz single-mode fiber [3].

M. Zhang et al., *Novel Optical Fiber Sensing Technology and Systems*, Progress in Optical Science and Photonics 28, https://doi.org/10.1007/978-981-99-7149-7_4

4.1 Noise Modulation Based Brillouin Optical Coherence Domain Reflection Technique

The Brillouin optical coherent domain reflectometry (BOCDR) technique is based on the integration of continuous optical signals, using continuous pump light modulated by periodic frequency to interact with reference light to generate correlation peaks for sensing [4]. In the classical sinusoidally modulated BOCDR distributed fiber optic sensing technique, to increase the sensing distance of the system, the modulation frequency of the optical source must be increased, however, this will lead to the decrease of the spatial resolution of the system, therefore, the system has the problem that the sensing distance and spatial resolution cannot be simultaneously combined. At the same time, to obtain high spatial resolution, it is necessary to increase the modulation depth of the optical source, and when the modulation depth is too large, the Rayleigh scattering spectrum will overlap with the Brillouin scattering spectrum, which introduces a large amount of noise and reduces the signal-to-noise ratio of the system, and even causes the Brillouin frequency shift to be unobservable, so the spatial resolution of the system cannot be increased indefinitely due to the Brillouin frequency shift in the optical fiber.

4.1.1 *Sensing Mechanism*

Noise modulation based on Brillouin optical coherent domain reflectometry uses a noise signal instead of a periodic sinusoidal signal to modulate the frequency of the laser source. Due to the aperiodicity of the noise signal, no periodic correlation peaks are generated in the fiber to be measured, thus overcoming the problem that the sensing distance of conventional BOCDR is limited by the periodicity of the modulated signal, while taking into account the sensing distance and spatial resolution [5]. This method generates only one correlation peak in the fiber, and the optical path difference between the pump light and the reference light is zero, which is equivalent to the zero-correlation peak in a conventional sinusoidal modulated BOCDR system. By adding a variable optical delay line (PODG) to the reference optical path, the optical delay line can be adjusted to change the optical range of the reference light along the way, so that this single correlation peak can be scanned and positioned in the fiber to be measured, thus achieving distributed measurements.

The optical paths of the pump light and the reference light in the noise-modulated BOCDR system are shown in Fig. 4.1. The optical path of the pump light is $L_1 = 2z + l_1 + l_2 + l_3$, and the optical path of the reference light is $L_2 = l_4 + l_5 + l_6$, where l_1 and l_2 are the optical paths experienced by the pump light before and after passing through the optical circulator, respectively, l_3 is the optical path from the optical circulator to the initial position of the fiber to be measured, z is the initial position of the fiber to be measured to the position of the point to be measured, l_4 and l_6 are the optical paths experienced by the reference light before and after passing through the

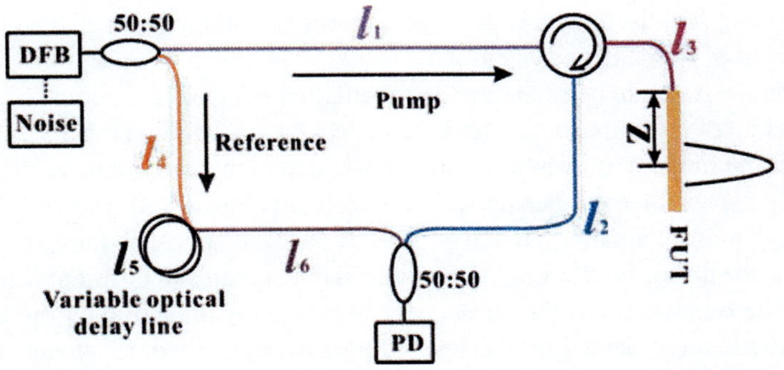

Fig. 4.1 Schematic diagram of the optical path of pump light and reference light in a noise-modulated BOCDR system

variable optical delay line, respectively, and l_5 is the optical path regulated on the variable optical delay line. l_4 and l_6 are the light paths of the reference light before and after passing through the variable optical delay line, respectively, and l_5 is the light path adjusted on the variable optical delay line. To make the pump light and the reference light in the second fiber coupler to achieve coherent beat frequency, the two optical paths L_1 and L_2 should be equal, when the l_5 in L_2 changes, to ensure that $L_1 = L_2$, only z can be changed in L_1. The position of the correlation peak is changed, and the continuous scanning of the correlation peak in the fiber to be measured can be achieved by continuous adjustment of the variable optical delay line.

When the noise signal is used to modulate the optical source, no periodic correlation peaks are generated in the fiber to be measured due to the acyclic nature of the noise signal, so the sensing distance is not limited by the periodic correlation peaks. Compared with the conventional BOCDR technique based on periodic sinusoidal modulation, the sensing range is significantly extended.

Noise-modulated distributed fiber optic sensing technology is based on the measurement of Brillouin frequency shift by coherent beat frequency of Stokes light and reference light generated by Brillouin scattering effect in the pump light in the fiber to be measured. Therefore, the spatial resolution of the system depends entirely on the coherence length of the Stokes light and the reference light, which is the coherence length of the optical source after modulation by the noise signal, or inversely proportional to the linewidth of the optical source after modulation by the noise. The spectral envelope of the noise-modulated laser is mostly in the form of a Gaussian-like feature, so the spatial resolution Δz is expressed as

$$\Delta z = L_c = \frac{c}{2n\Delta v} \tag{4.1}$$

where L_c is the coherence length of the optical source, n is the refractive index of the fiber, c is the speed of light in vacuum, and Δv is the linewidth of the optical source after noise modulation. By changing the linewidth of the optical source, the coherence length of the optical source can be tuned to achieve high spatial resolution measurement of the sensing system.

After analysis, it can be seen that the modulation depth of the optical source and the bandwidth of the noise source used both affect the linewidth of the optical source, the greater the modulation depth of the optical source, the wider the bandwidth, the greater the spread of the laser optical source linewidth caused, meaning that the coherence length is smaller. However, due to the limitations of the experimental conditions, the noise signal with good acyclic characteristics of different bandwidths could not be obtained, and the subsequent experiments only studied the effect of different modulation depths on the linewidth of the optical source under the same noise bandwidth, using a noise source bandwidth of 20 MHz.

4.1.2 Parameter Selection and Characteristic Analysis

The modulation depth of the laser optical source affects the linewidth of the detected optical source, and the Brillouin scattering threshold in the fiber is closely related to the linewidth of the optical source [6]. The variation of the Brillouin scattering threshold when the modulation intensity of the optical source is varied from −10 to −3 dBm is studied experimentally, as shown in Fig. 4.2. From Fig. 4.2a, it can be seen that the higher the modulation intensity of the optical source, the higher the Brillouin scattering threshold. Figure 4.2b shows the variation of the Brillouin scattering spectrum by changing the incident light power at the same modulation depth. With the increase of incident light power, the backward Stokes light intensity increases; after exceeding the Brillouin scattering threshold, the Stokes light intensity increases sharply.

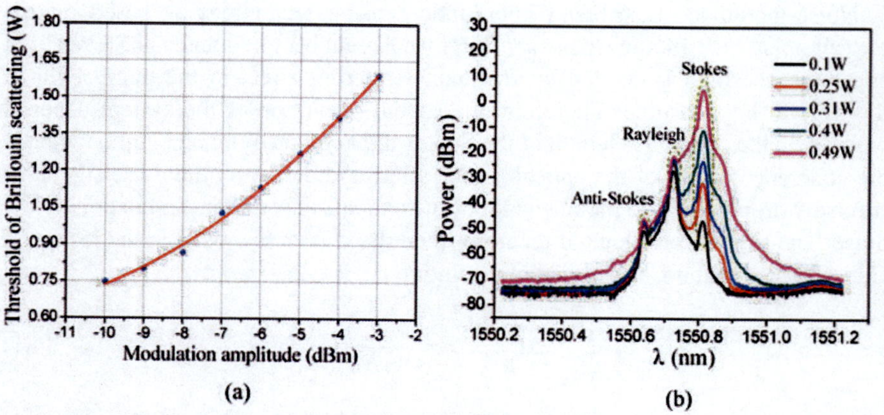

Fig. 4.2 Brillouin scattering characteristics. **a** variation of Brillouin scattering threshold with the modulation intensity of the optical source, **b** scattered light spectra at different incident light powers

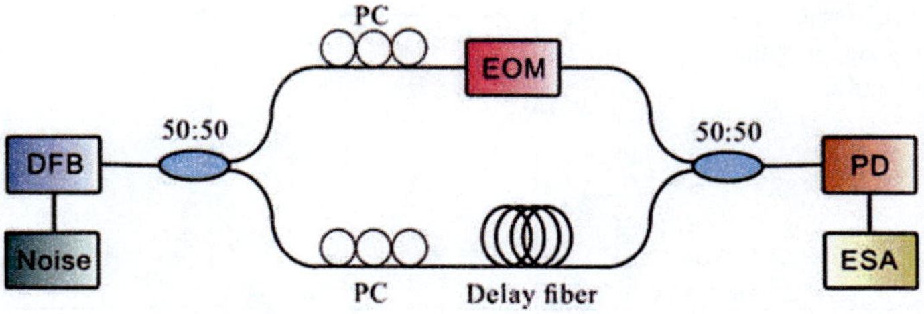

Fig. 4.3 Experimental setup for measuring the linewidth of optical sources by delayed self-difference method

The spatial resolution of the noise-modulated BOCDR system is equal to the coherence length of the optical source, which is inversely proportional to the linewidth of the optical source, therefore, to obtain the theoretical spatial resolution of the system, the linewidth of the optical source after modulation needs to be measured. After the laser source is modulated by noise, the linewidth of the optical source is broadened. Due to the accuracy limitation of the spectrometer, the broadened linewidth of the optical source cannot be measured directly by the spectrometer. Figure 4.3 shows the experimental setup of the delayed self-extrinsic method for measuring the linewidth of the optical source.

The laser signal modulated by the noise signal is divided into two ways by the first 50:50 optical coupler, one way through PC1 and modulated by EOM to produce a carrier-suppressed single-sideband optical signal, the other way through PC2 and then delayed by PODG (optical delay range of about 0–20km), the two signals are outlier tapped in another 50:50 optical coupler, The optical signal is converted into an electrical signal by the PD and received by the ESA for analysis, and the line width of the optical source is half the bandwidth of the beat spectrum signal. Figure 4.4 shows the self-heterodyne beat frequency signal when the modulation intensity of the optical source changes from −13 to −−2 dBm. As the modulation intensity of the optical source increases, the beat spectrum broadens, which means that the line width of the optical source increases. Table 4.1 shows the optical source line width change under different modulation intensity (Fig. 4.5).

4.1.3 Distributed Temperature Measurement

The system structure of the noise-modulated Brillouin optical coherence domain reflection system is shown in Fig. 4.6. A distributed feedback semiconductor laser with a central wavelength of 1550 nm is used as the laser source, and a 20 MHz arbitrary waveform generator is used as the noise signal source. The noise signal source is connected to the DFB laser through a coaxial cable with an impedance of 50 Ω. The modulated laser signal passes through a 50:50 fiber coupler and is divided into two channels, one for the pump light and the other for the reference

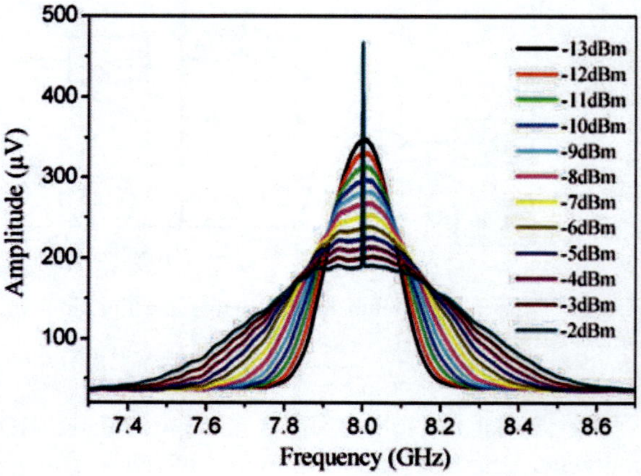

Fig. 4.4 Self-differential beat frequency signal with different modulation intensity

Table 4.1 Variation of line width of optical source under different modulation intensity

Modulation intensity of the optical source/dBm	Full width at half maximum of the beat spectrum/MHz	Line width of the modulated optical source/MHz
−13	166.681	83.841
−12	187.312	93.656
−11	212.008	106.004
−10	240.132	120.066
−9	271.483	135.742
−8	308.509	154.255
−7	353.793	176.897
−6	408.509	204.255
−5	472.379	236.190
−4	549.474	274.737
−3	645.326	322.663
−2	766.316	383.158

light. The pump light is amplified by a high power EDFA1 after passing through PC1, and the amplified pump light passes through the optical circulator into the FUT. In order to suppress the Fresnel reflection at the end of the fiber, the end of the fiber is immersed in refractive index matching fluid. The weak Stokes light is amplified by EDFA3 and then filtered by a tunable filter to remove the mixed Rayleigh scattered light and the spontaneous radiation noise of the amplifier, and then fed into another 50:50 fiber coupler. The reference light is amplified by EDFA2 in order to generate a strong coherent beat signal with the filtered Stokes light, and the amplified reference light is also fed into the 50:50 fiber coupler. The reference light and the Stokes light in the 50:50 fiber coupler are coherently tapped, and the tapped signal is converted into an electrical signal by the PD, which is received by

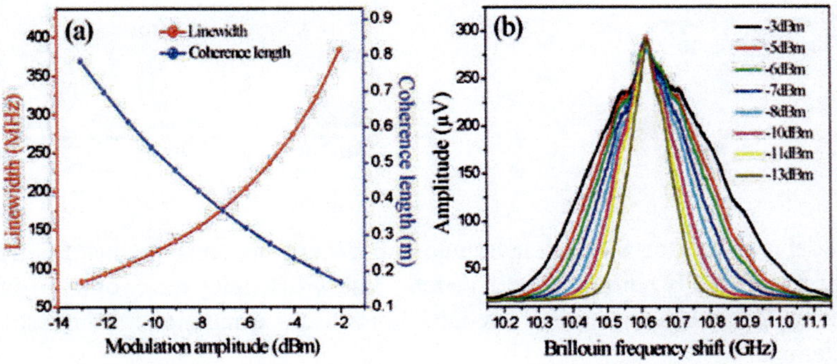

Fig. 4.5 Effect of different modulation intensities on optical source and Brillouin gain spectrum **a** variation of optical source linewidth and coherence length under different modulation intensities, **b** variation of Brillouin gain spectrum under different modulation intensities

a fast spectrometer with a bandwidth of 0~26.5 GHz and averaged to suppress the noise, and then the Brillouin gain spectrum is obtained. The Brillouin gain spectrum and the corresponding Brillouin frequency shift can be obtained at different locations by adjusting the variable optical delay line of the reference way to make the relevant peaks scan in the fiber to be measured, thus realizing the distributed measurement of fiber temperature.

The DFB laser used in the experiment is characterized by narrow spectral linewidth, stable wavelength, and tunability, which can meet the requirements of this sensing system perfectly. The arbitrary waveform generator with a bandwidth of 20 MHz can generate a period-free noise signal to ensure that there is only one correlation peak in the FUT in the sensing measurement, which enables accurate measurement of Brillouin frequency shift. A tunable filter with adjustable center wavelength and bandwidth (minimum bandwidth of 4 GHz) is used to effectively suppress Rayleigh scattered light mixed with Stokes light. The variable optical delay line used to adjust the optical range of the reference light consists of a programmable

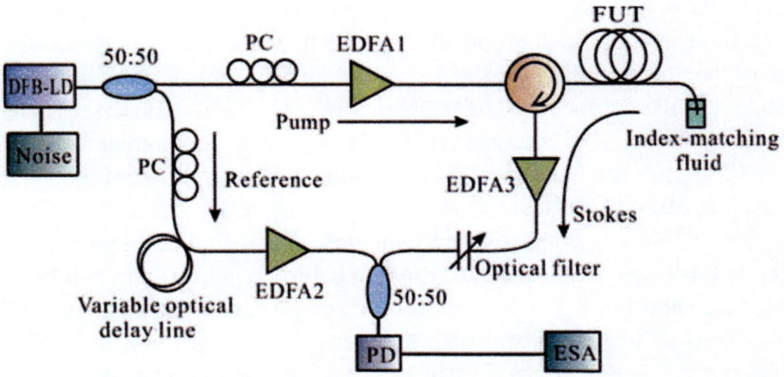

Fig. 4.6 Experimental setup of sensing system based on noise-modulated BOCDR

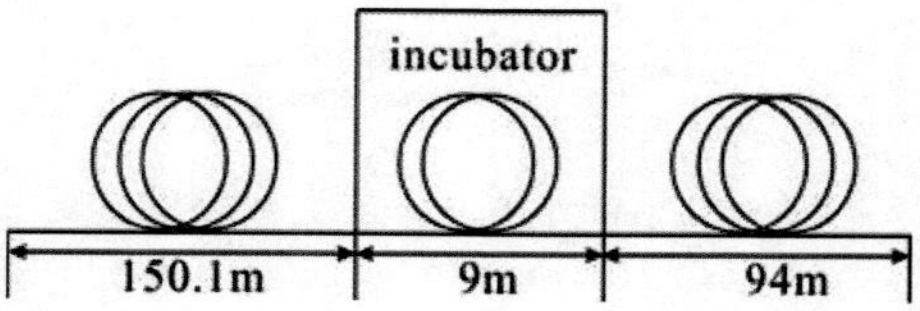

Fig. 4.7 Structure of the fiber in the experiment

optical delay generator with a delay range of 0–20 km and an accuracy of 5 m and a miniature electrically adjustable optical delay line with a delay range of 0–16 cm and an accuracy of 0.03 mm in series to achieve long-distance and high spatial resolution measurements.

Figure 4.7 shows the structure of the FUT, comprised of a standard single-mode fiber (SMF) with the length of about 253.1 m. In the experiment, the FUT was placed on a table in the laboratory where the temperature was 23.6 °C, and a 9-m section of the fiber was placed in the incubator. Figure 4.8 shows the timing, autocorrelation curves and frequency spectrum of the noise signal from the arbitrary waveform generator at 20 MHz. From the spectrum, the bandwidth of the noise signal is 20 MHz, and the timing and autocorrelation curves show that the noise signal has very good acyclic characteristics. The noise signal strength used in the experiment is −10 dBm, which means that the modulation strength of the optical source is −10 dBm.

Figure 4.9 shows the spectra of the signals obtained during the experiment. The curve (a) shows the backscattered spectrum at 26 dBm, which contains the anti-Stokes, Rayleigh and Stokes light. Curve (b) is the backscattered spectrum after amplification and filtering, in which only the Stokes light is included, and the Rayleigh scattered light and anti-Stokes light are well suppressed. Curve (c) is the reference spectrum after amplification, the center wavelength is equal to Rayleigh scattered light, and the intensity is comparable to that of Stokes light, so that the coherent beat frequency signal with high signal-to-noise ratio can be obtained. Curve (d) is the spectrum of the beat frequency signal of Stokes light and the reference light, and the intensity at the two peaks in the figure is equal, while the frequency difference is the Brillouin frequency shift.

The PODG of the reference light is adjusted so that the optical range of the reference light is equal to the optical range of the scattered light generated at a position of the 9m fiber placed in the thermostat, which is the position of the FUT to be measured in the thermostat. The temperature inside the thermostat was changed from 20 to 45 °C in 5 °C intervals for the temperature measurement experiment. Figure 4.10a shows the measured Brillouin gain spectra at different temperatures, from which it can be seen that the Brillouin gain spectrum shifts toward the high frequency direction as the temperature increases. The Brillouin frequency shifts at different temperatures were obtained from the Brillouin gain spectra and fitted, and the results are shown in Fig. 4.10b. From the figure, it can be seen that the Brillouin frequency shift increases linearly with the increase of temperature, and the slope of the fitted curve is 1.05 MHz/°C, which means that the experimentally measured temperature coefficient is 1.05 MHz/°C, which is consistent with the theoretical

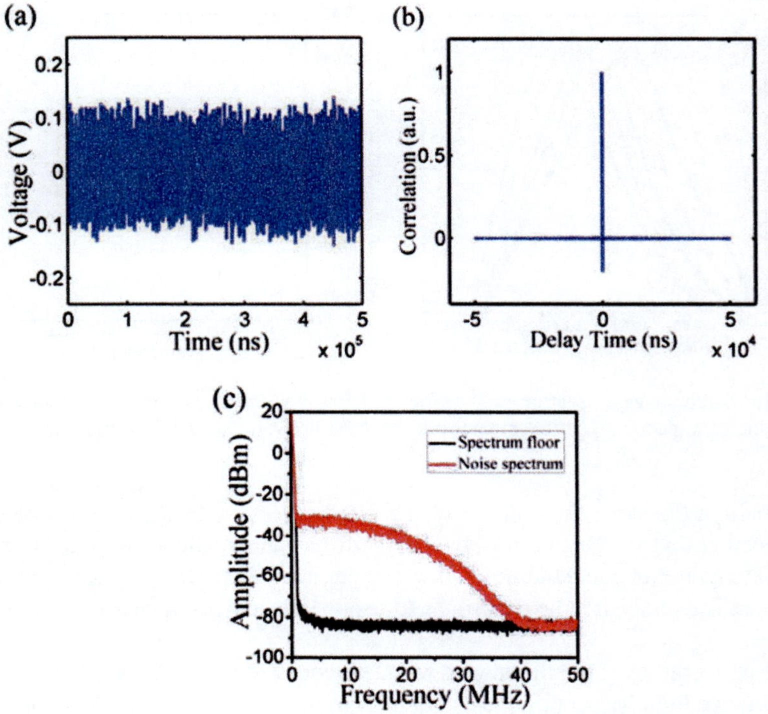

Fig. 4.8 Noise signal characteristics. **a** Timing; **b** Autocorrelation curve; **c** Frequency spectrum

Fig. 4.9 Spectrograms of different signals during the experiment based on noise-modulated BOTDR

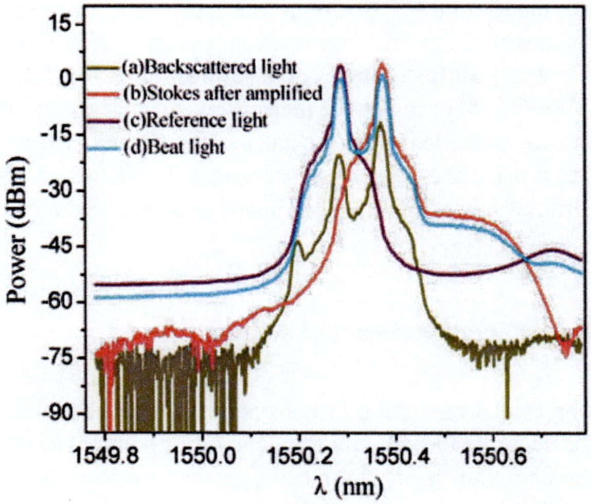

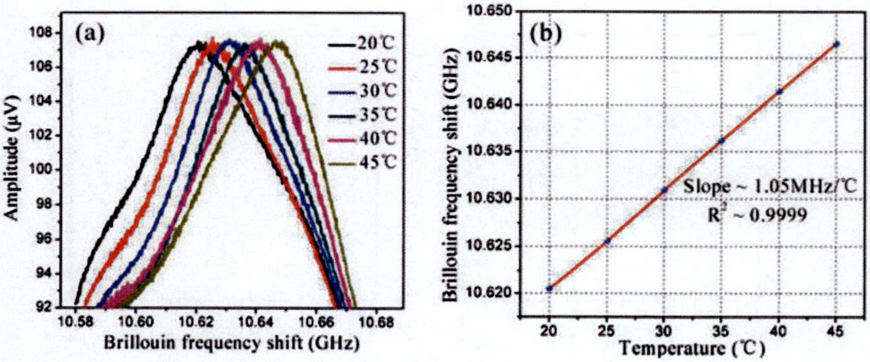

Fig. 4.10 Brillouin gain spectrum and Brillouin frequency shift with respect to temperature. **a** Brillouin gain spectrum at different temperatures, **b** Brillouin frequency shift versus temperature

value, and at the same time, the fit of the fitted curve is 0.9999, which is close to 1, indicating that the Brillouin frequency shift has an excellent linear relationship with the change of temperature, and the experimental results are consistent with the theoretical analysis. The experimental results are consistent with the theoretical analysis.

The temperature of the thermostat was kept constant at 50 °C, and the PODG of the reference light was continuously adjusted to scan the relevant peaks at different positions of the FUT, so as to obtain the Brillouin gain spectrum at different positions of the whole fiber as a distributed measurement of temperature. Figure 4.11a shows the distribution of Brillouin gain spectra at different positions of 253.1 m FUT, from which the Brillouin frequency shift at different positions is obtained, and the frequency shift variation curve shown in Fig. 4.11b is obtained. The frequency shift of the 9 m fiber inside the thermostat is significantly different from that of the rest of the fiber outside the thermostat, and the difference between the frequency shift inside and outside the thermostat is about 27 MHz, which corresponds to the temperature difference between the fiber inside and outside the thermostat (26.4 °C).

4.1.4 Performance Analysis

The spatial resolution of the system is defined as the fiber length corresponding to the decrease in Brillouin frequency shift from 90 to 10% of the peak [7]. The data of the fiber gain spectrum in and near the position in the thermostat in Fig. 4.11a were analyzed to derive the spatial resolution of the system, as shown in Fig. 4.12a. Also, the experimental analysis of the measurement resolution that can be obtained by the system at different modulation intensities was performed. Figure 4.12b shows the variation of Brillouin frequency shift near the thermostat obtained at the modulation

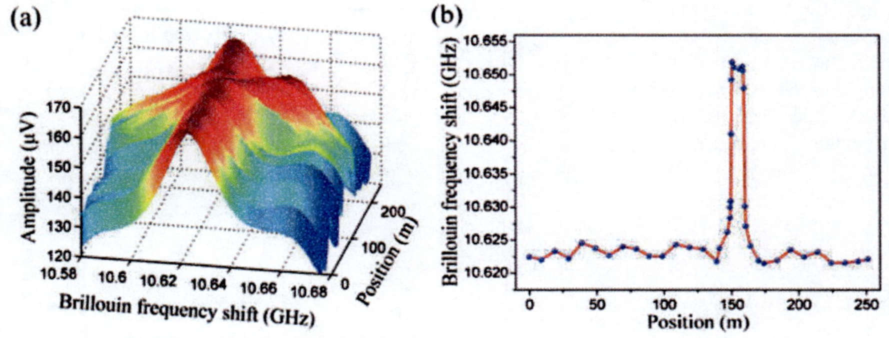

Fig. 4.11 Distribution of Brillouin gain spectrum and Brillouin frequency shift at different positions of the FUT **a** Brillouin gain spectrum distribution, **b** Brillouin frequency shift variation curve

intensity of −3 dBm. The spatial resolutions corresponding to Fig. 4.12a and b are 0.636 m and 0.19 m, respectively.

Figure 4.13 shows the comparison of the measured spatial resolution and the calculated coherence length under different modulation amplitude. From Fig. 4.13, we can see that the measured spatial resolution is almost the same as the coherence length of the modulated laser under different noise modulation amplitude. Therefore, it can be considered that this scheme can achieve relatively high measurement accuracy.

The temperature distribution over the length of 253.1 m fiber was successfully measured by modulating the semiconductor laser with a noise signal, and the experimental results were obtained with a temperature measurement accuracy of ± 1.3 °C and a minimum spatial resolution of 19 cm. The spatial resolution obtained in the experiment is in error with the theoretically calculated spatial resolution [8], which is mainly due to the following reasons: (i) the nonlinear effect in the fiber affects

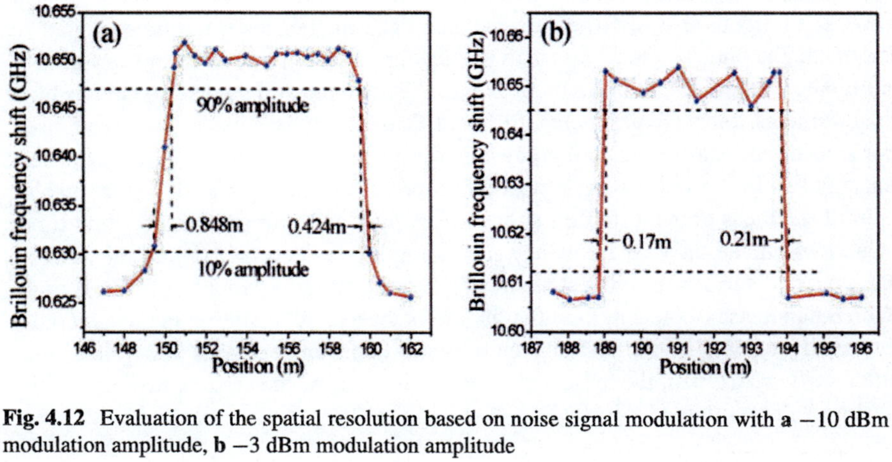

Fig. 4.12 Evaluation of the spatial resolution based on noise signal modulation with **a** −10 dBm modulation amplitude, **b** −3 dBm modulation amplitude

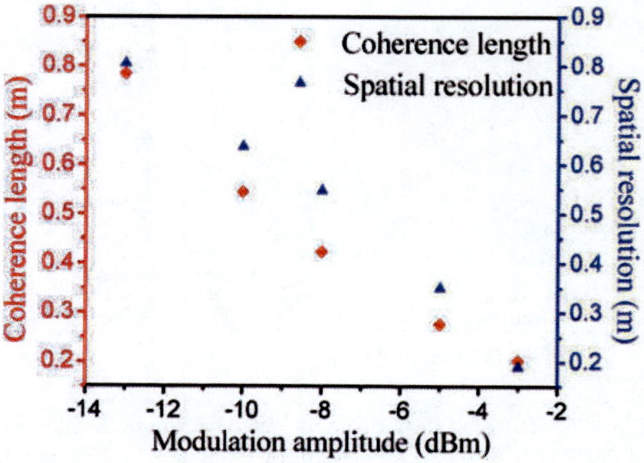

Fig. 4.13 The comparison of the measured spatial resolution and the coherence length of the modulated laser diode under different modulation amplitude

the coherence length of the pump light after the pump light enters the FUT; (ii) the Stokes light has to pass through a narrow-band filter for noise reduction before the coherent beat frequency with the reference light, which causes the linewidth of the Stokes light to change; (iii) the polarization correlation of Brillouin scattered light also affects its coherence length.

Compared with the traditional Brillouin light coherent domain reflection technology, which modulates the frequency of the optical source periodically, the BOCDR system, which uses the noise signal to modulate the frequency of the optical source, can theoretically solve the limitation of the periodic correlation peak on the measurement distance, and realize the long-distance, high-precision distributed fiber optic sensing measurement. At the same time, compared with other Brillouin scattering-based sensing systems, expensive pulse signal generators, high-frequency signal microwave sources, and high-performance electro-optical absorption modulators are unnecessary, and the device is simplified and the cost is reduced. The spectrometer is directly used as the data acquisition and processing device, saving the complicated data processing process, and the operation is simple, and the measurement time of the method is significantly reduced. The method uses the noise signal to modulate the optical source, and uses the acyclic characteristic of the noise signal to eliminate the periodic correlation peaks, so the disorder of the noise signal has high requirements. In the experiments, different Gaussian-type noise signals were used for sensing measurements, and the results were significantly different, with the best sensing results for noise signals with arbitrary waveforms. The high spatial resolution of the measurement implies short Stokes light and reference light action time, which will make the beat frequency signal amplitude decrease, so that the SNR of the system deteriorates, at the same time, when the system spatial resolution is high, with the increase of the length of the fiber to be measured, resulting in the generated Stokes light in the whole fiber scattered light proportion decreases, the Stokes light intensity is weakened, the SNR deteriorates; Eventually, the sensing distance of the system will be limited by its SNR.

4.2 Brillouin Optical Coherence Domain Reflectometry Based on Pseudo-Random Sequence Modulation

Zadok from Bar-Ilan University, Israel, proposed a BOCDA sensing system based on pseudo-random phase coding for Brillouin coherent domain analysis. However, the system needs to detect the signal double-ended incidence, the structure is complex, and when there is a break in the fiber under test (FUT), the system will not be able to achieve the measurement at all. In this section, we propose a random sequence modulation-based BOCDR system, which only requires pump light to be incident from a single end of the fiber making the system simple.

4.2.1 *Sensing Mechanism*

The principles of BOCDR based on random sequence modulation are the same as those of BOCDR based on the noise signal modulation, in that the Brillouin gain spectrum is obtained by coherently tapping the pump light with the reference light after frequency modulation, and then achieving the measurement of Brillouin frequency shift. The difference is that the modulation signal used to modulate the laser, the system uses a random sequence. The code length of the random sequence is adjusted so that the corresponding length is much longer than the length of the FUT, so that the random signal is also acyclic with respect to the length of the FUT, and only one correlation peak is generated in the fiber. The positioning of the system is similar to that of the noise-modulated sensing system in that the correlation peak is scanned in the fiber by adjusting the variable optical delay line of the reference light, thus achieving distributed sensing measurements. The sensing distance of BOCDR based on random sequence modulation is related to the code length and code rate of the random sequence, and the spatial resolution depends entirely on the coherence length of the optical source after modulation, which is inversely proportional to the line width of the optical source:

$$d_m = \frac{1}{2} M v_g T \tag{4.2}$$

$$\Delta z = \frac{v_g}{2\Delta v} \tag{4.3}$$

where M and T are the code length of the random sequence and the duration of a single random code, v_g is the speed of light in the fiber, and Δv is the linewidth of the optical source after modulation by the random sequence. The linewidth of the modulated optical source is related to the code rate $1/T$ and amplitude of the random sequence, and the increase of the code rate and signal amplitude will cause the linewidth of the optical source to increase, resulting in the reduction of the coherence length of the optical source and the achievement of high accuracy measurement. As shown in

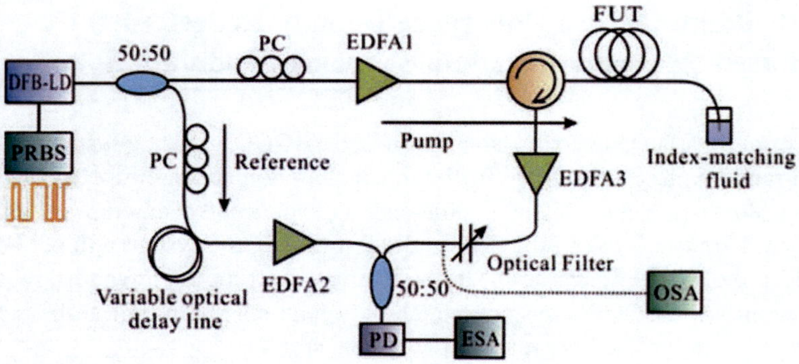

Fig. 4.14 Experimental setup of sensing system based on BOCDR with random sequence modulated laser optical source

Eq. (4.2), the sensing distance is also affected by the code rate $1/T$. Increasing the code rate leads to a decrease in the sensing distance while increasing the measurement accuracy, and the temperature range can be extended by increasing the code length. At present, the code length of pseudo-random code generator can be up to $2^{31}-1$, and the theoretical length of sensing fiber is 21977.4~439548 km when using code rate of 0.5~10Gbit/s.

The experimental setup of BOCDR sensing system based on random sequence modulated laser optical source is shown in Fig. 4.14. A distributed feedback semiconductor laser with a central wavelength of 1550 nm is used as the laser source, and a 10 Gbit/s random code generator (Centellax, TG2P1A) with a 13 GHz clock source (Centellax, TG1C1A) is used to generate the random sequence required for the experiment. The random code generator is connected to the DFB laser via a coaxial cable with an impedance of 50Ω. The modulated laser signal is divided into two paths through a 50:50 fiber coupler, one as the pump light and the other as the reference light, as shown in Fig. 4.14.

The pump light is amplified by the high power EDFA1 after passing through PC1, and the amplified pump light enters the fiber to be measured through the optical circulator. The backward Stokes light generated by the pump light in the fiber to be measured is output by the optical circulator. The weak Stokes light is amplified by the EDFA3 and then filtered by a tunable filter to remove the mixed Rayleigh scattered light and the spontaneous radiation noise of the amplifier, and then fed into another 50:50 fiber coupler. The reference light is amplified by EDFA2 in order to generate a strong coherent beat signal with the filtered Stokes light, and the amplified reference light is also fed into the 50:50 fiber coupler. The reference light and the Stokes light in the 50:50 fiber coupler occur in the coherent beat frequency, the resulting beat frequency signal by the photodetector into electrical signals, bandwidth 0~26.5 GHz fast spectrometer to receive this signal, and the average processing to suppress noise, and then obtain the Brillouin gain spectrum, measurement of the Brillouin gain spectrum of the central peak, you can get the Brillouin frequency shift. The Brillouin gain spectrum and the corresponding Brillouin frequency shift are obtained at different locations by adjusting the variable optical delay line of the reference light to scan the relevant peaks in the fiber to be measured, thus realizing the distributed measurement of fiber temperature.

The DFB laser used in the experiment also needs to have the characteristics of narrow linewidth, stable wavelength, and tunable to meet the experimental requirements. A total of 2^7–1, 2^{10}–1, 2^{15}–1, 2^{23}–1, 2^{31}–1 code lengths are available for the random code generator to meet the different sensing ranges required, the code rate can be varied from 0.5 to 10 Gbit/s, and the output signal peak-to-peak value is 300 mV. By adjusting the code rate and using amplifiers or attenuators in the electric domain to amplify or reduce the signal amplitude, The coherence length of the optical source can be tuned from meters to millimeters to meet different experimental requirements. A tunable filter (Yenistar, XTM50Ultrafine) with adjustable central wavelength and bandwidth of 4 GHz is used to suppress the Rayleigh scattered light mixed with Stokes light. The variable optical delay line for adjusting the optical range is used by a programmable optical delay generator with a delay range of 20 km and an accuracy of 5 m and a miniature motorized adjustable optical delay line with a delay range of 16 cm and an accuracy of 0.03 mm in series to achieve long-distance measurement of pump light and high-resolution measurement of the system.

4.2.2 Distributed Temperature Sensing

The experimental system was built according to the experimental scheme given in Fig. 4.14. A 250 m single-mode fiber was used as the sensing fiber. 1 m of the fiber was placed in a thermostat, and the rest of the fiber was placed on the experimental platform outside the thermostat, and the temperature of the experimental room was about 24 °C (Fig. 4.15).

The code rate, code length and peak value of the random sequence used in the experiment are 1.5 Gbit/s, 2^{15}–1, 300 mV, and the duty cycle is 1/2. According to the calculation of Eqs. (4.2) and (4.3), the length of the fiber that can be measured in the experiment is about 2 km, and the spatial resolution of the system is 58cm. Figure 4.16 shows the timing, spectrum and corresponding eye diagram of the random sequence, It can be seen from the figure that the random sequence has a good signal-to-noise ratio and the bandwidth of the signal is 1.5 Gbit/s.

The signal spectra measured in the experiment are shown in Fig. 4.17. The curve (a) is the backscattered spectrum at an incident light power of 25 dBm, which contains anti-Stokes, Rayleigh and Stokes light. Curve (b) shows the backscattered spectrum

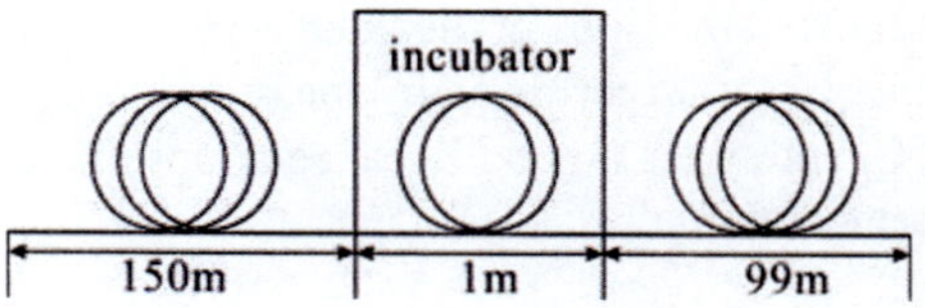

Fig. 4.15 Structure diagram of the sensing fiber of the system

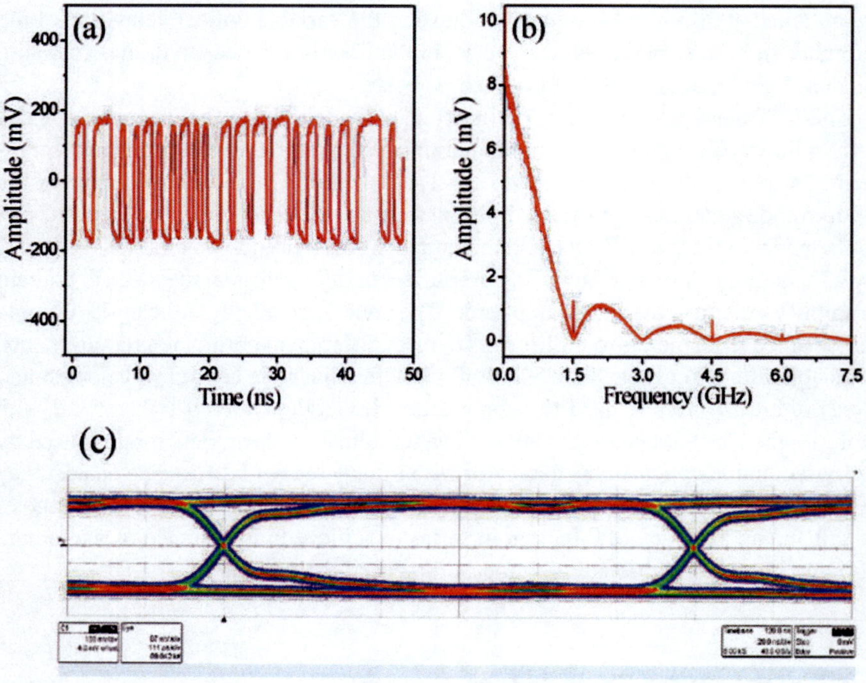

Fig. 4.16 Random sequence characteristics **a** timing; **b** spectrum; **c** eye diagram

after amplification and filtering, and it can be seen that the Rayleigh scattered light and anti-Stokes light are suppressed in this spectral component. Curve (c) shows the amplified reference spectrum with the same central wavelength as Rayleigh scattered light and the same intensity as Stokes light, in order to obtain a coherent beat signal with high signal-to-noise ratio. Curve (d) is the spectrum of Stokes light and the beat frequency signal of the reference light.

The variable optical delay line of the reference light was adjusted so that the relevant peak was located at a certain position of the 1m fiber in the thermostat. The Brillouin gain spectrum and Brillouin frequency shift were observed by changing the temperature in the thermostat from 21 to 51 °C in 10 °C intervals. The Brillouin gain spectra at different temperatures are shown in Fig. 4.18a, and the Brillouin gain spectra move toward the high frequency direction as the temperature increases, and the corresponding Brillouin frequency shift increases with the increase of temperature. The Brillouin frequency shift at different temperatures corresponding to Fig. 4.18a is fitted, and the fitted curve is shown in Fig. 4.18b, from which it can be seen that the slope of the fitted curve is 1.13 MHz/°C, and the degree of fit is 0.9916, that is, the experimentally measured temperature coefficient is close to 1.13 MHz/°C, and it has a good linear relationship between the Brillouin frequency shift and temperature.

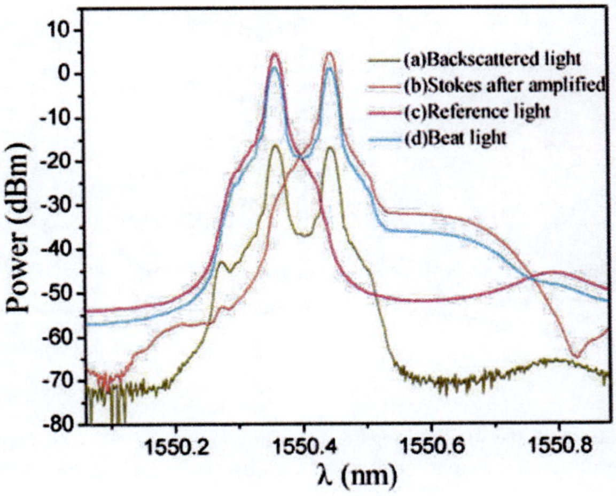

Fig. 4.17 Spectrograms of different signals during the experiment based on BOCDR with random sequence modulated laser optical source

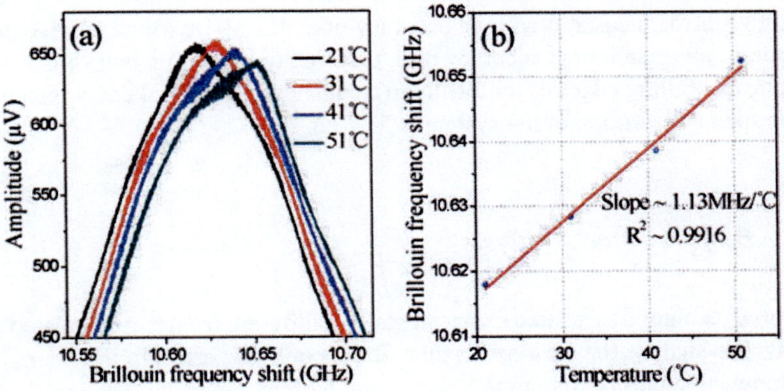

Fig. 4.18 Brillouin gain spectrum and Brillouin frequency shift versus temperature **a** Brillouin gain spectrum at different temperatures; **b** Brillouin frequency shift versus temperature curve

In order to verify the distributed measurement performance of the experimental system, the Brillouin gain spectrum at different locations of the FUT was observed by adjusting the variable optical delay line so that the correlation peaks were scanned in the FUT. For this purpose, the temperature of the thermostat was kept constant at 50 °C, and the variable delay line was continuously adjusted to obtain the Brillouin gain spectra at different positions of the fiber. Figure 4.19a shows the Brillouin gain spectra at different locations of the fiber. The deepest red part of the figure is the location of the 1 m fiber in the thermostat, which has a significant Brillouin gain spectrum shift compared with the red part of other areas. From the Brillouin gain spectrum at different positions in Fig. 4.19, the Brillouin frequency shift at different

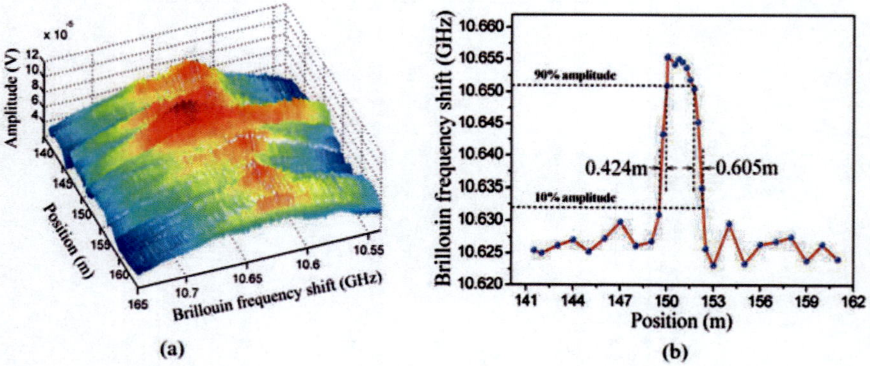

Fig. 4.19 Distribution of the Brillouin gain spectrum along the FUT and spatial resolution **a** Brillouin gain spectrum at different locations of the fiber; **b** Spatial resolution (0.514 m)

positions of the fiber can be obtained. The relationship between the fiber position and Brillouin frequency shift is shown in Fig. 4.19b, The difference between the Brillouin frequency shift inside and outside the thermostat is about 29 MHz. The Brillouin frequency shift is measured with an accuracy of ± 1.8 MHz, which corresponds to a temperature measurement accuracy of ± 0.96 °C. In the figure, half of the sum of the rising and falling edges of the Brillouin frequency shift is 51.4 cm, which means that the spatial resolution of the system is 51.4 cm.

4.2.3 Performance Analysis

The spatial resolution measured experimentally is different from that calculated theoretically. We analyze the reasons for this: ① the nonlinear effect in the fiber affects the coherence length of the pump light; ② the narrow-band filter used to filter the backward scattered light causes the change of the line width of the Stokes light, which affects the coherence length; ③ the polarization correlation of the Brillouin scattered light also has some influence on the coherence length; The modulation bandwidth of the laser used in the experiment is 2.5 GHz, and the code rate of the random sequence is 1.5 Gbit/s. It is known from Nyquist's theorem in the electric domain that the bandwidth of the signal after modulation will be much smaller than 1.5 GHz, which will definitely affect the measurement resolution of the system.

The BOCDR sensing technology using random sequence modulated optical source can solve the problem that the existing BOCDR system cannot take into account the sensing distance and spatial resolution, and the theoretical measurement distance can be up to 439,548 km with the highest spatial resolution of 1 cm. The system is expected to achieve real-time measurement in practical applications because of the significantly shortened measurement time without complicated data processing.

However, the system requires a high modulation bandwidth of the laser used. If an all-optical true random number generator can be used as the sensing optical source, the sensing distance and spatial resolution of the system will enable long-range and high-precision measurements. The authors' team has been engaged in the research of chaotic signals, and one of the most important applications of chaotic signals is the generation of random numbers. The true random signals are generated by using broadband chaotic signals with infinite code length and good random characteristics, and the random number prototype has been developed, but the generated random numbers are electrical signals, which cannot be used directly in the optical domain. The all-optical random number prototype has been successfully developed, and the optical random signal generated by this prototype can be used as the sensing source of the BOCDR system based on random sequence modulation, which can eliminate the electronic bottleneck problem of the current system and realize the long-distance and high-precision measurement.

Achieving high spatial resolution measurement also means that the coherence length between Stokes light and reference light is reduced, the coherence beat time is decreased, and the signal-to-noise ratio of the obtained signal is reduced. Moreover, the loss of signal at long distance transmission will be very large, so that the obtained Stokes light intensity is very low. These will become the limiting factor of the system performance, therefore, how to reduce the noise to obtain a high intensity signal becomes one of our main research problems in the future.

4.3 Brillouin Optical Coherence Domain Analysis Technique Based on Physical Random Code Modulation

The technique based on stimulated Brillouin scattering (SBS) allows for implementation of distributed fiber sensors with long-range and high-accuracy in temperature or strain measurement [9]. The distributed sensing technology has been found extensive applications in the fields of structural health monitoring, leakage detection along pipelines, aerospace engineering and geo-technical engineering.

The most extensive application of the SBS-based is Brillouin optical time-domain analysis (BOTDA) [10]. In the BOTDA system, a pump pulse and a counter-propagating continuous-wave (CW) light are injected into a fiber under test (FUT) to perceive the temperature/strain along the FUT by measuring temperature- and strain-dependent Brillouin gain spectrum (BGS). The typical sensing range of the BOTDA system can achieve up to 30–50 km [11]. However, it is very difficult to further extend the sensing distance of the BOTDA system due to the low signal-to-noise ratio (SNR) at the far end of the FUT. The increase of the pump peak power can prolong the sensing range, but the utilization of high pump pulse will lead to the broadening of the BGS resulted from the pump deletion and modulation instability [12]. Optical pulse coding has been utilized to extend the distance range of the BOTDA system where the total pump energy is increased by a factor equal to the pulse number in the pulse sequence without raising the peak power of it. For example, unipolar codes, such as Simplex [13], Colour Simplex [14], Time/frequency [15], Return-to-zero (RZ) [16] and Complementary [17] codes, have been employed to increase the sensing length. Besides, bipolar coding using Golay codes [18–20] has been proposed as a new breach in long distance sensing, showing an obvious advantage in the measuring range compared with unipolar codes.

Another significant SBS-based configuration is Brillouin optical correlation-domain analysis (BOCDA) [21], where both CW pump and probe waves with frequency modulation by sine waves are counter-propagated in a FUT to detect temperature/strain variation. The sine-based BOCDA suffers from a trade-off between the measurement range and the spatial resolution [22]. To solve this problem, optical pulse coding such as a binary pseudo random bit sequence (PRBS) [23–25] or perfect Golomb [26] codes has also been applied to high spatial-resolution BOCDA system. However, the pulse sequence regardless of PRBS or perfect Golomb codes has multiple periodic correlation peaks in the whole FUT, which gives rise to the ambiguous measurement of the sensing range. To avoid this ambiguity, on the one hand, the technique combining time-domain and correlation-domain analysis [2, 26] is employed in order that a large number of correlation peaks could be simultaneously interrogated along the fiber at the cost of complexity and high cost. On the other hand, a new pulse coding scheme allowing only one correlation peak in the FUT is urgently needed.

In this chapter, we propose a novel BOCDA system based on true random codes which are non-periodic. This ensures that there is only one correlation peak in the whole FUT, and the sensing distance will not be affected by multiple correlation peaks like other encoding methods. Since the system is optimized from the part of light source, its performance is improved without changing the rest part of this system. In this way, our system is less complex than others. The distributed sensing along a 1.1-km-long fiber is experimentally verified with a spatial resolution of 3.9 cm.

4.3.1 Sensing Mechanism

We propose a new technique for Brillouin optical coherence domain analysis based on physical random code modulation, and the high spatial resolution is demonstrated by theoretical simulations and experiments. The spatial resolution of 3.9 cm on a 1.1 km single-mode optical fiber was achieved experimentally by using a physical random code with 3 Gbit/s rate. The numerical simulation shows that a further higher spatial resolution of 1 cm can be achieved using physical random codes with higher code rates of 10 Gbit/s. In addition, the maximum uncertainty of the local Brillouin frequency shift is $\pm$ 1.1 MHz and the temperature coefficient is 1.15 MHz/°C.

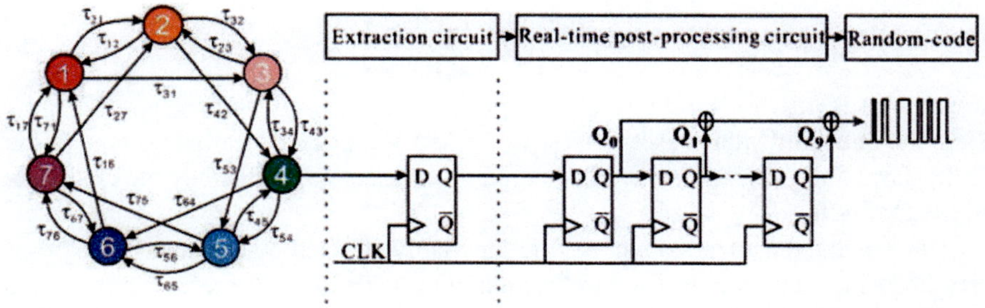

Fig. 4.20 Schematic diagram of true random code generation using the autonomous Boolean network oscillator

1. **High-speed physical random code generation**

In the proposed true random coding scheme for BOCDA, true random codes are generated by using autonomous Boolean network oscillator as the physical entropy source, shown in Fig. 4.20. The autonomous Boolean network oscillator consists of seven nodes with three inputs and outputs assembled in a bidirectional ring topology with nearest-neighbor coupling and self-feedback. Each node represents a logic gate. Point 1 performs exclusive XNOR operation, and the rest perform the inverse of the XOR operation. There is a delay time on each transmission link of the node, and the time it takes to propagate a signal from node j to node i is denoted by τ_{ij},$(i, j =$ 1,2,3,4,5,6,7). When the delay time in the network is set to be not exactly the same, the oscillator enters the random oscillation state, so-called Boolean chaos phenomenon [27]. After the generated chaotic signal is further sampled and quantized by the extraction circuit composed of a D flip-flop under the control of a clock (CLK), a real-time random code sequence is obtained. To optimize the randomness of the extracted random code sequence, a real-time post-processing circuit consisting of ten linear feedback shift registers (LFSRs) is needed. The whole random code generation system is implemented on a single field-programmable gate array (FPGA).

To better understand and observe the structure of autonomous Boolean network oscillators, the true random code sequence generation based on the autonomous Boolean network oscillator can also be numerically simulated by the theoretical model [28]. The evolution in time of the autonomous Boolean network oscillator can be described using the following Boolean delay equations:

$$x_n(t) = f_n[t, x_1(t - \tau_{n1}), x_2(t - \tau_{n2}), \cdots, x_n(t - \tau_{nn})] \tag{4.4}$$

Here each Boolean variable x_i depends on time t and the variable x_j at the previous moment. The Boolean variable xi is set as $x_i \in B = \{0, 1\}$, $(1 \leq i \leq n)$ and the function f_n represents the logical operators. τ_{ij} $(1 \leq i, j \leq n)$ represents the delay time it takes for the transmission from x_j to x_i.

The entropy extraction circuit (consisting of a clock-controlled D flip-flop) is used to sample and quantize the chaotic signal to generate a 0/1 binary sequence, and the sampling and quantization functions are defined as follows:

$$s_n(t) = \begin{cases} 0, & x_n(t) < C \\ 1, & x_n(t) > C \end{cases} \tag{4.5}$$

Here the threshold value is set as $C = V_{cc}/2$, and V_{cc} is the operating voltage of the D flip-flop. The generation rate of true random codes is determined by the clock rate of the D flip-flop.

The generated true random codes are inevitably affected by the metastable property of D flip-flop, which leads to the randomness decline. Therefore, a real-time post-processing circuit is added at the end of the random number generator to eliminate the residual bias in the random sequence. The characteristic polynomial of the real-time post-processing circuit consisting of ten LFSRs is depicted as $f(x) = x^9 + x^6 + x^5 + x^3 + x^2 + 1$. The formula of the Shannon entropy is given as follows:

$$H(X) = -\sum\nolimits_{x_n=\{0,1\}} p_X(x_n) \log_2 p_X(x_n) \tag{4.6}$$

where X and p_X (x_n) represent the random code sequence and the corresponding probability density, respectively. The Shannon entropy of the random code sequence increases from $H = 0.999501$ bit/sample to $H = 0.999999$ bit/sample via the real-time post-processing operation. This result shows that the residual bias is eliminated and the randomness is enhanced.

2. **Theoretical analysis of high spatial resolution**

The true random code sequences are generated in the optical domain by using the electrical true random codes extracted from the autonomous Boolean network oscillator to modulate the laser diode directly. The optical true random code sequences as the pump and probe waves are simultaneously injected into the sensing fiber from both ends, respectively. When the pump and probe waves satisfy the SBS phase-matching condition, which is, $\nu_{\text{pump}} - \nu_{\text{probe}} = \nu_{\text{B}}$ (ν_{B} being the Brillouin frequency shift (BFS)), the SBS interaction between pump and probe waves occurs at the meeting location. The generated beat signal will periodically modulate the refractive index of the sensing fiber to form a Brillouin dynamic grating (BDG) [29]. The magnitude of the corresponding acoustic wave is proportional to the temporal cross correlation between the complex envelopes of the pump and probe waves given as:

$$Q(t, z) = \frac{1}{2\tau_B} \int_0^t \exp\left(\frac{t^1 - t}{2\tau_B}\right) A_{pump}\left(t^1 - \frac{z}{v_g}\right) A_{probe} * \left[t^1 - \frac{z}{v_g} + \theta(z)\right] dt^1 \tag{4.7}$$

where A_{pump} and A_{probe} denote the complex envelope of the pump and probe waves, respectively. ν_g is the group velocity of light in the sensing fiber, and τ_{B} is the lifetime of the acoustic wave. The position-dependent temporal offset θ (z) is defined as θ (z) $= (2z - L)/\nu_g$, where L is the fiber length.

For $t \gg \tau_{\text{B}}$, the expectation value of the acoustic field magnitude at position z is

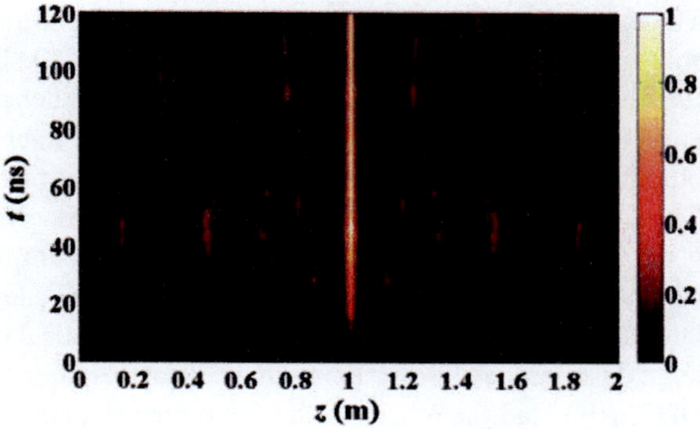

Fig. 4.21 The spatio-temporal two-dimensional distribution of the acoustic wave field *Q* (*t*, *z*)

$$\overline{Q(t,z)} = \frac{1}{2\tau_B}\int_0^t \exp\left(\frac{t^1 - t}{2\tau_B}\right)\overline{A_{pump}\left(t^1 - \frac{z}{v_g}\right)A_{probe} * \left[t^1 - \frac{z}{v_g} + \theta(z)\right]}dt^1$$
$$= c(\theta(z)) \tag{4.8}$$

where c (θ (z)) is the cross correlation function of the pump wave and the probe wave. In particular, when θ (z) = 0, c (θ (z)) is the auto-correlation function as the pump wave and the probe wave. At the center of the sensing fiber, the BDG is steadily and permanently generated within the correlation peak width, and the acoustic wave will be confined within the correlation peak width. Therefore, the spatial resolution of BOCDA based on true random codes is theoretically determined by the correlation peak width of the auto-correlation function of the true random code sequence. Actually, the correlation peak width is determined by the code duration of true random code sequences.

Figure 4.21 shows the spatio-temporal two-dimensional projection distribution of the acoustic field *Q* (*t*, *z*) generated by the optical true random code sequence with the rate of 10 Gbit/s as the pump and probe waves, respectively. Taking into account the duration of the simulation, 2-metersensing fiber is sufficient for our research. At the intermediate position of the sensing fiber, i.e., $z = L/2$, the acoustic wave is steadily and permanently generated with the correlation peak of approximately 1 cm, which corresponds to the theoretical spatial resolution of the proposed BOCDA system based on true random codes.

4.3.2 Distributed Temperature Measurement

The experimental setup of BOCDA system based on true random codes is illustrated in Fig. 4.22. The laser diode is directly modulated by high-speed true random codes extracted from the autonomous Boolean network. The optical true random codes with high extinction ratio are obtained through an electrical amplifier. The center wavelength and threshold current of the laser diode are 1550 nm and 20 mA, respectively.

The output of the laser source passes through the optical isolator (ISO1) and then is split into two branches by an optical coupler (10/90). The upper as the probe branch (90%) is modulated in a suppress-carrier, double-sideband format by the electro-optic amplitude modulator (EOM), which is driven by a microwave signal generator which outputs sine-wave with the modulation frequency that is near the BFS of the FUT. Then, the probe wave is launched into the FUT through a programmable optical delay generator (PODG), an erbium-doped optical fiber amplifier (EDFA1), a polarization scrambler (PS) and an isolator (ISO2). The PODG can be composed of the different length cascading fibers and two programmable optical delay lines, one of which has a large delay range of 0~20 km with the delay resolution of 30 cm (General Photonics ODG-101), and the other has the more precise delay range of 0~168 mm with the delay resolution of 0.3 μm (General Photonics MDL-002). The concrete process by controlling the PODG to locate the hot spot in the FUT is given as follows. First, by adjusting the length of the cascading fibers, the initial position of the correlation peak is located at the end connecting the OC of the FUT. Next, the scanning of the correlation peak is accomplished by cooperatively adjusting these two programmable optical delay lines. The probe wave is amplified to 10 dBm by the EDFA1 to meet the experimental requirements. The PS is placed after the EDFA1 to eliminate the effect of polarization on the SBS interaction of the BOCDA system when two branches of light meet at the center of the FUT. The lower as the pump branch (10%) is launched into the opposite end of the FUT via an optical circulator (OC) through a polarization controller (PC2) and an erbium-doped optical fiber amplifier (EDFA2). The pump wave is amplified to 28 dBm by the EDFA1 to meet the experimental requirements. After undergoing the SBS amplification by the pump wave along the FUT, the probe wave is filtered by the optical band-pass filter via the OC. The optical power of the filtered Stokes wave is recorded by a digital optical power meter (OPM) with an integrating sphere sensor. The range of frequency sweeping is set from 10.5 to 10.7 GHz, and the sweeping step is set as 1 MHz, and the scanning time of 1 MHz is 300 ms. These data are strictly corresponding to the acquisition speed of the OPM, which makes the measurement result more accurate. The structure of FUT is made up of a 1.1 km single-mode fiber (G.656) in which a 1 m fiber near 1.03 km is placed in a fiber thermostat.

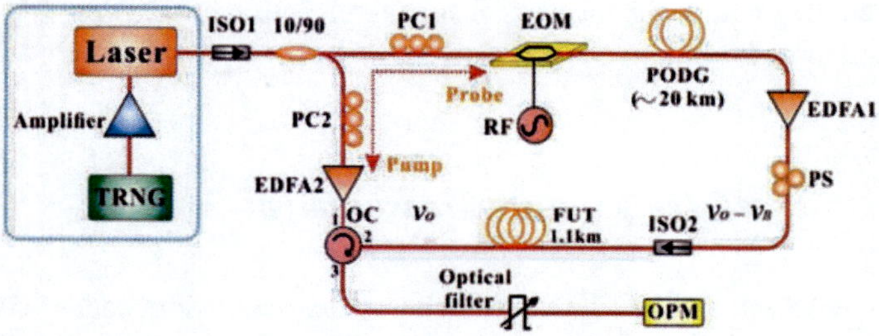

Fig. 4.22 Experimental setup of BOCDA system based on true random codes

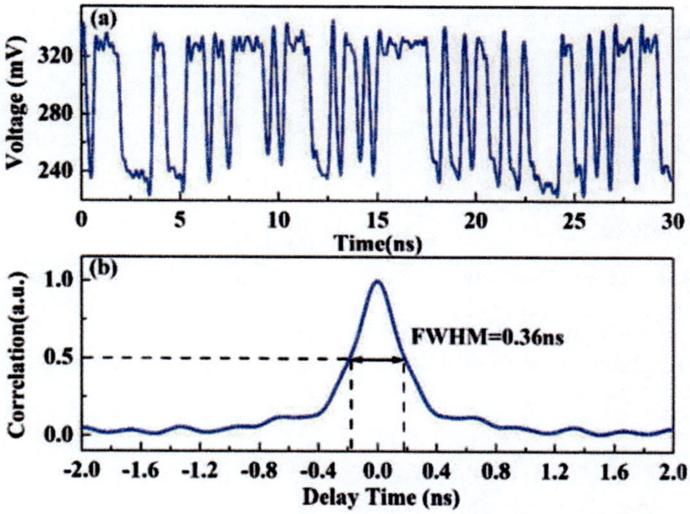

Fig. 4.23 True random codes from the autonomous Boolean network oscillator. **a** Time series; **b** Auto-correlation curve

In order to describe the characteristics of true random codes, the time sequence diagram and auto-correlation curve of true random codes are shown in Fig. 4.23. A high-speed random number generator based on autonomous Boolean network, is used to produce stable random number sequence, shown in Fig. 4.23a. The code rate of the random code generator we used is 3 Gbit/s, and the code width is 0.33 ns.

It can be seen from the time sequence of the oscillating signal that the time sequence signal of Boolean network changes from high level to low level, showing phase oscillation. The high and low levels of the generated random codes are about 330 mV and 240 mV, respectively. This is because that in practice, the autonomous Boolean network is realized by utilizing Field-programmable gate arrays (FPGAs). The generated signal is output by GX transceiver. The typical values of the high and low levels of the GX transceiver are about 320 mV and 200 mV, respectively. In this case, the 10 us time series of the true random code generated by the experiment are recorded to calculate the auto-correlation curve, shown in Fig. 4.23b. The full width of half maximum (FWHM) of the correlation peak of the auto-correlation curve is 0.36 ns, that corresponds to the spatial resolution which is measured at 3.6 cm.

The relationship of the BGS with temperature is illustrated in Fig. 4.24. Figure 4.24a shows the temperature dependence of the BGS in the FUT. First, the single correlation peak of the laser is located in the fiber thermostat by adjusting PODG in the probe path. Next, the temperature of the fiber thermostat is changed from 25 to 55 °C in succession. A phenomenon that the center frequency of the BGS moves from 10.622 to 10.656 GHz can be obviously observed. BGS line width is 30 MHz. From these BGS, we can get the temperature-dependence of BFS and the uncertainty of the estimated BFS, as shown in Fig. 4.24b. According to the fitting curve, the sensitive temperature coefficient of this system is 1.15 MHz/°C.

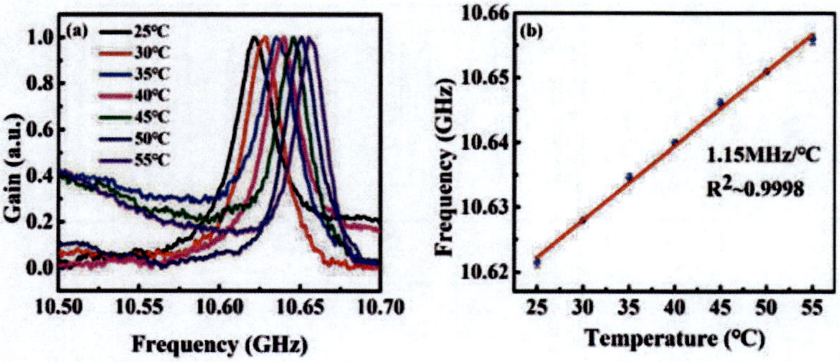

Fig. 4.24 The relationship of the BGS with temperature. **a** Temperature-dependence of the BGS along the FUT; **b** temperature coefficient of the BFS along the FUT

In addition, in measuring the Brillouin gain spectrum and Brillouin frequency shift with respect to temperature, temperature of the fiber thermostat is changed from 25 to 55°C in succession. 7 typical temperature points are selected, namely 25, 30, 35, 40, 45, 50 and 55 °C (an interval of 5 °C). At each temperature point, the BGS is obtained by averaging 30 repeating measurements. Supposed that the center frequency of each measured BGS, i. e., the BFS, is expressed as f_i ($i = 1, 2,..., 30$). The standard deviation δ_f of the BFS can be calculated according to the following formula:

$$\delta_f = \sqrt{\frac{\sum_{i=1}^{n} (f_i - \overline{f})}{n(n-1)}} \tag{4.9}$$

where $\overline{f} = \sum_{i=1}^{n} \frac{f_i}{n}$ is the average of the BFS, and n is the times of measurements, which is 30 in the experiment. Among these representative temperature points, the maximum standard deviation of the BFSs is considered as the measurement uncertainty of the BFS. Therefore, the maximum uncertainty value of the local BFS is $\pm$ 1.1 MHz with the temperature coefficient of 1.15 MHz/°C.

A three-dimensional plot of BGS measured along the FUT is shown in Fig. 4.25. The BFS at the hot spot section is clearly distinguished. In the experiments, the temperature of the hot-zone is set at 55 °C, and the room temperature is maintained constant at 25 °C. While the center frequency shift of BGS at room temperature is approximately 10.622 GHz, it's approximately 10.656 GHz at the thermostat. And there is about 34 MHz BFS at the hot spot section, which matches well with the real temperature variation of 30 °C.

Figure 4.26 describes the distribution of BFS measured along the FUT. The spatial resolution of the true random codes BOCDA system can be measured by the value of 10~90% of the rise and fall time equivalent length in meter for the hot spot section, where the rise and fall time equivalent length are 3.8 cm and 4 cm, respectively. Therefore, when the fiber in the thermostat box is 1 m long, the spatial resolution is

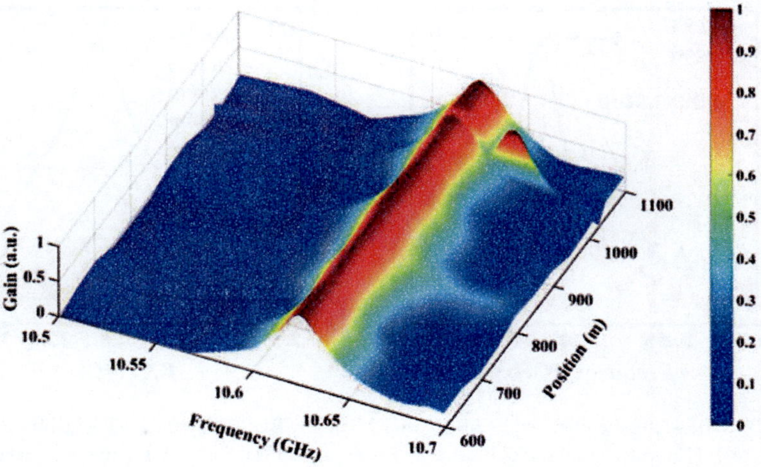

Fig. 4.25 Measured distribution of the BGS along the FUT

Fig. 4.26 Measured distributions of the BFS along the FUT

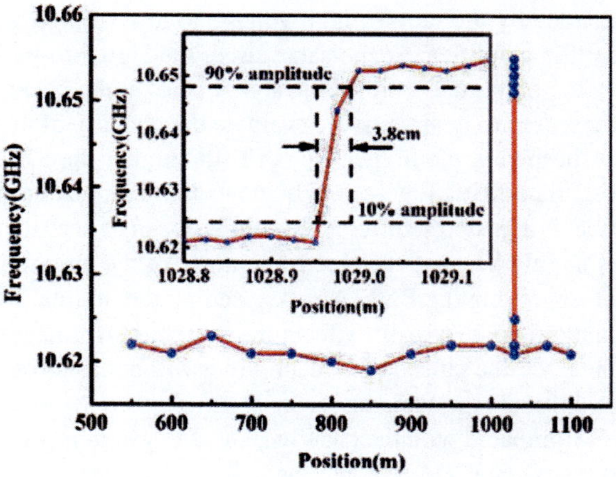

approximately 3.9 cm. In the experiment, we measured the BGS at about 50 positions of the sensing fiber and the spatial resolution obtained by the experiment is almost consistent with the above theoretical value.

4.3.3 Analysis of Influence Factors

Due to the limitation of experimental equipment, experiment currently reaches the transmission distance of 1.1 km and the spatial resolution of 3.9 cm. On the one hand, the transmission distance is mainly affected by the extinction ratio (EXT) which is

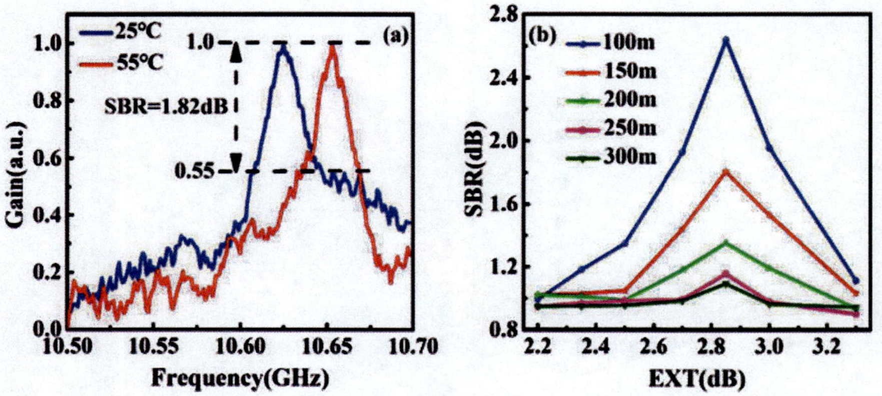

Fig. 4.27 BGS along the fiber and its SBR with EXT **a** The BGS along the FUT. The blue and red lines represent the correlation peaks located at 500 m of the FUT with the ambient temperature of 25 °C and at the middle of the 1-m heated fiber with the temperature of 55 °C, respectively; **b** the relationship curve between SBR and EXT at different locations

defined by the ratio of high voltage to low voltage of true random codes amplified by the amplifier. At the same time, the signal-to-background ratio (SBR) defined as the amplitude ratio of the signal peak to the background peak is experimentally measured to quantitatively evaluate the equality of the extracted BGS.

In theory, the higher the EXT, the higher the SBR. Figure 4.27a shows how to calculate SBR. Figure 4.27b shows the relationship between SBR and EXT when fiber is adjusted at different locations. It can be seen that there is an optimal EXT. This is mainly because the electronic noise of the amplifier is introduced in the process of improving the EXT. After exceeding the optimal EXT, the type distortion of true random codes greatly affects the system performance. In order to further improve the EXT, the pulsed optical amplifier without electronic noise is proposed to be used to optimize the system.

Compared with the chaotic BOCDA system researched by our group before [30], the new BOCDA system has many advantages. The auto-correlation trace of the chaotic signal has two kinds of residual side peaks except the main correlation peak. One is resulted by a weak amplitude auto-correlation of the chaotic signal occurring at the delay time of the external cavity, referred to as the time-delay signature (TDS) [31]. The other kind of side-lopes like the "coding noise" of PRBS exits in the entire trace. The SBR of single feedback chaotic laser source is low due to the existence of TDS and thick base noise. The random change of signal intensity also leads to the instability of SBR. The characteristics of the proposed true random code are non-periodic, and its signal intensity is stable. There is no TDS like chaotic lasers. The system has a higher SBR. The other kind of residual side peaks can show up, which will be the main limitation in our proposed scheme, although there is no TDS-induced off-peak amplification. Such residual off-peak SBS amplification contributes an additional noise mechanism, which can largely limit the performance of our proposed BOCDA system. To further suppress other residual off-peak amplification, the technique of the temporal gating will be employed in BOCDA system in the future work. From the experimental results, we get a better temperature coefficient and a longer sensing distance.

In conclusion, attributed to the better randomness and unpredictability of the true random codes, a high-spatial-resolution distributed fiber sensor for BOCDA based on true random codes has been proposed and achieved. The 1 cm high spatial resolution has been demonstrated by the numerical simulation with the 10Gbit/s rate true random codes. However, due to the limitation of experimental instruments, we achieved the spatial resolution of 3.9 cm over a 1.1 km single mode fiber by using true random codes with 3 Gbit/s rate. Besides, the maximum uncertainty of BGS is ± 1.1 MHz and the temperature coefficient is 1.15 MHz/°C. The research of BOCDA systems based on true random codes will promote the further study and application of true random code, and will also provide support for the research and development of the distributed optical fiber sensing system with high spatial resolution.

References

1. Zadok A, Antman Y, Primerov N, et al. Random-access distributed fiber sensing L J. Laser &. Photonics Rev 2012;6(5):L1–L5.
2. Elooz D, Antman Y, Levanon N, et al. High-resolution long-reach distributed Brillouin sensing based on combined time-domain and correlation-domain analysis. Opt Express. 2014;22(6):6453–63.
3. Cohen R, London Y, Antman Y, et al. Brillouin optical correlation domain analysis with 4 millimeter resolution based on amplified spontaneous emission. Opt Express. 2014;22(10):12070–8.
4. Hotate K, Zou W, Mizuno Y, et al. Proposal of Brillouin optical correlation-domain reflectometry (BOCDR). Opt Express. 2008;16(16):12148–53.
5. Zhang M, Bao X, Chai J, et al. Impact of Brillouin amplification on the spatial resolution of noise-correlated Brillouin optical reflectometry. Chin Opt Lett 2017;15(8):080603.
6. Chai J. Brillouin optical correlation-domain reflectometry based on non-periodical signal. Taiyuan: Taiyuan University of Technology; 2015.
7. Garus D, Geinitze, Krebber K, et al. Sensor fibres and a new concept for distributed temperature and strain sensing by Brillouin scattering. Opt Fiber Sens. 1996.
8. Hayashi N, Mizuno Y, Nakamura K. Suppression of ghost correlation peak in Brillouin optical correlation-domain reflectometry. Appl Phys Expr. 2014;7(11): 112501.
9. Zhou D, Dong Y, Wang B, et al. Single-shot BOTDA based on an optical chirp chain probe wave for distributed ultrafast measurement. Light Sci Appl. 2018;7:32.
10. Kurashima T, Horiguchi T, Tateda M. Distributed-temperature sensing using stimulated Brillouin scattering in optical silica fibers. Opt Lett. 1990;15(18):1038–41.
11. Diaz S, Mafang SF, Lopez-Amo M, et al. A high-performance optical time-domain Brillouin distributed fiber sensor. IEEE Sens. J. 2008;8(7):1268–72.
12. Minardo A, Bernini R, Zeni L, et al. A reconstruction technique for long-range stimulated Brillouin scattering distributed fibre-optic sensors: experimental results. Meas Sci Technol. 2005;16:900–8.
13. Soto MA, Bolognini G, Pasqualef D, et al. Simplex-coded BOTDA fiber sensor with 1 m spatial resolution over a 50 km range. Opt Lett. 2010;35(2):259–61.
14. Floch SL, Sauser F, Llera M, et al. Colour simplex coding for Brillouin distributed sensors. SPIE Fifth Eur Workshop Opt Fibre Sen. 2013.

15. Floch SL, Sauser F, Soto MA, et al. Time/frequency coding for Brillouin distributed sensors. 22nd Int Conf Opt Fiber Sens 2012.
16. Liang H, Li W, Linze N, et al. High-resolution DPP-BOTDA over 50 km LEAF using return-to-zero coded pulses. Opt Lett. 2010;35(10):1503–5.
17. Mao Y, Guo N, Yu KL, et al. L-cm-spatial-resolution Brillouin optical time-domain analysis based on bright pulse Brillouin gain and complementary code. IEEE Photonics J. 2012;4(6):2243–8.
18. Zan MSDB, Horiguchi T. A dual golay complementary pair of sequences for improving the performance of phase-shift pulse BOTDA fiber sensor. J Lightwave Technol. 2012;30(21):3338–56.
19. Soto MA, Floch SL, Thevenaz L. Bipolar optical pulse coding for performance enhancement in BOTDA sensors. Opt Expr. 2013;21(14):16390–7.
20. Yangz, Soto MA, Thevenazl. Increasing robustness of bipolar pulse coding in brillouin distributed fiber sensors. Opt Expr 2016;24(1):586–97.
21. Hotate K, Hasegawa T. Measurement of Brillouin gain spectrum distribution along an optical fiber using a correlation-based technique-proposal, experiment and simulation. IEICE Trans Electron, 2000.
22. Song KY, Hotate K. Enlargement of measurement range in a Brillouin optical correlation domain analysis system using double lock-in amplifiers and a single-sideband modulator. IEEE Photonics Technol Lett. 2006;18(3):499–501.
23. Denisov A, Soto MA, Thevenazl. Going beyond 1000000 resolved points in a Brillouin distributed fiber sensor: theoretical analysis and experimental demonstration. Light Sci Appl 2016;5:e16074.
24. Lopez-Gil A, Martin-Lopez S, Gonzalez-Herraez M. Phase-measuring time-gated BOCDA. Opt Lett. 2017;42(19):3924–7.
25. Shlomi O, Preter E, Ba D, et al. Double-pulse pair Brillouin optical correlation domain analysis. Opt Expr. 2016;24(23):26867–76.
26. London Y, Antman Y, Preter E, et al. Brillouin optical correlation domain analysis addressing 440000 resolution points. J Lightwave Technol. 2016;34(19):4421–9.
27. Rosin DP, Rontani D, Gauthier DJ. Ultra-fast physical generation of random numbers using hybrid boolean networks. Phys Rev E. 2013;87(4): 040902.
28. Song KY, Zou WW, He ZY, et al. All-optical dynamic grating generation based on Brillouin scattering in polarization maintaining fiber. Opt Lett. 2008;33(9):926–8.
29. Zhang JZ, Zhang MT, Zhang MJ, et al. Chaotic Brillouin optical correlation-domain analysis. Opt Lett. 2018;43(8):1722–5.
30. Zhang JZ, Feng CK, Zhang MJ, et al. Brillouin optical correlation domain analysis based on chaotic laser with suppressed time delay signature. Opt Expr. 2018;26(6):6962–72.
31. Ghil M, Zaliapin I, Coluzzi B. Boolean delay equations: a simple way of looking at complex systems. Physica D. 2008;237(23):2967–86.

Chapter 5
Chaotic Microwave Photon Sensing

Chaotic signals have important applications in communication systems and radar systems because of their good randomness and correlation characteristics. It is easy to generate chaotic lasers with bandwidth of 10 GHz by using semiconductor lasers. Due to its many outstanding characteristics, such as high speed, resist multipath fading, low power consumption, high penetration, high positioning accuracy and high security, Ultrawideband (UWB) technology has been widely used in short-range wireless communication, secure communication, radar detection, intelligent traffic positioning and other commercial, civil and military fields. Chaotic microwave photon sensing technology with chaotic laser as carrier is a new technology that integrates chaotic UWB technology and optical fiber transmission technology, and the short transmission distance of UWB technology is overcome. This technology can be well combined with optical radio technology to achieve remote signal transmission and control, and to compensate for the short coverage radius of UWB signals in applications.

5.1 Photogenerated Chaotic Ultrawideband Microwave Signals

5.1.1 Introduction

These outstanding characteristics of ultrawideband (UWB) technology have made it have many important and broad application prospects in many fields, such as high-speed wireless communication, wireless sensor network, through-wall radar, intelligent transportation and precision positioning [1]. For example, in the field of high-speed wireless communication, UWB systems can use an ultra-wide frequency band of gigahertz to achieve a transmission rate of 100 ~ 500 Mbit/s, which is

M. Zhang et al., *Novel Optical Fiber Sensing Technology and Systems*, Progress in Optical Science and Photonics 28, https://doi.org/10.1007/978-981-99-7149-7_5

much higher than existing communication systems such as 3G and Bluetooth [2]. UWB technology has been used for years in the military field, in 2002, the federal communications commission (FCC) removed the restrictions of UWB technology in the civilian field and allowed commercial applications of UWB technology, which became a milestone in the development of UWB technology [3].

Ultrawideband (UWB) signals are generally defined as signals with − 10 dB absolute bandwidth exceeding 500 MHz, or with a relative bandwidth beyond 20%, which is defined as 2 $(f_H - f_L)(f_H + f_L)$, where f_H and f_L are the high and low frequency cutoffs of the − 10 dB radiant point [4].

UWB signals are usually in impulse radio (IR) or multiband (MB) form. In addition, chaotic radio is also applied by researchers in the field of UWB technology researchers, as chaotic signals are wide-band and noise-like signals [5]. Chaotic-UWB is applied in the field of wireless communication, and becomes the physical layer candidate scheme of IEEE 802.15.4a standard. However, it is very difficult to generate high-frequency and broadband chaotic signals in the electrical field, which limits the scope of application of this communication mode [6].

IR-UWB, MB-UWB and noise-like Chaotic-UWB signals can only be effectively transmitted over a short distance from a few meters to tens of meters, due to the rapid attenuation of high-frequency signals during transmission and the limitation of UWB signal transmission power. In fact, UWB signal generation, coding, and control of UWB signals are often expected to be performed over long distances [7].

To promote the wide application of UWB technology and enable the local UWB system to interconnect with wired networks and other wireless networks, researchers have proposed and validated a new technology—UWB-over-fiber technology [10] since 2003, which integrates UWB technology [8] and fiber transmission technology [9]. In this technique, UWB signals are first generated and modulated at a central station and then transmitted to an indoor access point (AP) via optical fibers, where the UWB signal in the optical domain is converted to the electrical domain as the downlink signal to radiate to the free space, while current mature wireless communication technology (such as WLAN) can be adopted for uplink [11]. This hybrid optical UWB wireless communication technology not only uses the high-rate characteristics of UWB technology for the long-distance distribution of downlink signals, but also uses existing low-cost wireless communication technology as the uplink to transmit data. In addition to utilizing the huge bandwidth and low-loss transmission of optical fiber, UWB-over-fiber technology leverages the flexible and convenient advantages of UWB wireless communication. Finally, technology has attracted considerable interest in wireless communication because it meets people's needs for high-speed, high-capacity, long-distance and low-cost access [12].

The UWB microwave signals generated directly in the optical domain are the core problem of UWB over fiber technology. In this way, additional electrical-optical conversion can be avoided, and using optical methods to generate UWB signals has many other advantages, such as low loss, wide bandwidth, anti-electromagnetic interference, etc. [13].

At present, many institutions at home and abroad have carried out research on UWB signal generation in the optical field, and have made important progress. These methods fall roughly into four categories based on how they work (the working principle).

(1) The UWB pulse signal is generated based on the principle of photonics microwave delay filtering [14]. In this method, a negative coefficient is introduced by photonics microwave delay filtering to make the phase difference of the two pulses π. The last two pulses are combined in the detector to produce UWB pulse signal. Zhou Bingkun Group, Tsinghua National Laboratory for Information Science and Technology, Tsinghua University [15], Yao Jianping Group, Microwave Photonics Laboratory, University of Ottawa in Canada [3], Ning Tigang Group, Key Laboratory of All-Optical Networks and Modern Communication Networks, Ministry of Education, Beijing Jiaotong University [16], Perry Group, Research institude for Networks and Communication Engineering, Dublin City University in Ireland, based on Mach–Zehnder interferometer [17], Lin Jintong Group, Key Laboratory of Information Photonics and Optical Communication, Ministry of Education in China, Beijing University of Posts and Telecommunications, based on highly nonlinear photonic crystal fiber [6], Sun Junqiang Group, Wuhan National Laboratory for Optoelectronics College, Huazhong University of Science and Technology, used a periodically poled lithium niobate (PPLN) waveguide [18], Yariv Group, Department of Electronic Engineering, California Institute of Technology in USA, Zadok et al., College of Engineering, University of Bahilan in Israel, used high nonlinear fiber joint optical bandpass filter [19, 20], Bolea et al. ITEAM Research Institute, Universidad Politecnica de Valencia in Spain, generated photonic UWB pulse based on an Electro Absorption Modulator [21].
(2) The UWB pulse signal is generated based on the phase modulation-intensity modulation (PM-IM) conversion principle. Phase modulation to intensity modulation conversion can be achieved in the optical domain using scattering devices or optical discriminators. The groups of Shizhong Xie at Tsinghua National Laboratory for Information Science and Technology based on optical bandpass filter [22], Jianye Huang at the Photonics Systems Laboratory, University of Hong Kong based on highly nonlinear optical fibers [23], Yikai Su at the Advanced Optical Communication Systems and Networks, Shanghai Jiao Tong University using optical microring resonators [24], Xinliang Zhang at Wuhan National Laboratory for Optoelectronics, Huazhong University of Science and Technology based on semiconductor optical amplifiers and dense wavelength division multiplexers, Xinhuan Feng et al. at the Institute of Photonics Technology, Jinan University realized UWB photogenerated signals based on Bragg grating with polarization-maintaining fibers [7].
(3) The UWB pulse signal is generated based on the principle of frequency shaping and frequency time domain mapping. First, the spectrum is shaped into the corresponding pulse shape of single or double period by the filter in the optical domain, and then the mapping from frequency domain to time domain is realized

by using the dispersive element. LaRochelle's group at the Center for Optics, Photonics and Laser, Université deaval, has realized the optically-generated UWB signal using single-mode fiber combined with fiber Bragg grating [25].

(4) The UWB pulse signal is generated based on the nonlinear response characteristics of a semiconductor laser or semiconductor optical amplifier. Yu et al. at the Department of Photonics Engineering, Technical University of Denmark based on the relaxation oscillation effect of semiconductor laser response [26], Yonglin Yu at the Wuhan National Laboratory for Optoelectronics, Huazhong University of Science and Technology realized UWB optically generated signals based on nonlinear modulation response characteristics of semiconductor optical amplifiers [27].

The above research is mainly focused on generating UWB impulse pulse signals that satisfy the first derivative or higher derivative of Gaussian pulse of FCC indoor wireless communication spectrum radiating mask (FCC indoor mask), and its frequency range is limited to 3.1 ~ 10.6 GHz.

However, the rapidly developing environments such as wireless body-area networks, intelligent transportation system (22 ~ 29 GHz) and 60 GHz high band wireless personal–area networks require UWB signals with different frequency bands, different bandwidths and smooth power spectrum. Therefore, the FCC of the United States has specifically formulated the UWB radiation mask standard for different application objects.

Especially in the future, cognitive UWB (C-UWB) technology and UWB-over-fiber technology fusion more need to directly generate the center frequency can be dynamically adjusted flexibly, bandwidth can be controlled in the optical domain. It will be increasingly difficult to meet the needs of different environments in the future.

It can be seen that generating UWB signals with tunable center frequency, controllable spectrum bandwidth and conforming to FCC radiation template for different application layers in the optical domain will greatly expand the application field of UWB-over-fiber technology.

Recently, a series of optical generation schemes for tunable UWB signals have been proposed. In February 2010, Khan et al. of Purdue University proposed a proposed UWB arbitrary waveform generation technology based on photonics microcavity, achieving a wide range of tuning of UWB center frequency tuning. However, the expensive and complex control device reduces the stability of the system itself, and the spectrum noise is extremely large [28]. In January 2010, Peled et al. of Tel Aviv University, Israel, proposed a noise-like ultra-wideband signal generation method based on stimulated Brillouin scattering to amplify spontaneous radiation. The method has the advantages of adjustable signal center frequency and variable bandwidth, but it requires high-power laser signal input and a long distance (> 3.5 km) highly nonlinear fiber, and the signal quality needs to be further improved [29].

Under optical injection/optical feedback conditions, semiconductor lasers can generate chaotic lasers with random fluctuations in intensity. Using random noise characteristics of chaotic lasers with such intensity in the time domain, researchers have developed chaotic laser applications in the fields of time domain reflectometry measurement [30] and high-speed random number generation [31]. Chaos Laser and fiber sensing team found that Chaos laser has a natural wide spectrum characteristic of center frequency of 3 ~ 5 GHz and fractional bandwidth of 80 ~ 120%. Therefore, we are inspired to take advantage of the wideband characteristics of chaotic laser in the frequency domain to directly generate UWB signals in the optical domain and generate chaotic ultra-wideband microwave signals different from the traditional impulse pulse form [32], thus expanding the application of chaotic laser in the field of UWB technology.

5.1.2 Generating UWB Microwave Signals Using an Optically Injected Semiconductor Laser

This section describes the generation of tunable UWB microwave signals by externally injected chaotic semiconductor lasers. This scheme is based on the generation and control method of wideband chaotic laser with mixed optical feedback perturbation semiconductor laser by optical injection, and integrates chaotic wireless communication technology and microwave photonics technology to generate UWB microwave signals with tunable center frequency, controllable spectrum bandwidth and smooth power spectrum in the optical domain.

1. **System model and numerical simulation**

Figure 5.1 shows a system model for generating UWB pulse signals using an externally injected chaotic laser. The chaotic oscillation source is composed of both a mirror with adjustable reflection coefficient and a slave LD, and the master LD provides externally injected light. When the feedback intensity is appropriate, the slave LD will enter the chaotic state from the steady state, thus emitting the chaotic laser, and the spectral characteristics of the chaotic laser can be controlled by injecting light. Chaotic laser pulses are incident on an external modulator, so a non-return-to-zero code sequence is loaded into the external modulator to directly modulate the chaotic laser by on–off keying, thereby generating Chaotic-UWB pulses in nanosecond order. When the pulse duration T is satisfied T > 1/2ΔF (ΔF is the − 10 dB bandwidth of the original chaotic signal), the bandwidth of the Chaotic-UWB pulse signal will be the same as the ΔF of the original continuous chaotic signal.

Under the condition of optically injected external optically feedback LD, the theoretical model of the system can be described by a set of rate equations for electric complex amplitude E and carrier density N:

$$\frac{dE}{dt} = \frac{1+i\alpha}{2}\left[\frac{g(N-N_0}{1+\varepsilon|E|^2} - \tau_p^{-1}\right]E + \frac{k_f}{\tau_{in}}E(t-\tau)\exp(-i2\pi\upsilon_s\tau) + \frac{k_f}{\tau_{in}}E_j\exp(i\Delta\upsilon t), \tag{5.1}$$

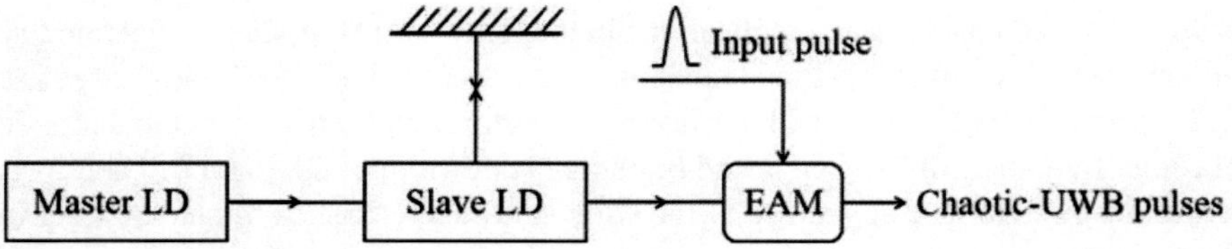

Fig. 5.1 Chaotic-UWB pulse generation system model by external optical injection chaotic laser. *LD* laser diode, *EAM* electroabsorption modulator

$$\frac{dN}{dt} = \frac{I_s}{qV} - \frac{N}{\tau_N} - \frac{g(N - N_0)}{1 + \varepsilon |E|^2} |E|^2 \tag{5.2}$$

where k_j and k_f denote the feedback and injection coefficient, $|E_j|$ is the optical field amplitude of the injection laser, $\Delta\nu = \nu_j - \nu_s$ is the optical frequency detuning between the master LD and the slave LD, α is the line width enhancement factor. The feedback delay $\tau = 218$ ns corresponds to the length of the optical cavity in the experiment, transparency carrier density $N_0 = 0.445 \times 10^6\ \mu m^{-3}$, threshold current small signal gain coefficient $g = 1.414 \times 10^{-3}\ \mu m^3/ns$, carrier lifetime $\tau_N = 2.5$ ns, photon lifetime $\tau_P = 1.17$ ps, round trip time in laser intracavity $\tau_{in} = 7.38$ ps, gain saturation parameter $\varepsilon = 5 \times 10^{-5}\ \mu m^3$, and active layer volume $V = 324\ \mu m^3$. In the simulation, a non-return zero sequence of 2 ns code length and 500 Mbit/s code rate is used to modulate chaotic laser to generate UWB chaotic laser pulse sequence. By changing the injection coefficient k_j, the bias current of the slave laser I_s, the optical frequency detuning $\Delta\nu$ and the linewidth enhancement factor α in the chaotic oscillation region of the laser, and analyzing the waveform, spectrum and frequency spectrum of the output pulse.

Figure 5.2 shows the frequency spectrum, time sequence and spectrum diagram of the generated UWB signal under parameters $\alpha = 5.0$, $\Delta\nu = 6$ GHz, $I_m = I_s = 3.88\ I_{th}$, $k_j = 0.08$, $k_f = 0.011$. In Fig. 5.2a, UWB pulse spectrum bandwidth reaches 6 GHz, and FCC-regulated indoor spectrum masking is well matched. The spectral characteristics of the chaotic pulse signal are effectively improved by adding external optical injection, which makes the spectrum flatter, improving the spectrum bandwidth and energy utilization rate in the FCC spectrum masking range. In Fig. 5.2c, the spectral component of the pulse signal is richer, which is the direct cause of UWB signal bandwidth enhancement.

We discussed the effects of parameters on UWB pulse generation by optically injected chaotic laser. The external and internal parameters are studied respectively.

(1) The influence of external parameters

In order to analyze the influence of parameters in the simulation model on the spectral characteristics of UWB signal and its internal relationship, three external parameters: optical injection intensity, injection detuning and bias current of the laser were adjusted when other parameters are constant. The simulation results are shown in Figs. 5.3, 5.4 and 5.5.

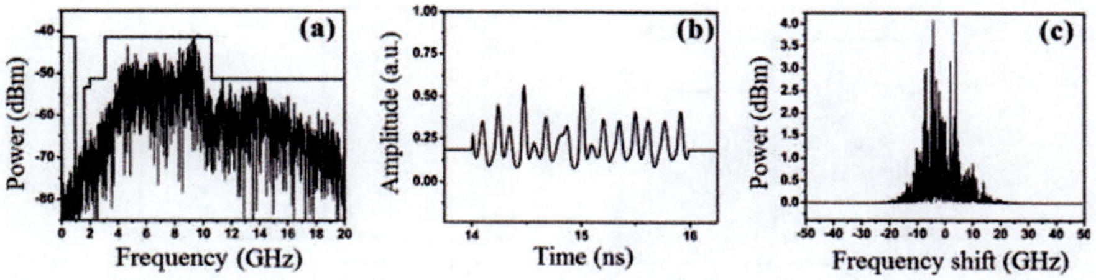

Fig. 5.2 UWB signal when $\alpha = 5.0$, $\Delta\nu = 6$ GHz, $I_m = I_s = 3.88\ I_{th}$, $k_j = 0.08$, $k_f = 0.011$. **a** RF spectra, **b** waveforms, **c** spectrogram

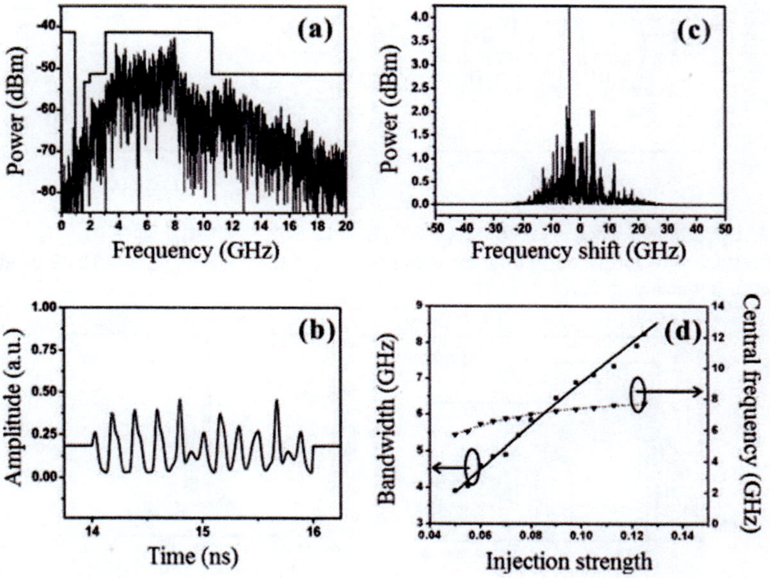

Fig. 5.3 UWB signal when $\alpha = 5.0$, $\Delta\nu = 6$ GHz, $I_m = I_s = 3.88\ I_{th}$, $k_j = 0.065$, $k_f = 0.011$. **a** RF spectra, **b** waveforms, **c** spectrogram, **d** Dependence of the UWB signal's $-$ 10 dB bandwidth and central frequency on k_j

To reduce the injection intensity to 0.065, set $\alpha = 5.0$, $\Delta\nu = 6$ GHz, $I_m = I_s = 3.88\ I_{th}$, $k_j = 0.065$, $k_f = 0.011$, Fig. 5.3a shows the simulation results of the pulse spectrum. At this time, the frequency spectrum bandwidth of the pulse signal is reduced to 4.8 GHz and centered at 6.55 GHz. Figure 5.3b and c show the spectrum and time series, respectively. In Fig. 5.3d, the relationship between UWB signal bandwidth and optical injection intensity shown by the solid line.

When the injection intensity increases from 0.05 to 0.12, the bandwidth changes from 3.9 to 8.2 GHz. When the intensity of the injected light increases, the beating between the master and slave LD will be intensified, so the amplitude of the difference frequency increases, which corresponds to the enhancement of the energy distribution

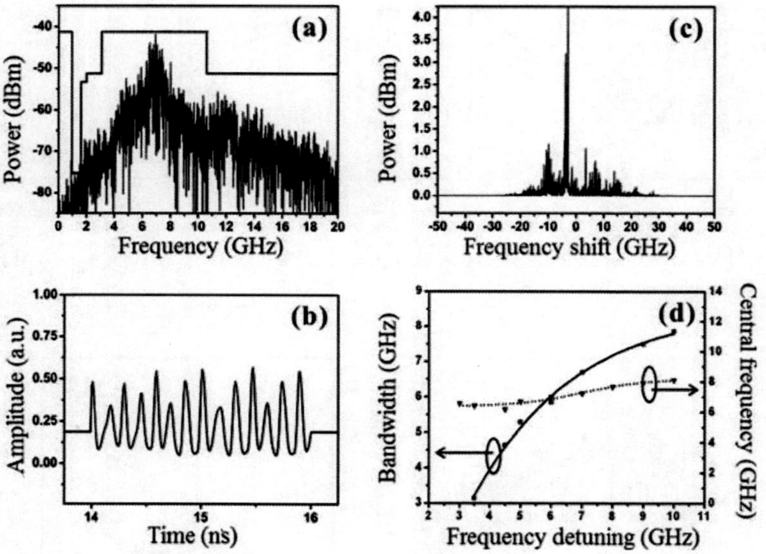

Fig. 5.4 UWB signal when $\alpha = 5.0$, $\Delta\nu = 3.5$ GHz, $I_m = I_s = 3.88\ I_{th}$, $k_j = 0.08$, $k_f = 0.011$. **a** RF spectra, **b** waveforms, **c** spectrogram, **d** Dependence of the UWB signal's − 10 dB bandwidth and central frequency on $\Delta\nu$

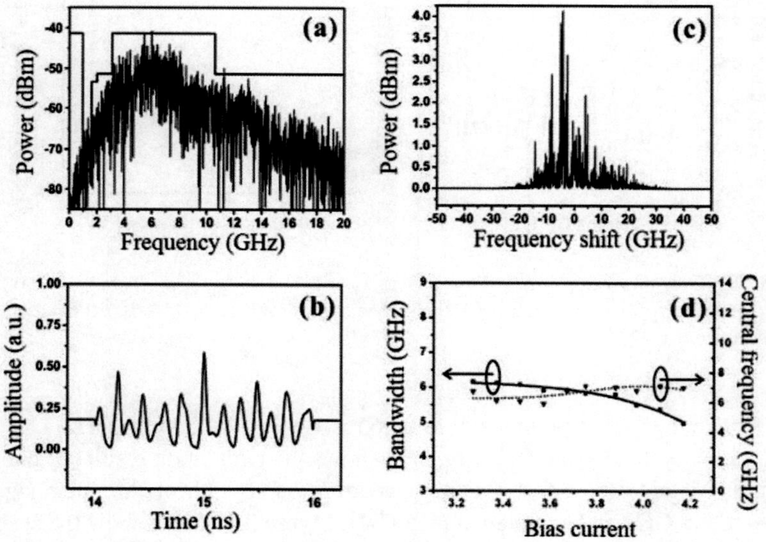

Fig. 5.5 The UWB signal when $\alpha = 5.0$, $\Delta\nu = 6$ GHz, $I_m = 3.88\ I_{th}$, $I_s = 4.07\ I_{th}$, $k_j = 0.08$, $k_f = 0.011$. **a** RF spectra, **b** waveforms, **c** spectrogram, **d** Dependence of the UWB signal's − 10 dB bandwidth and central frequency on I_s

near the difference frequency band in the frequency domain, thus broadening the spectrum of the UWB signal.

Figure 5.4a–c shows the simulation results of reducing the frequency detuning to 3.5 GHz when other parameters are constant. When the master and slave LD is reduced to 3 GHz in Fig. 5.4a, the frequency spectrum bandwidth of the pulse signal is reduced to 3.1 GHz and the center frequency to 6.4 GHz. When the output frequency of the master and slave laser is reduced, the spectral components of the chaotic laser are reduced, and the beating and mixing between spectral components is weakened, so the UWB pulse spectrum bandwidth is reduced. In Fig. 5.4d, the relationship between UWB signal bandwidth and frequency offset shown by the solid line. When the injection detuning increases from 3.5 to 10 GHz, the bandwidth changes from 3.1 to 7.9 GHz, and the frequency of periodic oscillations will increase. Intercoupling between high-frequency periodic oscillations and chaotic oscillations make the bandwidth wide.

Figure 5.5a shows the simulation results of UWB signal frequency spectrum by increasing the bias current of slave LD to 4.07 I_{th}. The frequency spectrum bandwidth is reduced to 5.4 GHz and center frequency at 7.1 GHz. Figure 5.5b and c shows the corresponding spectrogram and waveforms. When bias current of slave LD is increased from 3.27 I_{th} to 4.07 I_{th}, the bandwidth of the UWB signal is reduced from 6.16 to 4.98 GHz in Fig. 5.5d' solid line. This is because the energy distribution on the spectrum is formed by the beat frequency caused by the interaction between the intrinsic oscillation mode of the laser, the sideband mode caused by the relaxation oscillation and the external optical injection mode. The increase of bias current cause the relaxation oscillation frequency increasing, so the sideband mode of relaxation oscillation present a good match to the injection mode. At this time, the interaction between those two modes will narrow the bandwidth of the output signal.

Above analysis the − 10 dB bandwidth of UWB signal varies with the three external parameters. In addition, the external parameters will also affect the center frequency of UWB signal frequency spectrum. The dependence of the optical injection intensity, injection detuning and the bias current of the slave LD on center frequency are shown in the dashed lines in Figs. 5.3d, 5.4d and 5.5d. The change of external parameters affect the redistribution of UWB signal energy in the frequency domain or the enhancement of high frequency band energy (Figs. 5.3a and 5.4a), or the inhibition of low frequency band (Fig. 5.5a), so the center frequency changes slightly within 5 ~ 8 GHz.

(2) The influence of internal parameters

Among many internal parameters of LD, linewidth enhancement factor is an extremely important parameter, which represents the linewidth broadening and chirp characteristics of LD caused by carrier density fluctuations, as expressed in the following equations:

$$\alpha = -\frac{\partial x'/\partial N}{\partial x''/\partial N} \tag{5.3}$$

where x = x′ + ix″ denote the electrical polarizability, N is the carrier density. In fact, the linewidth enhancement factor essentially describes the coupling strength between gain and refractive index. The linewidth enhancement factor of LD with different material system and structure type is different. Linewidth enhancement factor lead the output of semiconductor laser show a variety of nonlinear dynamics phenomena, so it is necessary to figure out the influence of linewidth enhancement factor on the spectrum characteristics of chaotic UWB signals.

Figure 5.6a~c shows the frequency spectrum bandwidth is 5 GHz and center frequency is 6.6 GHz of the pulse signal, when the linewidth enhancement factor is reduced from 5.0 to 4.5. Figure 5.6d shows the dependence of the UWB signal's − 10 dB bandwidth and central frequency on α. When the linewidth enhancement factor increases from 2.0 to 7.5, the bandwidth of pulse signal changes from 4.1 to 7.5 GHz.

The reason why the pulse signal bandwidth will widen with the increase of the linewidth enhancement factor is that in the case of external light injection, the existence of the linewidth enhancement factor will lead to a large number of unstable dynamic characteristics. When α increases, the constant change of the main oscillation mode and sideband mode in spectrum, and the spectral components become richer. The strengthen intercoupling between these two modes make the output spectrum of chaotic laser broad. As shown in Fig. 5.6d's dotted line, the linewidth enhancement factor also effect the center frequency. The variation of parameter leads the redistribution of UWB signal energy in the frequency domain, resulting in

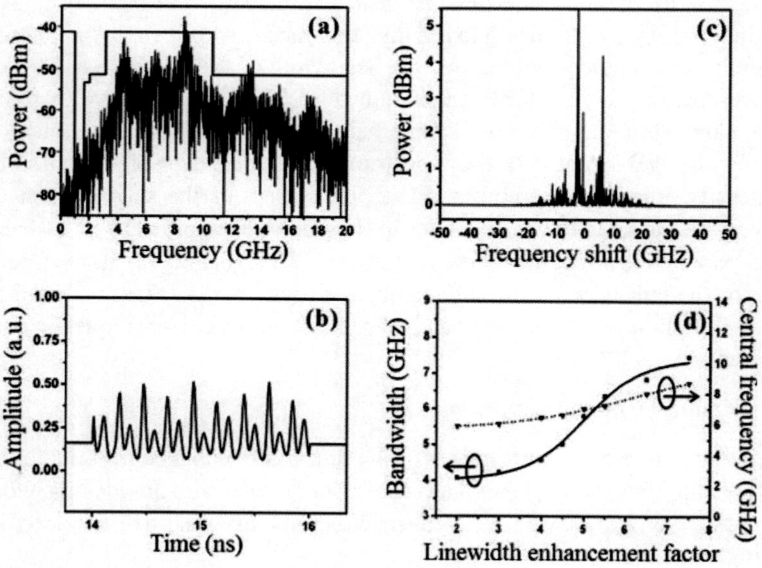

Fig. 5.6 The UWB signal when $\alpha = 4.5$, $\Delta\nu = 6$ GHz, $I_m = I_s = 3.88\ I_{th}$, $k_j = 0.08$, $k_f = 0.011$. **a** RF spectra, **b** waveforms, **c** spectrogram, **d** dependence of the UWB signal's − 10 dB bandwidth and central frequency on α

the enhancement of high frequency band energy and a small increase in the center frequency of UWB signal.

In addition, as shown in Fig. 5.7, the team studied the influence of fiber dispersion and loss on the spectrum of the generated chaotic UWB signal when it is transmitted in the fiber. In the simulation, setting optical fiber loss coefficient at 0.2 dB/km, dispersion coefficient at 17 ps/(km nm). Since the radiation power limit of UWB signal in FCC indoor communication spectrum template is mainly aimed at the transmission of UWB microwave signal in space, even if only consider optical fiber loss, to realize UWB-over-fiber, the power of UWB signals generated in the optical domain must higher than the power limit of the communication spectrum template in FCC. Figure 5.7a shows the generated UWB signal power spectrum before optical fiber transmission. The overall profile meets the requirements of FCC indoor communication spectrum template, but the power is higher than − 41.3 dBm. The power spectrum of the UWB signal after transmitted 1, 2.5, and 5 km through the optical fiber are in Fig. 5.7b~d. The chaotic UWB signal is less affected by optical fiber dispersion when transmitting in the optical fiber less than 5 km, and the spectrum profile is almost unchanged, only decay proportionally in amplitude.

The spectrum of UWB signals transmitted 15 km and 20 km in optical fibers respectively in Fig. 5.7e and f. The long-distance optical fiber dispersion lead the chaotic UWB pulse signal is broadened and distorted, and the spectrum of UWB signal is changed. In the low frequency interval less than 4 GHz, the energy distribution is homogenized and the spectrum profile becomes flat. In the vicinity of 8 and 12 GHz, the signal energy decreases and the spectrum profile becomes certain concave. In the high frequency part of 14 GHz, the signal energy increases and the spectrum profile has obvious protrusion.

Therefore, in chaotic optical borne UWB wireless system, the optical fiber dispersion will redistribute the spectrum energy of UWB signal when the optical fiber transmission larger than 15 km, resulting in the spectrum profile of UWB signal no longer strictly conforming to the power spectrum template of FCC indoor communication spectrum template. Therefore, it is necessary to introduce a dispersion management mechanism to compensate for the effect of optical fiber dispersion on UWB signal.

2. Experimental setup and results

Figure 5.8 shows the experimental setup for generating UWB microwave signals from external light injection chaotic laser. A commercial (DFB)-LD is used as a slave LD and a chaotic laser oscillator is constructed by an external fiber loop feedback system. Under appropriate feedback light intensity, the (DFB)-LD will enter chaos state from steady state. Another DFB-LD is employed as the master LD, injecting continous light into the slave LD through an optical fiber coupler to control the spectrum bandwidth and shape of the chaotic signal. The power and polarization state of the injection light are controlled by the VA2 and PC2, respectively. Maximum coupling efficiency is achieved when the polarization states direction of the injected light and the feedback light are the same as the slave LD. The output chaotic laser passes an optical isolator and injects into a 10 GHz bandwidth electrical absorption modulator (EAM). In the experiment, a nonreturn-to-zero codes with 2.5 ns length and 400 Mbit/

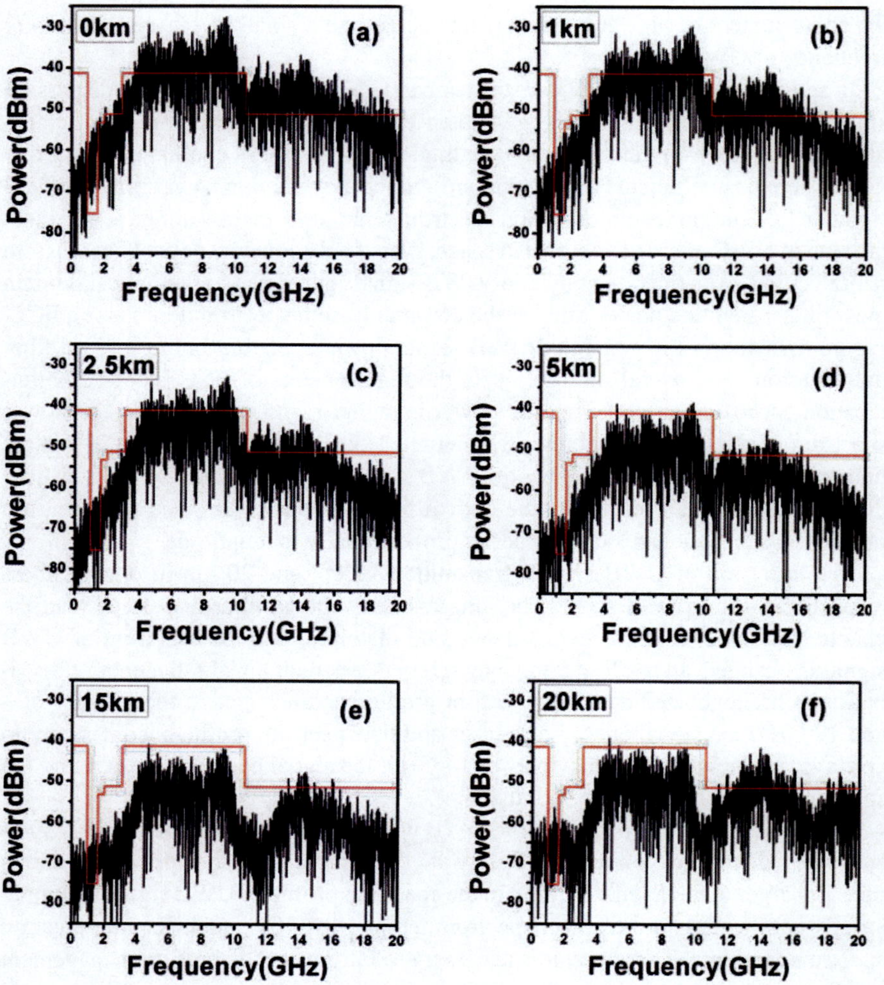

Fig. 5.7 Variation trend of RF spectra characteristics of optically generated chaotic UWB signal during transmission in optical fiber: **a** 0 km, **b** 1 km, **c** 2.5 km, **d** 5 km, **e** 15 km, **f** 20 km

s rate is used to modulate the chaotic light with a on–off keying to generate a series of chaotic laser pulse sequences on the order of nanoseconds. The generated UWB chaotic laser pulses were transformed into a chaotic ultra-wideband microwave pulses by a 50 GHz bandwidth ultra-fast photodetector (u^2t XPDV2020). Real-time high-speed oscilloscope (Lecroy 8600A) with 6 GHz bandwidth and spectrum analyzer with 26.5 GHz bandwidth (Agilent N9200A) were used to detect the generated UWB signals.

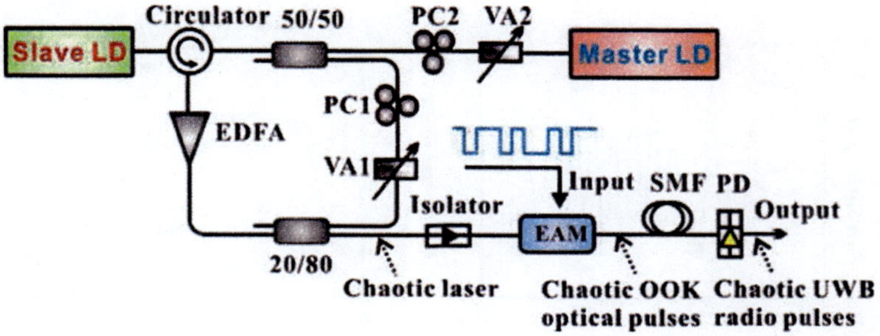

Fig. 5.8 Experimental setup for generating UWB microwave signals based on an optically injected chaotic semiconductor laser

In the experiments, the bias current of the slave laser is set to 1.5 times the threshold current, and its output wavelength was stabilized at 1553.457 nm by a precise temperature controller. The master LD wavelength could be adjusted by another temperature controller to adjust the optical frequency detuning between the master and slave LD. The length of the fiber ring feedback cavity was 43.6 m. The bandwidth of chaotic-UWB radio pulses will be the same as that of the original chaos when the pulse duration $T > 1/2\Delta F$ (ΔF is the − 10 dB bandwidth of the original chaotic signal). In the experiments, the − 10 dB bandwidth of the original chaotic signal is about 10 GHz, corresponding to the pulse duration limitation T about 0.05 ns, and the code length of the modulated signal is 2.5 ns, which is much longer than the minimum limit length of the chaotic pulse sequence. Therefore, the spectrum of chaotic laser can be conveniently adjusted by changing the injected power and the optical frequency detuning, thus, the spectrum of the generated UWB signal is controlled.

In the experiment, the injection power is adjusted to − 7.43 dBm, and the injection frequency detuning is 8.6 GHz. The chaotic UWB microwave signal that meets the indoor communication spectrum template specified by the Federal Communications Commission is obtained. The results are shown in Fig. 5.9. In Fig. 5.9a, the chaotic-UWB pulse sequence within the 2.5 ns duration comprises about ten pulses that have a FWHM of about 100 ps. In Fig. 5.9b, the measured spectrum (the blue line) is centered at 6.88 GHz, with a fractional bandwidth of about 116%, the power spectrum template of FCC indoor communication spectrum template is illustrated by the red line, and the gray line is the noise floor of spectrum analyzer. In this case, the resolution bandwidth (Res BW) of the spectrum analyzer is set to 3.0 MHz and video bandwidth (VBW) to 3.0 kHz. The spectrum of the generated chaotic-UWB pulses present a good match to the spectra of the FCC indoor mask and the power spectrum is smooth. It should be noted that the waveforms of each obtained UWB chaotic pulse train within 2.5 ns period are different (this is determined by the quasi-random characteristics of chaotic systems). The UWB chaotic pulse train in Fig. 5.9a is only a representative sample. The spectrum of UWB signal obtained in Fig. 5.9b is the average value of UWB chaotic pulse train in a period of time.

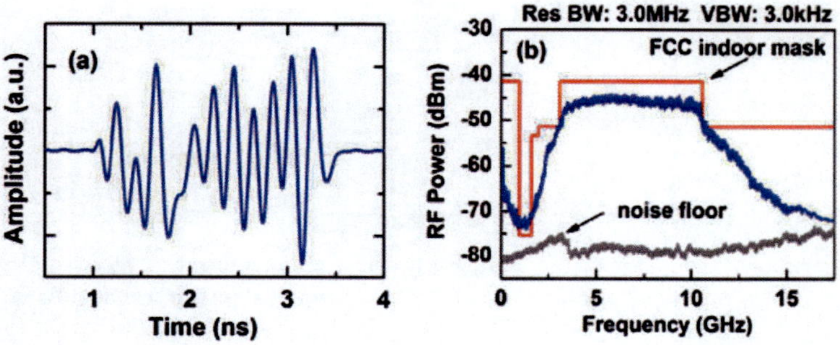

Fig. 5.9 The experimentally generated photonic UWB pulses. **a** Waveforms and **b** RF spectra

Figure 5.10 shows the spectrogram of the photogenerated UWB signal obtained in the experiment, the black line is the output spectrum of the slave laser during free oscillation, the green line is the output spectrum of the main laser, and the blue line is the output spectrum of the slave laser after external injection. The blue shift occurs when the slave laser moves from the long wavelength to the short wavelength direction, and the output frequency detuning is measured to be 8.6 GHz. The energy of the oscillation is transferred and redistributed towards the high frequency since the injected light and the beat effect from the slave laser wavelength occur.

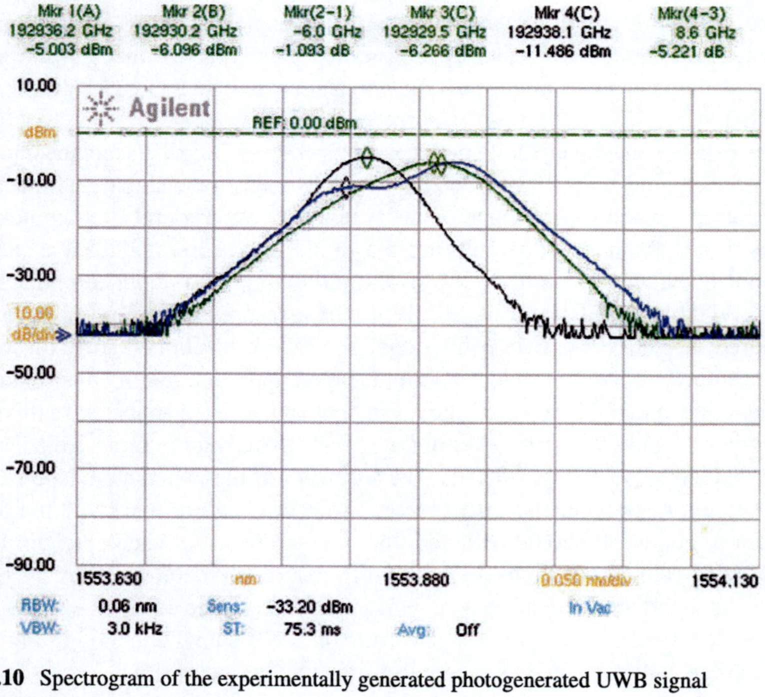

Fig. 5.10 Spectrogram of the experimentally generated photogenerated UWB signal

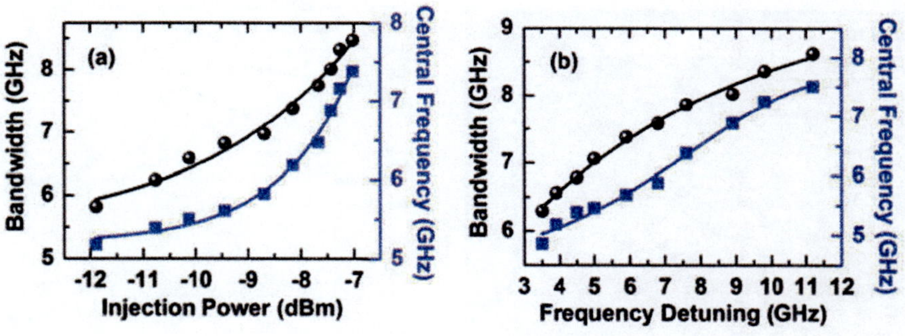

Fig. 5.11 Dependence of the UWB signal's bandwidth and central frequency on injection power (**a**) and optical frequency detuning (**b**)

Figure 5.11 shows the influence of injection power and injection frequency detuning on the center frequency and bandwidth of UWB signal. As shown in Fig. 5.11a, the central frequency increase from 5.25 to 6.88 GHz and bandwidth increase from 5.86 to 8.02 GHz with the injection power increasing from − 11.89 to − 7.43 dBm at 8.9 GHz frequency detuning. As shown in Fig. 5.11b, While the injection power is fixed on − 7.43 dBm, the central frequency increase from 5.82 to 6.88 GHz and bandwidth increase from 6.26 to 8.02 GHz with the frequency detuning increasing from 3.5 to 8.9 GHz. Thereby, the spectrum bandwidth and center frequency of generated UWB signals can be effectively adjusted by adjusting the injected power and frequency detuning.

Compared with UWB signals generated by optically feedback semiconductor lasers [33], UWB signals generated by optically injected chaotic semiconductor lasers have smoother spectrum, higher power and efficiency, easier adjustment of center frequency and bandwidth, and meet the requirements of FCC's indoor communication spectrum template [34], which is mainly attributed to additional external injected light. Under appropriate injection conditions, the frequency beat effect occurs between the external injected light and initial chaotic laser field, and the mutual coupling of higher frequency periodic oscillation, chaotic oscillation and high frequency broadband periodic oscillation is stimulated, thus the flatter spectrum and wider bandwidth of the output chaotic laser field. Therefore, the center frequency and bandwidth of UWB signals can be adjusted flexibly by using a chaotic semiconductor laser.

With the further increase of injected power and frequency detuning, the center frequency and bandwidth of UWB signal obtained in the experiment will continue to increase. Figure 5.12a shows the UWB signal generated when the injected power is − 5.6 dBm and the frequency detuning is 11.5 GHz. At this time, the − 10 dB bandwidth of UWB signal is 12.6 GHz, and the spectrum contour has exceeded the range of the communication spectrum template in room FCC. The spectrum is shown by the blue curve in Fig. 5.12b. However, the UWB radiant mask standards developed by FCC for different application object environments can be produced using this method.

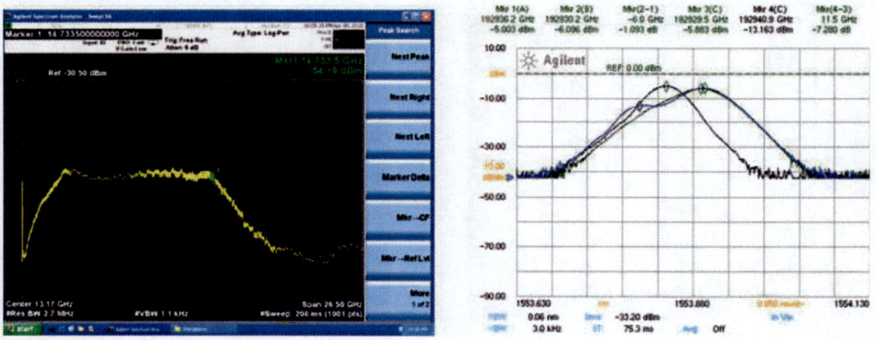

Fig. 5.12 The UWB signal when the injected power is − 5.6 dBm and the frequency detuning is 11.5 GHz **a** RF spectra and **b** spectrogram

Since the UWB system occupies an extremely wide frequency band, it overlaps with narrowband systems with approved frequency use rights, such as GPS, 3G, WLAN, etc. Though the transmitting power of the UWB system has a very strict limit, users are still worried that the UWB system will increase its noise base, cause interference, and reduce the performance of the system. So researchers introduce cognitive radio technology into the UWB system, hoping to perceive the surrounding electromagnetic environment, to adjust the frequency band used by the UWB system to form a notch to reduce the interference to the narrowband system, which is called cognitive UWB [35]. We studied the RF spectrum notch of UWB signals generated by optical injection chaotic LD to avoid overlapping with existing narrow-band systems on the RF spectrum, which shows in Fig. 5.13.

As shown in Fig. 5.13a, in order to avoid RF spectrum overlap with the UII system (such as 802.11a) around 5 GHz, the frequency detuning between the injected light

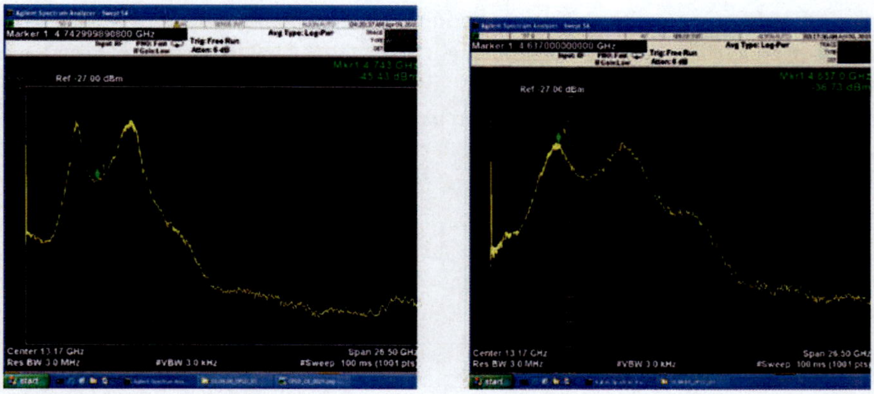

Fig. 5.13 Optical injection of chaotic LD generates RF spectrum notch of UWB. **a** RF spectrum notch of UWB signal obtained in the experiment and **b** Notch effect of UWB signal RF spectrum under different frequency detuning

and the chaotic LD is adjusted to realize the notch of the generated UWB signal in this frequency band.

By adjusting the injected power and frequency detuning, RF spectrum notch at different frequency bands can be realized. As shown in Fig. 5.13b, the notch occurs at around 6.7 GHz of UWB signal RF spectrum.

In conclusion, the RF spectrum notch of UWB signal in a specific frequency band can be conveniently realized by adjusting the frequency detuning and injected light power. This means that the method proposed in this section has the potential to be the preferred scheme for the spectrum dynamic adjustment and notch of C-UWB system in the future.

5.1.3 Widely Tunable Ultra-Wideband Signals Generation Utilizing Optically Injected Semiconductor Laser

In order to obtain widely tunable ultra-wideband (UWB) signals, we proposed an approach to generate UWB signals based on the nonlinear period-one oscillation mixing effect of optical pulse injection semiconductor laser, and obtained the UWB pulse signal output with the center frequency from 3.73 to 19.53 GHz.

The experimental setup of photonic tunable UWB microwave signals is shown in Fig. 5.14. The master laser diode (MLD) and the slave laser diode (SLD) are distributed feedback semiconductor lasers with a relaxation oscillation frequency of 4 GHz when biased at 28 mA. The power and polarization state of the injection laser can be adjusted by a variable attenuator (VA) and a polarization controller (PC) respectively. The frequency detuning of the master–slave lasers is still regulated by the precision temperature controller. The MLD is modulated by an electronic signal source (Agilent E8257D) to generate the gain-switched optical pulse signals. The gain-switched optical pulse signals are injected into the SLD through an optical circulator, and the UWB signals are generated by using the nonlinear dynamic response characteristics of the SLD. The UWB output signals are amplified by an erbium-doped optical fiber amplifier (EDFA) and divided into two outputs. One is measured the corresponding optical spectrum by an optical spectrum analyzer (Agilent 86140B), and the other is converted into UWB microwave signals via a bandwidth photodetector and measured by a 26.5 GHz electrical spectrum analyzer (Agilent N9020A) and a 50 GHz sampling oscilloscope (Agilent 86100B).

In the experiment, the repetition rate of the gain-switched optical pulses is 932 MHz, and its spectrum is shown in Fig. 5.15a. The power spectrum of the gain-switched pulses is a regular microwave frequency comb with a frequency spacing of 932 MHz, and the first harmonic have a 16 dB larger amplitude than other high-order harmonics at least. By adjusting the injected power and frequency detuning, the semiconductor laser can enter the state of single-period oscillation, double-period oscillation, four-period oscillation and chaos. In the experiment, we adjust the optical power of the injected SLD to $P_{inj} = 0.220$ dBm, and the optical frequency detuning

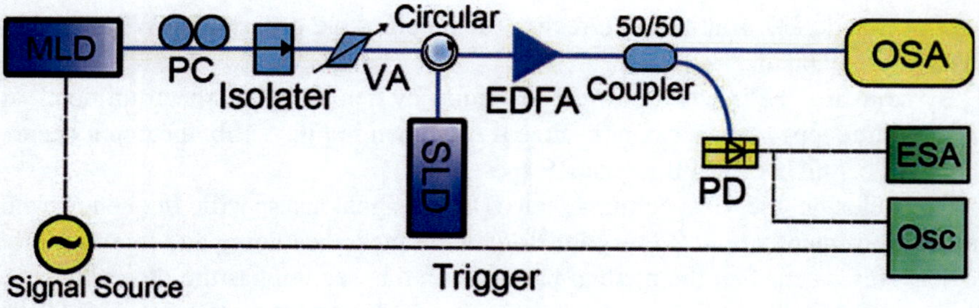

Fig. 5.14 Schematic setup for photonic tunable UWB signal generation. *PD* photodedtector, *OSA* optical spectrum analyzer, *ESA* electrical spectrum analyzer, *Osc* oscilloscope

to $\Delta\nu = -3.7$ GHz ($\Delta\nu = \nu_{inj} - \nu_s$, ν_{inj} is the optical frequency of the injected laser, ν_s is the optical frequency of the SLD), which drives the SLD in a single-period oscillation state. Because the frequency beating and reciprocal mixing lead to the energy redistribution of the optical pulses signal in the frequency domain, the UWB signal spectrum is shown in Fig. 5.15b. It can be seen that the center frequency of the UWB signals is 3.73 GHz (corresponding to the optical frequency detuning of the two lasers), and the − 10 dB bandwidth is 5.59 GHz. We measured that the fifth harmonic of the UWB signals (around 4 GHz) has a single side band (SSB) phase noise of about − 84 dBm/Hz at an offset frequency of 10 kHz. Compared with the incident gain-switched optical pulses, the obtained SSB phase noise is higher due to the overlap of the relaxation frequencies between both laser diodes.

In addition, we also investigated the optical and power spectrum characteristics of the optical UWB signals at different injection strengths and slave laser wavelengths, and the results are shown in Fig. 5.16. Due to the interaction between the injection field and the slave laser field, it makes the oscillation wavelength of the SLD red-shift or blue-shift to a certain extent (it is found in the experiment that the red-shift and

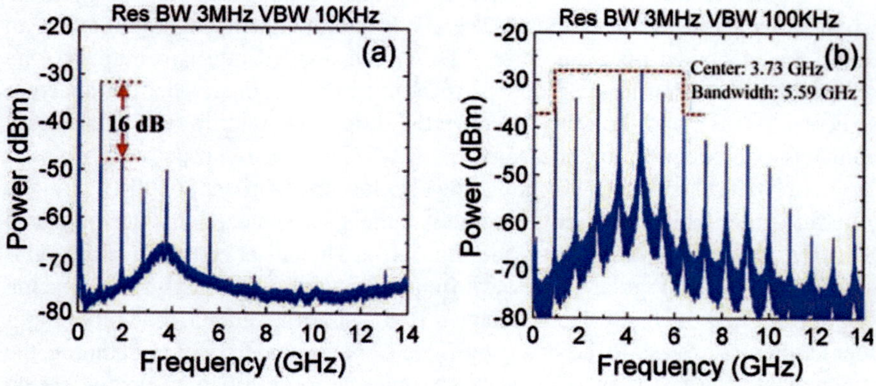

Fig. 5.15 **a** Power spectrum of gain-switched pulse train output from MLD. **b** Power spectrum of optical UWB pulse train output from SLD with $\Delta\nu = -3.7$ GHz, $P_{inj} = 0.220$ dBm

blue-shift depend on the relationship between the wavelength of the injected laser and the SLD: when the wavelength of the injected laser is greater than the wavelength of the SLD, the output wavelength of the slave laser will be blue-shifted; on the contrary, the output wavelength of the slave laser will be red-shifted. Oppo and others have given a more detailed theoretical simulation explanation for the frequency pushing effect of the laser under the optical injection) [36]. For the convenience of description and discussion, the wavelength values and frequency detuning shown in Fig. 5.16a are the stable output values after the interaction of light fields. When injected power P_{ing} is 0.220 dBm, the detuning is changed from − 3.7 to 19.5 GHz by setting the wavelength of the slave laser λ_s at 1553.477 nm. The optical spectrum is shown in Fig. 5.16a-1. Figure 5.16b-1 is the power spectrum corresponding to the UWB signals. It can be seen that the center frequency of the UWB microwave signals is 19.53 GHz, and the − 10 dB bandwidth is 5.59 GHz.

When other conditions remain unchanged, we reduce the injected power. As P_{inj} = − 0.629 dBm, the wavelength of the SLD will change from 1553.477 to 1553.500 nm due to the reduced injected power, and the frequency detuning of the two lasers will change from − 19.5 to − 16.7 GHz. The spectrum is shown in Fig. 5.16a-2. Accordingly, the center frequency of the obtained UWB signals changes to 16.74 GHz, and the result is shown in Fig. 5.16b-2. As P_{inj} = − 0.629 dBm, we tune the slave laser's wavelength λ_s from 1553.500 to 1553.542 nm. Due to the frequency-pushing

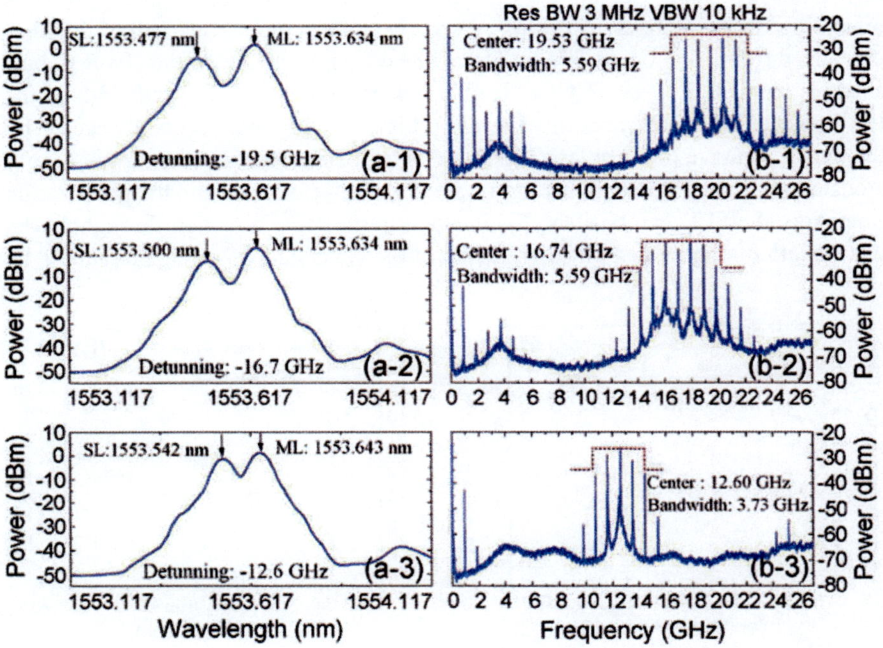

Fig. 5.16 **a** Optical spectrum and **b** corresponding power spectrum of the optical UWB signal: (a-1) and (b-1) P_{inj} = + 0.220 dBm, λ_s = 1553.477 nm; (a-2) and (b-2) P_{inj} = − 0.629 dBm, λ_s = 1553.500 nm; (a-3) and (b-3) P_{inj} = − 0.629 dBm, λ_s = 1553.542 nm.

effect, the injected laser is red-shifted in the cavity of the slave laser. The wavelength changes from 1553.634 to 1553.643 nm, and the frequency detuning changes to − 12.7 GHz, as shown in Fig. 5.16a-3. Figure 5.16b-3 shows the spectrum of the corresponding UWB signals. It can be seen that the center frequency of the UWB signals changes to 12.60 GHz, and the bandwidth changes to 3.73 GHz.

In the experiment, we tried to adjust the frequency detuning by changing the wavelength of the injected laser to achieve the widely tunable center frequency of the UWB signals. It is found that when the wavelength of the MLD is adjusted by the temperature controller, other parameters of the generated optical pulses will also change (such as polarization state, intensity, pulse shape, etc.), resulting in chaotic oscillation or injection locking state of the SLD and affecting the quality of the generated UWB signals.

Figure 5.17a shows the impact of the change of injection power from − 0.799 to 0.220 dBm on the center frequency and bandwidth of the generated UWB signals, when the injected laser wavelength λ_{inj} is 1553.634 nm. At this time, with the increase of injected power, the center frequency of the UWB signals increases from 15.82 to 19.53 GHz, while the bandwidth remains at 5.6 GHz. When the injected power changes in this range, the SLD will oscillate in a single-period state. When the injected power is not in this range, other nonlinear oscillations will occur in the SLD, resulting in the disappearance of the UWB pulse signals.

Figure 5.17b shows the changing trend of the center frequency and bandwidth of the UWB signals with the wavelength of the SLD when the injected power is maintained at $P_{inj} = 0.220$ dBm. As can be seen, when the wavelength of the SLD is reduced from 1553.643 to 1553.477 nm, the center frequency of the UWB signals changes from 3.73 to 19.53 GHz. In this process, the − 10 dB bandwidth of the UWB signals fluctuates randomly from 3.73 to 5.59 GHz, which is mainly caused by the random change of the polarization state of the SLD in the process of wavelength adjustment. In fact, if the polarization state of the injected laser and the polarization state from the SLD can be stably maintained in the same direction throughout the wavelength tuning process, the bandwidth of the generated UWB signals should be

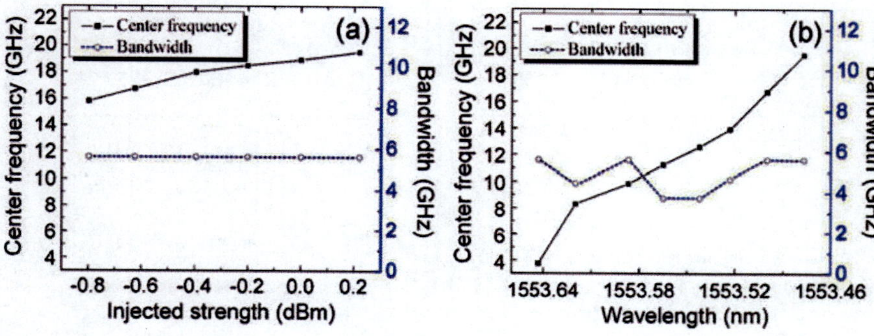

Fig. 5.17 **a** Center frequency and bandwidth varied as functions of the injection strength changed when $\lambda_{inj} = 1553.634$ nm. **b** Center frequency and bandwidth varied as functions of slave laser's wavelength changed when $P_{inj} = 0.220$ dBm

kept at 5.6 GHz. However, this also means that by changing the polarization state of the injected laser through the polarization controller, the − 10 dB bandwidth of the generated UWB signals can be adjusted to a certain extent.

The influence of the transmission of the photonic UWB signals in the optical fiber on its spectrum is the fundamental problem to realizing the UWB-over-fiber technology. For this reason, we have studied the transmission performance of the generated UWB signals in the optical fiber, and the results are shown in Fig. 5.18. The blue curve in the upper part of the figure is the spectrum of the generated UWB signals, and the yellow curve in the lower part is the spectrum obtained after the UWB signals are transmitted for 10 km. It can be seen that the spectrum silhouette of the UWB signals is unchanged after they are transmitted for 10 km in the optical fiber, but the overall only decrease in amplitude due to loss. This feature is crucial for its application in the UWB-over-fiber system.

Due to the bandwidth limitation of the spectrum analyzer, the variation range of the obtained UWB's center frequency can only be observed up to nearly 20 GHz. However, the center frequency of the generated UWB signals by the single-period oscillation mixing method is expected to reach a higher frequency band because some researchers have obtained the single-period state with oscillation frequency exceeding 90 GHz in the experiment [24, 37]. Therefore, it is expected to use this method to generate UWB signals in the 22 ~ 29 GHz frequency band for automotive anti-collision radar and the 60 GHz frequency band for ultra-wideband WPAN systems.

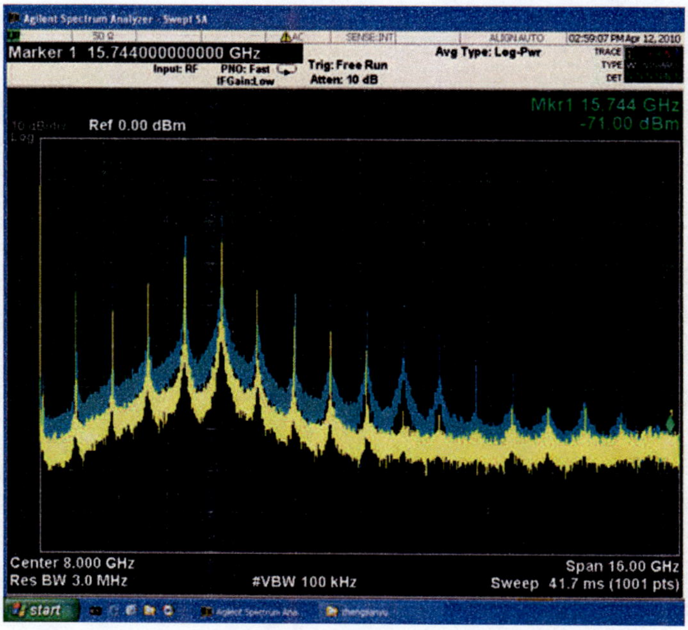

Fig. 5.18 Spectrum comparison of ultra-wideband signal before and after 10 km optical fiber transmission

5.1.4 Direct Modulation of Optical Feedback Semiconductor Lasers to Generate Chaotic UWB Signals

In this section, a simple and cost-effective solution is proposed and demonstrated—a directly current-modulated optical feedback semiconductor laser to generate a chaotic UWB pulse signal without an additional electro-optical modulator. The resulting chaotic UWB pulse signal not only has a controlled frequency bandwidth and tunable center frequency, but also achieves 34.08 km of optical fiber transmission without any dispersion compensation and with little change in the spectral shape, which is a significant improvement in the transmission distance and performance of the UWB signal compared to our previous work.

1. **Experimental setup and experimental results**

Chaotic lasers are a particular form of laser output instability, although a defined rate equation can also describe the dynamic characteristics of the laser, the laser output (light intensity, wavelength, phase) is no longer steady in the time domain, a noise-like random variation, where the feedback light perturbs the semiconductor laser and its output evolves from low-frequency undulation to chaos [38], and from doubly periodic to chaos. In the experiments, the bias current and feedback intensity of the semiconductor laser was adjusted to appropriate conditions so that the laser produced chaotic oscillations and then outputted a continuous chaotic laser, which was then directly internally modulated using a random sequence to produce a chaotic pulse sequence of 264 trains, each containing dozens of chaotic pulses with randomly fluctuating amplitudes. As described above, the experimental setup for generating chaotic pulsed microwave signals from a directly modulated semiconductor laser is shown in Fig. 5.19.

In the experiments, the threshold current of the laser was 16 mA, and the feedback loop length of the fiber was 5.6 m. To generate a spectrum that meets the requirements of the FCC indoor spectrum communication template, the bias current, feedback strength and modulation rate are optimized. The timing diagram of a directly modulated semiconductor laser random code and the frequency spectrum, timing diagram

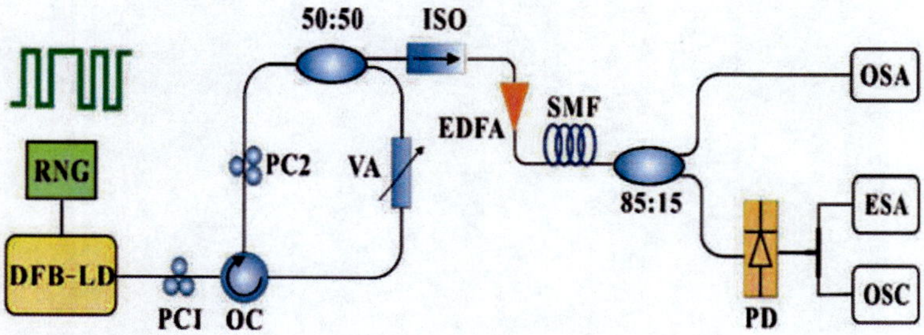

Fig. 5.19 Experimental setup for generating chaotic pulsed microwave signals from a directly modulated semiconductor laser

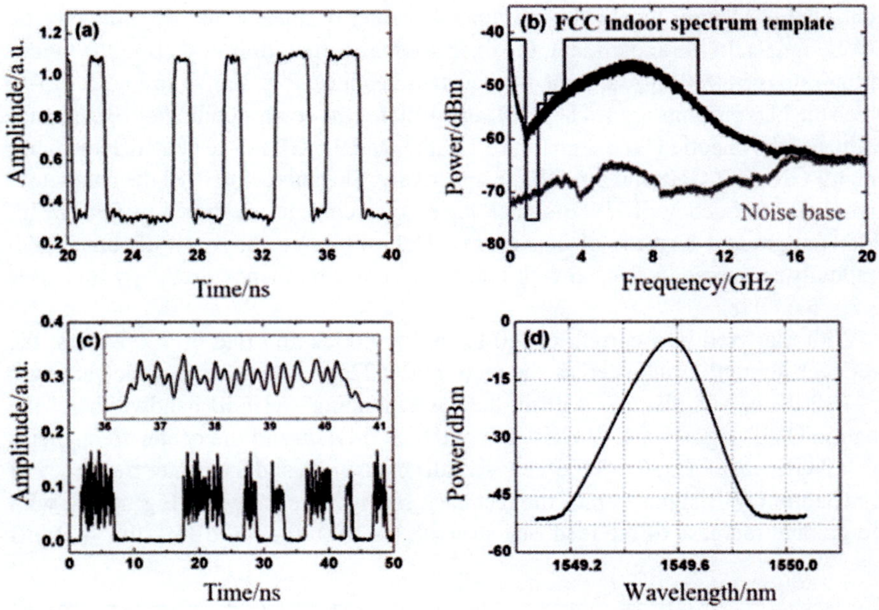

Fig. 5.20 Output characteristics of random code and chaotic UWB. **a** Timing diagram of the random code from the directly modulated laser; **b** spectrum diagram of the experimentally generated chaotic UWB pulse signal; **c** timing diagram; **d** spectrum diagram

and spectrum of a chaotic UWB pulse signal at a bias current of 32 mA, feedback strength of − 4 dBm and modulation rate of 960 Mbit/s are shown in Fig. 5.20.

Figure 5.20a shows a timing diagram of a random code for a direct semiconductor laser. In Fig. 5.20b, the upper curve shows the spectrum of the generated chaotic UWB pulse signal with a − 10 dB bandwidth up to 9.6 GHz, a relative bandwidth of 155% and a center frequency of 6 GHz; the middle curve shows the FCC indoor spectrum communication template, and the lower curve shows the spectrum meter noise base, it can be seen that the spectrum of the experimentally generated chaotic UWB pulse signals better The spectrum of the experimentally generated chaotic UWB pulses meets the requirements of the FCC indoor spectrum communication template. The chaotic UWB pulse train from the directly modulated optical feedback semiconductor laser is shown in Fig. 5.20c. The inset shows that the length of the chaotic UWB pulse train is 4 ns, containing about 20 pulses, so the half-height full width of each pulse is 100 ps. The central wavelength of the distributed feedback semiconductor laser was stabilized at 1549.572 nm with a − 20 dB linewidth of 0.2 nm by a temperature controller, as shown in Fig. 5.20d.

2. **Effect of bias current, feedback strength and modulation Rate on − 10 dB bandwidth and center frequency**

The effects of bias current, feedback strength and modulation rate on the spectral characteristics of the generated chaotic UWB signals are investigated to achieve a controllable spectral bandwidth and tunable center frequency output of the chaotic UWB signals. In the experiment, the fixed feedback strength is − 4 dBm, the modulation rate is 960 Mbit/s, and the bias current is adjusted as shown in Fig. 5.21a–c, when the bias currents are 1.4 I_{th}, 1.7 I_{th}, and 2.0 I_{th}, the corresponding − 10 dB bandwidths of the chaotic UWB signals are 7.8, 8.7, and 9.6 GHz. The center frequencies are 4.9 GHz, 6.2 GHz and 6.6 GHz, respectively. The curves of − 10 dB bandwidth and center frequency of UWB signal versus bias current are shown in Fig. 5.21d. With the gradual increase of bias current (1.1 ~ 2.0 I_{th}), the − 10 dB bandwidth gradually increases (5.6 ~ 9.6 GHz), and the center frequency gradually increases (3.8 ~ 6.6 GHz).

With the fixed bias current of 2.0 I_{th} and a modulation rate of 960 Mbit/s, the feedback strength is adjusted as shown in Fig. 5.22a~c, when the feedback strength is − 12 dBm, − 8 dBm, − 4 dBm, the corresponding − 10 dB bandwidths of the chaotic UWB signals are 8.0GHz, 8.75 GHz, 9.6 GHz, and the center frequencies are 4.2 GHz, 5.7 GHz, 6.6 GHz. The − 10 dB bandwidth and the center frequency of the chaotic UWB signal versus the feedback strength are shown in Fig. 5.22d, with the gradual increase of the feedback strength (− 14.0 ~ − 4.0 dBm), the − 10 dB

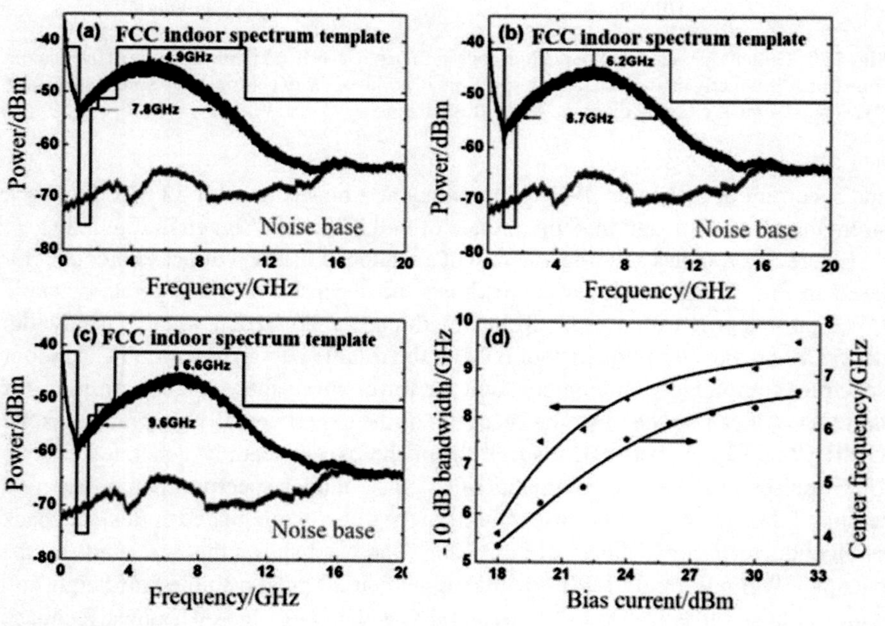

Fig. 5.21 Experimental spectrum and variation of bandwidth and center frequency with bias current. **a~c** Experimental spectra at bias currents of 22 mA, 27 mA and 32 mA, respectively; **d** Experimental spectrum of − 10 dB bandwidth and center frequency with bias current

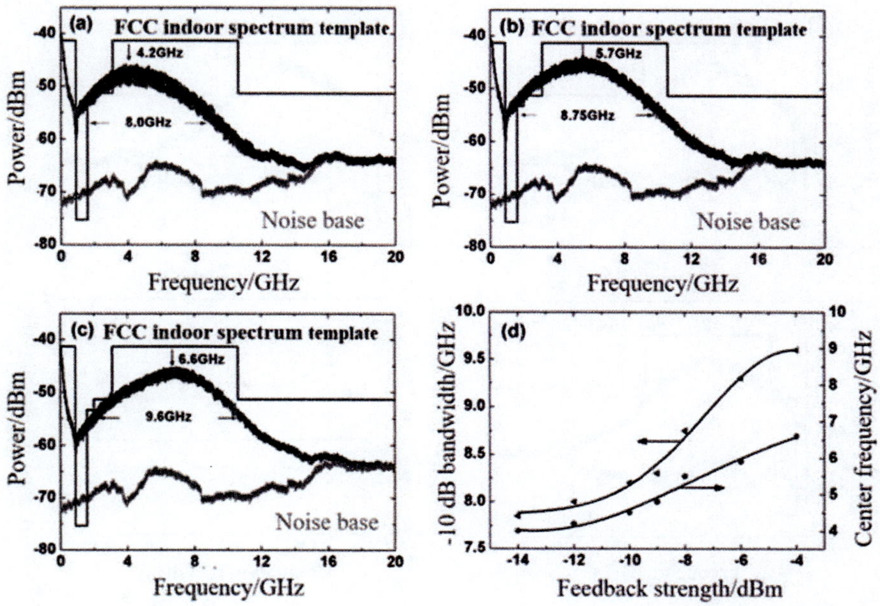

Fig. 5.22 Chaotic UWB signal output characteristics. **a**~**c** Experimental spectra at − 12 dBm, − 8 dBm and − 4 dBm feedback strengths, respectively; **d** Experimental curves of − 10 dB bandwidth and center frequency with feedback strength

bandwidth gradually increases (7.8 ~ 9.6 GHz), and the center frequency gradually increases (4.0 ~ 6.6 GHz).

The fixed bias current of 32 mA, feedback strength of − 4 dBm, varying the modulation rate, when the modulation rate is 360 Mbit/s, 720 Mbit/s, and 960 Mbit/s, respectively, the spectrum is shown in Fig. 5.23a–c. The curves of − 10 dB bandwidth and center frequency of UWB signal versus modulation rate are shown in Fig. 5.23d; with the gradual increase in modulation rate (240 ~ 960 Mbit/s), there is a slight increase of − 10 dB bandwidth and almost no chance of center frequency. As the modulation rate increases, the width of the chaotic pulse sequence gradually narrows, resulting in a slight broadening of the spectrum of the chaotic UWB signal with little change in the center frequency.

If the bias current and feedback strength are further increased, the − 10 dB bandwidth and center frequency will continue to increase, but the spectrum of the resulted chaotic UWB pulse signal will exceed the limits of the FCC indoor spectrum communication template. Directly modulated optical feedback semiconductor lasers give rise to vibrant, dynamic characteristics, such as periodic oscillations, low-frequency oscillations, and chaos [37–39]. In this section, the directly modulated optical feedback semiconductor laser generates a chaotic UWB signal; our study was limited to the feedback strength level (in the range of − 14 to − 4 dB), bias current level (in the range of 1.1–2.0 I_{th}), and modulation rate level (in the range of 240–960 Mb/s).

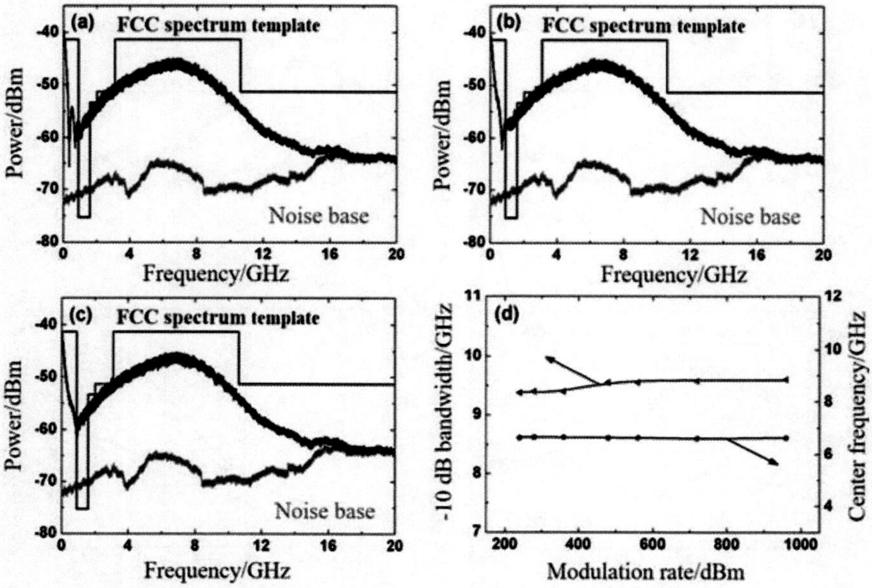

Fig. 5.23 Chaotic UWB signal output characteristics. **a~c** Experimental spectra at modulation rates of 360 Mbit/s, 720 Mbit/s, and 960 Mbit/s, respectively; **d** Experimental curves of the − 10 dB bandwidth and center frequency of the UWB signal spectrum with modulation rate

If we extend the limitation of the three parameters, the output will transition from chaotic oscillation to other dynamic characteristics.

3. **Transmission characteristics of chaotic UWB signals**

The transmission characteristics of the generated chaotic UWB signals in optical fibers are further investigated, and the trend of the chaotic UWB signal spectrum shape with the transmission distance is examined. As shown in Fig. 5.24, the bias current is 32 mA, the feedback strength is − 4 dBm, and the modulation rate is 960 Mbit/s. The chaotic laser pulses are amplified by EDFA and then transmitted over long distances; It can be seen that the spectral shape of the chaotic UWB signal remains almost unchanged after 10.14 km, 23.94 km and 34.08 km of fiber transmission respectively, but the energy is attenuated in different magnitudes due to the fiber transmission loss. This is because the change of spectrum shape mainly depends on the shape change of the chaotic UWB pulses and the experiments for the direct modulation of the optical feedback DFB laser output light line width is relatively narrow, in the fiber transmission, the narrow line width of the chaotic UWB pulses are less affected by fiber dispersion. Therefore, the fiber dispersion effect does not cause significant distortion of the pulse shape of the UWB signal, which makes the spectrum of the chaotic UWB signal almost unchanged, which means that the chaotic UWB signal generated under this system is very little affected by fiber dispersion. In addition, the absence of discrete spectral lines in the power spectrum of the chaotic UWB signal indicates that there is no need to introduce jitter

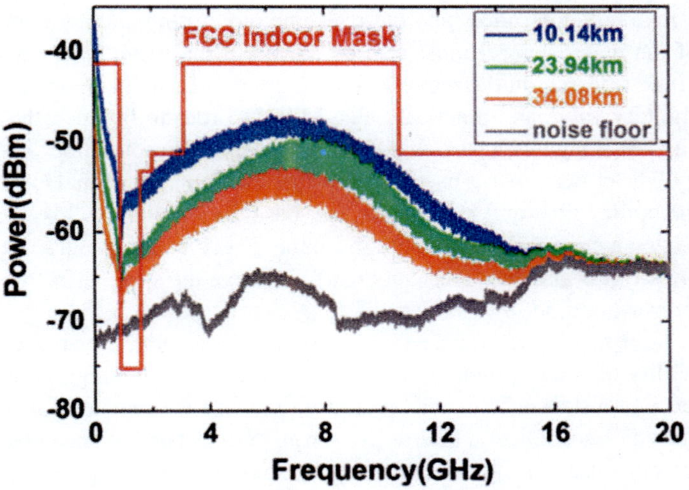

Fig. 5.24 Power spectrum variation tendency of chaotic UWB pulse signals in different distance fiber transmission

or optimize the modulation format to mitigate the negative effects caused by the discrete power spectrum, which reduces the cost and complexity of the system.

5.2 Chaotic Ultra-Wideband Microwave Photon Long-Range Ranging Technology

5.2.1 Introduction

As mentioned above, UWB radar has many advantages. However, broadband detection signals use traditional coaxial cables and waveguide transmission lines, which have large transmission losses and short operating distances. The distance between the radar antenna and the control center is limited to a few meters to tens of meters, which limits the application of broadband radar in military and civilian fields. As we all know, the radar antenna is a radiation source in the military, and it is also the target of enemy attacks. If the distance between the antenna and the control center is too close, once the antenna is attacked by aircraft or missiles, it will inevitably affect the control center, destroy the entire radar system, and endanger the safety of commanders and operators. In addition, for radar systems erected in harsh environmental conditions such as snow peaks and islands, operators must operate and maintain equipment in harsh environments. There are also the same problems in industry: for example, in the dark coal storage bunker with large dust and harsh environment, the measurement of the material level in the coal bunker by microwave ranging radar cannot be remotely controlled, and the real-time monitoring and equipment maintenance of the material level must also be solved in the harsh field environment. Therefore, the research on the remote technology of radar and antenna is a subject of great military value and practical value [37]. In order to broaden the effective coverage of UWB signals, researchers have proposed radio-over-fiber UWB radio technology. The optical fiber is used as a low-loss medium. The signal is transmitted through the optical fiber and reaches the user base station at a low loss. Then, it is converted into an electrical signal through a photodetector, and then transmitted and received by a UWB antenna. This can greatly improve the space effective range of UWB microwave radio [38].

Chaotic laser ranging radar based on chaotic laser has many advantages, such as low probability of interception, strong anti-interference ability, high precision and independence of distance. Moreover, it is easy to realize the remote control of radar by using chaotic laser. Chaotic laser is not sensitive to dispersion and does not need dispersion compensation. It can transmit tens of kilometers in optical fiber to the remote antenna end to realize the remote control of radar [39].

This section will introduce a chaotic UWB microwave photonic remote ranging sensing system. This system combines the previous work of chaotic UWB signal generation and transmission, and uses the chaotic laser generated by optical feedback and optical injection semiconductor laser as the detection signal of radar. The optical fiber remote control of 24 km is realized, and the distance measurement of single and double targets in 8.03 m free space is realized. The single target spatial resolution of 3 cm and the double target spatial resolution of 8 cm is obtained. This system combines optical UWB radio technology and chaotic correlation ranging technology, which can not only realize the remote control of UWB ranging radar, but also obtain high spatial resolution.

5.2.2 Experimental Device and Principle

Figure 5.25 shows the device diagram of chaotic UWB microwave photonic remote ranging system. The central station chaotic source is composed of optical feedback and optical injection semiconductor laser. The generated chaotic laser signal is divided into two channels through an 80:20 fiber coupler: one as the detection signal; the other way as a reference signal. The detection signal without any dispersion compensation is transmitted through 24.15 km SMF1 to the remote antenna. At the remote antenna end, the intensity of the optical signal is controlled by adjusting the tunable optical attenuator. The chaotic laser signal is converted into the corresponding chaotic UWB electrical signal by PD1, which is amplified by AMP1 and transmitted by the UWB antenna (TX Antenna). The detection signal is partially reflected after encountering the space target. The reflected signal is received by the same type of antenna (RX Antenna), then amplified by AMP2, and converted into the corresponding optical signal by performing intensity modulation on the laser. The optical signal is amplified by an optical amplifier, transmitted by 24.15 km SMF2, and

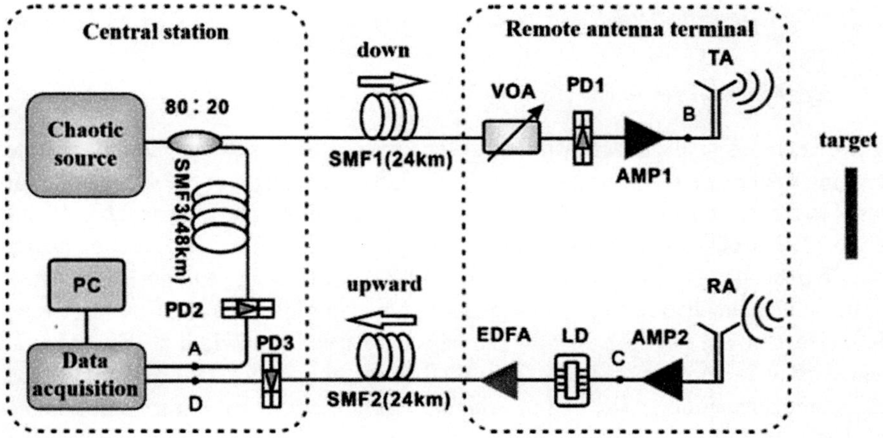

Fig. 5.25 Experimental installation of chaotic UWB microwave photon distance measurement system

returned to the central station. After PD3, the chaotic laser signal is converted into a UWB chaotic electrical signal and the reference signal transmitted by 48 km SMF3 is cross-correlated to obtain the distance information of the target. The reference signal and the reflected signal are collected and stored by the real-time oscilloscope, and then the computer performs the correlation operation and displays the results. The power spectrum of each signal is observed by the signal analyzer.

Because the chaotic signal has a sharp δ function correlation form, the position of the reflected event, that is, the target position can be determined by cross-correlation between the reference signal and the target reflected signal. The reference signal is the function X(t) of time t, and the functional relationship between the detection signal carrying the target position information and time is $a \cdot X(t - \tau)$, where a is the loss coefficient of the microwave signal in the air, and τ is the propagation time of the detection microwave signal in the air. Then the relationship between the cross-correlation operation of the two signals:

$$X(t) * a \cdot X(t - \tau) \approx a \cdot \delta(\tau) \tag{5.4}$$

where the peak position of the correlation curve corresponds to the time τ, then the distance between the target position and the antenna is $c \cdot \tau/2$, where c is the electromagnetic wave velocity in vacuum (the electromagnetic wave velocity in air is approximately considered as the electromagnetic wave velocity in vacuum).

The bandwidth of PD1 and PD2 (NewFocusModel1554-B) used in the experiment is 12.5 GHz, the bandwidth of PD3 is 40 GHz, and the bandwidth of AMP1 (A-INFOLA1018N3209) and AMP2 (A-INFOLA1018N2420) is 1 ~ 18 GHz.

5.2.3 Experimental Results and Analysis

1. **Experimental results of ranging**

The detection signal characteristics of the chaotic UWB microwave photonic remote ranging system are shown in Fig. 5.26. The output wavelength of the semiconductor laser is stable at 1549.770 nm. When the bias current of the laser is 22 mA (1.4 times the current), the feedback optical power is – 11.55 dBm, the injection power is 1.76 dBm, and the frequency detuning is 7.1 GHz, the timing and power spectrum of the generated chaotic detection signal are shown in Fig. 5.26a and b, respectively. Its average power is – 6.2 dBm. It can be seen that the bandwidth of the detection signal is wide and flat, and the – 10 dB bandwidth reaches 18 GHz. Fig. 5.26c is its autocorrelation curve, and the sidelobe level is very low. Figure 5.26d is the peak distribution of the signal time series. It can be seen from the figure that the distribution belongs to Gaussian-like distribution. The chaotic signal time series amplitude is evenly distributed on both sides of 0 V, indicating that the chaotic signal has good randomness. The UWB ranging radar with this detection signal has excellent anti-interference.

The experimental results of single target ranging of chaotic UWB microwave photonic remote ranging radar are shown in Fig. 5.27a. The target is a metal plate with a size of 35 cm × 35 cm. The cross-correlation curves of the detection signal

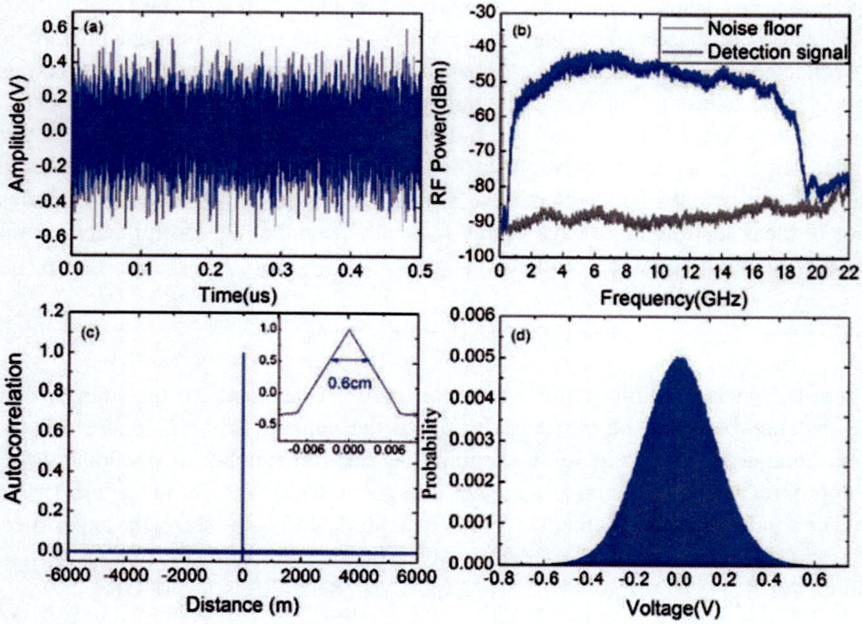

Fig. 5.26 Detection signal characteristics of chaotic UWB microwave photon remote ranging radar **a** Waveforms; **b** RF spectra; **c** Autocorrelation curve; **d** Peak distribution

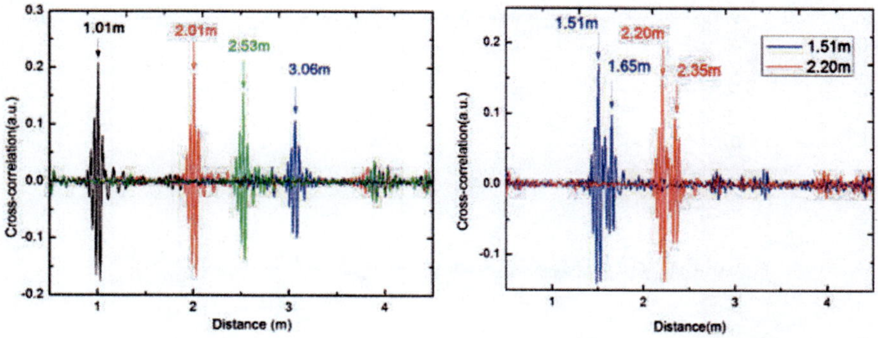

Fig. 5.27 Experimental results of single target ranging for chaotic UWB microwave photon remote ranging radar. **a** single target space ranging and **b** dual-target space ranging

and the reference signal are shown when the distance between the target and the antenna is 1.01 m, 2.01 m, 2.53 m and 3.06 m, respectively. As shown in the figure, the position of the target is clearly marked by the correlation peak. As the distance between the antenna and the target becomes farther, the attenuation of the UWB signal in the free space increases, and the peak value of the cross-correlation signal becomes lower.

Furthermore, the distance measurement of two targets at the same time is experimentally studied. Figure 5.27b is the experimental result of space ranging of two targets with the same size. The target in the experiment is two 35 cm × 35 cm metal plates. The cross-correlation curve of the blue curve in the figure is the result of two targets at 1.51 m and 1.65 m from the chaotic radar transmitting antenna. The red curve in the diagram is the measurement results of two targets at 2.20 m and 2.35 m from the chaotic radar transmitting antenna respectively. As shown in the figure, the positions of the two targets are clearly marked by the correlation peaks.

Figure 5.28 shows the experimental results of dual-target ranging with different sizes. In this experiment, the two targets are no longer two metal plates of the same size. One of the targets is still a 35 cm × 35 cm metal plate, and the other target is the air shower chamber, and the metal plate is placed in the air shower chamber channel. The transmitting antenna and receiving antenna are facing the target metal plate, and the curve in the figure is the ranging experiment results of the system: among them, the first peak represents the distance of the air shower chamber, which is 3.89 m; The second peak represents the distance of the target metal plate at 4.09 m. Because the reflected signal of the metal plate is stronger than that of the air shower chamber, the correlation peak of the metal plate placed in the air shower chamber channel is higher than that of the air shower chamber.

2. **Spatial resolution**

The spatial resolution of the chaotic signal correlation ranging technology is determined by the full width at half maximum of the chaotic signal autocorrelation curve. The narrower the full width at half maximum of the main peak, the higher the spatial

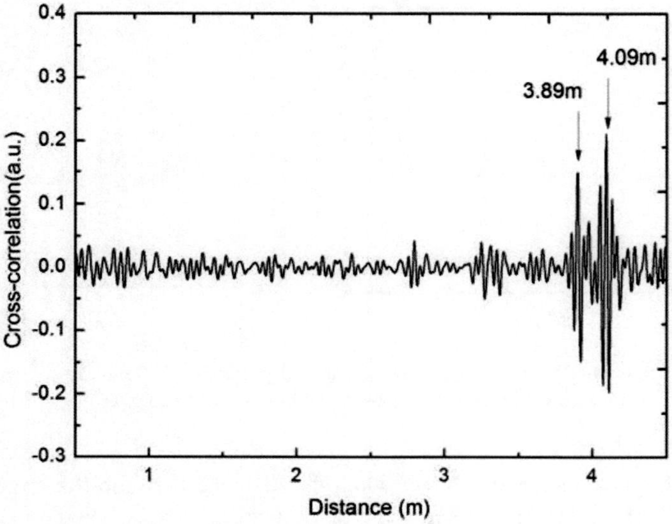

Fig. 5.28 Experimental results of dual-target ranging with different sizes

resolution of the system. In addition, the Wiener-Khinchin theorem shows that the signal autocorrelation function and its power spectrum are a pair of Fourier transform pairs, so the full width at half maximum of the autocorrelation function mainly depends on the bandwidth of the chaotic signal. The wider the bandwidth of the signal, the narrower the full width at half maximum of the autocorrelation function, and the higher the spatial resolution. The autocorrelation curve of the detection signal in this experiment is shown in Fig. 5.29 and its FWHM is about 0.6 cm. The chaotic signal generated by the experiment has a wide bandwidth, which can reach 18 GHz, and its distance resolution can reach 0.3 cm. However, since the spatial resolution of the system is limited by the 6 GHz bandwidth of the real-time oscilloscope, the limit resolution of the system is 0.6 cm.

However, in the experiment, the actual range resolution of the system is greater than 0.6 cm due to the interference of other signals in space, the dispersion of the detected signal in the connecting transmission cable, and the bandwidth of other electronic devices used. In order to determine the actual range resolution of the ranging radar system, the method used in reference [27] is adopted: In the experiment, the two antennas are aligned with each other, and there is no obstacle between the transmitting antenna and the receiving antenna, so that the signal transmitted by the transmitting antenna is directly received by the receiving antenna. The cross-correlation curve of the received signal and the transmitted signal is shown in Fig. 5.30a, and its full width at half maximum is 6 cm. Therefore, the spatial ranging resolution of the system is 3 cm, that is, the actual range resolution of the system is 3 cm.

Further, as shown in Fig. 5.30b, we place the target at a distance of about 1.49 m from the radar and use the chaotic radar system to detect its distance. The experimental results are shown in the red curve, and then the target is moved back 3 cm.

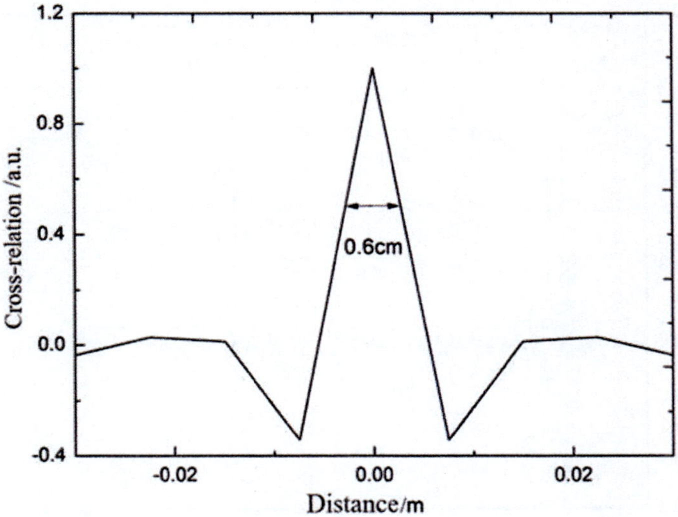

Fig. 5.29 The autocorrelation curve of the detected signal

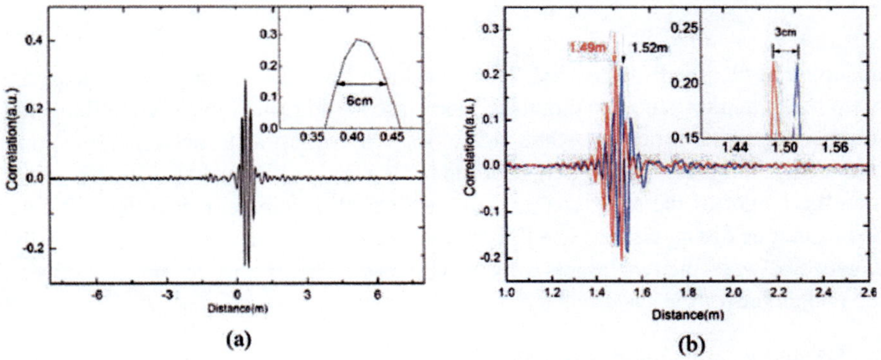

Fig. 5.30 Experimental test result. **a** The cross-correlation curve between two antennas 0.5 m apart; **b** experimental results of spatial range resolution of single target

The chaotic radar system detection experimental results are shown in the blue curve. This is also the minimum distance of the single target that the ranging system can distinguish in the experiment, thus verifying that the single target range resolution of the ranging radar system is 3 cm. The illustration in the figure is the amplification of two cross-correlation peaks.

In order to analyze the resolution of the system between two targets in multi-target ranging, the distance between two targets is changed during the experiment to make it gradually smaller. The test results of the distance resolution between the two targets are shown in Fig. 5.31. When the distance between the two targets is 8 cm, the two peaks are still obvious. If the distance between them is reduced, the second peak can

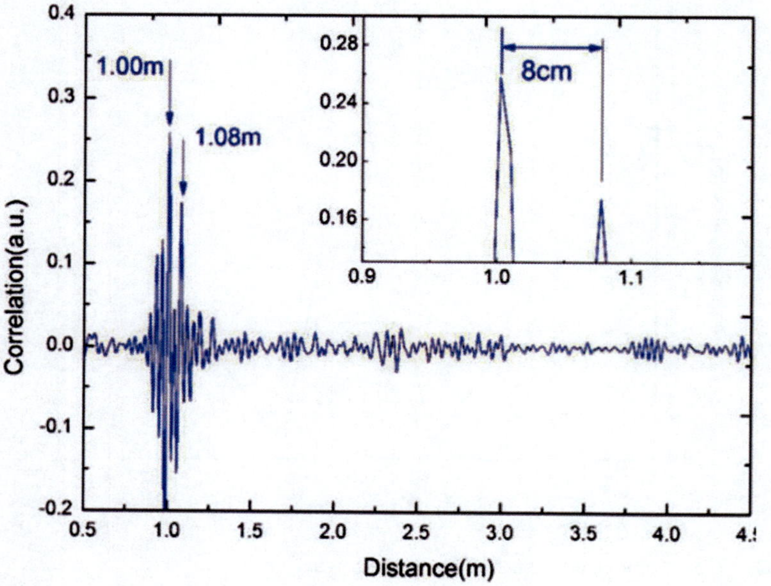

Fig. 5.31 The test results of the distance resolution between the two targets

no longer be clearly distinguished. This is because when the two targets are close, the main peaks marking the two targets will be close to their side lobes. When their side lobes are at the same position, their side lobes will overlap with each other, resulting in the correlation curve. The height of the peak marking the target position is lower than the height of the side lobe, and the main peak is submerged in the side lobe and cannot be distinguished. The illustration in the figure can clearly distinguish the distance between the two targets is 8 cm. Therefore, the distance resolution between the two targets of the system is 8 cm.

3. **Dynamic range measurement**

In order to evaluate the actual measurable range of the chaotic detection signal with an average power of − 6.2 dBm under the conditions of the existing experimental device, the peak noise level (PNL) was analyzed by measuring the single target with spatial distances of 0.69 m, 1.57 m, 2.02 m, 2.51 m, 3.11 m, 3.63 m, 4.11 m, 4.60 m, 5.24 m, 5.80 m, 6.54 m, 7.08 m, 7.61 m and 8.18 m, respectively.

$$\mathrm{PNL} = 10\lg\frac{C_1}{C_2} \tag{5.5}$$

In the formula, C1 is the peak of the reflection peak, and C2 is the peak of the highest noise except the reflection peak. By analyzing the relationship between the peak noise level and the spatial transmission distance, the measurement dynamic range of the chaotic signal under this power is obtained. The results of the experiment

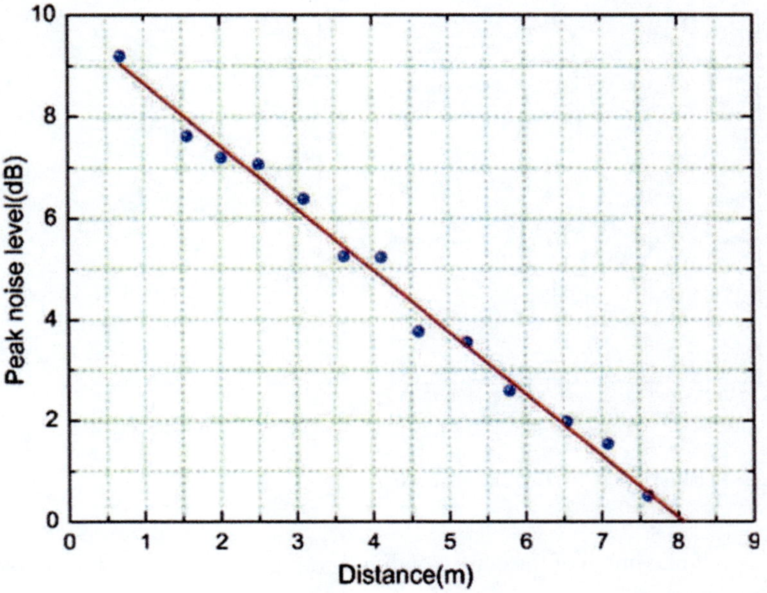

Fig. 5.32 Dynamic range measurement

are shown in Fig. 5.32. The results show that the detection signal of this power can achieve a spatial measurement distance of 8.03 m under the existing experimental device.

4. **The influence of bandwidth of data acquisition system on system resolution**

The bandwidth of the detection signal generated by the chaotic signal source of the chaotic UWB microwave photonic remote ranging system introduced in this section is 18 GHz, but the maximum bandwidth of the real-time oscilloscope for collecting data is 6 GHz, which will limit the ranging accuracy of the system. For this reason, we studied the distance resolution of the system under different acquisition bandwidths. When the acquisition bandwidth of the oscilloscope is set to 1 GHz, the FWHM of the auto-correlation curve of the detection signal is 5.30 cm, that is, the limit resolution of the system is 5.30 cm; when the acquisition bandwidth is set to 4 GHz, the FWHM of the auto-correlation curve of the detection signal is 0.98 cm; when the acquisition bandwidth continues to increase to 6 GHz, the FWHM of the auto-correlation curve of the detection signal is 0.60 cm.

The experimental results are shown in Fig. 5.33. Figure 5.33a is the auto-correlation curve of the measured detection signal when the acquisition bandwidth of the high-speed oscilloscope is set to 1 GHz, 3 GHz, 4 GHz and 6 GHz, respectively. Figure 5.33b is the FWHM between the acquisition bandwidth of the oscilloscope and the auto-correlation curve of the measured detection signal. It can be seen that the wider the bandwidth of the collected detection signal is, the narrower the full

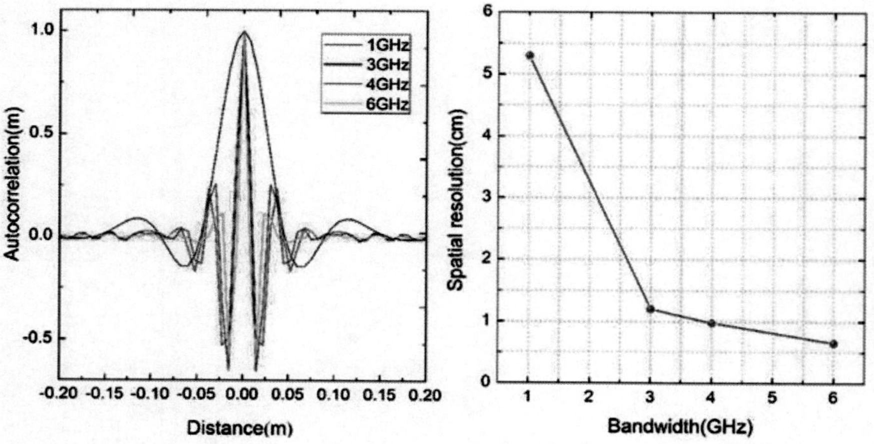

Fig. 5.33 Relationship of FWHM and bandwidth

width at half maximum of the corresponding correlation curve is, and the higher the ranging accuracy is.

5. **Influence of Signal Distortion on Ranging**

The interference of other electromagnetic wave signals in space, the dispersion of the detection signal itself, and the bandwidth and noise of other experimental devices will affect the ranging performance of the system. However, due to the strong anti-interference ability of the chaotic detection signal, although affected by the above factors, the system can still measure a single target. The distance resolution can reach 3 cm. As shown in Fig. 5.34, the gray curve is the noise base of the signal analyzer. The red power spectrum curve in Fig. 5.34 corresponds to the power spectrum of point A in Fig. 5.25 system device, that is, the power spectrum of the signal generated in the optical domain of the central station, and its bandwidth can reach 18 GHz. The black line in Fig. 5.34 is the power spectrum of the chaotic detection signal of point B, that is, the detection signal power spectrum of the radar system. Because the bandwidth of the low noise amplifier is 1 ~ 18 GHz, the frequency component of the chaotic signal below 1 GHz is filtered out. The blue spectral line in Fig. 5.34 is the power spectrum of the C-point signal in the system, that is, the output signal generated by the microwave amplifier after the target reflection signal received by the receiving antenna is mixed with the electromagnetic wave signal in the air. Due to the transmission loss in the air, the higher the signal frequency, the greater the attenuation, so the high frequency energy of the signal received at point C is much lower than the energy of the low frequency signal.

The green spectral line in Fig. 5.34 is the power spectrum of the detection signal at point D in the system, that is, the collected target reflection signal is limited by the 10 GHz laser modulation bandwidth, and the bandwidth of the echo signal will be greatly reduced. In the experiment, the distance information of the target is obtained

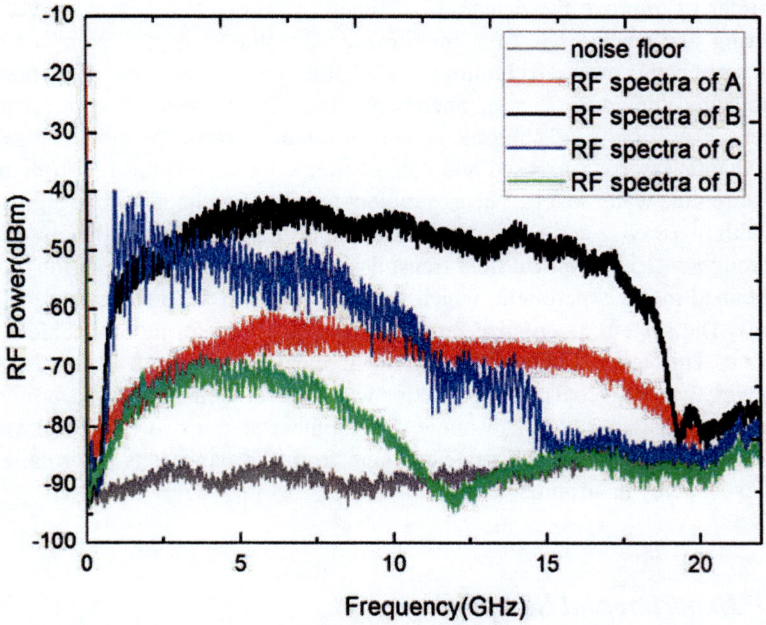

Fig. 5.34 Signal power spectrum curves at different locations in the system. Red: power spectrum of chaotic signal generated in optical domain of central station; Black: power spectrum of radar detection signal; Blue: the output signal of the mixed signal of the target reflection signal received by the receiving antenna and the electromagnetic wave signal in the air after microwave amplifier; Green: power spectrum of the collected target reflection signal

by cross-correlation operation between the black corresponding reference signal and the green corresponding detection signal in Fig. 5.34, and the remote ranging is realized. It can be seen from the figure that although the distortion of the detection signal is serious, the system also has good ranging performance due to the good anti-interference performance of the chaotic signal.

5.3 Chaotic Microwave-Photonic for Remote Water-Level Monitoring

With the rapid development of science and technology, the ecological balance has been seriously damaged due to people's irrational development of the natural environment and economic development, and the frequency of disasters has continued to rise, seriously threatening the survival of mankind. To do a good job of flood prevention and disaster reduction, improving detection and prevention means is the premise.

In order to improve the detection accuracy of non-contact remote water-level monitoring and realize real-time monitoring of water-level, the team combined optical carrier microwave technology and chaotic correlation ranging technology, and integrated the advantages of microwave water-level sensor and optical water-level sensor. A microwave-photonic sensor for remote water-level monitoring based on the chaotic laser is proposed and demonstrated, which enables real-time monitoring of remote water-levels. The experiment generated a chaotic probe signal with bandwidth of 18 GHz and real-time monitoring of the remote water-level was realized through 24.15 km optical fiber transmission, and the spatial resolution of 2 cm was obtained in the experiment, which was very close to the theoretical resolution (1.6 cm). The system can realize remote water-level monitoring and achieve high resolution. This technology has important application prospects of flood control project early warning, coal mine production water-level monitoring, flood monitoring stations in mountain valleys, resource development stations in oceans or islands and other fields, which can liberate workers from the dangerous environment of water-level survey environment.

5.3.1 Experimental Setup

An experimental setup of microwave-photonic sensor for remote water-level monitoring based on chaotic laser is shown in Fig. 5.35. The remote water-level sensor system consists of three parts: the central office, the remote antenna unit and the fiber link. Firstly, at the central office, the ultra-wideband chaotic laser is generated by using an optically injected semiconductor laser with optical feedback. And the generated chaotic laser is split into two branches by a single mode fiber coupler with a coupling ratio of 80:20: one is the probe light and the other is the reference light. The power of the detected optical signal is controlled by a variable optical attenuator (VA). The optical signal with appropriate power is transmitted by 24 km SMF1 to the remote antenna unit, converted from photo detector1 (PD1, 12.5 GHz) to electrical signal, amplified by AMP1 (1 ~ 18 GHz) and transmitted by the ultra-wideband transmitting antenna (TA, 1 ~ 18 GHz). Due to the strong reflecting characteristics of water, the detection signal will be reflected when it touches the water's surface. The reflected signal is collected by a receiving antenna (RA, 1–18 GHz), and amplified by another low noise amplifier (AMP2, 1–18 GHz). After that, the transformed optical signal is amplified by an erbium doped fiber amplifier (EDFA), transmitted back towards the central office over single mode fiber (SMF2, 24 km) and then detected by photo detector3 (PD3, 40 GHz). A chaotic laser in the other branch is transmitted over single mode fiber (SMF3) which is compensating fiber, and detected by photo detector2 (PD2, 12.5 GHz) within the central office as a reference signal.

The height of the water-level is calculated by cross-correlation between the probe signal and the reference signal, and the remote water-level sensing system is calibrated by a determined water-level height before the experiment begins.

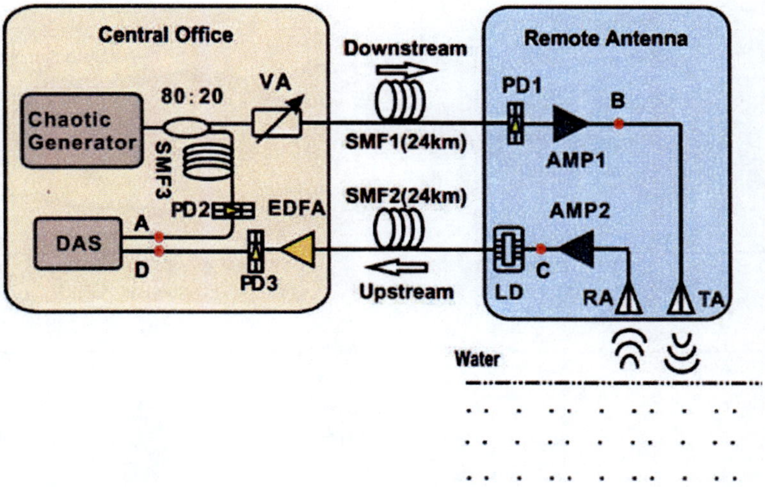

Fig. 5.35 Experimental setup of microwave-photonic sensor for remote water-level monitoring based on chaotic laser

5.3.2 Detection Signal Characteristics and Transmission Characteristics of Remote Water-Level Sensor

In the experiments, the slave LD is biased at 1.4 times of its threshold current, and its output wavelength is stabilized at 1549.770 nm. Figure 5.36 shows the detection signal features of the remote water-level sensor system. The time series of the transmit signal is shown in Fig. 5.36a. The waveform has a noise-like feature and no obvious periodicity is shown in the figure. In Fig. 5.36b, the blue curve is the power spectra of the chaotic detection signal obtained experimentally when the output power of the chaotic laser is − 0.79 dBm. The gray curve is the noise floor of the signal analyzer. As seen in the Fig. 5.36, it can be seen that the bandwidth of chaotic signal is up to 18 GHz. The power spectral density is rather flat and smooth in the range from 1 GHz to near 19 GHz. The signal below 1 GHz is filtered by AMP1. Figure 5.36c is the autocorrelation trace of the time series with a correlation length of 20 μs. The illustration is a magnification of its main peak. It can be clearly seen that its half-height and full width are 1.6 cm, which can reach the ideal ranging accuracy of 1.6 cm. This is mainly limited by the bandwidth of the oscilloscope. The noise level of the autocorrelation curve is very low, which has significant advantages as the detection signal of the ranging system.

Figure 5.36d shows an amplitude histogram of the chaotic signal and that the chaotic signal generated in this way makes a good randomicity. The anti-interference ability in sensing detection is guaranteed.

Figure 5.37 demonstrates the transmission characteristics of power spectra at different points in the system shown in Fig. 5.37. The gray curve is the noise floor of the signal analyzer. The red power spectrum in Fig. 5.37 corresponds to the power

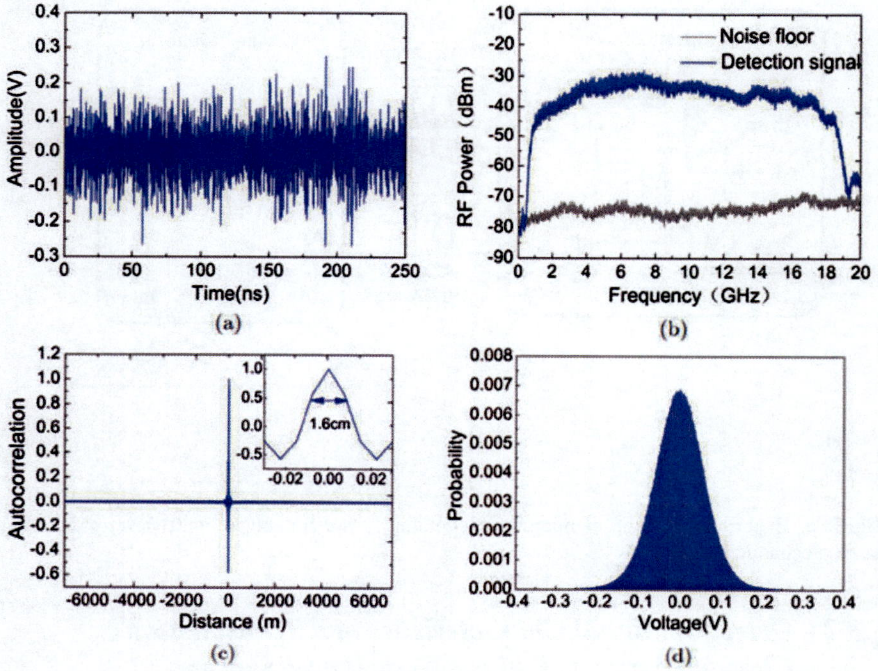

Fig. 5.36 Characteristics of transmit signal. **a** Time series, **b** power spectra of the chaotic detection signal obtained experimentally with detuning frequency f = 7.1 GHz, the optical power of injection P = 1.76 dBm, and feedback strength − 11.55 dBm. The gray curve is the noise floor. **c** Autocorrelation trace of the chaotic state and **d** Amplitude histogram of time series

spectrum of the reference signal, that is, the power spectrum of the chaotic laser generated in the optical domain of the central station, with a bandwidth up to 18 GHz. The black curve in the figure is the power spectra of the detection signal, that is, the detection signal power spectrum of the radar system. Since the bandwidth of the low noise amplifier is 1 ~ 18 GHz, the chaotic signal frequency components below 1 GHz are filtered out. The blue spectral line of the Fig. 5.37 is the output signal of the mixed signal of the target reflection signal and electromagnetic wave signal in the air received by the receiving antenna in the system after the microwave amplifier. Due to the transmission loss in the air, and the higher the signal frequency, the greater the attenuation, so the received signal high-frequency energy is much lower than the energy of the low-frequency signal. The green spectral line of the Fig. 5.37 is the power spectrum curve of the final detection signal of the system, that is, the reflected signal of the collected target. As can be seen from the Fig. 5.37, after the spatial transmission of the signal, the distortion is relatively large, that is, the signal distortion is relatively serious.

Figure 5.38a shows the measuring results of the microwave-photonic sensor for remote water-level monitoring based on the chaotic laser. The illustration in the upper right corner of the figure shows a simple schematic of water-level measuring.

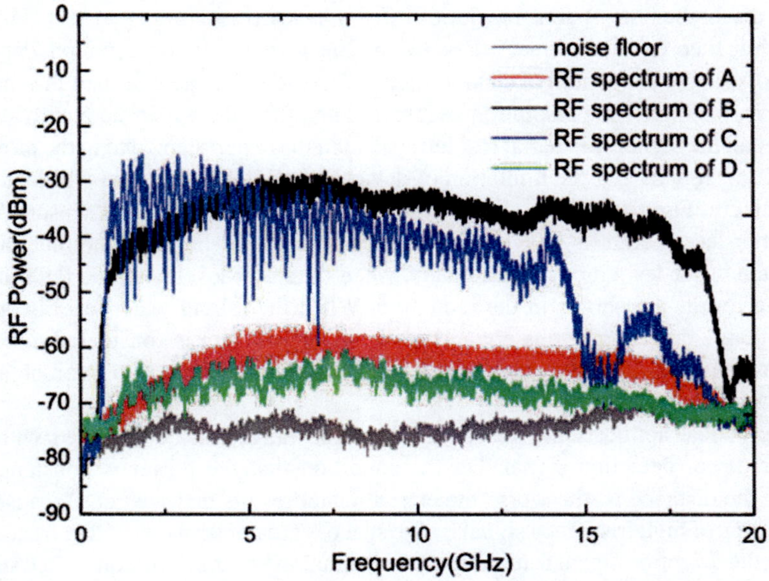

Fig. 5.37 The power spectra feature of different points in the system. Points A, B, C and D are marked in this figure by four dots

The height of the measured water-level can be obtained by measuring the distance between the water's surface and the antenna. With the water-level of 0.99 m, 1.12 m and 1.44 m away from the radar antennas, respectively, the height of the water's surface is indicated clearly by the peaks of cross-correlation traces. The ranging results are shown in the figure. As you can see from the Fig. 5.38, the sidelobe and noise levels are very low.

For the correlation theory of ranging, the resolution of chaotic signal is determined by the FWHM of the autocorrelation curve. The narrower the FWHM of the main

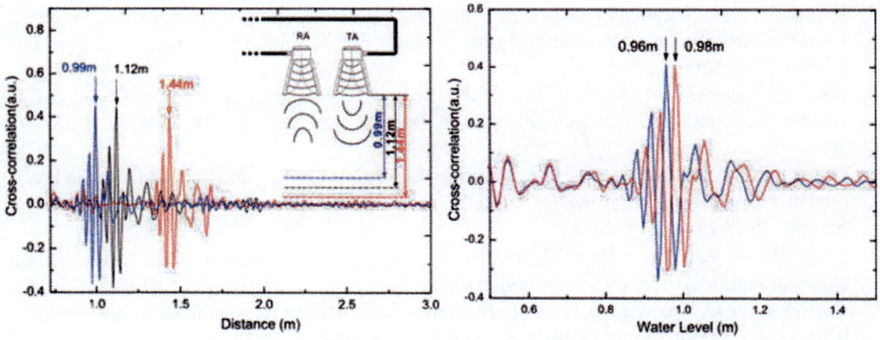

Fig. 5.38 Experimental test results. **a** Measuring results of remote water-level sensor. **b** The spatial resolution of remote water-level sensor

peak, the higher the spatial resolution of the system. According to the Wiener–Khinchin theorem, the signal's FWHM of the autocorrelation function depends primarily on the bandwidth of chaotic signal. The wider the bandwidth of the chaotic signal is, the higher the resolution becomes. Therefore, the wider the bandwidth of the signal, the higher the spatial resolution. But in the experiment, due to the effect of interfering signals and the limitation of devices, the actual resolution will be larger than 1.6 cm. Figure 5.38b shows the resolution of remote water-level sensor. When the water-level height is 0.96 m, the ranging results are shown in the blue curve; when the water-level drops 2 cm, that is, when the water-level height is 0.98 m, the ranging results are shown in the red curve. When the falling water-level height is lower than 2 cm, the range is not accurate. Therefore, 2 cm is considered to be the range resolution of the system, which is very close to the range resolution obtained by the theoretical FWHM.

In practical applications, the longer detection range can be achieved with the greater power detection signal. On the condition that the power is constant, the farther the distance is, the worse the experimental results that we get. Because the attenuation of high frequency signal in free space is greater than that of low frequency signal, the detection signal is high frequency signal. After transmission in free space, the signal will have a certain degree of distortion. In the actual test, the lower the water-level, that is, the distance between the height of the water and the antenna, the worse the water-level test effects.

References

1. Ieello GR, Rogersoo GD. Ultra-wideband wireless systems. IEEE Micro Mag. 2003;4(2):36–47.
2. Roy S, Foerster JR, Somayazulu VS, et al. Ultrawideband radio design: the promise of high-speed, short range wireless connectivity. Proc IEEE. 2004;92(2):295–311.
3. Pan SL, Yao JP. Optical generation of polarity and shape-switchable ultrawideband pulses using a chirped intensity modulator and a first-order asymmetric Mach-Zehnder interferometer. Opt Lett. 2009;34(9):1312–4.
4. Federal Communications Commission. First report and order in the matter of revision of part 15 of the commission's rules regarding ultra-wideband transmission systems. Tech. Rep. ET-Docket 98-153, FCC 02-48
5. Han SM, Popov O, Dmitriev AS. Flexible chaotic UWB communication system with adjustable channel bandwidth in CMOS technology. IEEE Trans Microw Theory Tech. 2008;56(10):2229–36.
6. Zhang FZ, Wu J, Fu SN, et al. Simultaneous multi-channel CMW-band and MMW-band UWB monocycle pulse generation using FWM effect in a highly nonlinear photonic crystal fiber. Opt Express. 2010;18(15):15870–5.
7. Feng XH, Li ZH, Guan B, et al. Switchable UWB pulse generation using a polarization maintaining fiber Bragg grating as frequency discriminator. Opt Express. 2010;18(4):3643–8.
8. Wan MY, Yee C, Yee ML. Wireless ultra-wideband communications using radio over fiber. In: IEEE conference on ultra-wideband systems and technologies; 2003.
9. Zeng F, Yao JP. An approach to ultrawideband pulse generation and distribution over optical fiber. IEEE Photonics Technol Lett. 2006;18(7):823–5.

10. Yao JP, Zeng F, Wang Q. Photonic generation of ultrawideband signals. J Lightwave Technol. 2007;25(11):3219–35.
11. Salkintzis AK, Fors C, Pazhyannur R. WLAN-GPRS integration for next-generation mobile data networks. IEEE Wirel Commun. 2002;9(5):112–24.
12. Juan YS, Lin FY. Demonstration of ultra-wideband (UWB) over fiber based on optical pulse-injected semiconductor laser. Opt Express. 2010;18(9):9664–70.
13. Chen HW, Wang TL, Li M, et al. Optically tunable multiband UWB pulse generation. Opt Express. 2008;16(10):7447–52.
14. Chang QJ, Tian Y, Ye T, et al. A 24-GHz ultra-wideband over fiber system using photonic generation and frequency up-conversion. IEEE Photonics Technol Lett. 2008;20(19):1651–3.
15. Zhao Y, Zhen XP, Zhang HY, et al. UWB-over-fiber transmission system using a dual-output Mach-Zehnder modulator. Chin Opt Lett. 2010;8(5):454–6.
16. Li J, Ning TG, Pei L, et al. Optical ultra-wideband pulse generation and distribution using a dual-electrode Mach-Zehnder modulator. Chin Opt Lett. 2010;8(2):138–41.
17. Shams H, Anandarajah AK, Perry P, et al. Electro-optical generation and distribution of ultrawideband signals based on the gain switching technique. J Opt Commun Netw. 2010;2(3):122–30.
18. Wang J, Sun QZ, Sun JQ, et al. All-optical UWB pulse generation using sum-frequency generation in a PPLN waveguide. Opt Express. 2009;17(5):3521–30.
19. Zeng F, Yao J. Ultrawideband impulse radio signal generation using a high-speed electrooptic phase modulator and a fiber-Bragg-grating-based frequency discriminator. IEEE Photonics Technol Lett. 2006;18(19):2062–4.
20. Zadok A, Wu XX, Sendowski J, et al. Photonic generation of ultra-wideband signals via pulse compression in a highly nonlinear fiber. IEEE Photonics Technol Lett. 2010;22(4):135–7.
21. Bolea M, Mora J, Ortega B, et al. Optical UWB pulse generator using an N tap microwave photonic filter and phase inversion adaptable to different pulse modulation formats. Opt Express. 2009;17(7):5023–32.
22. Wang SG, Chen HW, Xin M, et al. Optical ultra-wide-band pulse bipolar and shape modulation based on a symmetric PM-IM conversion architecture. Opt Lett. 2009;34(20):3092–4.
23. Zhou EB, Xu X, Lui KS, et al. A power-efficient ultra-wideband pulse generator based on multiple PM-IM conversions. IEEE Photonics Technol Lett. 2010;22(14):1063–5.
24. Liu F, Wang T, Zhang Z, et al. On-chip photonic generation of ultra-wideband monocycle pulses. Electron Lett. 2009;45(24):1247–9.
25. Abtahi M, Mirshafiei M, Magne J, et al. Ultra-wideband waveform generator based on optical pulse-shaping and FBG tuning. IEEE Photonics Technol Lett. 2008;20(2):135–7.
26. Yu XB, Gibbon TB, Pawlik M, et al. A photonic ultra-wideband pulse generator based on relaxation oscillations of a semiconductor laser. Opt Express. 2009;17(12):9680–7.
27. Lv H, Yu YL, Shu T, et al. Photonic generation of ultra-wideband signals by direct current modulation on SOA section of an SOA-integrated SGDBR laser. Opt Express. 2010;18(7):7219–27.
28. Khan MH, Shen H, Xuan Y, et al. Ultrabroad-bandwidth arbitrary radiofrequency waveform generation with a silicon photonic chip-based spectral shaper. Nat Photonics. 2010;4:117–21.
29. Peled Y, Tur M, Zadok A. Generation and detection of ultra-wideband waveforms using stimulated Brillouin scattering amplified spontaneous emission. IEEE Photonics Technol Lett. 2010;22(22):1692–4.
30. Wang AB, Zhang MJ, Xu H, et al. Location of wire faults using chaotic signal. IEEE Electron Device Letter. 2011;32(3):372–4.
31. Li P, Wang YC, Zhang JZ. All-optical fast random number generator. Opt Express. 2010;18(19):20360–9.
32. Zhang MJ, Liu TG, Wang AB, et al. Photonic ultrawideband signal generator using an optically injected chaotic semiconductor laser. Opt Lett. 2011;36(6):1008–10.
33. Zhang MJ, Zheng JY, Wang AB, et al. Chaotic ultra-wideband over fiber link based on optical feedback laser diode. Microw Opt Technol Lett. 2013;55(7):1504–7.

34. Zhang MJ, Liu M, Wang AB, et al. Photonic generation of ultrawideband signals based on a gain-switched semiconductor laser with optical feedback. Appl Opt. 2013;52(31):7512–6.
35. Lansford J. UWB coexistence and cognitive radio. International workshop on ultra-wideband system; 2004.
36. Velanas P, Bogris A, Argyris A, et al. High-speed all-optical first-and second-order differentiators based on cross-phase modulation in fibers. J Lightw Technol. 2008;26(18):3269–76.
37. Liu Y, Kikuchin, OJ. Controlling dynamical behavior of a semiconductor laser with external optical feedback. Phys Rev E. 1995; 51(4):2697–700.
38. Takiguchi Y, Liu Y, Obsubo J. Low-frequency fluctuation induced by injection-current modulation in semiconductor lasers with optical feedback. Opt Lett. 1998;23(17):1369–71.
39. Wang XF, Xia GQ, Wu ZM. Theoretical investigations on the polarization performances of current-modulated VCSELs subject to weak optical feedback. J Opt Soc Am B Opt Phys. 2009;26(1):160–8.

Chapter 6
Distributed Fiber Optic Raman Thermometer and Applications

The distributed fiber optic Raman sensing technology uses the principle of optical time domain reflectometer combined with the temperature effect of Raman scattered light to achieve distributed temperature measurements along the fiber. In 1985, J.R. Darkin (UK) completed the first sensing measurement based on the OTDR principle and Raman scattering effect of optical fiber, which opened the prelude of distributed fiber optic sensing based on Raman effect by domestic and foreign research scholars. In 1987, Chongqing University developed the first prototype of fiber optic temperature measurement in China, which initially realized the functions of fiber optic temperature measurement. The first prototype of fiber-optic temperature measurement was developed by Chongqing University in 1987, which initially realized the functions of fiber-optic temperature measurement, and then China University of Metrology, Beijing Aviation Aerospace University, Shandong University, Taiyuan University of Technology, and Tianjin University have carried out the research on distributed fiber optic Raman measurement temperature measurement techniques. This chapter firstly introduces the distributed fiber-optic Raman sensing technology, the new demodulation method and the development of a new distributed fiber optic Raman sensing technology, and incorporate the research results into the fiber optic Raman thermometer in the development of the prototype. Finally, the developed thermometer is applied to a practical engineering project.

6.1 Novel Fiber Optic Raman Demodulation Techniques

6.1.1 Spontaneous Raman Scattering Effect

Optical fiber is a low-loss, high-speed optical transmission medium, but because the medium itself has a certain degree of inhomogeneity, the light wave in the fiber will interact with the mechanism of molecular or atomic, resulting in backscattering. This

M. Zhang et al., *Novel Optical Fiber Sensing Technology and Systems*, Progress in Optical Science and Photonics 28, https://doi.org/10.1007/978-981-99-7149-7_6

backscattering includes three types: Rayleigh scattering, Brillouin scattering, and Raman scattering. Typical backscattering in optical fibers is illustrated and described in Sect. 3.2.1.

For the cause of Raman scattering effect, there are two main explanations: classical theory and quantum theory.

The classical explanation of the cause of Raman scattering is based on the interaction of matter with electromagnetic radiation. When monochromatic light with electric vector E is incident on a substance, the distribution of positive and negative charges of molecules or atoms of the substance will change or form an electric dipole moment.

In a certain incident light range, the inductive dipole moment M per unit volume and the incident electric E is proportional to the incident vector E:

$$M = \alpha E \tag{6.1}$$

where α is the polarizability tensor. Under a certain condition, the vibrating electric dipole will re-radiate electromagnetic waves. Usually, only the scattering of visible light is considered, and the mass of the nucleus is so different from that of the electron that only the electron can keep up with the vibration of visible light. Therefore, the light scattering from matter is only the contribution of electrons. The frequency and wave vector of the scattered light are different from the incident light.

According to the classical electromagnetic theory, the incident light electromagnetic field-induced dipole moment is

$$\boldsymbol{M}(t) = \sum e_i \boldsymbol{r}_i(t) \tag{6.2}$$

A charge with acceleration $\boldsymbol{r}_i$(t) radiates electromagnetic waves, and the radiation intensity is proportional to $\left|M(t)^2\right|$, if the electric field component E in the electromagnetic field varies in the following form

$$\boldsymbol{E} = \boldsymbol{E}_0 \cos \omega_L t \tag{6.3}$$

where ω_L is much larger than the atomic vibration frequency and is comparable to the vibration frequency of the electron. Then the induced dipole moment M can be written as the number of steps of the electric field E:

$$\boldsymbol{M} = \alpha \boldsymbol{E} + \frac{1}{2!}\beta \boldsymbol{E}^2 + \frac{1}{3!}\gamma \boldsymbol{E}^2 + \cdots + \frac{1}{n!}\delta \boldsymbol{E}^2 \tag{6.4}$$

where α is the electron polarization rate of the order of $10^{-40}\mathrm{CV}^{-1}\mathrm{m}^2$, β is the hyperpolarization rate, which is of the order of $10^{-50}\mathrm{CV}^{-1}\mathrm{m}^2$, γ, δ is the higher order rank tensor, γ of order $10^{-60}\mathrm{CV}^{-1}m^2$. where the electron polarization rate α depends on the distribution of charges in the system, i.e. $\alpha = \alpha(p)$. If the atomic coordination changes during the vibration, the parameter p, which characterizes the charge distribution, also changes, i.e. α changes, where Q is the generalized phonon.

$$\alpha = \alpha_0 + \left(\frac{\partial \alpha}{\partial Q}\right)_0 Q + \frac{1}{2!}\left(\frac{\partial^2 \alpha}{\partial Q^2}\right)_0 Q + \frac{1}{3!}\left(\frac{\partial^3 \alpha}{\partial Q^3}\right)_0 Q + \cdots \tag{6.5}$$

where the primary term of Q determines the primary Raman effect, the second term of Q determines the secondary Raman effect, and so on analogously.

If the atoms in the molecule vibrate with ω_q frequency, then, from $Q = Q_0\cos\omega_q t$, the electron polarization rate in the primary Raman effect varies with time as

$$\alpha(t) = \alpha_0 + \left(\frac{\partial \alpha}{\partial Q}\right) Q_0 \cos \omega_q t \tag{6.6}$$

Substituting Eq. (6.5) into Eq. (6.1), we have

$$\begin{aligned} M(t) &= \alpha_0 E_0 \cos \omega_L t + \left(\frac{\partial \alpha}{\partial Q}\right)_0 Q_0 E_0 \cos \omega_L t \cos \omega_q t \\ &= \alpha_0 E_0 \cos \omega_L t + \frac{1}{2}\left(\frac{\partial \alpha}{\partial Q}\right)_0 Q_0 E_0 \cos(\omega_L - \omega_q)t \cos(\omega_L + \omega_q)t \end{aligned} \tag{6.7}$$

Equations (6.7) show that the vibration of the induced dipole moment M has not only the incident light frequency ω_L, , but also ($\omega_L \pm \omega_q$) two new frequencies symmetrically distributed on both sides of ωL. They originate from the modulation of the electron polarization rate by atomic vibrations α modulation. The former corresponds to frequency-invariant elastic light scattering, such as Rayleigh scattering, and the latter corresponds to frequency-inelastic light scattering with a change in frequency, such as Raman scattering. The decrease in frequency $(\omega_L - \omega_q)$ is called the Stokes frequency, while the increase in frequency $(\omega_L - \omega_q)$ is called the anti-Stokes frequency.

If the electron polarization rate is considered as a scalar quantity, the electron polarization rate when the atoms in the medium are all in equilibrium position is α_0, then the change in the electron polarization rate due to the vibration of the diatomic molecule is $\Delta\alpha$, there is

$$\alpha = \alpha_0 + \Delta\, \alpha \tag{6.8}$$

The diatomic molecular vibration is optically modeled as a plane wave with frequency ω_q and wave vector q. The electron the amount of change in polarization rate can be expressed as

$$\Delta\, \alpha = \Delta\, \alpha \cos(\omega_q t - qr) \tag{6.9}$$

And the incident light wave frequency is still ω_L, the wave vector is k_L plane electromagnetic wave as $E = E_0 cos(\omega_L t - k_L r)$, then the induced dipole moment is

$$|M| = (\alpha + \triangle \alpha)|E| = \left[\alpha 0 + \triangle \alpha \cos\left(\omega_q - qr\right)\right]|E_0| \cos(\omega_L t - k_L r)$$
$$= \alpha_0 E_0 \cos(\omega_L t - k_L r) + \frac{1}{2} \triangle \alpha_0 E_0 \left\{\left[\cos(\omega_L + \omega_q)t - (k_L + q)r\right]\right.$$
$$\left. + \left[\cos(\omega_L - \omega_q)t - (k_L - q)r\right]\right\} \quad (6.10)$$

where $(\omega_L t - k_L r)$ is the corresponding characteristic quantity of Rayleigh scattering, and $(\omega_q \pm \omega_L)$ and $(k_L \pm q)$ are the corresponding characteristic quantities of Raman scattering, i.e., the frequency ω_s and wave vector k_s of Raman scattered light respectively represent the energy conservation and momentum conservation laws followed in the inelastic Raman scattering process.

The classical theory can describe the Raman scattering process, but the quantitative description of Stokes light, anti-Stokes light does not match with the experimental facts. To solve these problems, a comprehensive description can be made by quantum theory [1, 2].

The scattering phenomenon can also be explained by the collision of light quanta (particles) with molecules of the medium. According to quantum theory, monochromatic light with frequency ν_0 can be considered as a quantum of light with energy $h\nu_0$, h is Planck's constant. When a photon acts on a molecule, both elastic and inelastic collisions can occur. In elastic collisions, no energy is exchanged between the photon and the molecule, and the photon only changes its direction of motion, the photon only changes its direction of motion, but not its frequency. This elastic scattering process corresponds to Rayleigh scattering. In the inelastic collision process, photons energy exchange occurs with molecules, the photon not only changes its direction of motion, but also transfers part of its energy to the molecule, which is transformed into the vibration or rotation energy of the molecule, or the photon gets energy from the vibration or rotation of the molecule. In both processes, the frequency of the photon changes. The process of photon energy gain corresponds to the increasing frequency of Raman-anti-stokes scattering. The process of photon energy gain corresponds to Raman-anti-Stokes scattering with increasing frequency, and the process of photon energy loss corresponds to Raman-Stokes scattering with decreasing frequency.

In an inelastic collision, the energy and momentum of the molecules of the medium are changed, which means that an energy level jump occurs the structure is shown in Fig. 6.1. In Fig. 6.1, E1 and E2 represent the two vibrational energy levels of the molecule. When the incident light interacts with the molecule in the energy level E_1, the molecule will jump from the energy level E_1 to the imaginary energy level Es, by due to the instability of the energy levels, the molecule jumps from the imaginary energy level Es to the energy level E_2, where a photon of energy $h\nu_s$ is scattered Stokes photon. When the incident photon interacts with a molecule at energy level E_2, the molecule jumps from energy level E_2 to imaginary energy level E_a, and due

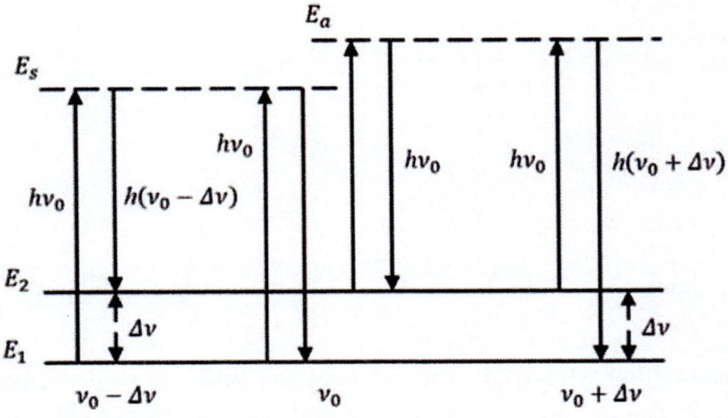

Fig. 6.1 Schematic diagram of scattering energy level structure

to the instability of the energy level, the molecule will jump from the virtual energy level E_a to the energy level E_1, and then scatter the photon with energy $h\nu_a$. The frequency of Stokes light can be expressed as

$$\nu_s = \nu_0 - \frac{E_2 - E_1}{h} \tag{6.11}$$

Meanwhile, the frequency of the anti-Stokes light can be expressed as

$$\nu_{as} = \nu_0 + \frac{E_2 - E_1}{h} \tag{6.12}$$

The initial energy level of the molecule is at energy level E_1 when Stokes light is produced, and the initial energy level of the molecule is at E_2 when anti-Stokes light is produced. Since the intensity of the Stokes light is several orders of magnitude larger than that of the anti-Stokes light, and it makes up for the shortcomings of the classical Raman interpretation [3, 4].

6.1.2 Single-Ended Structure Temperature Demodulation Method

Temperature demodulation method based on single-ended structure is a common temperature demodulation method, according to the results of the device, single-ended temperature demodulation structures can be further divided into single- way and dual-way demodulation devices [5]. The common single-ended demodulation system structure is shown in Fig. 6.2.

For a single demodulation system, a pulsed laser incident a pulsed light to a wavelength division multiplexer (WDM) and inject it into the sensing fiber, the

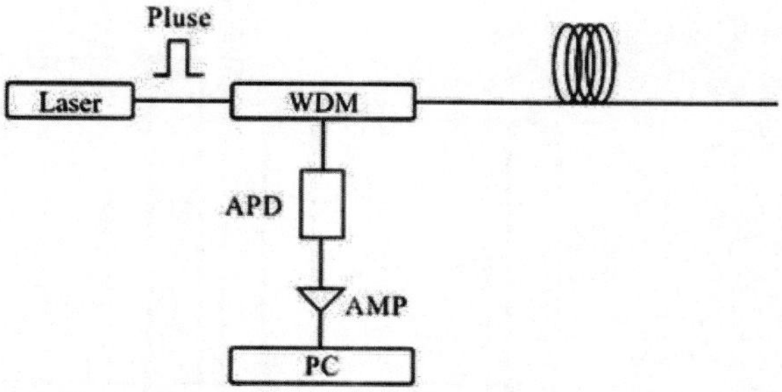

Fig. 6.2 Temperature demodulation method based on single-ended structure

Raman scattered light, Rayleigh scattered light and Brillouin scattered light in the fiber are scattered into the WDM, the Raman scattered anti-Stokes light is filtered out by the WDM and enters the avalanche photodetector (APD) in turn and an amplifier. The amplified electrical signal is then transmitted to the data acquisition card (DAC) and then demodulated by the corresponding algorithm to obtain the temperature change information along the sensing fiber.

The scattered light intensity of the three types of scattered light in optical fibers can be expressed by the following expression:

$$I_r(L) = M_r K_r V_r^4 I_0 \exp(-2\alpha_0 L) \tag{6.13}$$

$$I_s(T, L) = M_s K_s V_s^4 I_0 R_s(T) \exp[-(\alpha_0 + \alpha_s)L] \tag{6.14}$$

$$I_a(T, L) = M_a K_a V_a^4 I_0 R_a(T) \exp[-(\alpha_0 + \alpha_s)L] \tag{6.15}$$

where, $I_r(L)$ is the amplified backscattered Rayleigh light intensity; $I_s(T, L)$ and $I_a(T, L)$ are the amplified Raman-Stokes and anti-Stokes light intensities with respect to temperature and position, respectively; M_r, M_s, and M_a are the amplification of Rayleigh, Stokes, and anti-Stokes scattered light by the light amplification system, respectively; K_r, K_s, and K_a are the scattering cross-section coefficients of Rayleigh, Stokes, and anti-Stokes scattered light, respectively; ν_r, ν_s, and ν_a are the optical frequencies of Rayleigh, Stokes and anti-Stokes scattered light, respectively; I_0 is the incident light intensity; α_r, α_s, α_a are the attenuation coefficients of incident light, Stokes and anti-Stokes scattered light, respectively; L is the position of the fiber; T is the temperature at the L position; $R_s(T)$ and $R_a(T)$ respectively are the Boltzmann factors of the Raman Stokes and Raman anti-Stokes scattered light in the fiber, related to the settlement number of energy levels, which can be expressed as

$$R_s(T) = \left[1 - \exp(\frac{-h \bigtriangleup \nu}{kT})\right]^{-1} \tag{6.16}$$

$$R_a(T) = \left[\exp(\frac{-h \bigtriangleup \nu}{kT}) - 1\right]^{-1} \tag{6.17}$$

where, h is the Planck constant, $\Delta\nu$ is the Raman shift, which is approximately equal to 13.2 THz, k is the Boltzmann constant.

The conventional self-demodulation method is divided into two steps, firstly, the whole sensing fiber is placed at the same temperature T_0 for calibration, and then the sensing fiber is placed at the measurement temperature T, and the Raman anti-Stokes scattered light intensity of the entire fiber is obtained in these two steps. The final demodulation equation is

$$I_a(T, L)/I_a(T_0, L) = \frac{Ra(T)}{Ra(T_0)} = \exp(\frac{h \Delta \nu}{kT})/\exp(\frac{h \Delta \nu}{kT_0}) \tag{6.18}$$

From Eq. (6.18), it can be seen this method cannot eliminate the effect of the incident light source power fluctuation and fiber loss.

In the actual measurement, due to the fluctuation of the light source and the unstable magnification of APD, the actual measured light intensity was

$$I_a(T, L) = (M_a + \Delta M_a)K_a v_a^4(I_0 + \Delta I_0)Ra(T)\exp[-(\alpha_0 + \alpha_a)L] \tag{6.19}$$

where ΔM_a is amount of change in magnification of APD during measurement, and ΔI_0 is the increment of incident luminous flux during measurement, substituting Eq. (6.19) into Eq. (6.18), we get

$$I_a(T, L)/I_a(T_0, L) = (M_a + \Delta M_a)(I_0 + \Delta I_0)R_a(T)/M_a^* I_a^* R_a(T_0) \tag{6.20}$$

It is clear from Eq. (6.20) that the measurement results vary with the magnification of APD and laser power. Based on this, a reference thermostat with a constant temperature of Tc was introduced at the front end of the sensing fiber, and the improved experimental setup is shown in Fig. 6.3, and the improved algorithm is

$$\frac{I_{a'}(T, L)}{I_{a'}(T_c, L)}/\frac{I_a(T_0, L)}{I_a(T_c, L)} = \frac{R_a(T)}{R_a(T_0)}/\frac{R_a(T_0)}{R_a(T_c)} \tag{6.21}$$

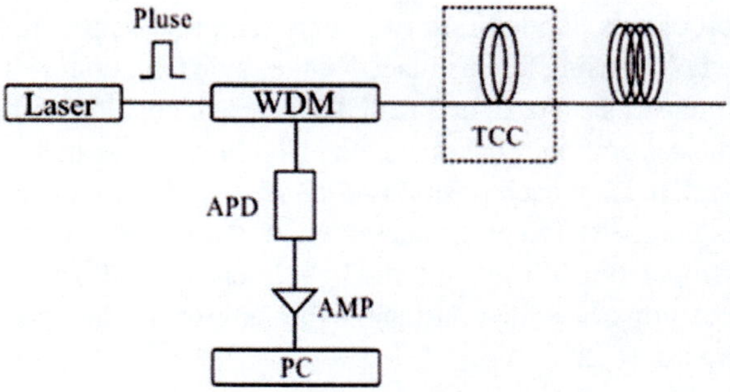

Fig. 6.3 Schematic diagram of single-way demodulation improved structure

Fig. 6.4 Experimental setup of two-way demodulation system

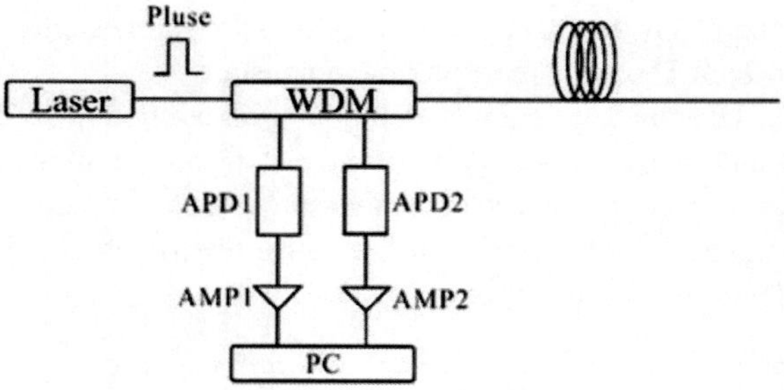

It is obvious from Eq. (6.21) that the change in laser power and the instability of the APD magnification will be automatically corrected by the reference thermostat [6].

In the single-way demodulation algorithm, only the inverse Stokes light in Raman scattering is used, and the biggest problem of this method is that it cannot eliminate the influence of fiber loss on the temperature measurement results, which ultimately leads to the stability and accuracy of temperature measurement are affected. Therefore, the engineers proposed a two-way demodulation algorithm and a system architecture to match it, this is shown in Fig. 6.4.

In the two-way demodulation system, there are two temperature demodulation methods. The first one is to use Raman backward Stokes scattered light and Raman-anti-Stokes scattered light, and the second one is to use Rayleigh backward scattered light and backward Raman-anti Stokes scattered light. In the demodulation mode of Rayleigh scattered light demodulation of anti-Stokes scattered light, the Rayleigh scattered light is stronger than the Raman-anti-Stokes scattered light by about 40 dB, which means that the damage of Rayleigh scattered light is much stronger than that of Raman-anti-Stokes scattered light, which greatly affects the lifetime of the system. The Raman-Stokes light demodulation method is widely used in dual demodulation systems because the light intensity of Raman-Stokes light and the light intensity of anti-Stokes light are close to each other, so there is no problem of affecting the system lifetime. Therefore, it is widely used in two-way demodulation systems, and the following section will focus on this method.

In the Raman-Stokes and anti-Stokes demodulation methods, the measurement process is divided into two steps. The first step is the calibration process, i.e., the whole sensing fiber is placed at the same temperature T0 to obtain the scattered light intensity Is (T0, L) and Ia (T0, L) at each position of the whole fiber. The first step is the calibration process, which is to obtain the scattered light intensity Is (T0, L) and Ia (T0, L) at each position of the fiber at the same temperature T0. The second step is the measurement process, where the sensing fiber is placed at the measurement temperature T The scattered light intensity Is (T, L) and Ia(T, L) are obtained at each position of the fiber during the measurement. Then the individual scattering intensity by Eq. (6.22).

$$\frac{I_a(T,L)}{I_s(T,L)}/\frac{I_a(T_0,L)}{I_s(T_0,L)} = \frac{R_a(T)}{R_s(T)}/\frac{R_a(T_0)}{R_s(T_0)} = \exp\left[-\frac{h\,\Delta\,\nu}{k}\left(\frac{1}{T}-\frac{1}{T_0}\right)\right] \tag{6.22}$$

Since only the temperature to be measured in Eq. (6.22) is an unknown parameter, the others are all known constants or already measured. Therefore, the temperature to be measured can be derived from Eq. (6.22) [7, 8].

In the Rayleigh scattered light demodulation Raman-inverse Stokes scattered light demodulation system, the whole temperature demodulation process is also divided into two steps i.e., the whole sensing fiber is calibrated at the same temperature T_0 and the sensing fiber is measured at the measurement temperature T. However, the Raman-Stokes scattered light intensity is not obtained in these two steps, but the Rayleigh scattered light intensity is obtained instead. The final demodulation equation is expressed as

$$\frac{I_a(T,L)}{I_r(L)}/\frac{I_a(T_0,L)}{I_r(L)} = \frac{R_a(T)}{R_a(T_0)} = \exp\left(\frac{h\,\Delta\,\nu}{kT}\right)/\exp\left(\frac{h\,\Delta\,\nu}{kT_0}\right) \tag{6.23}$$

Equations (6.22) and (6.23) are the demodulation equations for the temperature measurement of the two dual demodulation systems, respectively, and the two dual demodulation systems introduce Rayleigh scattered light and Raman-Stokes scattered light as reference light, respectively. Although this method the optical fiber loss can be effectively removed from the temperature measurement results, but due to the introduction of different central wavelength light, the speed of the two types of light propagation in the fiber is not the same. However, due to the introduction of different central wavelengths, the two types of light do not propagate at the same speed in the fiber, which introduces a new dispersion problem and affects the performance of the system.

In a distributed fiber optic Raman temperature measurement system, the end of the sensing fiber does not have the same refractive index as the air due to the fiber material. The same refractive index of the fiber material and air leads to strong Fresnel reflectometers at the end of the fiber. Figure 6.5a shows the Stokes and inverse Stokes signals received by the acquisition card. The Fresnel reflection peak at the end of the sensing fiber is obvious, and the positions of the reflection peaks of the Stokes signal and the anti-Stokes signal are misaligned. This phenomenon is caused by fiber dispersion. This phenomenon becomes more and more serious as the distance of the sensing fiber increases, which affects the application of dual demodulation system for long-distance sensing.

Therefore, an algorithm is needed to compensate for this. The Stokes and anti-Stokes backscattered light is calculated. at the end of the fiber due to the Fresnel reflection, L1 and L2, respectively, and obtain the difference $Lc = |L1\text{-}L2|$, and finally the position of the Stokes backscattered data point is panned, and the panning algorithm is

$$I_s(L) = I_s\left(L + \frac{L_c}{L_2}L\right) \tag{6.24}$$

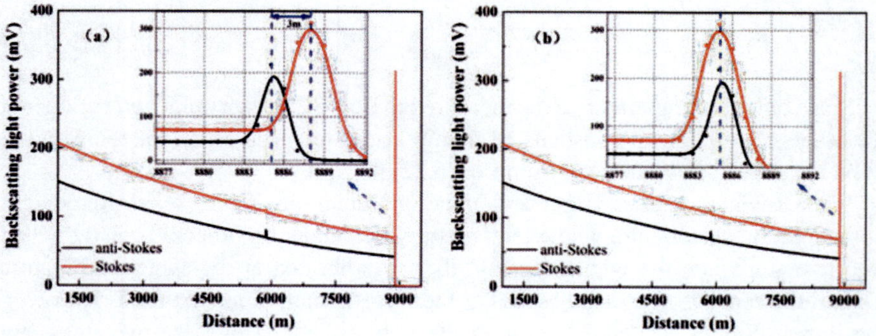

Fig. 6.5 Backscattered light intensity of anti-Stokes and Stokes. **a** Before dispersion compensation, **b** after dispersion compensation

where, Is (L) represents the light intensity of the backscattered Stokes light at L of the fiber. The dispersion can be compensated by Eq. (6.24) to improve the sensing accuracy over long distances in a dual demodulation system.

Temperature measurement experiments based on a single-ended dual demodulated distributed fiber optic temperature sensing system were conducted at a room temperature of 30.0 °C. A sensing fiber of length 13.0 km was used in the experiment. The sensing fibers to be tested were set at 3.07 km and 11.83 km distances, respectively. The temperature of the sensing fibers was controlled by a water bath and set at 40.0 °C, 45.0 °C, 50.0 °C, 55.0 °C, 60.0 °C, and 65.0 °C, respectively. and 65.0 °C, respectively. Figure 6.6 shows the temperature of the sensing fiber to be tested at 3.07 km and 11.83 km. It can be seen the temperature demodulation method based on the single-ended distributed fiber optic temperature sensing system can accurately detect the temperature along 13.0 km.

As shown in Fig. 6.7a, there is a significant temperature rise peak at the front end of the sensing fiber to be measured due to fiber dispersion. When the temperature of the sensing fiber is set at 75.0 °C, the measured temperature at the sensing fiber can reach up to 89.0 °C. The experimental results show that the temperature

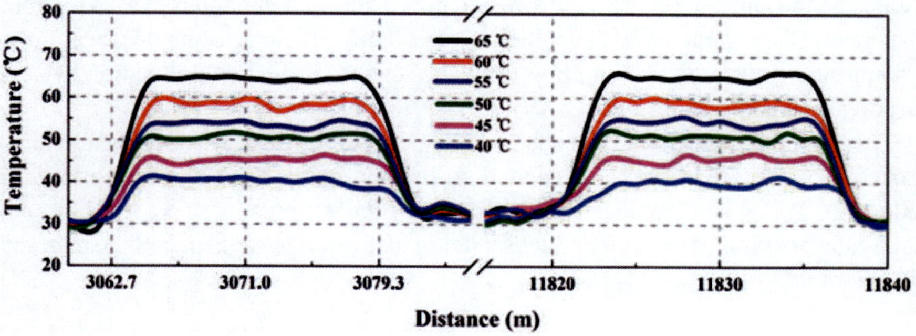

Fig. 6.6 Temperature measurement results along the reference fiber

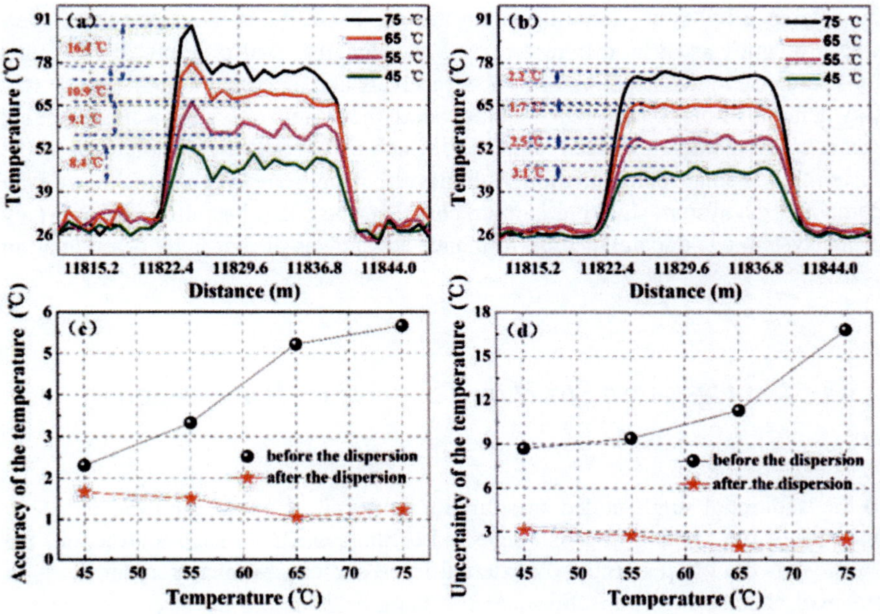

Fig. 6.7 Results of the measured temperature **a** The result of the measured temperature before dispersion compensation, **b** the result of temperature measurement after dispersion compensation, **c** temperature accuracy before and after dispersion compensation, **d** uncertainty of temperature before and after dispersion compensation

measurement accuracy is 13.8 °C, the relative error is 13.9%, and the temperature uncertainty is 16.4 °C. These temperature demodulation results greatly affect the temperature measurement accuracy of the distributed fiber optic sensing system, which is caused by the dispersion of the Stokes and anti-Stokes signals after long distance transmission without calibration.

To eliminate the errors caused by fiber dispersion, the Stokes signal needs to be compensated before the temperature measurement, using the temperature dispersion compensation algorithm mentioned above. Figure 6.7b shows the temperature demodulation results of the sensing fiber to be measured after the dispersion compensation. The experimental results show that the peak temperature rise of the sensing fiber disappears and the temperature profile is smoother than that before the dispersion compensation. Figure 6.7c and d represent the temperature measurement accuracy and temperature uncertainty, which show that the temperature accuracy is optimized from 5.6 to 1.2 °C, and the relative error is reduced from 7.4 to 1.6%. The temperature uncertainty was optimized from 16.4 to 2.4 °C. It can be seen the dispersion compensation algorithm has a significant effect on improving the accuracy of temperature measurement [9, 10].

The main difference between the single demodulation system and the dual demodulation system is that the wavelength division multiplexer filters out only the Raman-inverse Stokes scattered light, i.e. the system uses the same central wavelength as the reference light and signal light. On the one hand, it can overcome the dispersion problem caused by different central wavelengths, it also eliminates the need for symmetry in the selection of optical receivers and electrical amplifiers for two-way demodulation systems. On the other hand, the system only uses a single optical receiver and a single electrical amplifier, which can greatly reduce the cost of the system, so the single demodulation method is also widely used. Although the two-way demodulation system structure is more complex, the overall stability and accuracy of the system has unparalleled performance advantages of the single demodulation system.

6.1.3 Double-Ended Structure Temperature Demodulation Method

In the traditional single-ended structured temperature demodulation method, only one end of the sensing fiber is connected to the system. In such a structure, the sensing fiber is often affected by external force and causes measurement errors in practical engineering applications. At the same time, due to the structural characteristics of the single-ended structure, the single-ended structure-based temperature demodulation method must be pre-calibrated in practice, i.e., the entire sensing fiber needs to be placed in a constant temperature field or curve fitting methods are used to compensate for the different attenuation of the anti-Stokes and Stokes light. Once any component of the sensing system or the sensing fiber needs to be replaced, the entire system needs to be recalibrated before it can start operating to measure temperature. This greatly limits the engineering applications of distributed fiber optic Raman temperature measurement systems [11].

Therefore, the engineers proposed a temperature demodulation method based on a double-ended structure. This structure can obtain both forward and backward Raman scattering signals, and the structure device is shown in Fig. 6.8.

For the dual-ended demodulation system, the pulsed laser incident the pulsed light into the WDM and injected into the sensing fiber, the resulting backscattered light enters the Raman filter through the circulator, the filter filters out the required Stokes and anti-Stokes light and enters both the photoelectric converter and the electrical

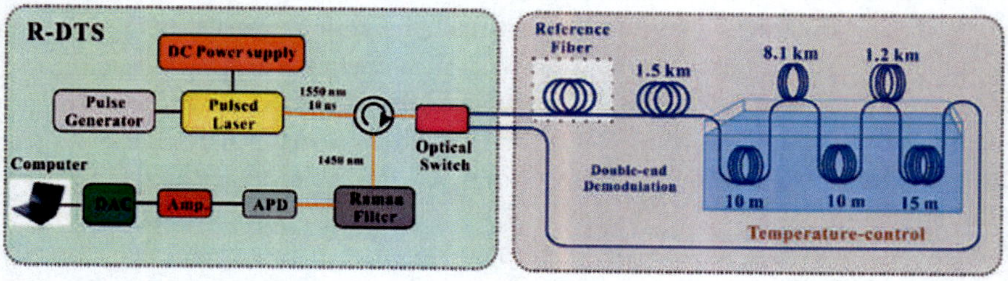

Fig. 6.8 Double-ended temperature demodulation structure device diagram

amplifier, the amplified electrical signal is captured by the data acquisition card and demodulated by different algorithms to obtain the temperature at the sensing fiber.

The double-ended structure temperature demodulation method firstly requires the ring structure to obtain the Raman backscattered light information in different loop directions. The ratio of Stokes light to anti-Stokes light can be expressed by Eqs. (6.25) and (6.26) to express:

$$R_1(T, L) = \frac{I_{s1}}{I_{a1}} = \frac{K_s}{K_a}(\frac{\nu_s}{\nu_a})^4 \exp(-\frac{h \triangle \nu}{kT}) \exp\left[\int_L^l (\alpha_a(L) - \alpha_s(L))dL\right] \quad (6.25)$$

$$R_2(T, L) = \frac{I_{s2}}{I_{a2}} = \frac{K_s}{K_a}(\frac{\nu_s}{\nu_a})^4 \exp(-\frac{h \triangle \nu}{kT}) \exp\left[\int_L^l (\alpha_a(L) - \alpha_s(L))dL\right] \quad (6.26)$$

where, Is1 and Ia1 represent the intensity of Stokes light and anti-Stokes light in the same direction, respectively, Is2 and Ia2 represent the Stokes light intensity and anti-Stokes light intensity in the other direction, respectively, Ks and Ka are the coefficients of the scattering cross section of Stokes and anti-Stokes light respectively, h is Planck's constant, Δν is the frequency shift of Raman scattering from the fiber, k is the is the Boltzmann constant, and T is the measurement temperature, αa and αs are the attenuation coefficients of the anti-Stokes light and Stokes light in the fiber, respectively, and L is the sensing position, and l represents the length of the sensing fiber. The Raman scattering intensity of each position in the sensing fiber loop is obtained by geometrically averaging the Raman scattering signals in different directions, which is calculated as follows:

$$R(T, L) = \sqrt{R_1(T, L)R_2(T, L)} = \frac{K_s}{K_a}(\frac{\nu_s}{\nu_a})^4$$

$$\exp(-\frac{h \triangle \nu}{kT}) \exp\left[\int_L^l (\alpha_a(L) - \alpha_s(L))dL\right] \quad (6.27)$$

From Eq. (6.27), it can be seen the attenuation coefficient $\int_L^l (\alpha_a(L) - \alpha_s(L))dL$ is independent of the position L after the detection of Raman scattering signals in different directions. Therefore, the double-ended temperature demodulation system can effectively eliminate the influence of local external physical perturbations on the temperature measurement results [12].

As already described in the previous section, the calibration process is an important factor affecting the practical application of distributed Raman temperature measurement systems. In the double-ended structure, by setting a reference fiber with a constant temperature in front of the sensing fiber, the Stokes and anti-Stokes signal strengths of the reference fiber can also be obtained from the geometric mean of the two signals in different directions. The Stokes and anti-Stokes signal strengths of the reference fiber can also be obtained from the geometric mean of the two signals in different directions, which is calculated as follows:

$$R_0(T_0, L_0) = \sqrt{R_1(T_0, L_0)R_2(T_0, L_0)} = \frac{K_s}{K_a}(\frac{\nu_s}{\nu_a})^4$$
$$\exp(-\frac{h \Delta \nu}{kT}) \exp\left[\int_L^l (\alpha_a(L) - \alpha_s(L))dL\right] \tag{6.28}$$

The temperature information along the sensing fiber can be obtained by combining Eqs. (6.27) and (6.28), and the specific calculation method is as follows the calculation is as follows:

$$\frac{1}{T} = \left[\ln(\frac{R(T, L)}{R_0(T_0, L_0)})(-\frac{k}{h \Delta \nu})\right] + \frac{1}{T_0} \tag{6.29}$$

In order to verify the proposed double-ended temperature demodulation algorithm, the experiments shown in Fig. 6.9 were carried out. A high-power pulsed laser operating at 1550 nm is used to generate a pulsed laser with a maximum peak power of 30 W, a pulse width of 10 ns, and a frequency of 6 kHz. The receiver consists of a low-noise avalanche photodiode, a transimpedance amplifier, and a high-speed data acquisition card. In order to measure the temperature along the sensing fiber, the sensing fiber is 13 km long and has a gradual refractive index of 62.5/125 μm multimode fiber. Considering the effect of Fresnel reflectometer of the fiber, the length of the fiber was changed to 13 km. A fiber ring of 80 m in length was placed at the end of the sensing fiber. The sensing fiber to be measured consists of two parts, the lengths are 15 m (FUT 1) and 16 m (FUT 2). The two sensing fiber ends are connected to the distributed fiber optic sensing system through optical switches, which alternate the forward and backward pulse signals in the sensing fiber. This allows the pulse signal to be sent alternately forward and backward in the sensing fiber.

The temperature measurement experiments were carried out at a room temperature of 30.0 °C. The sensing fiber used in the experiments was a 13.0 km common multimode fiber, and the sensing fibers to be tested were set at 3.07 km and 11.83 km, respectively. The length of each sensing fiber to be tested is about 15 m, as shown in Fig. 6.10. The temperature of the sensing fibers to be tested was controlled by a water bath and set at 40.0 °C, 45.0 °C, 50.0 °C, 55.0 °C, 60.0 °C, and 65.0 °C. The temperature measurement results are shown in Fig. 6.11, The actual room temperature fluctuation in the experiment was 4.2 °C. In addition, in a double-ended distributed fiber optic sensing system, the temperature demodulation method requires forward and backward Raman backscatter signals to demodulate the temperature information along the sensing fiber. Since the signal as the signal-to-noise ratio decreases along the sensing fiber distance, the measurement results show that the temperature resolution is worse near the end of the sensing fiber distance compared to the middle region of the sensing fiber length. The experimental results show that the temperature resolution based on the double-ended the temperature demodulation method based on a

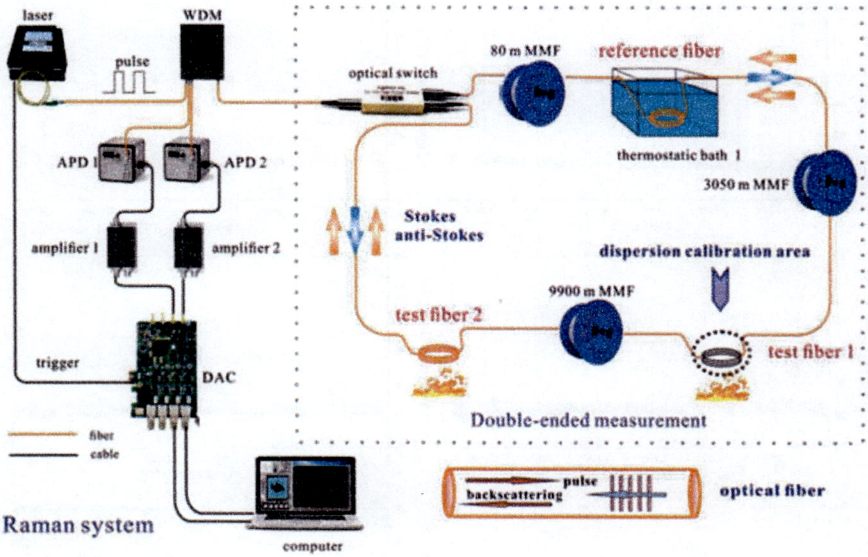

Fig. 6.9 Double-ended distributed fiber optic Raman sensing experimental setup

double-ended fiber optic sensing system can accurately detect the temperature of the 13.0 km sensing fiber. In addition, the double-ended distributed fiber sensing system with reference temperature can omit the calibration process before the temperature measurement [10].

Compared with the single-ended demodulation system, the double-ended demodulation system has two obvious advantages. On the one hand, the geometric averaging of the loop effectively reduces the impact of external local disturbances on the

Fig. 6.10 Double-ended structure with anti-Stokes signals in two different directions

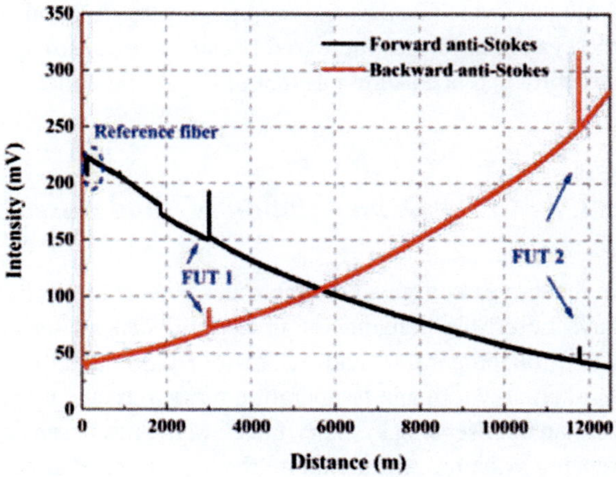

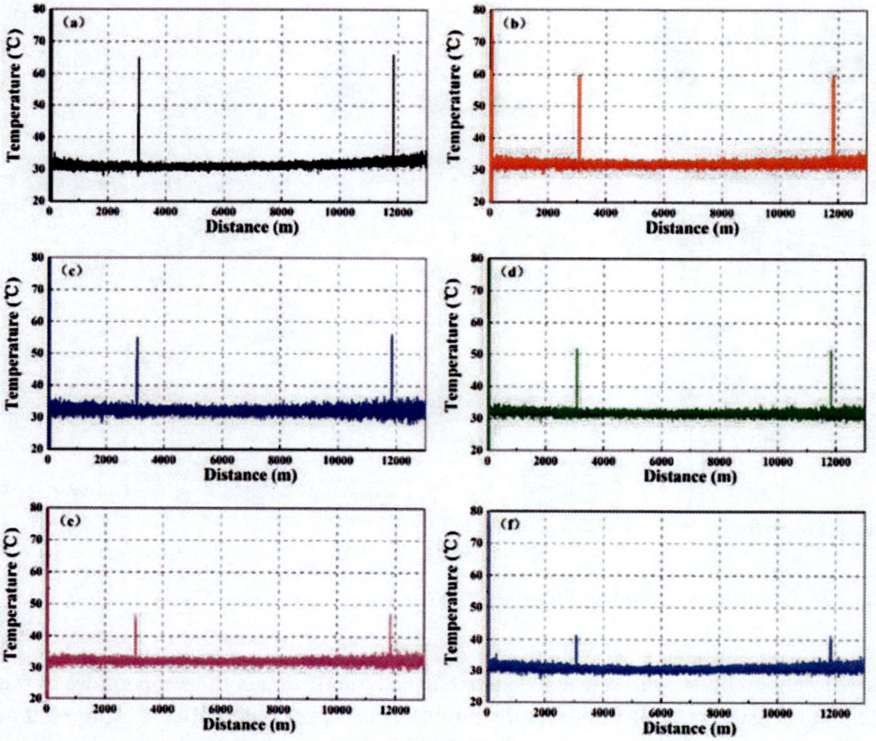

Fig. 6.11 Temperature measurement results of the double-ended structure of the fiber to be tested at different temperatures **a** 65 °C **b** 60 °C **c** 55 °C **d** 50 °C **e** 45 °C **f** 40 °C

temperature measurement accuracy. On the other hand, the double-ended temperature demodulation system can eliminate the calibration step in the single-ended structure, which makes the engineering application more convenient. The cost of the system is further increased because the sensing fiber used in the double-ended structure is twice as long as that of the single-ended system.

6.1.4 Crack Demodulation Method Based on Fiber Loss

In recent years, highway tunnels and other traffic infrastructure monitoring systems have raised the demand for fiber optic sensors for simultaneous measurement of structural temperature and crack width. Currently, there are three main types of structural crack width and temperature measurements using fiber optic sensing systems: Raman fiber sensing systems, fiber Bragg grating sensing systems, and Brillouin fiber sensing systems. Among them, the traditional Raman fiber sensing system uses the spontaneous Raman scattering effect for temperature detection. Fiber Bragg grating sensing system is a quasi-distributed sensor with high resolution and measurement accuracy.

However, it has some limitations in long-distance detection applications such as remote monitoring of tunnels. Brillouin fiber optic sensing system can extract temperature and strain information by Brillouin frequency shift, according to the demodulation principle, mainly can be divided into time domain demodulation system (BOTDR, BOTDA) and relevant domain demodulation system (BOCDR, BOCDA) [13–18].

Currently, Taiyuan University of Technology has proposed a Raman scattering-based crack measurement method for optical fiber loss, which the method mainly uses the sensing optical fiber to sense the change of external stress and the optical fiber loss changes with it. The optical fiber loss changes with the sensed stress change, and the structural crack width information is extracted by the Raman scattering loss property based on OTDR technique. Compared with the inverse Stokes signal, the sensing fiber will be more sensitive to the temperature. When external tension (caused by structural cracks) is applied to the sensing fiber, the sensing fiber will deform. This deformation affects the effective scattering area in the sensing fiber, which eventually leads to loss of light passing through the location. In addition, the loss characteristics of the sensing fiber will be affected by the different levels of tension applied in the sensing fiber. In addition, the loss characteristics will appear in different states when different levels of tension are applied to the sensing fiber. That is, as the external tension increases, the fiber loss will also increase. The schematic diagram of the measurement principle is shown in Fig. 6.12.

A diagram of the experimental setup for structural crack detection by distributed fiber optic sensing system is shown in Fig. 6.13. The device It mainly consists of a pulsed laser, a wavelength division multiplexer, a high-speed data acquisition card, an optical switch, and an avalanche photodiode. The pulsed laser from the pulsed laser is switched into the loop sensing fiber by the optical switch, and the different The Raman scattered light of different wavelengths is filtered by the WDM, and

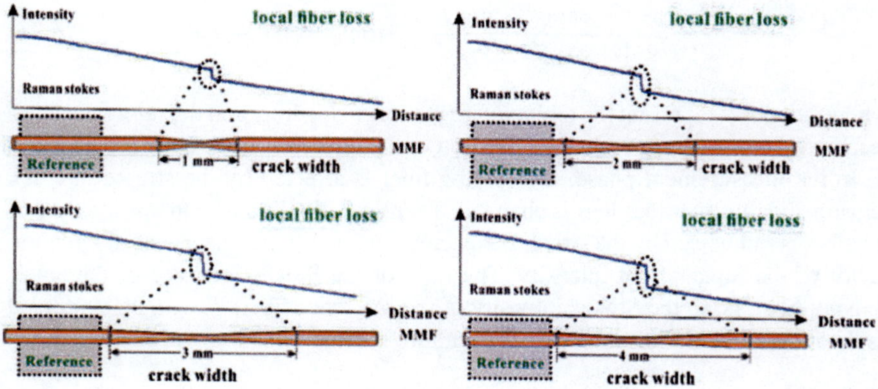

Fig. 6.12 Schematic diagram of crack width measurement based on local fiber loss

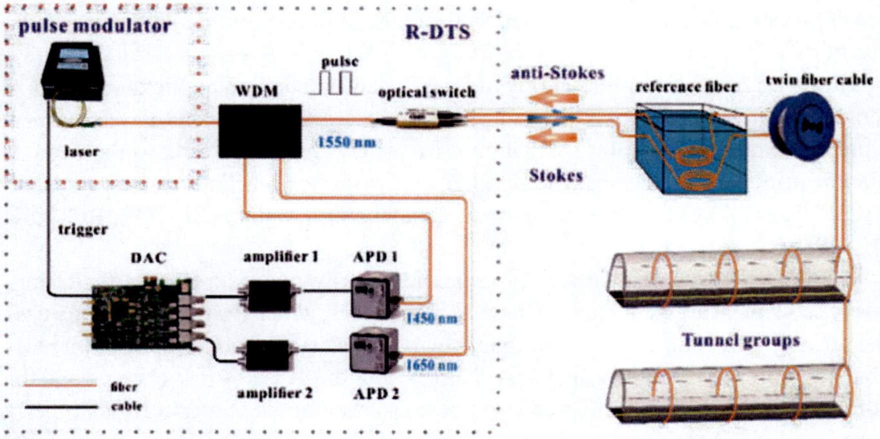

Fig. 6.13 Schematic diagram of crack demodulation structure based on fiber loss

then enters the APD to convert the optical signal into telecommunication signals. signal into telecommunication signal, and then it is captured by a high-speed data acquisition card.

When the sensing fiber is kept in its natural state without external forces, the Raman-Stokes scattering intensity obtained by the system can be expressed as Isc can be expressed as

$$I_{sc} = K_s v_s^4 \frac{1}{1 + \exp(h \bigtriangleup v / kT_c^r)^{-1}} (\alpha_0 + \alpha_s) L \tag{6.30}$$

In addition, the ratio of the Raman-Stokes intensity of the entire sensing fiber to the Raman-Stokes intensity of the reference fiber can be expressed as

$$\frac{I_{sc}}{I_{sc}^{re}} = \frac{1 + \exp(h \bigtriangleup v / kT_c^r)^{-1}}{1 + \exp(h \bigtriangleup v / kT_c)^{-1}} (\alpha_0 + \alpha_s)(L - L_r) \tag{6.31}$$

where Tc and Tr represent the temperature at the sensing fiber and the reference fiber, respectively, and Lr represents the position of the reference fiber.

In the measurement phase, the sensing fiber is affected by the structural crack tension and thus the fiber loss is changed. Therefore, the Raman-Stokes intensity of the fiber is influenced by the structural tension. Define S(L) as the tensor modulation factor of the Stokes light intensity, The ratio of the Stokes intensity of the entire sensing fiber, Is, to the Stokes intensity of the reference fiber, I_s^{re}, can be expressed as the ratio of the Stokes intensity of the entire sensing fiber I can be expressed as

$$\frac{I_{sc}}{I_{sc}^{re}} = S(L) \frac{1 + \exp(h \bigtriangleup v / kT_c^r)^{-1}}{1 + \exp(h \bigtriangleup v / kT_c)^{-1}} (\alpha_0 + \alpha_s)(L - L_r) \tag{6.32}$$

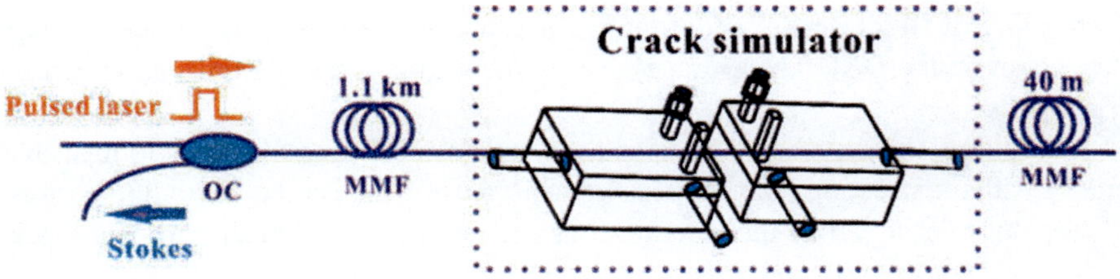

Fig. 6.14 Diagram of the experimental setup for measuring the width of structural cracks

where, T_c^r represents the temperature of the reference fiber. Therefore, the crack width information can be obtained by Eqs. (6.31) and (6.32) The demodulation equation of the crack width information can be expressed as

$$S(L) = \frac{I_{sc}}{I_s^{re}}\frac{I_{sc}^{re}}{I_{sc}} = \frac{1 + \exp(h \bigtriangleup \nu/kT^r)^{-1}}{1 + \exp(h \bigtriangleup \nu/kT)^{-1}}\frac{1 + \exp(h \bigtriangleup \nu/kT_c^r)^{-1}}{1 + \exp(h \bigtriangleup \nu/kT_c)^{-1}} \tag{6.33}$$

From Eq. (6.33), it can be seen when the distributed fiber Raman system obtains the temperature data along the sensing fiber The temperature data along the sensing fiber can be obtained by the distributed fiber optic Raman system, which can be used to monitor the cracks in the structure. In order to verify the feasibility of the proposed crack detection method, a crack width measurement was carried out as shown in Fig. 6.14. The crack width simulator (HWHR Instruments, A053F) was used to apply axial tension on the multimode sensing fiber to simulate the generation of structural cracks. The 1.1 km of multimode sensing fiber was connected to an optical circulator at one end. to the optical loop, and the other end to the fiber tensioning platform, and the end of the sensing fiber, 40 m fiber ring is connected to the other end of the optical the other end of the sensing fiber is connected to the other end of the fiber tensioner.

In the experimental stage, different levels of axial tension were applied to the multimode sensing fiber at 1.1 km, and then the Raman-Stokes signal is collected by DAC and demodulated by the above-mentioned demodulation steps to obtain the specific experimental results are shown in Fig. 6.15. Figure 6.15 shows the variation of the Stokes local loss at the 1.1 km position for different axial tensions (crack widths). From the experimental results, it can be seen, the local fiber loss varies considerably at 1.1 km where the structural crack is generated. At different axial tensions, the local fiber loss shows different states at different axial tensions. Specifically, as the crack width increases, the local fiber loss also increases. The experimental results are the same as the theoretical analysis. The relationship between the fiber loss coefficient and the crack width is shown in Fig. 6.16, which shows that the fiber loss coefficient remains in the range of 1.6~5.6 mm. This implies that the crack width information can be obtained from the local fiber loss using Raman-Stokes light, and the effective measurement range can be 1.6~5.6 mm, which is suitable for large-scale crack width monitoring of structures. In addition, the measurement time of this method is only dependent on the DAC The average measurement time is up to 1.04 s, as shown in Fig. 6.16b.

In addition, the distributed fiber optic sensing system uses fiber loss to measure cracks without affecting its own measurement performance and temperature performance. In order to verify the simultaneous measurement of temperature and crack

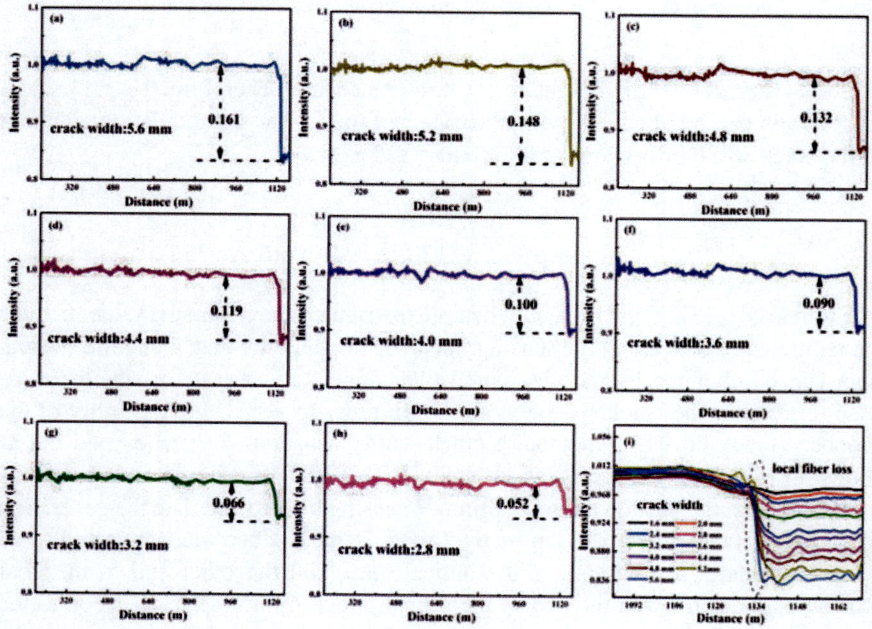

Fig. 6.15 The overall distribution of fiber loss results for different crack widths, the fiber axial tensions are **a** 5.6 mm, **b** 5.2 mm, **c** 4.8 mm, **d** 4.4 mm, **e** 4.0 mm, **f** 3.6 mm, **g** 3.2 mm, **h** 2.8 m, **i** overall distribution of fiber axial tension

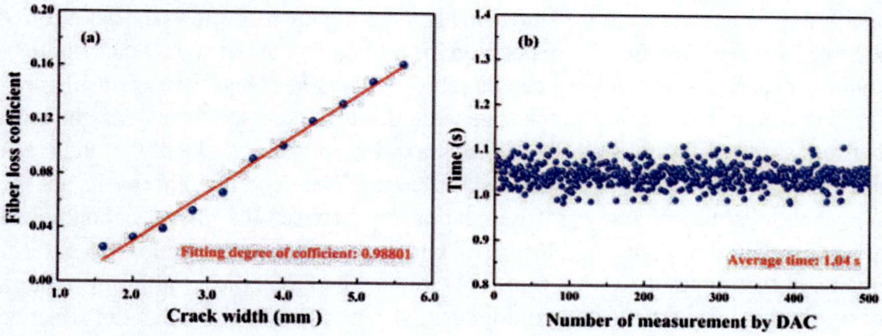

Fig. 6.16 Crack width characterization based on fiber loss effect. **a** Relationship between fiber loss coefficient and crack width information; **b** measurement time of the system

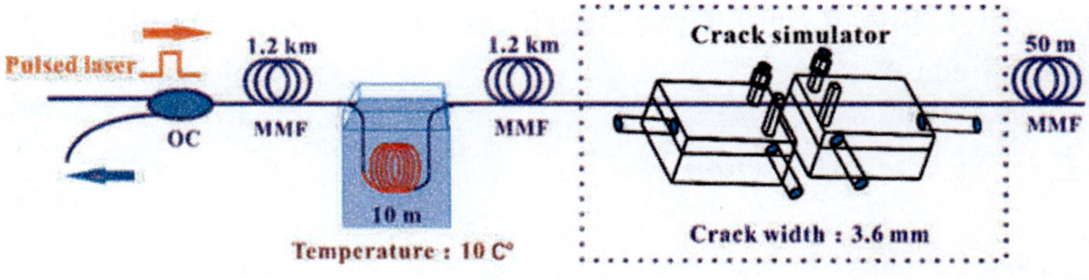

Fig. 6.17 Schematic diagram of the structure of the experimental setup

width using the same sensing fiber, the experimental setup is shown in Fig. 6.17. FUT 1(60 m) was placed at 1.2 The temperature was set to 10.0 °C.

The FUT2 (1 m) was placed at 2.4 km and stretched using a fiber stretching platform (stretching length of 3.6 mm) to generate axial tension. The results of temperature and fiber loss distribution along the fiber are shown in Fig. 6.18. The temperature distribution along the sensing fiber is shown in Fig. 6.18a, which can clearly distinguish the temperature distribution at FUT 1 (10.0 °C) and the ambient temperature (23.0 °C). Figure 6.18b shows the results of the fiber loss distribution.

From Fig. 6.18b, it can be seen the fiber loss coefficient at 2.4 km is 0.090. Based on the relationship between fiber loss and crack width in Fig. 6.16a, it can be observed that the detected crack width at 2.4 km is 3.6 mm. the experimental results show that the monitored crack width is consistent with the parameters set by the fiber stretching platform. This experiment shows that the fiber optic sensor can be used to detect temperature and a wide range of crack conditions with a single sensing fiber.

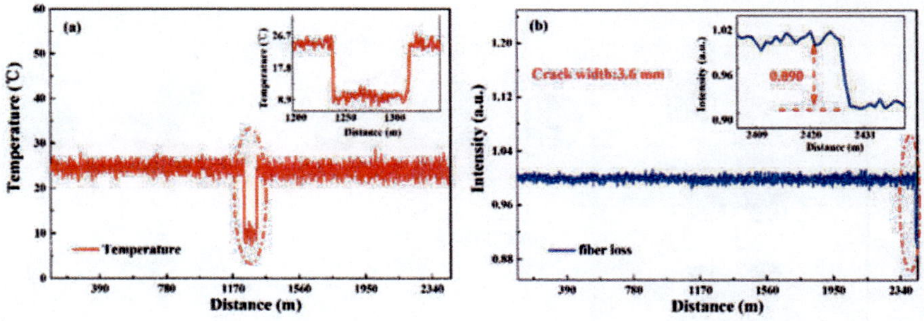

Fig. 6.18 Simultaneous temperature and crack width measurements using a single sensing fiber. **a** Temperature measurement results along the sensing fiber, **b** fiber loss distribution results due to crack width

6.2 High-Speed Real-Time Distributed Fiber Optic Raman Thermometer

6.2.1 System Integration

Section 6.1 describes in detail the physical mechanism of the distributed fiber optic Raman temperature measurement system, the principle of temperature demodulation, and the proposed new temperature demodulation method and new Raman distributed fiber optic sensing technology. To realize the application transformation of the research results, the research results are integrated into the development of the prototype of the optical fiber Raman thermometer, the team has developed a high-speed real-time distributed optical fiber Raman thermometer for engineering applications.

The experimental setup of the high-speed real-time distributed fiber Raman thermometer is shown in Fig. 6.19. A high-power pulsed laser with a wavelength of 1550 nm generates a narrow pulse width light signal into the multimode fiber (MMF), and its backward Raman scattering light is filtered out by a wavelength division multiplexer to 1450 nm Raman-anti-Stokes scattering light and 1650 nm Raman-Stokes scattering light and coupled to an avalanche photodetector. Due to the anti-Stokes optical signal being weak, an APD and a signal amplifier are used to convert and amplify the optical and electrical signals, the converted voltage is passed through a high-speed data acquisition card and a computer for data acquisition and temperature demodulation.

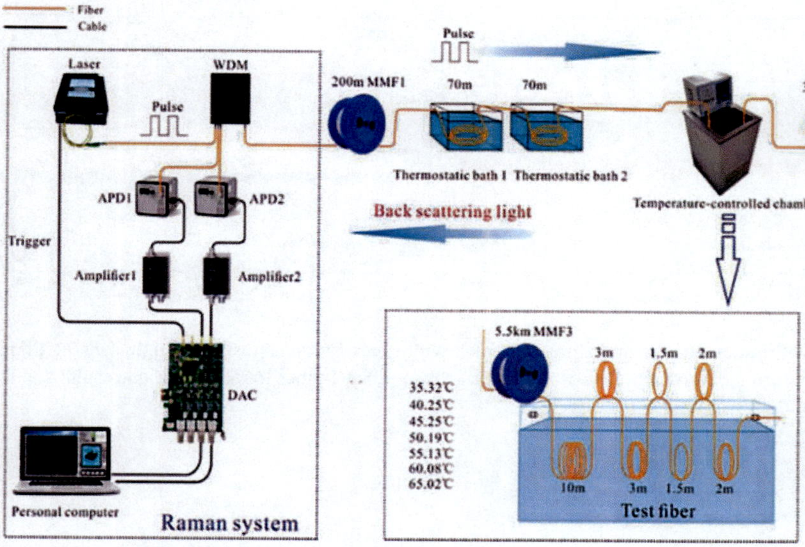

Fig. 6.19 High-speed real-time distributed fiber optic Raman thermometer set-up diagram

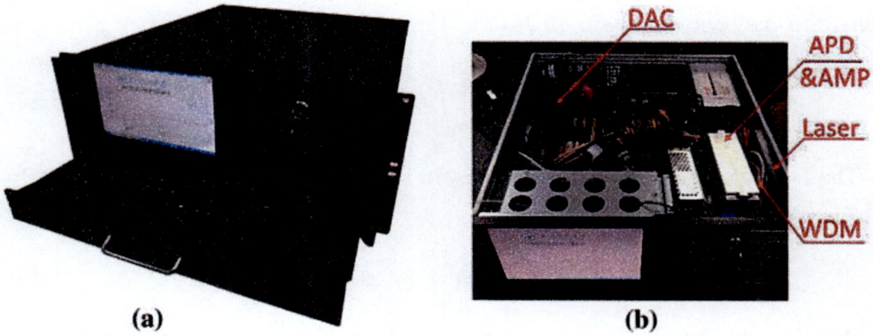

Fig. 6.20 High-speed real-time distributed fiber optic Raman thermometer. **a** Entity diagram of the thermometer, **b** internal device distribution

The appearance of the integrated high-speed real-time distributed fiber-optic Raman thermometer and the distribution of internal components are shown in Fig. 6.20.

6.2.2 *Wavelet Mode Maximum Denoising Method*

In a distributed fiber optic Raman temperature measurement system, the spatial temperature information is obtained using the principle of spontaneous Raman scattering and optical time domain reflection generated when light is transmitted in the fiber [19]. As the Raman scattering signal is nearly 30 dB weaker than the Rayleigh scattering signal, the Raman backscattering signal in the fiber is very weak [20]. And the avalanche photodetector, low-noise amplifier, and acquisition system will bring a lot of noise to the backscattering signal, so the acquired backscattering signal will often be drowned in noise. The traditional solution is cumulative averaging, i.e. increasing the number of cumulative averages to enhance the signal denoising capability, but increasing the number of cumulative averages means increasing the system data processing time, thus increasing the single measurement time, which does not meet the requirements of real-time monitoring of the object to be measured [21–23]. To solve this problem, the wavelet mode maximum method [24, 25] can be used to denoise the Raman signal. The wavelet mode maximum denoising method, as one of the wavelets transforms, inherits the advantages of the wavelet transform to overcome the shortcomings of traditional spectral analysis methods, and uses a window of variable size and position to analyze the signal, which can meet the requirements of signal time–frequency localization processing, and effectively eliminate the noise in the abruptly changing signal. The principle and implementation are as follows.

Family of wavelet functions $\varphi_{ab}(t) = |a|^{-0.5}\varphi(\frac{t-b}{a})$ is obtained from the wavelet basis functions φ(t) by means of the scale factor a and a position factor b obtained by scaling and displacement. For a continuous signal f (t) the continuous wavelet transform can be given by Eq. (6.34)

$$W_f(a, b) = |a|^{-0.5} \int f\ (t)\varphi^*(\frac{t-b}{a})dt \tag{6.34}$$

The family of wavelet functions is discretized, i.e. $a = a_0^m$, $b = na_0^m b_0$, usually a0 = 2, b0 = 1can be obtained

$$\varphi_{m,n}(t) = 2^{-0.5m}\varphi(2^m t - n)\, m, n \in Z \tag{6.35}$$

Based on this, Mallat used the multi-resolution analysis (MRA) method to unify the construction of wavelet bases and proposed the now widely used method of fast decomposition and reconstruction of Mallat wavelets. The specific operation principle of the Mallat decomposition and reconstruction for orthogonal wavelets is shown in Fig. 6.21. The original signal f is decomposed into high-frequency detail coefficients g1 by a high-pass (band-pass) filter H, and into a low-frequency approximation s1 by a low-pass filter L. The low-frequency approximation s1 is then decomposed into high-frequency detail coefficients g2 by a high-pass (band-pass) filter H, the low-pass filter L is then decomposed into low-frequency detail coefficients s2, and so on, until it is decomposed into the right number of layers. The wavelet reconstruction, on the other hand, does the opposite, by reconstructing the last decomposed high-frequency detail coefficients into the last decomposed low-frequency approximation coefficients through the inverse of the high-pass (band-pass) filter H' and the last decomposed low-frequency approximation coefficients through the inverse of the low-pass filter L'. The low-frequency approximation coefficients and the high-frequency detail coefficients from the previous decomposition are then combined into the low-frequency detail coefficients from the previous decomposition through the inverse L' of the low-pass filter and the inverse H' of the high-pass (band-pass) filter, and so on until the original signal f is reconstructed.

The wavelet mode extremum denoising method, as a kind of wavelet transforms, maintains the same basic decomposition reconstruction idea as Fig. 6.21, and the core of its denoising is to process the coefficients after wavelet decomposition [26]. The wavelet coefficients of the scattering intensity curve at the actual temperature of 30.10 °C are used to analyze the denoising implementation of the wavelet mode maximum, where the wavelet coefficients of this scattering curve are shown in

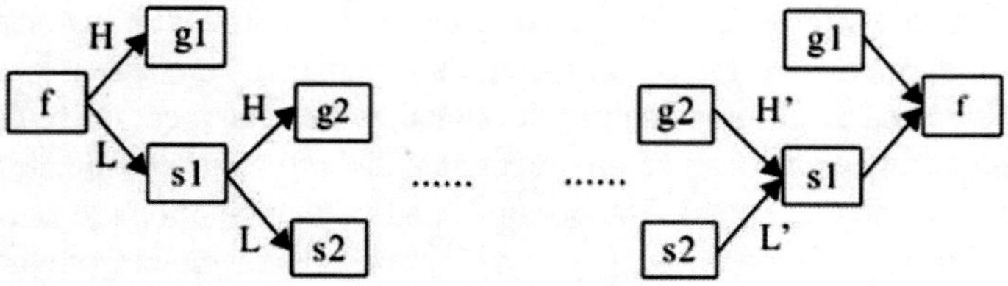

Fig. 6.21 Schematic diagram of wavelet decomposition reconstruction

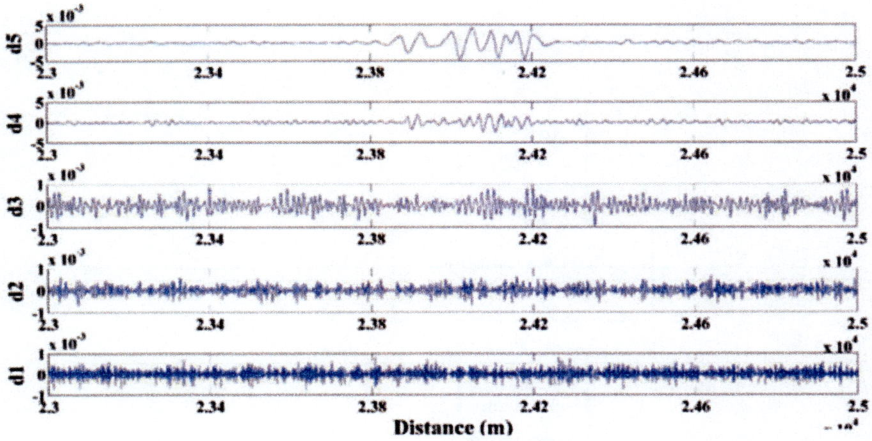

Fig. 6.22 Modal maxima corresponding to wavelet coefficients of different decomposition layers

Fig. 6.22. In Fig. 6.22, the number of wavelet decomposition layers from top to bottom is decomposition 5, 4, 3, 2 and 1 layers respectively, where the horizontal coordinate is the sensing distance and the vertical coordinate is the coefficient size. As can be seen from the graph, as the number of wavelet decomposition layers rises, the modal maximum of the temperature signal at detection area 1 gradually increases, while the noise signal gradually decreases. According to this characteristic of the signal and noise, the wavelet mode maxima denoising process is shown in Fig. 6.23.

The wavelet mode extreme value denoising method first decomposes the original signal by selecting the appropriate number of decomposition layers. The number of decomposition layers can be determined by the change of the modal maxima of the wavelet coefficients, i.e. the ratio of the wavelet modal maxima coefficients to the noise maxima coefficients is maximum at the maximum number of decomposition layers, and the ratio increases with the increase of the number of decomposition layers. After obtaining the coefficients of the maximum decomposition layer (d5), the modal maxima of the coefficients were extracted, and then the extracted modulus maximum value is judged by the threshold.

If it is greater than the threshold value, the modal maxima will be retained, otherwise, the modal maxima will be set to zero. The retained modal maxima and other coefficients are reconstructed to the previous layer (d4) and compared with the original coefficients of that layer (d4), if there is a modulo maximum of the corresponding symbol, the modulo maximum value is retained, otherwise, the value is zeroed, while the number of layers (d4) and other modulo maxima values that meet the threshold are retained, and then the modulo maximum value retained by both d5 and d4 is compared to the upper layer d3 of d4, and so on, until the first decomposition layer. At this time, the original coefficient, the retained modulo maximum value coefficient, and the change to zero coefficient are reconstructed to achieve wavelet mode maximum denoising.

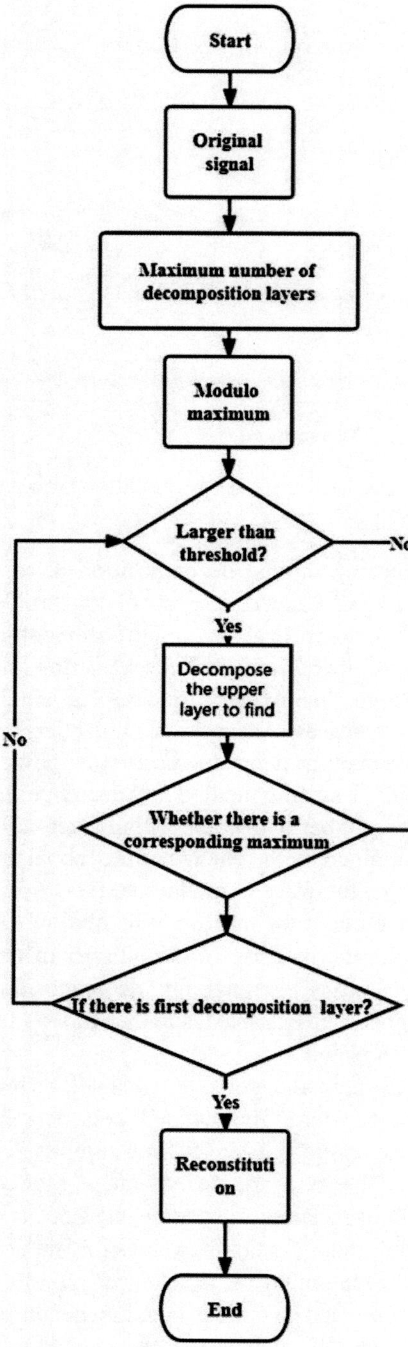

Fig. 6.23 Flow chart of wavelet mode extreme value denoising

As one of the wavelet transforms, the wavelet mode maximization method is extremely suitable for detecting sudden changes in Raman scattering intensity due to temperature changes, and can effectively reduce the cumulative averaging times of the system, thus reducing the measurement time of the system, while ensuring the temperature measurement accuracy.

To verify the effectiveness of wavelet mode denoising in Raman temperature measurement, the wavelet mode denoising method was used in Raman temperature measurement experiments, and the results are shown in Fig. 6.24. The graph shows the Raman-anti-Stokes scattering light intensity before and after wavelet mode denoising for a test interval of 65 °C and an incident light power of 14.6 mW, where the original signal is the Raman scattering signal after averaging 10,000 times. Figure 6.24a shows the full-fiber intensity distribution, where the small plot shows the local Raman scattering light intensity. Figure 6.24b shows the scattering intensity before and after wavelet denoising at the end of the fiber, with 5 layers of wavelet decomposition and the wavelet basis function "sym6". The wavelet decomposition layer is 5 and the wavelet basis function is "sym6". The noise is reduced from 2.64 to 0.65 mV, the signal after denoising is the same as the original signal, both in the temperature range and at the end of the fiber at room temperature, and the signal at the mutation point can be extracted well.

To demonstrate that the method can effectively reduce the fluctuation of temperature measurement results, three different test interval temperatures (TCC temperature) were set to 65 °C, the incident optical power was set to 14.6 mW, and the temperature of the reference thermostat tank was set to 26 °C. The signals before and after denoising were demodulated separately, where the number of wavelet decomposition layers was 5 and the wavelet basis function was"sym6". The demodulation results are shown in Fig. 6.25, where the original signal is a Raman scattering signal with a cumulative average of 10,000 times. Figure 6.25a–c shows the temperature measurement results at the 10 m fiber ring in each test section with the maximum temperature fluctuation after wavelet transform mode extremes, compared with the temperature measurement results without wavelet transform.

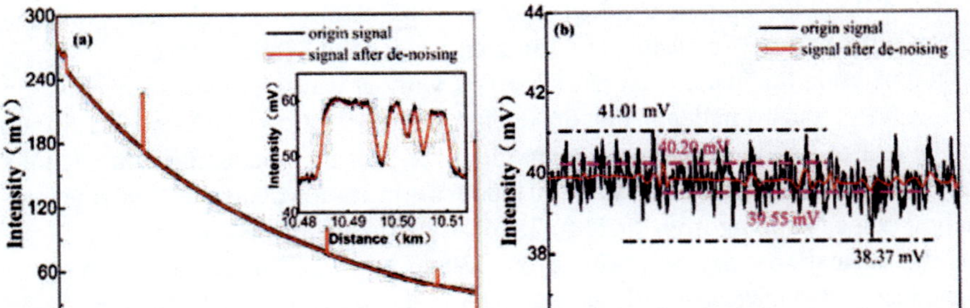

Fig. 6.24 Raman anti-Stokes light intensity diagram before and after wavelet mode extreme denoising. **a** Full fiber, **b** fiber tail

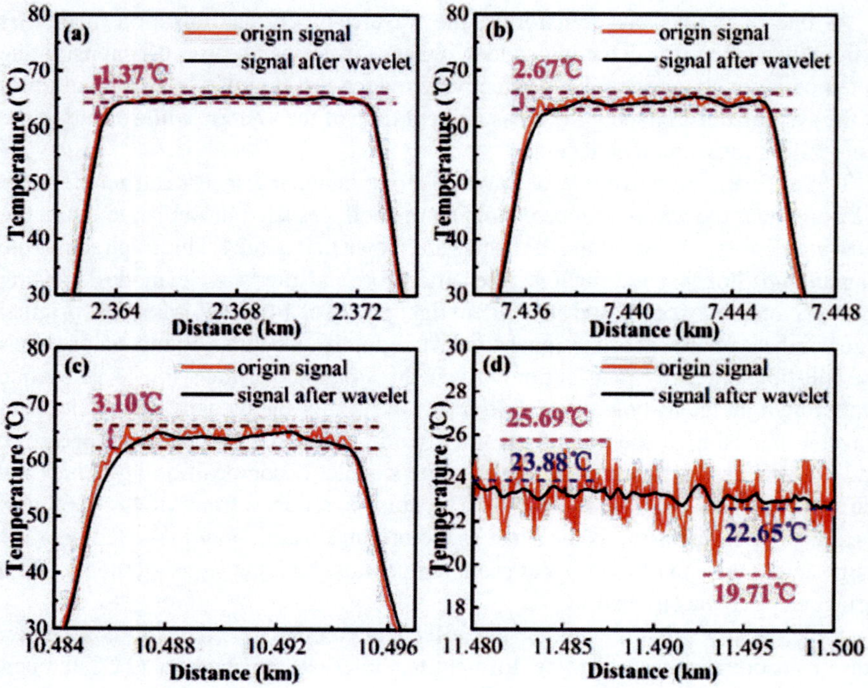

Fig. 6.25 Demodulation results of the original signal and the wavelet denoised signal at different sensing distances. **a** Test point 1, **b** test point 2, **c** test point 3, **d** end of fiber

Figure 6.25d shows the temperature demodulation of the original and wavelet-transformed signals at room temperature at the end of the fiber

As can be seen from Fig. 6.25, with the increase of sensing distance, the effect after wavelet mode extreme value denoising becomes more and more obvious. At test point 1 (about 2.36~2.39 km), the temperature fluctuation was 1.37 °C before wavelet denoising and 0.76 °C after wavelet denoising, and at test point 2 (about 7.43~7.47 km), the temperature fluctuation was 2.67 °C before wavelet denoising and 1.56 °C after wavelet denoising. In particular, at test point 3 (about 10.48~10.52 km), the temperature fluctuation can reach 3.10 °C without wavelet mode denoising, and after wavelet transform denoising, the temperature fluctuation is reduced to 1.60 °C. At the end of the fiber, it is clear from the graph that the temperature fluctuations become smaller after wavelet transform mode maximization denoising, with the fluctuation range changing from 19.71 to 25.69 °C before denoising to 22.65–23.88 °C, and the fluctuation range decreasing from 5.98 to 1.23 °C. This phenomenon is due to the increase of the sensing distance and the attenuation effect in the fiber, resulting in a gradual weakening of the signal strength, especially at the end of the fiber, where the signal is completely drowned in noise. At test point 1, the signal strength was 230.5 mV, the noise fluctuation was 1 mV before the wavelet denoising, and 0.1 mV after the denoising; at test point 3, the signal strength was 60.8 mV, and the noise fluctuation before and after the denoising remained the same as at test point 1. Especially at the end of the fiber, the signal strength was 40 mV due to the attenuation of the fiber and the room temperature, and the noise fluctuation before and after denoising was the same as at test points 1 and 2. Although the noise fluctuation is the same before and after the fiber, the signal strength gradually decreases as the distance rises, and the weakening of the strength leads to an increasing ratio of noise fluctuation to signal, which means that the demodulated signal caused by the noise fluctuation will also increase. Therefore, in the case of the same noise fluctuation change, the weaker the signal strength, the more obvious the denoising effect.

6.2.3 Early Warning Model Based on Real-Time Movement Method

The real-time moving method is based on a set of recent actual data to build an early warning model to predict future changes in temperature values along the fiber optic route, the early warning model consists of a primary moving average method and a second moving average method [27]. The temperature warning model consists of three main aspects, as shown in Fig. 6.26, and is implemented as follows.

(1) Fixed threshold alarm model: If the temperature information $T \geq Th$ (Th is the fixed alarm threshold set by the system) is demodulated by the distributed fiber optic Raman thermometer at each point along the fiber, the system displays the location of the alarm point.
(2) Temperature difference alarm procedure: distributed fiber Raman thermometer demodulation of the same position within a continuous period of the previous temperature data T1 and the next measurement of the temperature data T2 and take the absolute value of $Ta = |T1 - T2|$, if $Ta \geq Th$ (Th is the system set alarm threshold), the system displays the location of the alarm point.
(3) Temperature prediction warning procedure: Raman thermometer will measure the same position in continuous time 5 sets of temperature data, respectively set to $T_{t\text{-}4}$, $T_{t\text{-}3}$, $T_{t\text{-}2}$, $T_{t\text{-}1}$, T_t (T_t is the time t, Raman thermometer demodulation of temperature information, $T_{t\text{-}n}$ is the time t of the first n Raman thermometer demodulation of temperature information), one by one to compare the five sets of temperature data whether the time $T_{t\text{-}4}, < T_{t\text{-}3} < T_{t\text{-}2} < T_{t\text{-}1} < Tt$. If the condition is met, the system enters the second moving average prediction method, otherwise, it enters the primary moving average prediction method.

The one-time moving average forecasting method is:

$$T_{pr} = (T_{t-4} + T_{t-3} + T_{t-2} + T_{t-1} + T_t)/5 \tag{6.36}$$

T_{pr} is the temperature value for the customized prediction time.
The quadratic moving average method is:

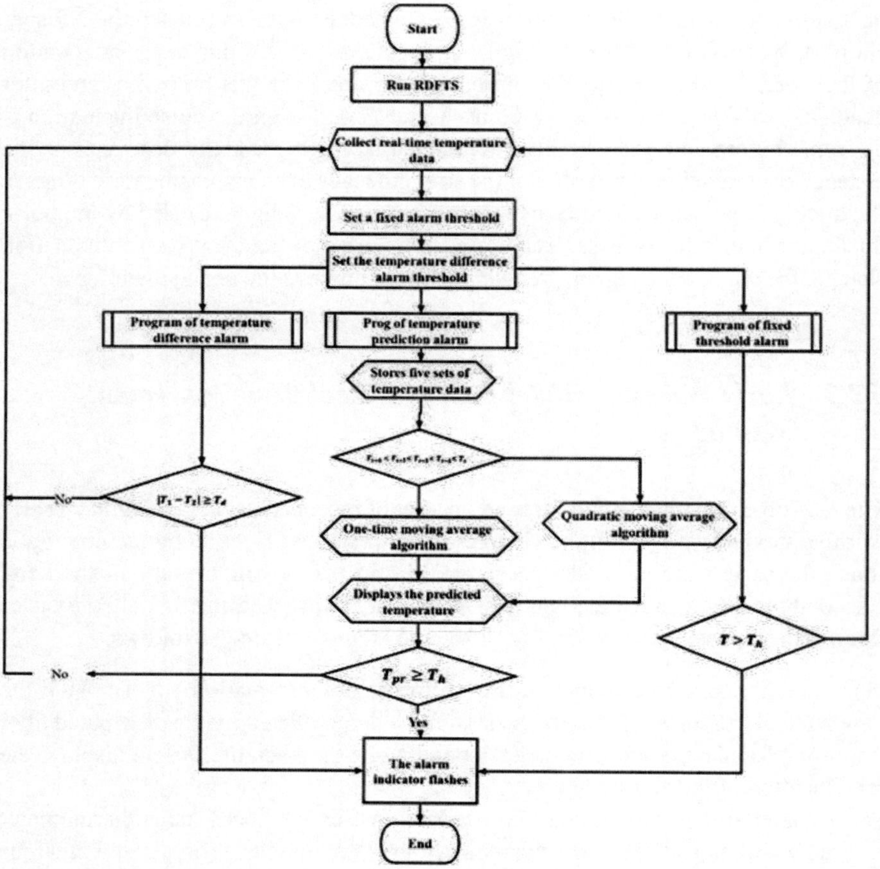

Fig. 6.26 Flow chart of temperature warning based on "moving average method"

$$T_{pr} = a_t + b_t(\frac{T_Y - T_N}{t_1}) \tag{6.37}$$

T_Y is the user-defined prediction time, T_N is the current time, t_1 is the time it takes for the Raman thermometer to cdemodulate one temperature time required for the data, and the time units of T_N, T_Y and t_1 must be the same. Actually

$$a_t = 2T_t^{(1)}T_t^{(2)} \tag{6.38}$$

$$b_t = \left[T_t^{(1)} - T_t^{(2)}\right]/2 \tag{6.39}$$

and

$$T_t^{(1)} = (T_{t-4} + T_{t-3} + T_{t-2} + T_{t-1} + T_t)/5 \tag{6.40}$$

$$T_t^{(2)} = \left[T_{t-4}^{(1)} + T_{t-3}^{(1)} + T_{t-2}^{(1)} + T_{t-1}^{(1)} + T_t^{(1)}\right]/5 \tag{6.41}$$

The system compares the above calculated full fiber predicted temperature data Tpr with the input fixed alarm threshold Th, if Tpr > Th, the Raman system shows the position of the alarm point.

In the moving average forecasting method described in reference [29], the set of historical data in the one-time moving average forecasting method is continuously updated as time moves forward, and each time the Raman thermometer demodulates a new set of historical data, the one furthest from the original set of historical data is removed. The quadratic moving average method is a second-time moving average of the primary moving average, then the prediction model is built based on the primary moving average and the quadratic moving average.

1. **Indoor early warning experiment**

In the indoor experiment, the fiber to be measured by the distributed fiber Raman thermometer was placed in a high-precision thermostat, and then the temperature of the thermostat was adjusted randomly and the early warning system started to collect data. The measured and predicted data are shown in Fig. 6.27.

With the black curve showing the actual temperature measured by the Raman thermometer and the red curve showing the predicted temperature by the Raman thermometer. Figure 6.27b shows a partial enlargement of detection area 1. Table 6.1 shows the advanced response times for different areas of the fiber to be measured. The experimental results show that the early warning model can predict the temperature trend along the fiber in advance.

To further determine the prediction accuracy based on the moving average prediction method, the absolute and relative errors between the predicted and measured temperatures were analyzed, and histograms of the prediction errors were derived. Figure 6.28a shows the average error between the predicted and measured temperatures. The results show that the average prediction error for the predicted temperature is 0.33 °C and the average relative error is 1%. Figure 6.28b shows the histogram of the prediction error rate of temperature. The prediction error is 80.49% in the range of −0.5~0.5 °C and 98.07% in the range of −1.0~1.0 °C. The results show that the predicted temperature and the measured temperature are in good agreement, and the experimental results can meet the requirements of the distributed fiber optic Raman temperature measurement system for the next stage of temperature prediction.

2. **Outdoor warning experiment**

To further verify the accuracy of the real-time mobile method for predicting the temperature of distributed fiber optic Raman thermometers, an armored fiber optic cable was used to simultaneously measure and predict the temperature of a large outdoor building for a total of 24 h. In the experiment, the sensing fiber optic cable was laid on the ground and fence of the outdoor building (Taiyuan University of Technology's Museum of Learning) for monitoring and predicting the outdoor building temperature. There are two ways of laying the fiber optic cable in this experiment,

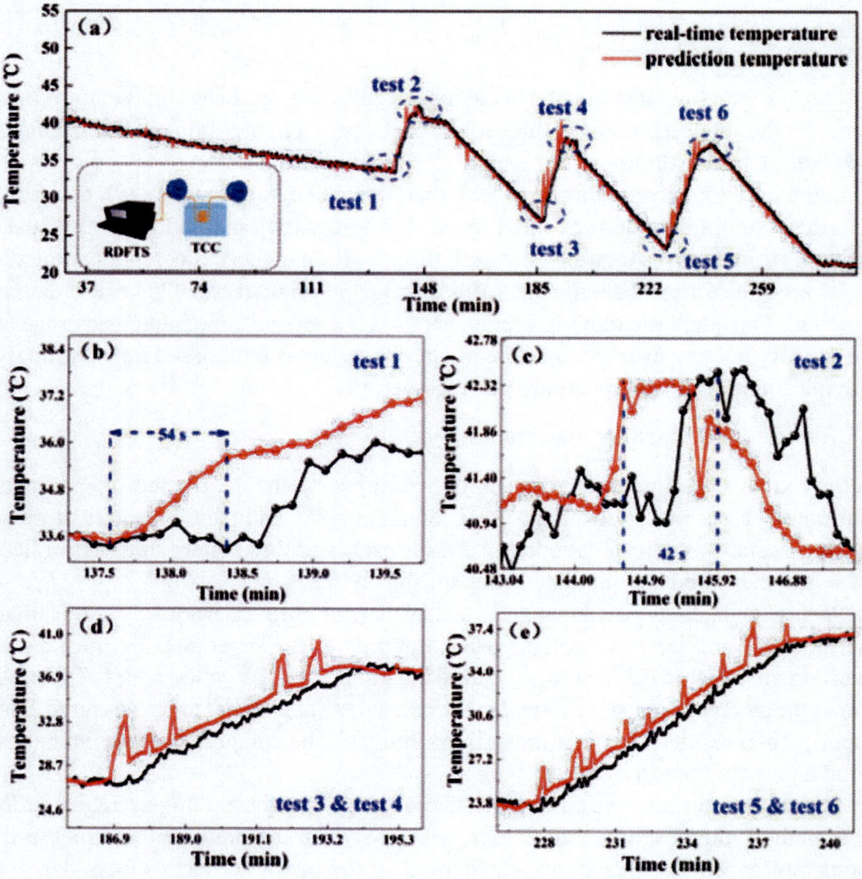

Fig. 6.27 Experimental results of predicted and measured temperatures **a** Whole sensing fiber, **b** local amplification of detection region 1, **c** local amplification of detection region 2, **d** local amplification of detection areas 3 and 4, **e** local amplification of detection areas 5 and 6

Table 6.1 Advance response times for different detection zones

Detection zones	Forecast time (min)	Actual time (min)	Pre-response time (sec)
Test 1	137.57	138.62	54
Test 2	145.23	146.27	42
Test 3	186.06	186.99	55
Test 4	193.83	194.41	34
Test 5	226.78	227.59	55
Test 6	242.44	243.13	41

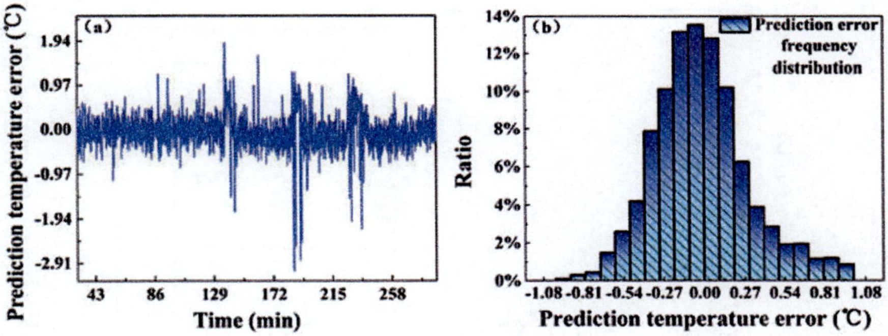

Fig. 6.28 Temperature prediction errors for indoor experiments. **a** Predicted and measured temperature errors, **b** temperature error histogram

as shown in Fig. 6.29. One of them is a flat structure, where the fiber optic cable is placed directly on the ground. The other is a twisted structure, where the sensing fiber is wrapped around the fence of the building. The distributed fiber optic Raman sensing system performs 24 h real-time temperature monitoring and prediction. The prediction time is 1 min set by the Temperature Early Warning Model (TEWM), which means that the prediction time is 1 min after the current time.

Figure 6.30 shows the experimental results predicted by the proposed temperature ahead warning model and the temperature results measured in real-time. The black curve shows the actual measured temperature at a single point of the building as a function of time, while the red curve shows the predicted temperature results at that point. The relationship between the predicted (red curve) and real-time measured

Fig. 6.29 Outdoor temperature prediction experiment and sensing fiber optic cable laying

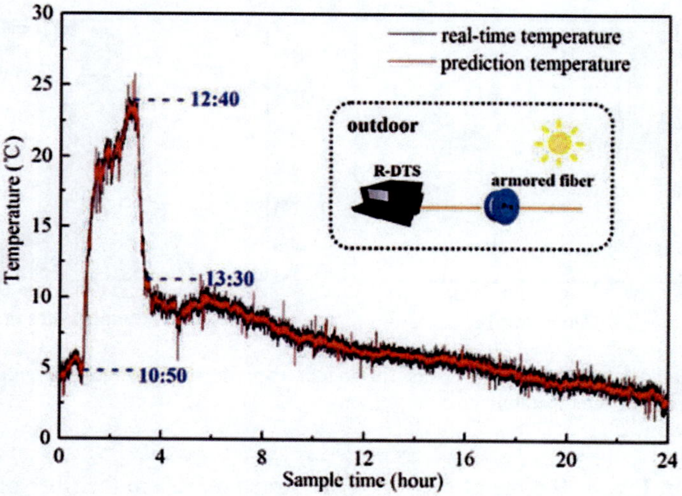

Fig. 6.30 Temperature ahead warning model predicted temperature results versus real-time measured temperature results

temperatures (black curve) can be obtained from Fig. 6.30. The red curve represents the temperature data predicted by TEWM. The black curve represents the actual measured temperature data after 1 min. As can be observed, the TEWM can predict the trend of the actual measured temperature.

Figure 6.31 shows the error plots of the predicted and measured temperatures for the outdoor experiments. From Fig. 6.31a, it can be seen that the average absolute error of the predicted values compared with the measured values is 0.31 °C, and the average relative error is 5%, of which the maximum absolute error is 3.02 °C. Figure 6.31b shows the histogram of prediction errors, in which the proportion of prediction errors in the range of 0~0.1 °C is 26.42%, and the proportion of prediction errors ≤ 0.5 °C is 84.68%. The results of the outdoor prediction experiments showed that the predicted temperature values based on the moving average prediction method were in good agreement with the actual measured values.

6.2.4 Main Technical Specifications of the Thermometer

1. **Temperature measurement accuracy**

The accuracy of temperature measurement is generally defined as the absolute difference between the actual temperature and the measured temperature [11]. For the experimental measurements, the incident optical power was set at 14.6 mW, the room temperature was 22.82 °C, the temperature of the constant temperature bath was set at 26.0 °C, and the temperature of the test zone (TCC) was set from 35.0 to 65.0 °C, and a test temperature point was taken at 5.0 °C intervals.

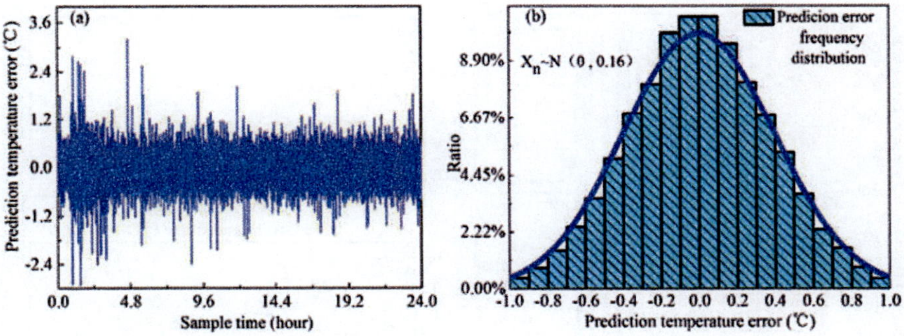

Fig. 6.31 Errors in predicted results for large buildings. **a** Predicted and measured temperature errors, **b** temperature error histogram

The cumulative averaging method, the dynamic noise floor method, the self-demodulation method, and the wavelet mode extreme value transformation method were used simultaneously in the experiment. The number of accumulated averages was 10,000, the number of layers of wavelet mode maximum decomposition was 5 and the wavelet basis function was "sym6".

The results of the experimental tests are shown in Fig. 6.32. Figure 6.32a-I and a-II show the temperature distribution and the accuracy (blue dots) and uncertainty (magenta bars) at test point Fig. 6.32b-I and b-II show the temperature distribution and accuracy and uncertainty at test point 2, Fig. 6.33 c-I and c-II are the temperature distribution and the accuracy and uncertainty of temperature measurement at test point 3 respectively. From Fig. 6.33a-I, b-I, and c-I, it can be seen that the temperature fluctuation of the system gradually increases with the increase of sensing distance. At test point 1 (about 2.36–2.39 km), the maximum temperature fluctuation of the system is 0.85 °C, at test point 2 (about 7.43–7.47 km), the temperature fluctuation of the system is 1.60 °C. At test point 3 (approximately 10.48–10.52 km), the system temperature fluctuations deteriorated further, with a temperature fluctuation of 1.66 °C. This is because, after the wavelet mode maximum denoising, the noise fluctuation remains essentially constant over the whole fiber, while the signal strength drops from 275 mV at the front end of the fiber to 40 mV at the end of the fiber. If the noise fluctuation remains constant, as the backscattering signal strength gradually decreases, the noise fluctuation has a greater impact on the demodulation results, resulting in a greater temperature fluctuation. The temperature uncertainty in Fig. 6.32 a-II, b-II, and c-II deteriorates as the sensing distance increases, which can also be explained by the above principle, i.e. the weaker the scattering signal, the greater the temperature uncertainty as the noise fluctuation remains constant.

From Fig. 6.32a-II, b-II, and c-II, it can be seen that the temperature measurement accuracy of the system gradually deteriorates with increasing distance. At test point 1, the temperature measurement accuracy of the system is positive and the maximum value is 0.52 °C. At test point 2, the temperature accuracy was negative with a maximum value of −1.00 °C. At test point 3, the accuracy deteriorated further from

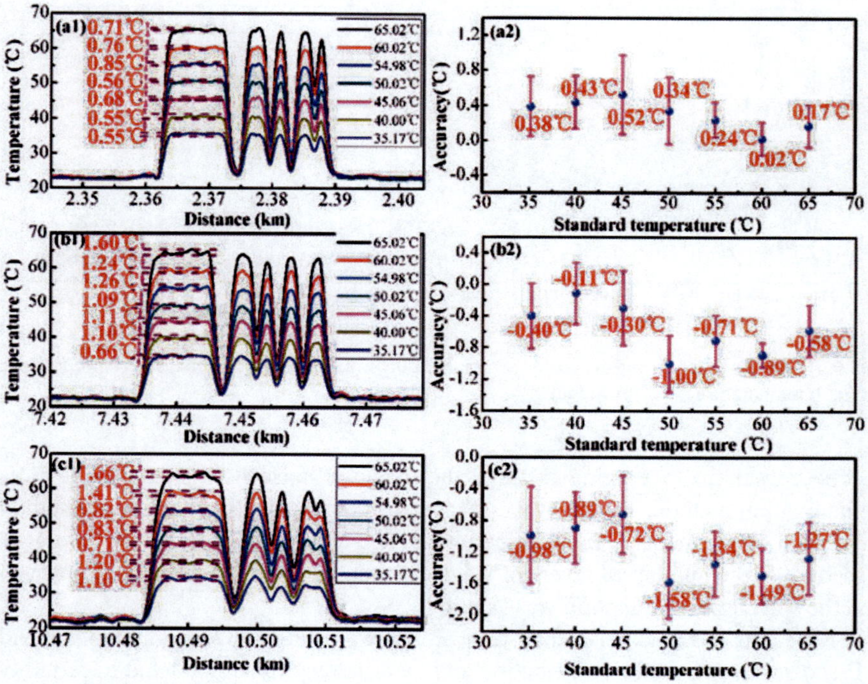

Fig. 6.32 Distribution of temperature measurement and accuracy and uncertainty of temperature measurement at different test points. **a** Test point 1, **b** Test point 2, **c** Test point 3

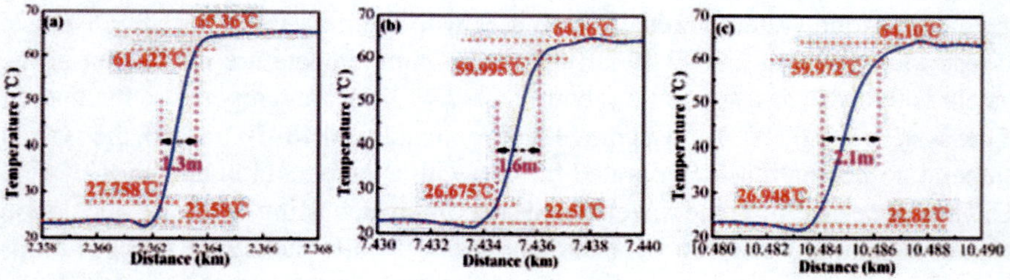

Fig. 6.33 Spatial resolution of the system. **a** Test point 1, **b** test point 2, **c** test point 3

the results at test point 2 but still reached −1.58 °C. This phenomenon is also because the effective information is drowned out by the noise as the distance increases, while the wavelet transform modal maximum has the effect of filtering out the noise, but it also filters out a certain amount of the useful signal.

As the sensing distance rises, the signal strength gradually decreases, with the signal strength at test point 3 decaying by nearly 4 dB compared to that at test point 1. In the case where the wavelet mode maximum filtered signal remains the same, the weaker the signal, the greater the impact of the filtered signal on the temperature measurement results, and it will be less than the actual value. In other words, as the sensing distance increases, the accuracy of the temperature measurement results will gradually deviate from and be smaller than the actual value due to the filtering of a small portion of the effective signal.

Table 6.2 shows the demodulation results of the wavelet mode maximum for the same group of scattering signals before and after denoising at test point 3 when the TCC temperature was controlled at 60.02 °C. As can be seen from the table, the demodulation temperature before denoising was higher than that after denoising by about 0.7 °C in all five measurements. In other words, after the wavelet denoising, the effect of the filtered signal on the temperature measurement results is about 0.7 °C at that temperature point. It can thus be verified that the deterioration of the system temperature measurement accuracy with the spatial resolution is due to the filtering of the wavelet transform removal of useful signals.

In summary, with the increase of the sensing distance, although the temperature accuracy, temperature fluctuation, and temperature uncertainty of the system gradually deteriorated, the system was still able to achieve a temperature accuracy of 1.58 °C, a temperature fluctuation of 1.66 °C and a temperature uncertainty of 0.61 °C in the interval of about 10.48 km.

2. **Spatial resolution**

There are two main definitions of spatial resolution, the direct measurement of spatial resolution for spatial resolution analysis and the spatial resolution of the optical fiber length corresponding to 10–90% of the temperature variation [28].

The direct measurement method provides an approximate measure of the spatial resolution of the system, which is defined by looking at the minimum length of the sensing fiber placed in the thermostat bath that can accurately measure the actual temperature. The spatial resolution of the system can therefore be derived from Fig. 6.33a–c. As can be seen from the graphs, the spatial resolution of the system decreases as the sensing distance increases. The minimum length at test point 1, test point 2 and test point 3 where the actual temperature can be accurately measured is 1.5 m, 2 m, and 3 m, respectively, so the corresponding spatial resolution for these three locations is 1.5 m, 2 m, and 3 m, in that order, under this definition. This measurement method can only determine the range of spatial resolution, i.e. between the maximum fiber ring length just before saturation and the maximum fiber ring length before saturation, and does not accurately measure the spatial resolution of the system. To accurately characterize the spatial resolution of a system, the length of the fiber corresponding to 10–90% of the temperature change is often used as the spatial resolution. The spatial resolution of the three measurement zones was analyzed and the results are shown in Fig. 6.33.

Table 6.2 Demodulation results before and after wavelet mode extreme denoising of the same group of scattering signals at test point 3 Unit:°C

	1	2	3	4	5
Before denoising	60.00	59.54	59.93	59.18	60.03
After denoising	59.30	58.88	59.21	58..49	59.31

As can be seen from Fig. 6.33, the spatial resolution of the system becomes progressively worse as the sensing distance increases. The spatial resolution of the system is 1.3 m and 1.6 m at test point 1 and test point 2 respectively, while at test point 3 the spatial resolution of the system is 2.1 m.

The theoretical spatial resolution l of the system is determined by the pulse width of the laser pulse (tp), the bandwidth of the amplifier and photodetector (Bpd), and the bandwidth of the acquisition card (Bd). That is, the spatial resolution lp determined by the laser, the spatial resolution lpd determined by the amplifier and photodetector, and the maximum spatial resolution ld determined by the acquisition card in Eq. (6.42).

$$l_p = \frac{t_p c}{2n},\ l_{pd} = \frac{c}{2B_{pd}},\ l_d = \frac{c}{2B_d} \tag{6.42}$$

where n is the refractive index of the fiber and c is the propagation speed of light in a vacuum. In this system, the pulse width of the laser is 10 ns, the bandwidth of the amplifier and a photodetector is 80 MHz and the bandwidth of the acquisition card is 100 MHz, so the theoretical spatial resolution of the system, which is determined by the pulse width of the laser pulse, the bandwidth of the detector and the bandwidth of the acquisition card, is 1.25 m. This is not consistent with the spatial resolution obtained in the experiments, because on the one hand, there is intra-mode dispersion in the optical fiber, and intra-mode dispersion causes pulse broadening, and the broadening size increases with the sensing distance; on the other hand, as the sensing distance increases, the signal-to-noise ratio of the system decreases and the spatial resolution deteriorates.

3. **Temperature stability**

To further investigate the temperature stability of the system, a repeatability measurement study was carried out on the system. The temperature of the test zone was set at 34.87 °C and 10 measurements were taken at 30-min intervals under these conditions, the specific results are shown in Table 6.3.

From Table 6.3, we can calculate that the average values of temperature measurement at test point 1, test point 2, and test point 3 are 35.14 °C, 34.26 °C and 33.79 °C respectively, and the accuracy of temperature measurement at the three test points are 0.27 °C, 0.60 °C and 1.08 °C respectively. The fluctuation of temperature measurement results of 10 measurement experiments is 0.9 °C, 0.72 °C, and 1.02 °C at test

Table 6.3 Measurement results for 10 m fiber ring at three test points (°C)

	1	2	3	4	5	6	7	8	9	10
Test 1	34.50	35.07	35.40	35.28	35.06	35.31	35.00	35.11	35.33	35.34
Test 2	33.94	34.11	34.21	34.39	33.89	34.35	34.18	34.65	34.58	34.36
Test 3	33.35	33.58	34.12	33.86	33.65	34.07	34.37	34.02	33.45	33.34

point 1, test point 2, and test point 3 respectively. From these 10 measurements, it can be seen that the system has good repeatability.

4. **Positioning deviation**

In addition to ensuring the accuracy of the temperature measurement, the distributed fiber optic Raman temperature measurement system also needs to accurately locate the location of temperature anomalies to eliminate safety hazards and ensure production safety at the first sign of an event. For this purpose, the positioning accuracy of the distributed fiber Raman temperature measurement system is tested.

To ensure the accuracy of the experimental results, a commercial optical time domain reflectometer (AOR-500S) was used to calibrate the length of the optical fibers, and when the measurement distance is 10 km, the uncertainty of the distance is up to 2.1 m, which is within acceptable limits. The lengths of the fibers were measured using the distributed fiber optic Raman temperature measurement system and compared with the calibration values of the AOR500S to obtain the positioning accuracy of the prototype.

The eight optical fibers to be tested were marked with serial numbers 1–8, and the experimental steps were as follows:

(1) The length of fiber 1 was measured by the AOR-500S and the length of fiber 1 was measured by the fiber Raman meter, and the measured lengths were recorded;
(2) Fuse fiber 2 to the end of fiber 1, and measure the length of the fused fiber was measured by the AOR-500S and the fiber Raman sample machine, and the measured lengths were recorded;
(3) Repeat step (2) above until all 8 trays of fibers are fused and the length data shown in Table 6.4 are measured by the instrument.

From the comparison of the measurement data of AOR-500S and the optical fiber Raman thermometer in Table 6.4, it can be seen that the results of the distributed optical fiber Raman thermometer to be measured are consistent with the length measurement results of AOR-500S, and the measurement error is within 2 m. Taking into account the systematic errors in the OTDR measurement process, the distributed optical fiber Raman temperature measurement system can accurately measure the exact location of the fiber, which means that the event point can be precisely located.

Table 6.4 Fiber length values from AOR-500S and fiber Raman measurements (m)

	1	2	3	4	5	6	7	8
AOR-500S	1298	2508	4350	5567	6785	7997	9211	11,210
Raman system	1296	2508	4350	5567	6784	7995	9210	11,209

6.3 Long-Distance and High-Precision Distributed Fiber Optic Raman Thermometer

6.3.1 System Integration and Multi-stage Thermostatic Control Technology

1. **System Integration**

As Fig. 6.34 shows the designed long- distance and high-precision distributed fiber optic Raman thermometer. To improve the performance index of temperature measuring system, the authors' team designed the key components independently, and through the combination of temperature control and bias control, the high precision and stable conversion from optical signal to electrical signal is realized, which lays a foundation for further improving the sensing distance and temperature measurement accuracy of the system. The long-distance high-precision distributed fiber Raman thermometer has the following advantages:

(1) Wide temperature measurement range, the temperature control of the key components of the system can ensure the stability of the system in a larger temperature range;
(2) Long measurement distance, for the photoelectric conversion module, bias voltage control is introduced to ensure the SNR of the system by adjusting its working bias and effectively improve the measurement distance of the system for different applications.
(3) High measurement precision, through the dynamic adjustment of temperature and bias voltage, the photoelectric conversion process is more stable, the temperature measurement accuracy of the system is more reliable and stable, and the temperature abnormal area can be warned in real time. Different alarm channels and different alarm thresholds can be set to reduce the probability of false negatives.

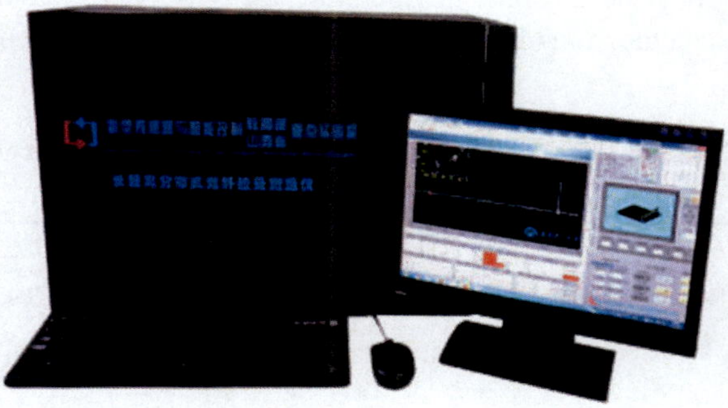

Fig. 6.34 Long-distance and high-precision distributed fiber optic Raman thermometer prototype

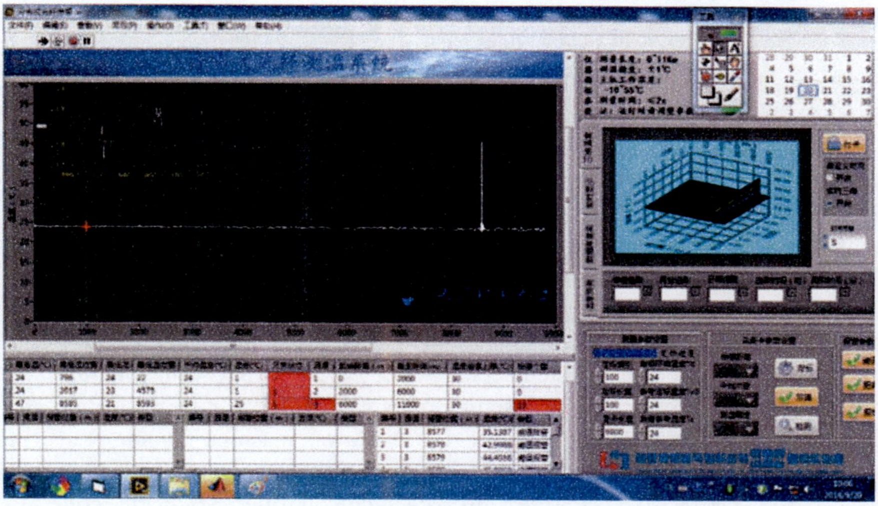

Fig. 6.35 Long-distance and high-precision distributed fiber optic Raman thermometer temperature demodulation interface

The software system of LabVIEW temperature demodulation for upper machine is designed for needle pair long distance high precision optical fiber Raman temperature measuring hardware system. The software interface of LabVIEW is shown in Fig. 6.35, and the software system contains the following main functions:

(1) Temperature display along the fiber. The relationship between fiber length and temperature is established, and the measured temperature distribution along the fiber is presented to the user by means of a waveform graph.
(2) Temperature abnormal point alarm. When there is an abnormal temperature point in the fiber, the alarm light of the system will light up, and the specific location of the alarm point will be highlighted to ensure the timeliness of the alarm.
(3) Historical data search. A database query system has been established to store the measured temperature distribution periodically, which is convenient for users to check later.
(4) Temperature 3D display module. The three axes are time, distance and temperature, which are used to display in real time or user-defined 3D map of temperature over time.

2. **Multistage thermostatic control technology**

(1) Overall design of the system

In distributed fiber Raman temperature measurement system, the parameters of each device are easily changed by the influence of ambient temperature, which seriously affects the stability and accuracy of temperature measurement system, especially the calibration fiber and APD [29, 30]. Long-distance and high-precision distributed fiber optic Raman thermometer with temperature control system for APD and calibrated fiber. As shown in Fig. 6.36, the temperature control system is mainly divided into four parts, including the incubator, upper machine module, lower machine module, and power supply. The thermostatic box is mainly used to place the calibration fiber and APD, and its built-in high-precision temperature sensor can detect the temperature of the thermostat. The temperature of thermostatic box can be adjusted dynamically by semiconductor thermoelectric cooler. The upper machine module uses the hardware resources of the Avnet V5 development board. The module is designed with FreeRTOS operating system and STemwin interface to achieve the functions of periodic update of temperature data, data storage, USB communication and thermostatic box temperature setting. data storage, USB communication and thermostatic box temperature setting [31]. The lower machine module reads the temperature sensors (ADT7410 and AD7793) periodically through the integrated bus and the serial interface bus using the STM32F4 microcontroller [32]. For the collected temperature data, on the one hand, the lower machine sends it to the upper machine through the local network bus of the controller, on the other hand, it adjusts the duty cycle loaded on the PWM wave of the driver board with the corresponding control algorithm, so as to change the working power of the TEC and realize the feedback control of the temperature. The power supply uses the Chaoyang power supply 4NIC-Q250KT, with three outputs of 12 V/8A, 5 V/5A and 3 V/10 W, supplying power to the TEC (12 V/5 V), the upper machine (12 V) and the lower machine (12 V) respectively.

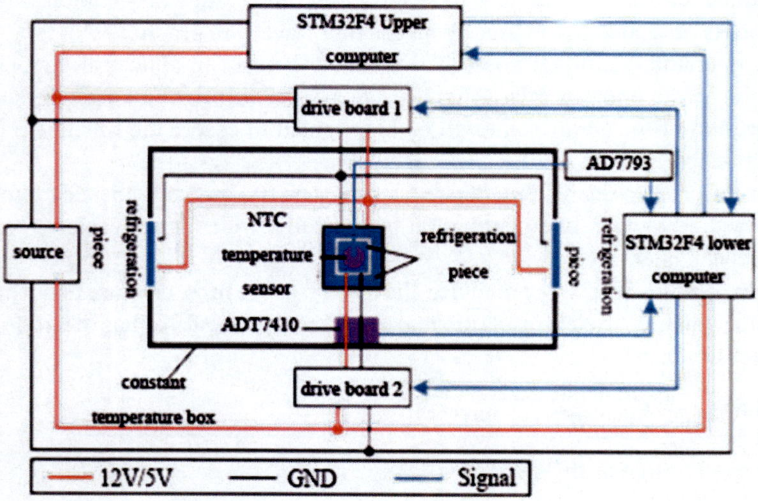

Fig. 6.36 Temperature control system structure diagram

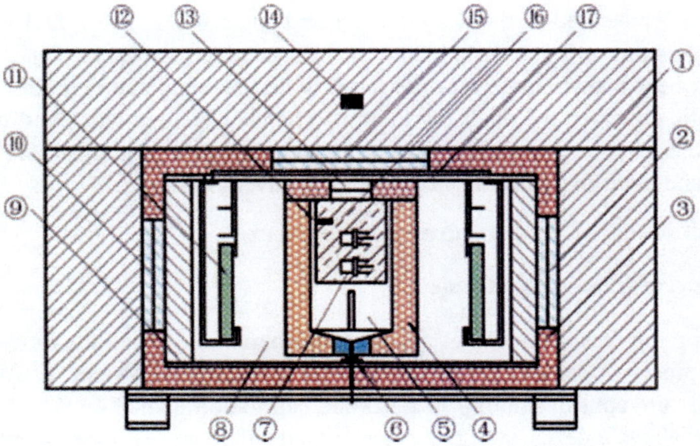

Fig. 6.37 Thermostatic box mechanical structure diagram. ① ⑨ Cooling fin; ② ⑩ **13 15** Refrigeration piece; ③ Heat-insulating cotton; ④ Thermal insulation foaming material; ⑤ Secondary constant temperature room; ⑥ ATD7410-1; ⑦ APD; ⑧ First-level constant temperature room; **11** Calibration fiber; **12** NTC temperature sensor; **14** ATD7410-3; **16** Heat conduction copper block; **17** ATD7410-2

(2) **Mechanical structure design**

The whole thermostatic box is divided into two layers, the internal and external nesting structure, on the one hand, is conducive to ensuring the stability of the temperature control of the inner thermostatic chamber, on the other hand, is conducive to reducing the overall volume of the thermostatic box, convenient for the later integration of fiber optic Raman thermometer. The specific mechanical structure is shown in Fig. 6.37, where the outer thermostatic chamber holds the calibration fiber and the inner thermostatic chamber holds the APD.

Each part of the thermostatic box is tested by high-precision temperature sensors. Because the calibration fiber is placed in the outer constant greenhouse and cooled by air conduction, in order to ensure the uniformity of the temperature of the calibration fiber, temperature sensors are placed at the top and bottom of the primary thermostatic chamber. In order to reduce the volume of the thermostat, the thermoelectric cooler is selected as the semiconductor thermoelectric cooler, which is mainly based on the Peltier effect of semiconductor materials, normal operation, one side of the cooling, one side of the heating, and can be achieved by changing the direction of the current cooling and heating surface conversion, is a better choice to achieve temperature control [33]. In order to achieve good temperature control effect during normal operation of TEC, it is necessary to conduct heat and cold generated by heating surface and cooling surface in a timely manner. Therefore, outer heat sink and cooling fan are installed on the heating surface of TEC, and inner heat sink is installed on the cooling surface. To ensure the uniform and stable temperature of the APD, a heat conduction copper block is installed at the cooling end of the cooling plate.

The APD is embedded in the heat conduction copper block, and a high-precision NTC temperature sensor is installed on the heat-conducting copper block. In order to prevent heat conduction from interfering with the temperature stability of the thermostatic chamber, the outer thermostatic chamber is filled with heat insulation foam and thermal insulation foam between the inner and outer heat sinks, and between the inner and outer thermostatic chamber, respectively.

3. **Circuit design of temperature control system**

(1) **Temperature sensing circuit**

Considering the feasibility and economy of temperature control, the long-distance and high-precision roman thermometer is set to ± 0.005 °C and ± 0.1 °C for the temperature control stability of APD and calibration fiber [34, 35]. In order to achieve the above indexes, thermistors and digital temperature sensors were selected as temperature sensors for the inner and outer thermostatic chambers respectively. ADT7410 is selected as the digital temperature sensor, and a negative temperature coefficient linear temperature sensor is selected as the thermistor, together with an A/D converter chip to achieve temperature measurement.

(2) **Drive circuit design**

In order to reduce the overall volume of the thermostatic box and facilitate the later integration of Raman thermometer, the semiconductor thermoelectric cooler is selected, which is often used in the cooling of CPU, infrared sensor, refrigerator and other electronic devices, has the advantages of small size, light weight, high reliability [33]. The outer thermostatic chamber of the thermostatic box is placed in the calibration fiber. In order to ensure the ambient temperature of the calibrated fiber, the TEC of 12 V and 8 A is selected for the outer thermoelectric cooler, and the size is 40 mm × 40 mm. In order to achieve a wide range of temperature control, the inner and outer thermoelectric cooler are selected double layer TEC, the maximum temperature difference can reach 80.0 °C.

In order to stabilize the temperature of the calibrated fiber and APD, it is necessary to control the working power of TEC by the single chip microcomputer. The output voltage of the single chip microcomputer is 3.3 V, and the driving current is usually a few hundred milliamps. Therefore, the driving circuit needs to be used to realize the conversion of the control instruction of the single chip microcomputer to the working power of the TEC. The H-bridge drive circuit is used to achieve power amplification, and the pulse wave of a certain width is output by the single chip microcomputer to adjust.

H-bridge circuit is named for its shape like the letter H. It is often used to drive DC motors. By controlling the on–off of H-bridge, the forward rotation, reversal and braking of the motor can be realized. Since the switching of TEC cooling and heating and the forward and reverse rotation of the motor are achieved by changing the current direction, the H-bridge circuit used for motor drive can also be applied to the TEC drive. Switching between the cooling and heating sides of the TEC can be achieved by forward and reverse rotation, thus increasing or decreasing the temperature. The control of the TEC drive circuit mainly includes two aspects, one is the control of the drive direction, and the other is the control of the drive power. For the drive direction, the microcontroller of the lower computer can control the level of the universal control input and output interface to achieve conversion, while the drive power needs to adjust the duty ratio of the PWM to control.

Pulse width modulation is a technique to control analog circuits by changing the duty ratio of the outgoing square wave. The control principle of the PWM waveform of the lower computer is shown in Fig. 6.38. The timer of the microcontroller is set to work in the upward counting PWM mode, where CNT is the counting value, CCRx is the threshold value, and ARR is the maximum count value. When the program is running, when the value of CNT is less than the value of CCRx, the I/O port outputs a low level, and CNT keeps adding itself according to the period T. When the value of CNT is greater than the value of CCRx, the I/O port outputs a high level. When the value of CNT increases to be equal to ARR, CNT is reset and starts counting from 0 again. Therefore, the frequency and duty ratio of the output PWM can be obtained as.

$$f = \frac{1}{ARR \times T'} duty\,\text{ratio} = 1 - \frac{\text{CRRx}}{\text{ARR}} \tag{6.43}$$

When T is fixed, the frequency of PWM can be adjusted by changing the value of ARR, and the duty ratio of PWM can be adjusted by changing the value of CCRx. Figure 6.39 shows the flow chart of the microcontroller to realize PWM duty ratio adjustment, where the PWM channel is selected as TIM5_CH1 and the corresponding GPIO is PH10, so it is necessary to initialize TIM5 and GPIOH. The timer output PWM also needs to configure ARR value, PSC (clock prescaler coefficient), initial CCR1 value, counting mode, and PWM mode. Set ARR value to 99, PSC to 83, initial CCR1 value to 90, up counting mode, and PWM1 mode.

The PWM frequency is calculated by Eq. (6.44), when the timer clock frequency is 84 MHz, the PWM frequency is 10 kHz, and the duty ratio is 90%. For the PWM

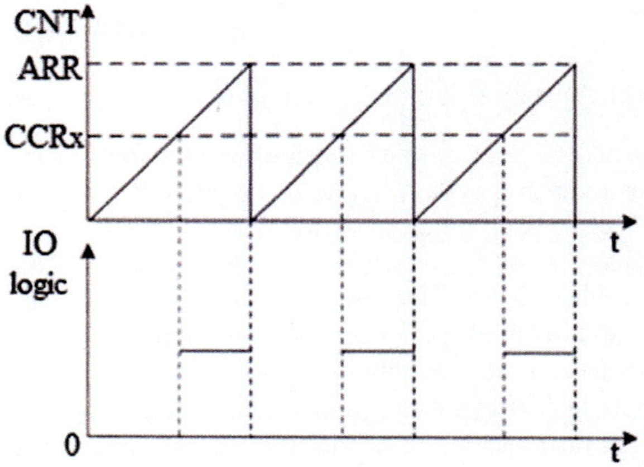

Fig. 6.38 PWM wave generation principle

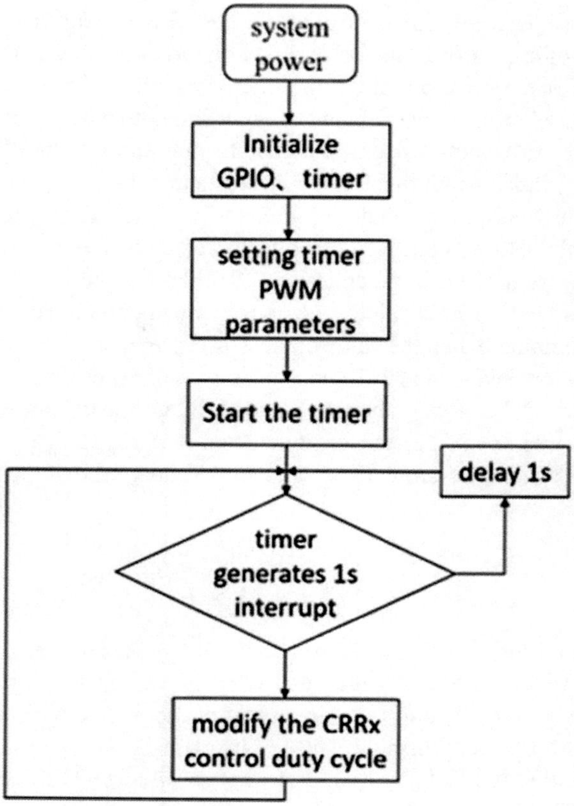

Fig. 6.39 PWM duty cycle adjustment flow chart

cycle control, the 1 s timer interrupt of TIM7 is also used, so the PWM duty ratio can be adjusted by checking the status of the interrupt flag bit and adjusting the value of CCR1.

$$f = \frac{f_{CLK}}{(PSC + 1)(ARR + 1)} \tag{6.44}$$

(3) **Testing and optimization of temperature control system**

After the software and hardware design of the upper and lower machine, the temperature control system has been equipped with temperature control and monitoring. Therefore, the whole system is tested and optimized. When controlling the temperature, it was found that the high-frequency on/off of the TEC driver circuit MOS generated a lot of electromagnetic interference, which seriously affected the accuracy and stability of the temperature measurement of the temperature sensor.

In order to ensure the control effect of temperature control system, it is necessary to control the temperature sensing circuit against electromagnetic interference. The main methods to reduce the influence of electromagnetic interference on electronic devices are shielding and filtering [35]. In this temperature control system, the ADT7410 and NTC are connected to the lower machine through wires, so they are easily affected by electromagnetic interference. Considering the limited role of filtering and the complexity of the design process, the wire connecting the temperature sensor and the lower computer was replaced with a shielded wire, and one end of the shield was connected to the lower machine and grounded. Figure 6.40 shows the temperature change curve of ADT7410 before and after replacing the shield, the set temperature of both temperature curves is 10 °C. It can be found that the fluctuation of temperature curve is ± 0.2 °C when using ordinary wire, and the fluctuation of temperature curve is ± 0.1 °C after replacing with shielded wire, so the shielded wire can effectively shield electromagnetic interference to ensure the accuracy of temperature measurement. After replacing all the wires between the temperature sensor and the lower machine with shielded wires, the temperature of the outer thermostatic chamber was set to 10 °C and the temperature of the inner thermostatic chamber was set to 5 °C by the upper machine, as shown in the temperature change curve in Fig. 6.41 and the interface of the upper machine in Fig. 6.42.

The red curve in Fig. 6.41 shows the temperature change of ADT7410-1, and the blue curve shows the temperature change of ADT7410-2.

The magenta color is the temperature change curve of NTC. At a room temperature of about 22.0 °C, the temperature of ADT7410-1 reached a stable state of (10 ± 0.1) °C after 1000 s. The temperature of ADT7410-2 reached a steady state of (10 ± 0.1) °C after 500 s, and the temperature of NTC gradually approached 5.0 °C after 1500 s, and finally stabilized at (5 ± 0.005) °C due to the power of TEC thermoelectric cooler. For the upper machine, the temperature data is updated timely, the touch response is accurate, the communication with PC and lower machine is normal, and the overall operation is stable and reliable. Therefore, the temperature control system has achieved the design goals in terms of control accuracy, stability and human–machine interaction, and can meet the requirements of the Raman thermometer.

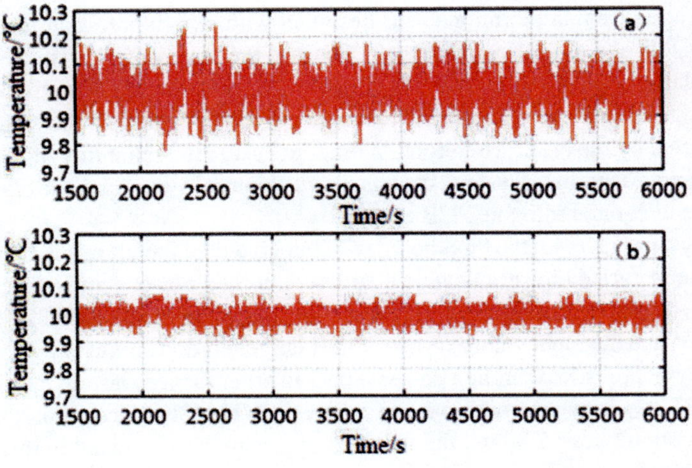

Fig. 6.40 ADT7410 temperature change curve comparison chart. **a** Ordinary wire, **b** shielded wire

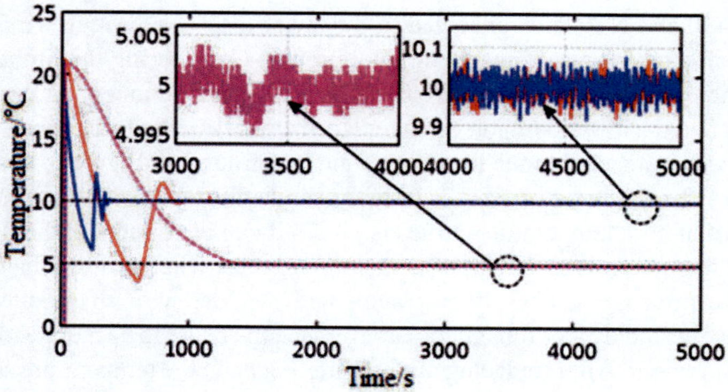

Fig. 6.41 Temperature control system temperature change curve

Fig. 6.42 The upper machine interface of the temperature control system after stabilization

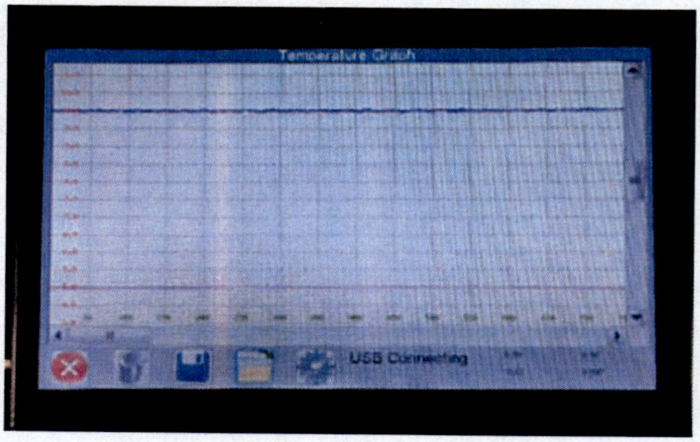

(4) **Temperature control and bias control system integration**

The fiber optic Raman thermometer is designed with a high precision temperature control system and a bias control system, because the bias control system is used to stabilize the gain of the APD by bias regulation under the action of the temperature control system, and the control process needs to be informed of the temperature data collected by the NTC. Therefore, in order to ensure the reliability of the Raman temperature measurement system, the temperature control system and the bias control system are integrated into one PCB board.

As shown in Fig. 6.43, the PCB board with integrated temperature and bias control system, the integrated lower machine board with unified power supply. Because the rated voltage of the cooling fan is 12 V, the input voltage of the STM32F407 core board is 5 V, and the input voltage range of the bias generation circuit is 2.8~5.5 V, the system board adopts 12 V unified power supply in order to meet various applications. The DC-DC step-down circuit is used to supply the STM32F407 core board and bias generation circuit after 5 V, and the voltage conversion chip selected in this book is LM2596.5.0.

Fig. 6.43 Integrated temperature and bias control system board

The LM2596 is a step-down power management IC with fixed or adjustable voltage outputs of 3.3, 5 and 12 V, and a maximum drive current of 3 A with over-temperature and current limit protection. Figure 6.44 shows the LM2596 voltage reducing circuit, Fig. 6.45 shows the internal structure of the LM2596. From Fig. 6.45, it can be seen that VIN and OUT are connected through a triode, and the LM2596 voltage reducing circuit is a typical BUCK voltage reducing circuit, which works as follows.

When the triode is on, VIN passes through the triode to the OUT pin, due to the presence of diode D3, VIN can only flow through L6 to capacitor C35 and the load, at this time VIN excites the inductor and provides energy to the capacitor and the load, the voltage across the capacitor is less than VIN. When the triode is off, VIN and OUT are disconnected. At this time, diode D3 conducts due to the current continuation of the inductor, and inductor L6 and capacitor C35 provide energy to the load at the same time, and the output voltage is equal to the voltage across capacitor C35. When the triode conducts and closes regularly, the energy conversion in the circuit reaches a balance and the output voltage is basically stable and less than the

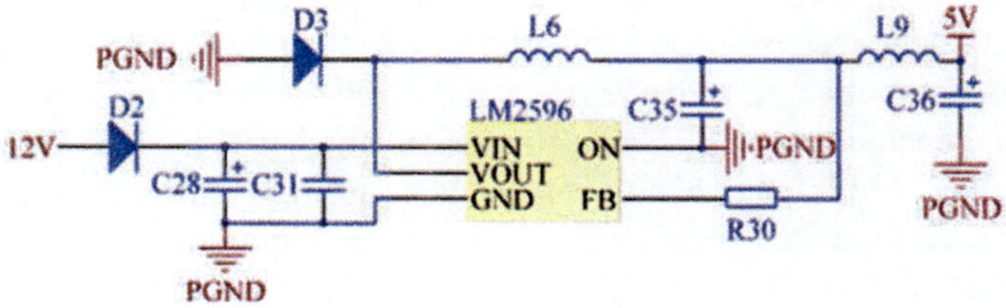

Fig. 6.44 LM2596 buck schematic

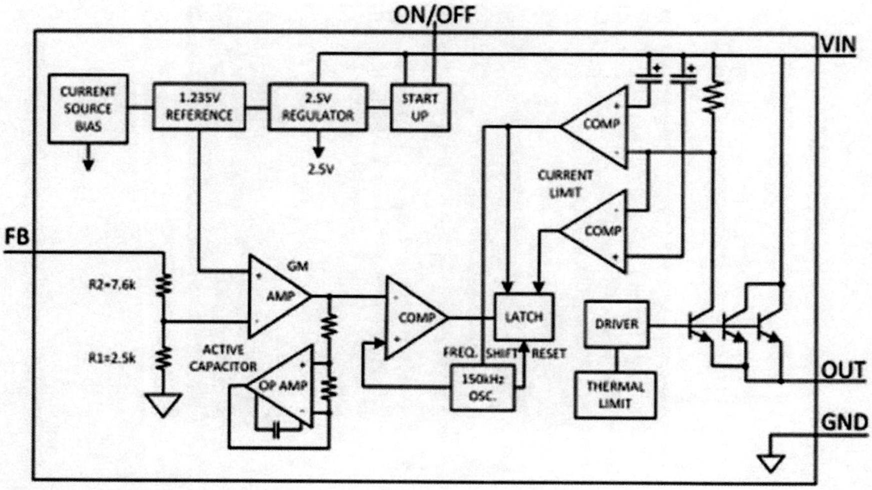

Fig. 6.45 LM2596 internal structure diagram

input voltage. PWM is often used to control the triode, and the output voltage can be adjusted by changing the duty ratio while the frequency remains constant.

After integrating the temperature control system and the bias control system, the optimal operating bias VB of the APD needs to be determined to ensure the normal operation of the bias compensation system and the demodulation effect of the Raman thermometer. For this purpose, the bias generation circuit was designed to be adjustable from 30 to 45 V, and the temperature control was applied to the Raman temperature measurement system to test its demodulation effect at different bias voltages. Because the minimum operating voltage of the amplifier circuit is 30 V and the breakdown voltage of APD is −40~44 V, the design is to adjust the output voltage of the bias control system at 30~39 V and to demodulate the amplified signal of APD.

In order to verify the proposed temperature control and bias system, the total length of the sensing fiber of the distributed fiber Raman thermometer is 9288 m, and the room temperature is about 25 °C. The temperature distribution curve along the fiber under different bias conditions is obtained by signal acquisition and processing, as shown in Fig. 6.46. In order to truly reflect the effect of APD bias on the final temperature measurement results, the temperature profile was not added with a denoising algorithm. It can be seen from the figure that the temperature fluctuation range of the temperature distribution curve decreases significantly when the bias voltage output from the bias generation circuit gradually increases from 30.34 to 38.65 V. Comparing the temperature at 9000 m, when the bias voltage is 30.34 V, the fluctuation range is 16.17 °C (maximum temperature 32 °C, minimum temperature 15.83 °C); when the bias voltage is 38.65 V, the fluctuation range is 3.52 °C (maximum temperature 25.61 °C, minimum temperature 22.09 °C). It can be concluded that with the increase of the bias voltage at both ends of the APD, the gain value of the APD increases significantly, the signal-to-noise ratio of the amplified signal is significantly improved, and the fluctuation range of the final demodulated temperature value is also significantly reduced. Therefore, in order to ensure the distributed fiber optic Raman thermometer to achieve better demodulation accuracy, the bias voltage of APD should be selected to a larger value, but because the breakdown voltage of APD is −40~44 V, in order to ensure the reliability of Raman temperature measurement system, this book design bias control system output reference bias value is 38.65 V, bias control range is 38.2~39.2 V. after temperature stabilization of temperature control system The upper computer interface is shown in Fig. 6.47.

The temperature of the thermostatic box is set by the upper machine. Considering the heating problem of APD, in order to ensure the constant temperature of APD, the temperature of three control links is set to 10 °C. The output voltage of the bias system is adjusted to 30.34 V, 33.21 V, 35.93 V and 38.65 V respectively, and the temperature value demodulated by Raman thermometer is recorded. The upper machine interface of the temperature control system after temperature stabilization is shown in Fig. 6.47. For different bias values, the temperature control effect of the temperature control system is basically the same. The temperature of control links 1

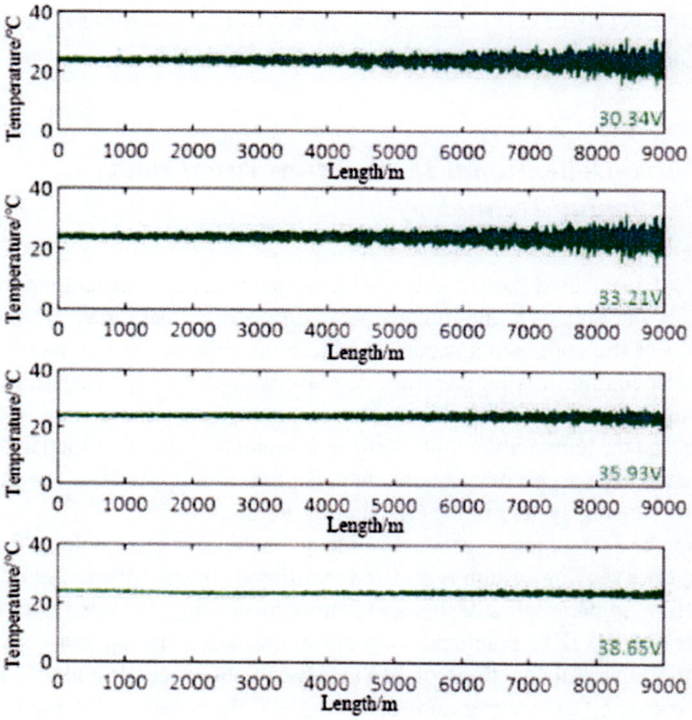

Fig. 6.46 Temperature distribution curves measured at different bias pressures

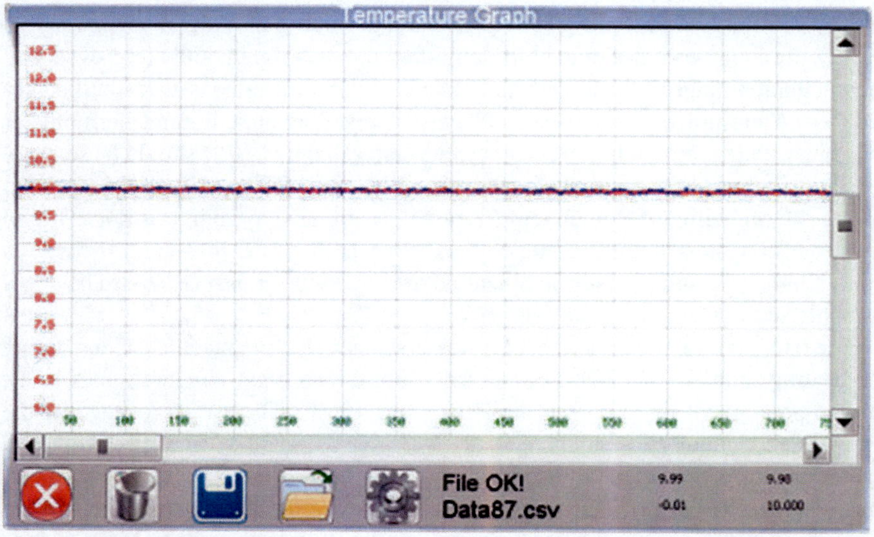

Fig. 6.47 The upper machine interface after setting the temperature to 10 °C

and 2 is stable at (10 ± 0.1) °C, and the temperature of control link 3 is finally stable at (10 ± 0.01) °C due to the fluctuation of APD working state.

6.3.2 *Three-Dimensional Temperature Visualization Positioning Technology*

Conventional distributed Raman fiber sensing systems display the temperature information along the fiber as a one-dimensional "distance-temperature" based curve or "list" [36]. For the application scenarios where the sensing fiber is laid intricately or has local bending, this temperature display method based on one-dimensional "distance-temperature" curve or "list" cannot quickly grasp the overall situation of the system, and the temperature mutation point cannot be quickly located in the actual three-dimensional space environment, which limits the application of distributed Raman fiber sensing system in the engineering field.

Based on the 3D temperature visualization and positioning technology, the Raman distributed fiber sensing system is realized by using the great advantages of SOLIDWORKS (SW) in parametric design and feature modeling, combined with the great advantages of LabVIEW graphical software in hardware driving and data acquisition, and the powerful functions of MATLAB in data processing algorithms. The mapping from the "1D temperature data" to the "3D visualization model" of the spatial fiber is completed. According to Reference [37], a visual positioning technology based on optical fiber 3D temperature display is proposed, which is based on the layout of optical fiber inside the tunnel and its 3D model temperature visualization display. The three-dimensional temperature visualization positioning technology is specifically implemented by the following steps.

1. **Tunnel visual modeling**

When establishing the tunnel model, SW divides the tunnel into two independent entities: tunnel and surrounding rock. The final 3D visualization model of the tunnel is obtained by intelligent assembly in SW. When establishing the tunnel model, the calculation model takes the single-hole double-lane as an example, and the tunnel length is selected as 700 m. In order to facilitate the subsequent modification of the tunnel parameters, intelligent dimensioning is adopted.

(1) After entering SW software, create a new part, first draw a sketch of the tunnel solid, and use the smart dimension tool to mark it, then stretch the tunnel solid sketch longitudinally, and remove the top of the stretch to get the tunnel solid model.
(2) Build the tunnel surrounding rock model, sketch the tunnel surrounding rock sketch, and make each parameter of the tunnel surrounding rock consistent with the tunnel entity through intelligent dimensioning.
(3) After entering the SW software, create a new assembly, insert the tunnel solid model and the tunnel surrounding rock model created above, and complete the assembly according to the actual spatial location.
(4) To obtain a sense of realism, the solid road surface of the tunnel was selected and its appearance was set to asphalt, while the transparency of the tunnel model was set to 0.8.

2. **Laying of fiber optic cable models in tunnels**

In order to monitor the distribution of temperature field in the tunnel and locate the temperature mutation area in the tunnel quickly, a longitudinal linin curve (LLC) is designed to lay out the sensing fiber [38, 39]. The longitudinal lining curve is divided into two parts: the lining curve and the longitudinal curve, the longitudinal curve is alternately distributed according to the direction of the tunnel, and the lining curve is laid in an arc-shaped way and close to the top of the tunnel lining structure. The longitudinal curve is used to monitor the internal temperature field of the tunnel, considering that in case of fire in the tunnel, the fire heat spreads vertically upwards, the longitudinal curve is not sensitive to the fire monitoring, while the liner curve can be used to monitor the fire heat, and at the same time, the fire distribution size and propagation direction can be distinguished.

Refer to the established tunnel model to draw a longitudinal lining cable model, create new parts in the SW, select the datum plane and draw a three-dimensional sketch of the optical fiber distribution, first, a set of longitudinal liner curves are created and used as the reference curves, and the fiber curves are obtained by replicating the reference curves. Perform a scan command on the resulting curve to obtain a longitudinal alignment model of the fiber distribution model (LLM). Finally, the tunnel visual model and the longitudinal liner model with fiber optic deployment were intelligently assembled in SW to obtain Fig. 6.48.

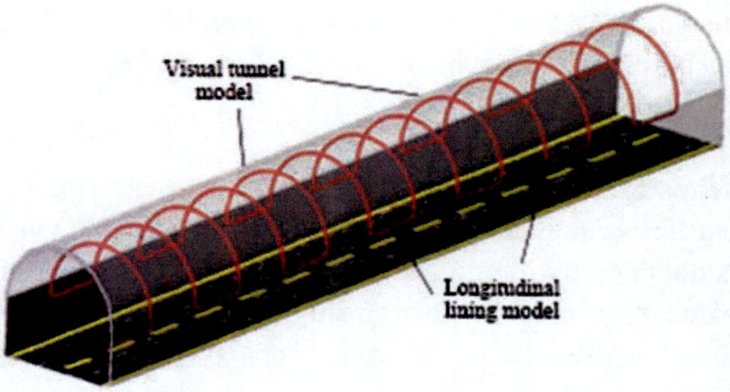

Fig. 6.48 Tunnel 3D visualization model and fiber optic deployment method

3. **Temperature 3D visualization positioning technology**

Temperature 3D visualization positioning technology by means of virtual instrument software LabVIEW sensor mapping. The established fiber distribution model is converted into a format recognized by virtual instrument LabVIEW, which contains the normal vector coordinate information of the fiber distribution model. In order to display the temperature data on the fiber optic model, we designed a sensor deployment program, which can read the vertex array of the fiber optic model in the fiber optic distribution direction and map the temperature data collected by the data acquisition card to the fiber optic 3D model.

The sensor layout program input the read vertex array into the programming function index array of the virtual instrument software LabVIEW, and set the index length to Ni. Through the index array, the points in the vertex array are read in turn according to the index book order i = 0, 1, 2, 3,..., M-1, the index length is M, and Ni = M·i (i = 0 ~ N-1), and the virtual temperature sensors are arranged in turn at N:

$$M = \min\left\{\frac{\nu\Delta T}{2}, \frac{\nu\tau}{2}, \frac{\nu\mu}{2}\right\} \tag{6.45}$$

where, v is the speed of light propagation in the optical fiber, ΔT is the pulse width of the light source in the Raman sensing system, and τ is the data acquisition.

The interpolation algorithm is applied to the temperature unknown points between the virtual temperature sensors. The unknown points close to the virtual temperature sensor are given a smaller weight value, and the farthest points are given a larger weight value. The interpolation algorithm is as follows.

$$G(l) = \sum_{i=1}^{m} \frac{W_i}{[d_i(x, y)]^r} \Big/ \sum_{i=1}^{m} \frac{W_i}{[d_i(x, y)]^r} \tag{6.46}$$

where, G (l) denotes the value of the attribute Z of the unknown point G, m denotes the number of observed points, Wi denotes the attribute value of the i th observed point, di (x, y) denotes the distance from the i th point to the P point. In general, the value of r is 2, and the larger the value of r, the smoother the change of the interpolation result.

In the interpolation algorithm, first determine whether the unknown point coincides with the virtual temperature sensor point: if it coincides, the value of the unknown point is the value of the virtual temperature sensor point; if it does not coincide, the weight coefficient of each unknown point to the virtual temperature sensor point is calculated, and the interpolation result of each grid node is calculated according to the inverse distance interpolation algorithm. Then, the corresponding rendering color value of each interpolation point is given by comparing the scale attribute of the color strip, so as to obtain the real-time 3D temperature field representation image of the whole model. The overall flow of the 3D temperature visualization positioning technique is shown in Fig. 6.49.

4. **Simulation results of 3D visualization positioning technology**

In order to verify the feasibility of the three-dimensional visualization positioning technology, the layout of the optical fiber in the tunnel satisfies the longitudinal lining

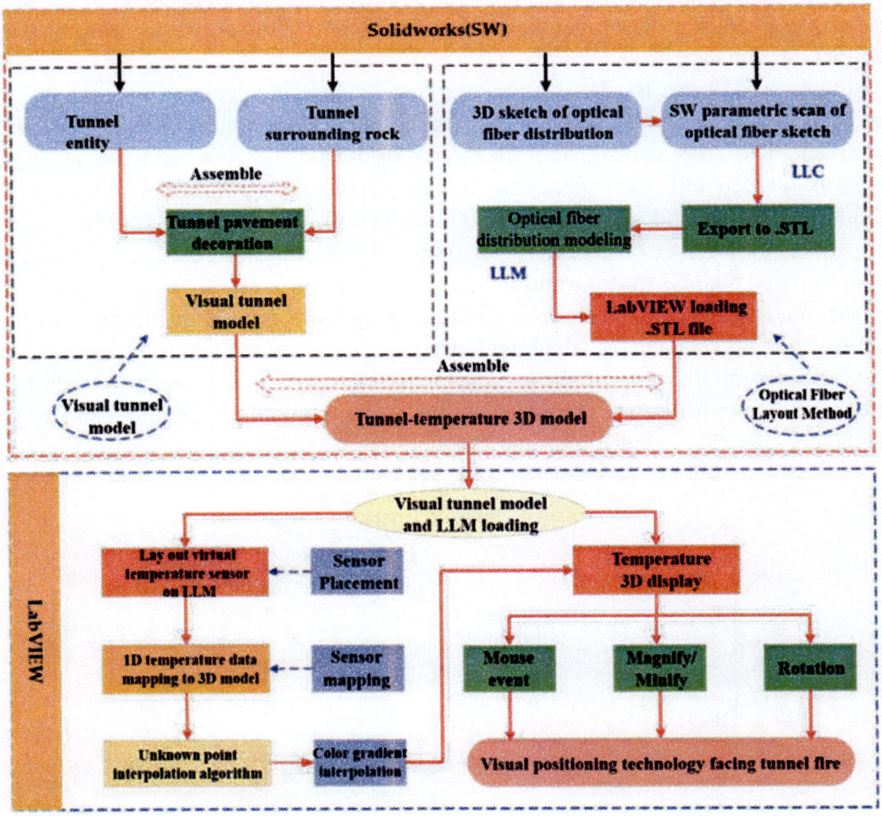

Fig. 6.49 3D temperature visualization and positioning technology flow chart

curve model, and the longitudinal lining curve model is composed of the longitudinal curve and the length of the lining curve of each group, and the length is 5 m and 18 m respectively.

Therefore, the length of the longitudinal lining curve model is 212 m. The LLM is drawn according to the actual ratio of 1:100. Figure 6.50a is a one-dimensional temperature curve, in which the abscissa represents the sensing distance, the ordinate corresponds to the temperature information along the fiber, and the blue elliptical dotted frame represents the temperature mutation area along the fiber. The layout model of the fiber in the tunnel is shown in Fig. 6.50b. The 1D temperature data is mapped to the longitudinal lining curve model of the optical fiber layout through the sensor layout program. Figure 6.50b shows the mapping results. The blue region represents the indoor temperature along the fiber, and the red region represents the temperature mutation region on the one-dimensional map. Therefore, assuming a fire occurs in the tunnel, the 3D temperature visualization positioning technology can precisely locate the fire source in the actual tunnel 3D spatial environment, this temperature display can not only have an intuitive and fast understanding of the temperature field in the tunnel, and can achieve rapid and precise location of the temperature change area.

Figure 6.51 shows the 3D temperature visualization and positioning technology and temperature warning operation system for tunnel fire monitoring. system. The total length of the sensing fiber in the experiment is 25 km, and the sensing fiber is divided into three channels, namely channel 1 (0~7500 m), channel 2 (7500~15,000 m) and channel 3 (15,000~25,000 m). It is assumed that the alarm thresholds of the three channels are set to 40.0 °C (blue warning), 50.0 °C (orange warning) and 80.0 °C (red warning), respectively.

As shown in Fig. 6.51, the red dotted frame shows the alarm information of the one-dimensional curve, which includes the number of temperatures exceeding the temperature alarm threshold in each channel, as well as the warning location, the maximum temperature, the minimum temperature and the average temperature in the channel. The temperature mutation regions 1 (Region1), 2 (Region2) and 3

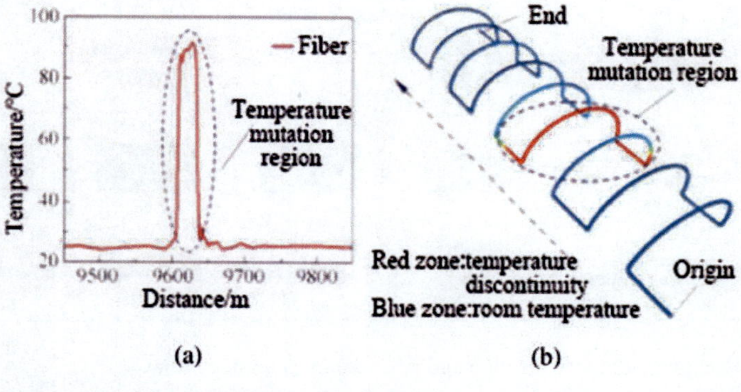

Fig. 6.50 3D temperature display technology mapping results. **a** 1D temperature profile, **b** laying model of fiber optic cable in the tunnel

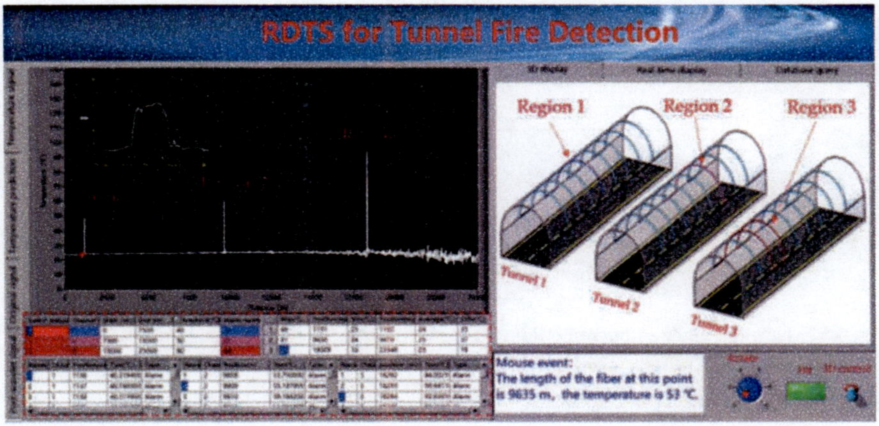

Fig. 6.51 3D temperature visualization and positioning technology and temperature warning platform for tunnel fire monitoring

(Region3) in the one-dimensional curve are mapped to Tunnel1, Tunnel2 and Tunnel3 in the tunnel group. In the tunnel group, different alarm thresholds are displayed in different alarm colors. Blue warning, orange warning and red warning represent the alarm thresholds of 40.0 °C, 50.0 °C and 80.0 °C, respectively. By combining the alarm information of the 1D curve with the 3D temperature visualization and positioning technology, it is possible to quickly and accurately locate areas of rapid temperature change.

6.3.3 The Main Technical Indexes of the Thermometer

1. **Sensing distance**

In order to study the system performance of the long-distance high-precision distributed optical fiber Raman thermometer, a long-distance temperature measurement experiment was designed. Figure 6.52 shows the setup of the temperature measurement experiment, in which the total length of the sensing fiber is 30 km. Four fiber rings are set up in the sensing fiber, the length of the first two rings is 20 m, the length of the third ring is 50 m, and the length of the fourth ring is 60 m. The set temperature of the thermostatic tank was changed, and the temperature of the four fiber optic rings to be measured was obtained by using the distributed fiber optic Raman system after the thermostatic tank was stabilized. Figure 6.53 shows the temperature distribution of the whole sensing fiber measured by the system when the ambient temperature is about 27 °C and the setting temperature of water tank is 50, 70 and 90 °C. It can be seen from the figure that the distributed fiber Raman temperature measurement system can accurately measure the temperature distribution along the 30.0 km sensing fiber.

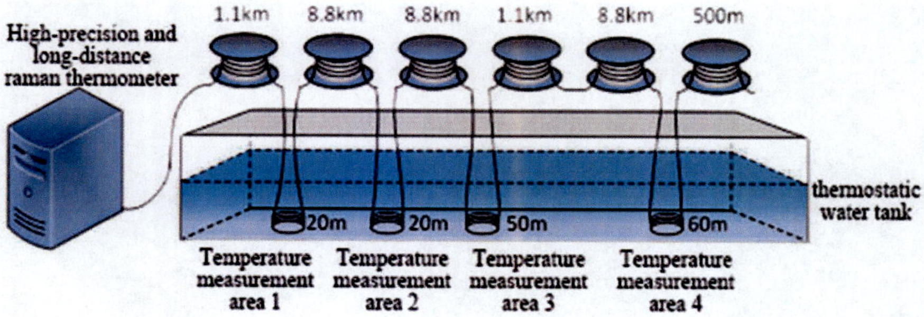

Fig. 6.52 New distributed fiber optic Raman thermometer testing device diagram

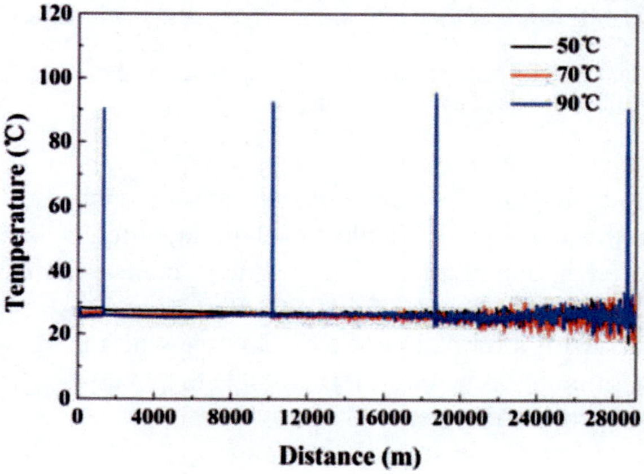

Fig. 6.53 Temperature distribution along the sensing fiber

2. **Spatial resolution**

To study the spatial resolution of the long-range high-precision distributed fiber Raman temperature measurement system, the temperature data of the four fiber rings to be measured were scaled up by setting the temperature of the thermostatic tank to 90 °C, and the temperature distribution in the local area of the fiber was obtained as shown in Fig. 6.54.

In order to calibrate the spatial resolution of the long-range high-precision distributed fiber Raman thermometer, the temperature profile of the temperature measurement area was analyzed, and the distance difference between 90 and 10% temperature change at the sensing fiber length of 1380 m was 3.1 m and 3.0 m, respectively. The spatial resolution of the high-precision long-range distributed Raman thermometer at 1380 m is about 3.05 m. The temperature curve of the temperature measurement area at 10.2 km is enlarged and analyzed to obtain Fig. 6.54b. The distance difference between 90% temperature change and 10% temperature change is 3.4 m, and the spatial resolution of the high-precision distributed fiber Raman thermometer at 10.2 km is about 3.4 m. The temperature curve of the temperature measurement area at 18.8 km is enlarged and analyzed to obtain Fig. 6.54c, where

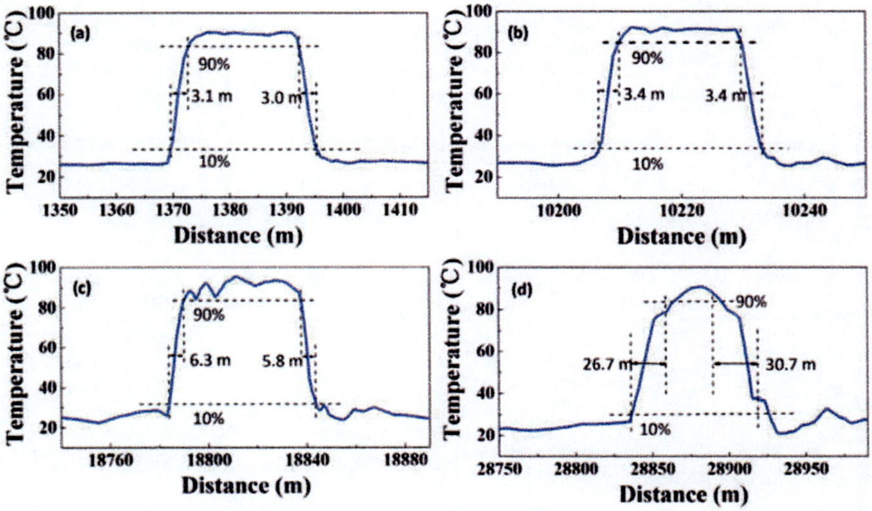

Fig. 6.54 Spatial resolution of long-range and high-precision distributed fiber optic Raman temperature measurement system

the distance difference between 90 and 10% temperature change is 6.3 m and 5.8 m respectively, and the spatial resolution of the high precision long-range distributed fiber optic Raman thermometer at 18.8 km is about 6.05 m. Finally, the temperature curve of the 28.9 km temperature measurement area is enlarged and analyzed to obtain Fig. 6.54d, in which the distance difference between 90% temperature change and 10% temperature change is 26.7 m and 30.7 m, respectively, and the spatial resolution of the long-range and high-precision distributed fiber Raman thermometer at 28.9 km is about 28.7 m.

3. **Temperature measurement range and accuracy**

To further investigate the temperature range and accuracy of the long-range and high-precision distributed fiber-optic Raman temperature sensing system at different temperatures, the temperature of the constant temperature bath was adjusted with the temperature control range: −30.0~90.0 °C and the temperature step value: 20.0 °C. The temperature data measured by the temperature measuring fiber in the water tank are enlarged, and the results are shown in Fig. 6.55. From Fig. 6.55, it can be seen that the measured temperature values of the fiber ring 1 and the fiber ring 2 are basically the same, and the curve is relatively stable, while the measured fiber ring 3 and the fiber ring 4 are limited by the length of the fiber and the signal-to-noise ratio of the signal, the measured temperature values are quite different and the curve fluctuates greatly.

In order to effectively calculate the temperature measurement error of the system, the temperature measurement data of the distributed fiber Raman temperature

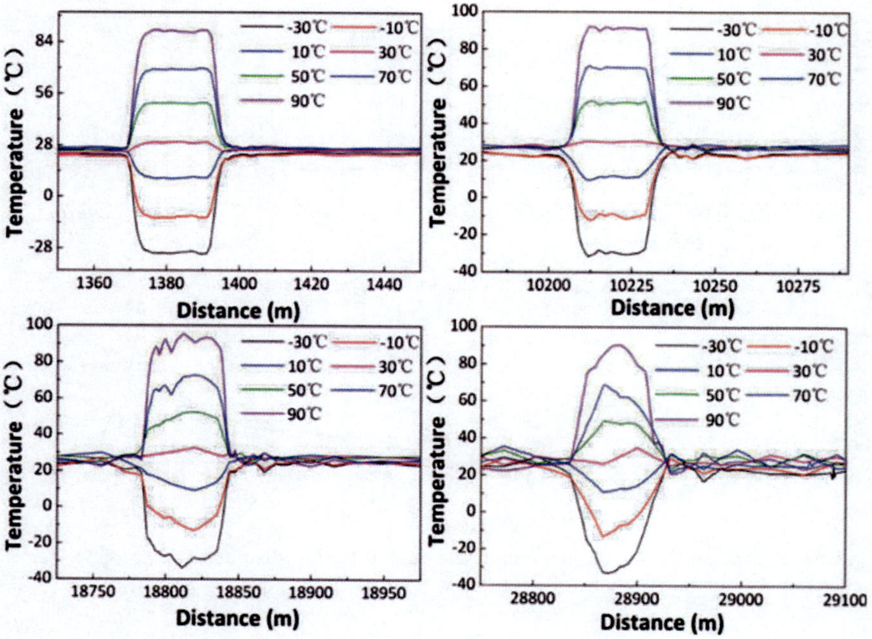

Fig. 6.55 Temperature distribution of the fiber ring to be measured along the sensing fiber

measurement system at different temperatures and the calibration temperature data of Agilent 34,410 A were compared, and the temperature measurement error curves of different temperature measurement points were plotted as shown in Fig. 6.56. In order to effectively calculate the temperature measurement error of the system, the temperature measurement data of the distributed fiber Raman temperature measurement system at different temperatures and the calibration temperature data of Agilent 34,410 A were used to draw the temperature measurement error curve of different temperature measurement points as shown in Fig. 6.56. From the figure, it can be seen that for the fiber ring 1 and fiber ring 2, the temperature measurement error remains constant for different temperatures and can be controlled within ± 1 °C, while for the fiber ring 3 and fiber ring 4, the temperature fluctuation is relatively large compared to different temperatures, in which the temperature measurement error of fiber ring 3 can reach ± 2 °C, while the temperature measurement error of fiber ring 4 is −6~2 °C. Considering the application of fire prevention for transportation infrastructure, the system can still effectively detect the presence of abnormal temperature points and provide effective alarms in actual production.

4. **Temperature stability**

The definition of the temperature stability of the Raman system is the most stable measurement of the sensor at a constant temperature over a period of time [40]. In the experiment, based on the traditional demodulation method and the proposed multi-stage constant temperature control and bias voltage control method, we compared

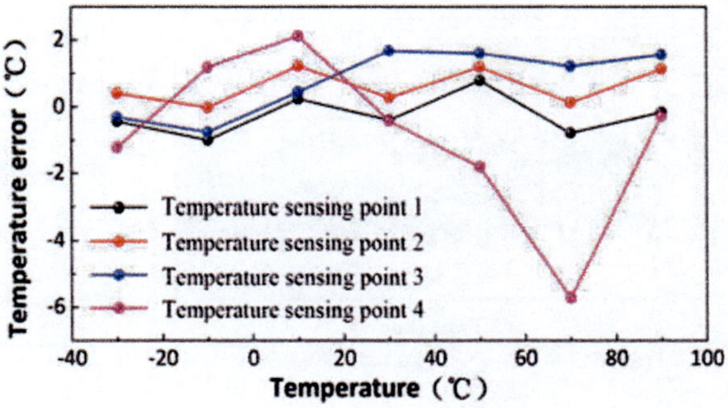

Fig. 6.56 Temperature measurement error curve of different temperature measurement points

the temperature range (within 4.5 h) and standard deviation. The experimental results of the system temperature fluctuation are shown in Fig. 6.57.

The four fiber rings to be measured were partially placed in a constant temperature water bath and maintained at a constant temperature of 30 °C. After the long-range and high-precision distributed fiber Raman temperature measurement system was stabilized (1 h), the temperature was measured continuously using the conventional demodulation-based method and the designed multi-stage thermostat control and bias control method, respectively. Figure 6.57 a–I, b–I, c–I and d–I represent the temperature stabilization trends without multi-stage thermostatic control and bias control algorithms at 1.38 km, 10.22 km, 18.80 km and 28.90 km respectively. Figure 6.57 a–II, b–II, c–II and d–II represent the temperature stability trends of the multi-stage thermostatic control and bias control algorithms at 1.38 km, 10.22 km, 18.80 km and 28.90 km respectively. The temperature signal at sensing distances of 1.38 km, 10.22 km and 18.80 km averaged 5000 times and contained 810 sets of data in 4.5 h. The temperature signal at sensing distances of 28.90 km averaged 30,000 times and contained 135 sets of data. The temperature signal at the sensing distance of 28.90 km was 30,000 times on average, containing 135 sets of data. The experimental results show that the measurement temperature stability of the long-range high-precision distributed fiber optic Raman temperature measurement system has been optimized. The temperature stability was optimized from ± 12.6 °C to ± 7.2 °C over a sensing distance of 28.9 km. Therefore, the multi-stage thermostat control and bias control system can provide reliable and highly stable temperature measurement performance.

5. **APD test analysis**

Different selection and control methods of APD will lead to changes in positioning accuracy, because APD is in Geiger mode, with high gain, high sensitivity, single photon response and other advantages [41–43]. Reference [44] studied the effect of operating temperature, dark current, bias voltage on the gain of APD and the gain variation in Geiger mode, starting from the gain conditions of APD. There are many internal factors affecting the magnitude of APD photocurrent, including dark current noise, scattering noise, thermal noise and a series of other factors, among which the

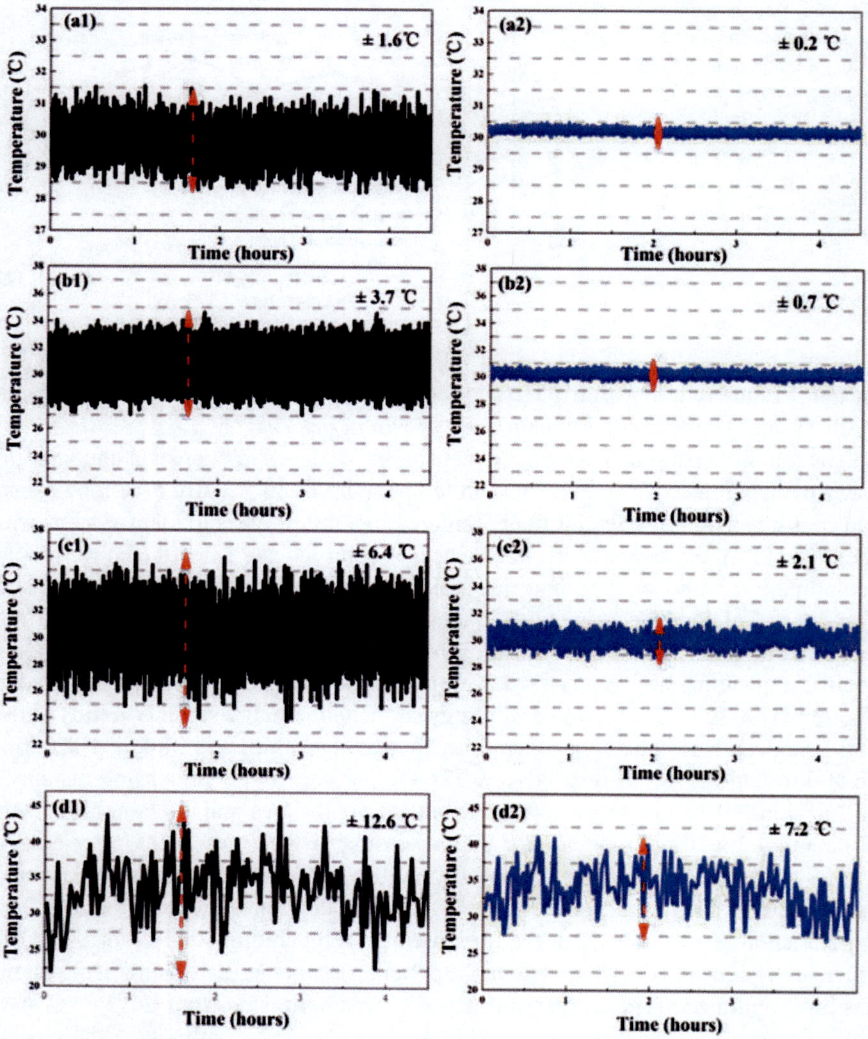

Fig. 6.57 System temperature measurement volatility comparison chart

dark current noise has a greater impact on the gain. The change of dark current of InGaAs-APD at different voltage and temperature is analyzed by experiment.

The InGaAs-APD test device diagram is shown in Fig. 6.58, and the physical device diagram is shown in Fig. 6.59. The APD test device is mainly composed of a printed circuit board and an experimental bracket. Among them, the printed circuit board contains three 1N4148 high-frequency switching diodes, two 5.1 kΩ resistors, a 100 Ω slide resistor, and multiplexer switches. In the experiment, the bias voltage at both ends of the APD was strictly controlled by using a 0–70 V adjustable precision power supply and multiple switches to test the APD characteristics. The bracket part includes multiple APDs and temperature probes to be tested, which can accurately monitor temperature changes. By controlling the on–off of multiple switches, the gain characteristics of multiple APDs can be directly detected. The GD5510Y InGaAs-APD produced by the 44th Research Institute of China Electronics Technology Group was used in the experiment.

(1) Test results and analysis

The APD dark current affects the magnitude of the photocurrent gain. Several InGaAs -APD of the same type were placed on a holder in a high-precision thermostat bath with alcohol solution. The operating temperature (±0.05 °C) of the InGaAs-APD was kept constant and the bias voltage was varied the variation of dark current was

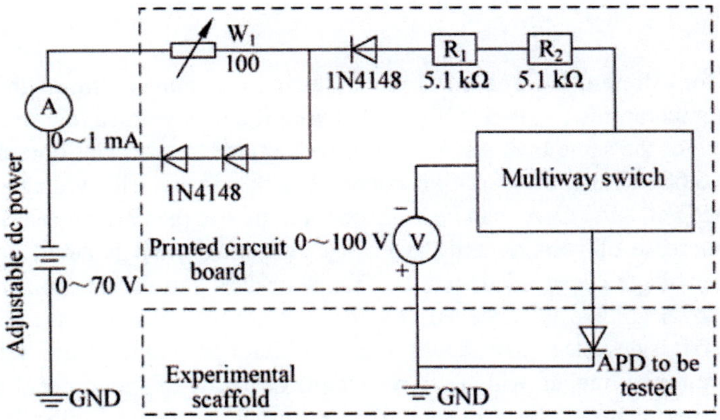

Fig. 6.58 InGaAs-APD test setup diagram

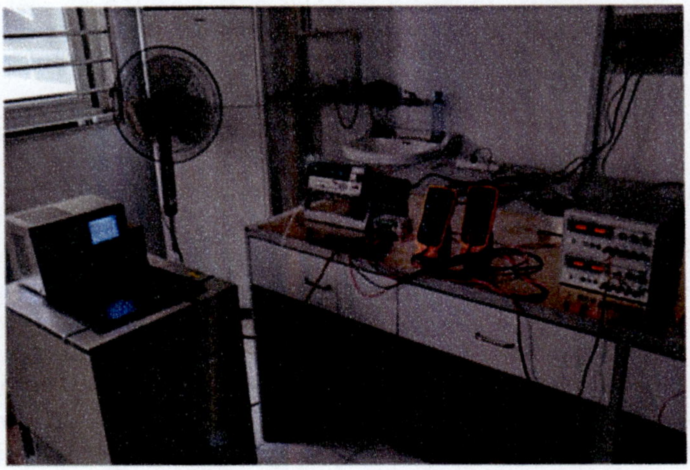

Fig. 6.59 APD test physical diagram

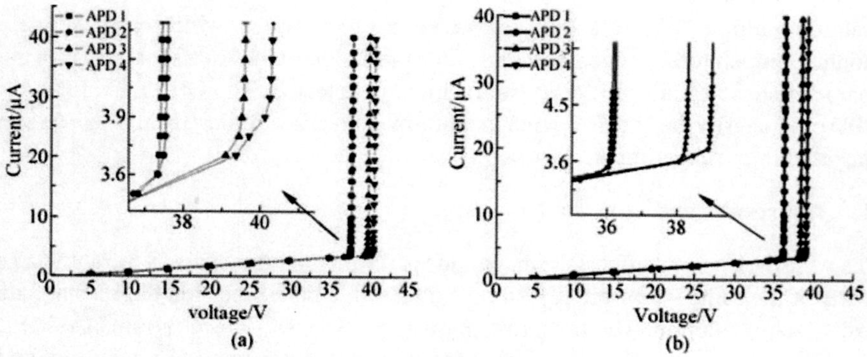

Fig. 6.60 Dark current versus voltage curves for different APD. **a** Dark current versus bias voltage at −10 °C, **b** Dark current versus bias voltage at −20 °C

observed for different InGaAs-APD. By adjusting the temperature of the bath at different temperatures (−30~5 °C) and changing the bias voltage, the trend of the dark current of the same InGaAs-APD at different temperatures was obtained.

Figure 6.60a shows the variation curve of dark current with voltage for four InGaAs-APD at −10 °C. As can be seen from the figure: the dark current increases with the increase of voltage, and the change of dark current is small before the breakdown voltage is exceeded, and the change of dark current is exponential after the breakdown voltage is exceeded. The dark currents of different APD are also different, so it is necessary to find two similar APD for pairing in distributed Raman temperature measurement, which can reduce the difficulty of data processing later.

In the experiments, all the APD dark current curves were tested with temperature to facilitate the matching of APD and data processing. Figure 6.60 shows the variation of dark current with voltage for a single APD at −10 °C and −20 °C respectively. It is found that the breakdown voltage varies for different APD at the same temperature, and the breakdown voltage varies for the same APD at different temperatures. Therefore, the use of APD requires attention to the control of APD temperature.

Figure 6.61 shows the variation of dark current for the same APD at different temperatures (−30~5 °C). It can be seen from the figure that the dark current of the same APD gradually decreases with decreasing temperature and at the same reverse bias. Therefore, in a distributed Raman system, the lower the temperature, the lower the effect of dark current on the distributed Raman system at a certain bias voltage.

The breakdown voltage curves of the two APD with temperature are given in Fig. 6.62, the breakdown voltage of the APD increases with increasing temperature. The temperature coefficient of breakdown voltage for APD No. 1 is 136.86 mV/°C and for APD No. 2 is 134.83 mV/°C. The average coefficient is approximately 135 mV/°C. The graph shows that temperature has a large influence on the breakdown voltage of the APD, so the temperature of the APD is precisely controlled by loading a two-stage temperature control device in the distributed Raman temperature measurement prototype system.

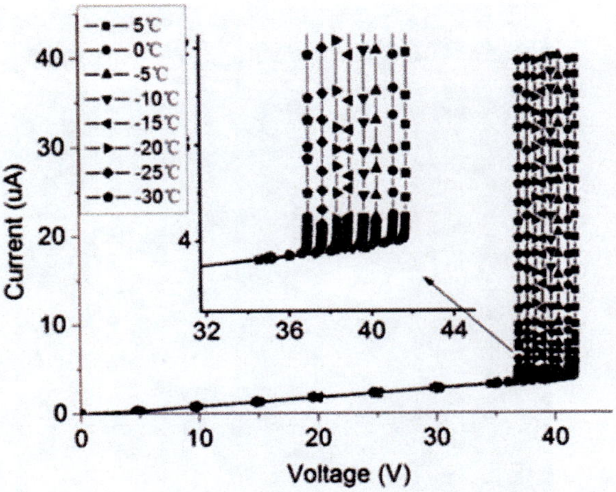

Fig. 6.61 Dark current versus temperature at different temperatures

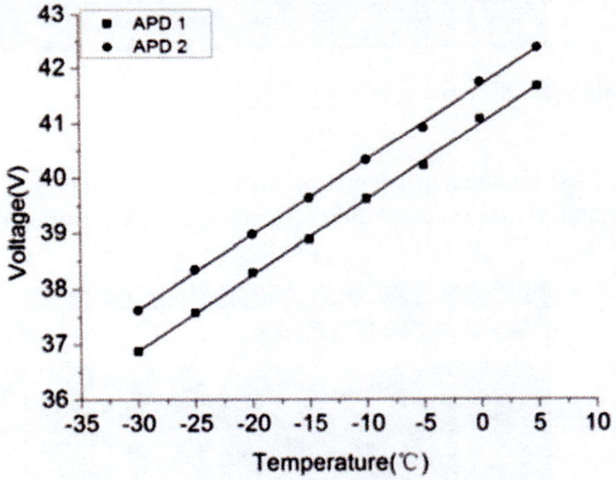

Fig. 6.62 APD breakdown voltage versus temperature

(2) **Analysis of APD noise characteristics**

Firstly, we analyze the source and distribution of the main noise of the APD when there is no external illumination. At this time, there are two main types of noise in the avalanche diode: dark current noise and scattered particle noise, of which scattered particle noise is also known as quantum noise, which is particularly small in the absence of light and can be ignored, so the main analysis of dark current noise. When no light enters, the APD itself has a substrate noise, set its scan bandwidth to 100 Mbit/s, through 1000 times the average, can get its substrate noise, from Fig. 6.63 can be seen, the APD itself has a low substrate noise, can be neglected.

The bias voltage is slowly increased at the incident side of the APD, as shown in Fig. 6.64, and the spectrum gradually rises from 0 to 60 Mbit/s. This indicates that the noise of the APD gradually increases from 0 to 60 Mbit/s under the influence

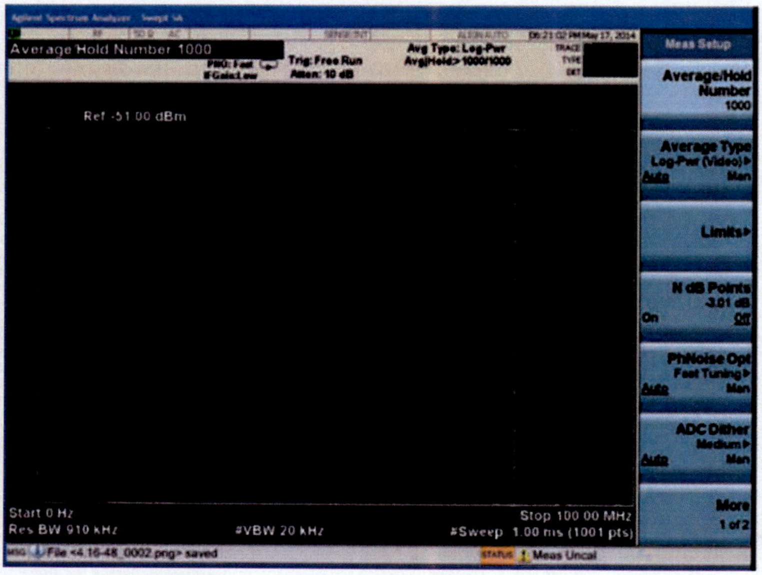

Fig. 6.63 Substrate noise of the APD itself

of the bias voltage, which is mainly dark current noise. After 60 Mbit/s the noise remains constant and therefore needs to be controlled below the 60 Mbit/s interval.

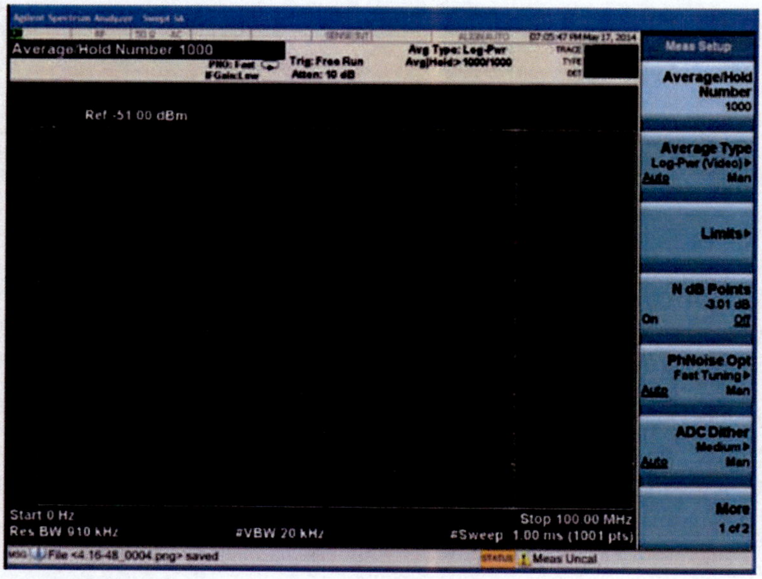

Fig. 6.64 APD bias noise

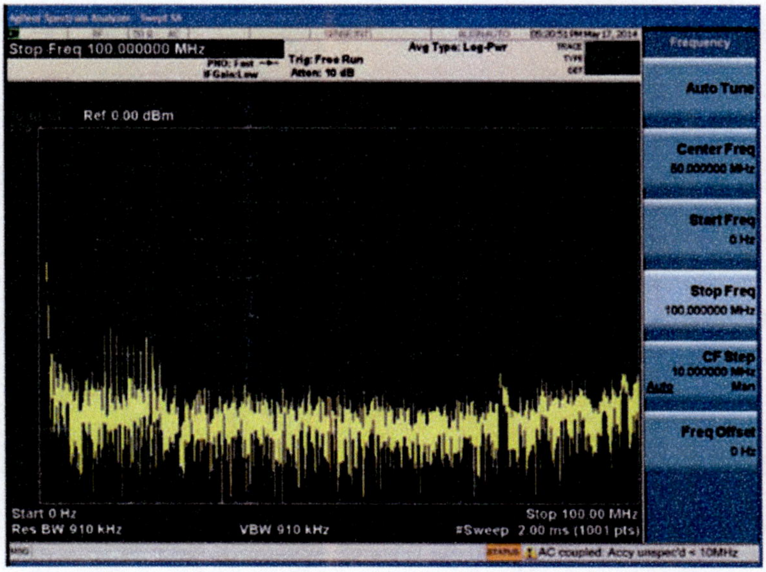

Fig. 6.65 Dark current noise spectrum

When connected to the photocurrent, the noise is further modified from the original base. In the experiment, the bias voltage at both ends of the APD was changed to control its dark current magnitude, while the spectrum meter signal was used to trigger the oscilloscope to collect its timing and spectrum, and the effect of noise on gain was observed by changing the current magnitude of the input dark current. Figure 6.65 shows that the noise is present at 10–20 Mbit/s for a dark current of 3.9 μA, and that the noise is present in the spectrum, but it is not always present, so it is part of the burst noise and needs to be removed when the final data is processed.

(3) Analysis of APD gain characteristics

When Raman backscattered light is incident on the APD, changes in incident optical power, external temperature and reverse voltage all have a significant effect on the gain change. The gain of the APD is limited by the size of the applied bias and breakdown voltage, and it is necessary to make the gain of the APD as large as possible in order to maximize its efficiency, so the applied bias should be close to the breakdown voltage to achieve high gain. Preliminary experiments have shown that the lower the input optical power, the greater the amplification effect on the APD, thus indicating that the APD is suitable for small signal amplification. The APD was connected to a low-power light source, and the gain was investigated at different bias and current levels. In the experiments, a pulsed laser is used to boost the pulsed light for the APD as described previously, and the input power is controlled by an optical attenuator The pulsed laser is synchronized by its own synchronization signal, enabling the two to be synchronized and collected. The pulses of the APD are slightly behind the synchronous signal due to its response time. In our experiments,

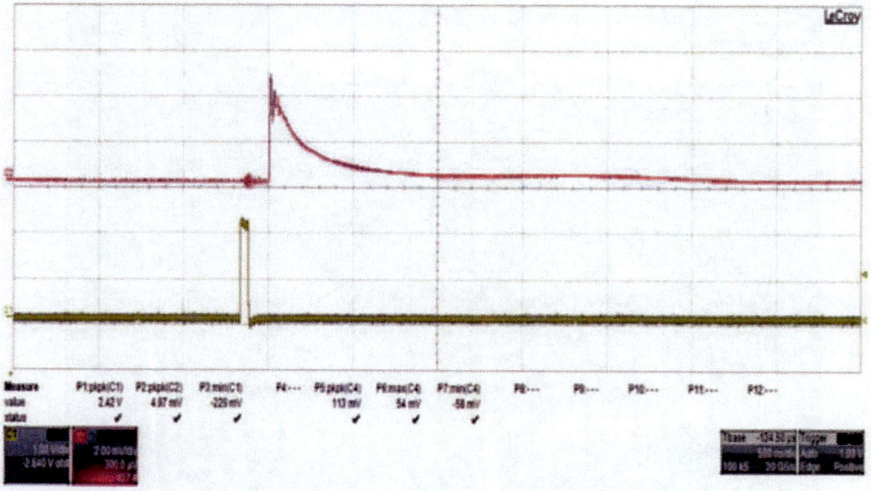

Fig. 6.66 Pulse synchronization timing diagram

we set the photocurrent at a fixed value to ensure the accuracy of the temperature at both ends of the APD (the temperature of the bath varied less than 0.1 °C), and then changed the bias voltage, gradually increasing the reverse bias voltage at both ends of the APD and moving closer to the breakdown voltage to observe the change in gain. As shown in Fig. 6.66, the amplitude of the starting light signal was obtained by averaging the light current of 0.1 μA for several times, and by continuously changing the reverse bias voltage loaded on both ends of the APD, the pulse amplitude of the APD at different bias voltages can be obtained and compared to obtain the pulse amplitude gain curve, i.e. the optical signal amplification.

It can be seen from the experiments that the APD gain increases with increasing bias voltage at the same temperature and increases exponentially in a certain interval, and that the pulse gain of the same APD signal varies at different temperatures. As the photocurrent increases, the APD gain becomes saturated. Because of the presence of multiplying noise, dark current noise and a number of other factors, the gain will no longer increase with the increase in photocurrent after a certain value, so there will be a maximum gain. The APD optical current is controlled at its inflection point, which is the optimal avalanche gain of the APD. In distributed fiber optic sensing, by adjusting the voltage and operating temperature of the APD, the APD photocurrent can be controlled at the limit of gain, thus achieving a large avalanche gain.

As shown in Fig. 6.67, the gain change curve when the input photocurrent is 0.1 μA, after experimental analysis: this model APD has the maximum gain at 10 °C, therefore, in the distributed Raman temperature measurement prototype, this APD can be controlled at around 10 °C, when the photocurrent is 1 μA, its gain can be maximized, through analysis, it can be seen that this interval is the bias size when the APD is in Geiger mode.

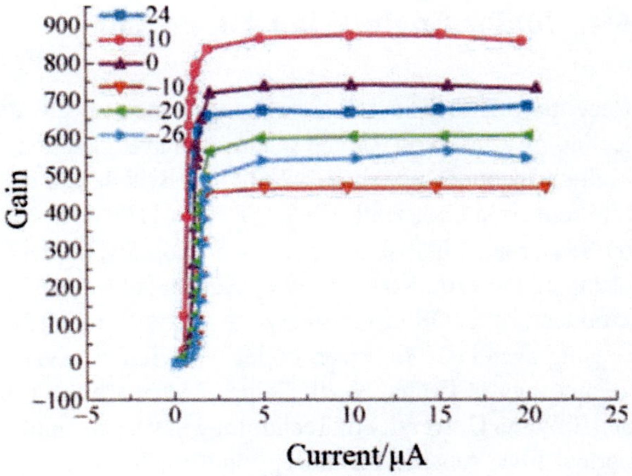

Fig. 6.67 APD gain variation curve

In the following, we need to analyze the Geiger model, its feasibility under the Geiger model and its advantages and disadvantages.

(4) **APD in Geiger mode analysis**

The APD works in Geiger mode, i.e. the bias voltage of the APD operates at a higher than the breakdown voltage of the avalanche diode, and the gain at this time will be several times higher than the gain when it works in general mode [45]. When the APD is operating in the general region, its bias voltage is about 2 V below the breakdown voltage, when the APD is operating in the safe region, the temperature and voltage changes basically will not damage the APD, the disadvantage is that the gain is small, the APD cannot be used effectively. When the APD is in Geiger mode, the current will quickly reach a large value with a slight change in voltage, which is generally less than 1 ps. The high responsiveness and high gain factor of Geiger mode is of great interest for distributed fiber optic Raman temperature measurement systems. The disadvantage is that when the APD is in Geiger mode, the APD is more affected by external temperature and voltage variations. Currently, Geiger mode is mainly used for single photon detection, single photon counting, radar imaging, quantum communication and other fields. If we can effectively control the operating mode of the APD in Geiger mode, it will be of great value for Raman temperature measurement systems.

The amplification of the APD in Geiger mode is significantly greater than the amplification of the APD before the multiplication interval. Most Raman temperature measurement products control the amplification interval of the APD in the area before the avalanche interval for safety and control conveniently, to avoid permanent damage to the APD due to insufficient control accuracy. By improving the control circuitry to precisely control the voltage and current across the APD within the avalanche interval, the APD gain can be significantly increased.

6.4 Major Engineering Application

Recently, a series of commercialized products of fully distributed optical fiber sensing technology based on Raman scattering have appeared. For example, the DTS-800, a high-end product launched by YORK in British, and distributed optical fiber temperature sensors such as Halo-DTS, DTS-SR, DTS-MR, DTS-LR and DTS-XR produced by Sensornet, a British company. Additionally, the DFS-1000 developed by Fujikura of Japan, the 5100 Series made by SensorTran of the United States and the LHD4CH produced by LIOS of Germany are all typical distributed optical fiber temperature sensing systems. Moreover, colleges such as University of Electronic Science and Technology of China, North China Electric Power University, Tsinghua University and Taiyuan University of Technology have also conducted researches on distributed optical fiber sensing technology and made some achievements. This has made a significant contribution to the independent research and development of distributed optical fiber temperature sensing system with long distance, high precision and high stability, which has greatly promoted the study on its practical application.

Distributed optical fiber temperature sensing system has unique advantages over electrical sensing system, which makes its application prospect very broad and is often competent for the tasks that the electrical sensing system is difficult to achieve, including electric power, chemical, infrastructure, transportation, coal mine, water conservancy, petroleum and other fields. For instance, it is capable of the monitoring of cable temperature and ampacity, temperature measurement in high temperature environment, safety monitoring of buildings, bridges, tunnels and subways, and temperature monitoring of aerospace industry.

In short, temperature change is the most important and direct manifestation of changes in object characteristics. Therefore, temperature monitoring has become a significant tool for fault diagnosis and accident warning. The demand for fully distributed temperature monitoring is very extensive, and many fields such as energy, gas pipeline, electric power and aviation all regard it as a necessary method of fault diagnosis and accident warning. However, the optical fiber sensing technology based on Raman scattering, which has the advantages of full distribution, long distance, and high temperature measurement accuracy, integrates signal transmission and sensing information on a continuous optical fiber, and can simultaneously obtain the distribution information of the measured object with time and space.

6.4.1 Application in Safety Monitoring of Long-Distance Natural Gas Pipeline

1. **Necessity of safety monitoring system for gas pipeline temperature**

With the continuous development of society and economy, people's living standard is improving, and the use of gas is becoming more and more widespread in life. The transportation of natural gas has high requirements on pipelines, and the laying and maintenance of long-distance pipelines is of great significance to the delivery of natural gas. Once there is a problem with the transportation pipeline, it will seriously affect people's lives and safety, and bring losses to the society and economy. Owing to gas pipelines are deeply buried underground, electrochemical corrosion, bacterial corrosion, stress corrosion, and stray current corrosion will occur to traditional electrical sensors, and it is basically impossible to perform regular manual inspections of deeply buried pipelines [46]. Therefore, it is necessary to install distributed optical fiber temperature safety monitoring systems on gas pipelines.

2. **Monitoring requirements for gas pipeline temperature safety monitoring system**

Long-distance gas pipelines are mostly laid in remote mountainous areas, and most of them are buried underground. In some extreme terrain areas with complex structures, dense mines and many goafs, geological disasters such as shallow surface folds and slippage can easily cause pipeline rupture, abnormal deformation, or gas leakage due to abnormal temperature and other reasons, which poses a major threat to the safety of long-distance gas pipelines. This requires that the monitoring sensor system has the capacity to implement macroscale, fully-distributed and high-precision temperature monitoring on large-scale and long-distance structures. Corrosion is one of the main causes of accidents in long-distance natural gas pipelines. Corrosion may not only thin the wall of the pipeline in a large area, resulting in excessive deformation or rupture, but also may directly cause pipeline perforation or stress corrosion cracking, resulting in gas leakage accidents. In this case, on the one hand, the sensing system should have good corrosion resistance, on the other hand, it should have the ability of real-time and long-term monitoring, that is, the excessive reaction of the gas pipeline (too high temperature, too low temperature, etc.) can be monitored for a long term. What's more, the system is required to achieve advanced warning of temperature and detect the abnormalities of gas pipeline transportation faster, so as to timely alert the gas pipeline management department [47, 48].

3. **Application cases**

Distributed optical fiber temperature measurement systems are widely used in large-scale projects such as forest fires, tunnels, and natural gas pipelines. Health monitoring of gas pipelines can detect potential problems and then enable timely repair to avoid catastrophic consequences. In cooperation with Shanxi CBM Jishu Co.,Ltd. (Natural Gas) and SHANXI GUOXIN ENERGY Co.,Ltd., the distributed optical fiber sensor developed by the author's research group on the "Gas Pipeline Network Digital System Development and Demonstration Project" has been applied to the safety monitoring of Qinshui natural gas pipeline in Shanxi Province. Figure 6.68 shows the construction of 10.0 km-level demonstration project of distributed optical fiber Raman temperature sensor developed by Taiyuan University of Technology in the pipeline section of "Nanda Valve Room—Guo Jialing" from Qinshui to Yangcheng, Shanxi Province.

In the field temperature test experiment of the pipeline, the pipeline is heated by hot air blowing at 1920 m along the pipeline to simulate the heating phenomenon

Fig. 6.68 Gas pipeline temperature field test

caused by the leakage of the gas pipeline, and the distributed optical fiber Raman thermometer is used for real-time temperature measurement. The temperature test results are shown in Fig. 6.69, which show that the distributed optical fiber Raman temperature measurement system can accurately detect and locate the temperature distribution information along the gas pipeline.

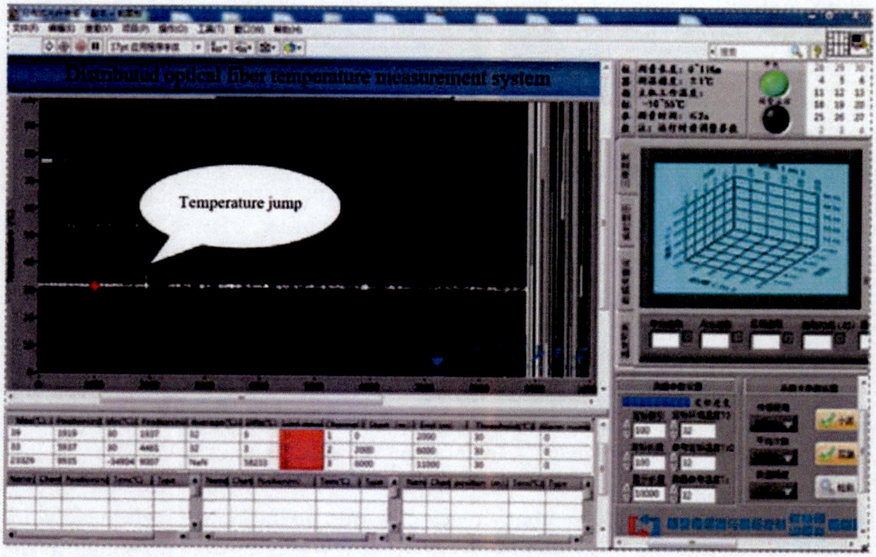

Fig. 6.69 Temperature test results at worksite

6.4.2 Application in Power Cable Fault Detection

In order to solve the impact of overhead transmission lines on urban appearance and expand the bearable range of power load demand in the development process, more and more urban power cables have begun to gradually replace overhead transmission lines through power renovation projects, and have been applied in urban power supply system. By retrofitting the cables into the ground, the floor space can be reduced, the circuit layout can be simplified, the operating cost can be reduced, and daily life can be facilitated [49]. What needs to be considered is how to solve the problem of heating caused by the operation of the cable. Heating phenomenon usually occurs when the cable is running, and excessive temperature may cause fire. Because the cables are below the ground and the environment is complex, it may make it difficult to carry out the rescue work, further magnifying the loss. In order to ensure the normal operation of cables and protect the property safety of the society, it has become an urgent need to study the application of temperature measurement technology in cable fault detection.

The current carrying capacity of the cable refers to the amount of current that the cable passes through during the process of electric energy transmission, and this index is directly related to the heating degree of the conductor when the current flows. Due to the differences of laying state, operating environment and other conditions, the insulation aging degree of cable conductors varies, which will affect the actual current-carrying capacity of the cable. Therefore, it is necessary to carry out highly accurate online measurement of the conductor temperature during the actual operation of the cable, and deduce the empirical formula of the influence of the actual current carrying capacity of the cable on temperature rise according to the actual production experience. On this basis, the allowable current carrying capacity of the cable is monitored, and at the same time, accurate real-time temperature measurement and control is performed on different parts of the cable and its accessories and the laying environment. This method can greatly improve the operation efficiency of medium and high voltage power cables, and fully guarantee the operation safety. The distributed optical fiber temperature monitoring system has a series of excellent characteristics that other traditional temperature measurement systems cannot achieve. For example, when the current is flowing, multi-point, comprehensive, real-time, and accurate distributed measurement can be achieved by online monitoring of the thermal effect of the cable, so as to solve the problems that were difficult to deal with in time after an emergency occurred on the cable and its accessories in the past [50, 51].

As the power transmission carrier of the power system, the operation state of the power cable is directly related to the safety and stability of the power grid, so it is extremely important to monitor its own running state. The high-voltage cable optical fiber distributed temperature monitoring system must ensure the completion of the following core functions: ① Monitor the surface temperature of the running cable in real time; ② Accurately locate the abnormal heating position of the running cable, and display the temperature status of the heating position; ③ Convert the real-time online surface temperature of the cable detected by the optical fiber distributed temperature monitoring system into the operating current carrying capacity of the cable through auxiliary software, so as to provide historical basis for customers to determine the maximum current carrying capacity of the cable and maximize the service life of the power cable [52].

6.4.3 Application of Coal Gangue Mountain Fire Monitoring

The internal environment of the coal gangue mountain is complex and changeable, so it is difficult to figure out the flow state of the air inside the coal gangue mountain, the distribution of the temperature field and the scattering law of the remnant coal, and thus it is hard to determine where fire caused by spontaneous combustion of coal will occur in the coal gangue mountain. Therefore, when monitoring the internal temperature field of the coal gangue mountain, the monitoring range should be as large as possible to ensure that the temperature information of all possible fire locations can be received. The investigation of the application of the optical fiber temperature measurement system in the temperature monitoring of multiple coal gangue mountains shows that for the traditional method, such as laying temperature measurement optical fiber along the upper and down crossheading of working face to the inside of the coal gangue mountain [53], as illustrated in Fig. 6.70. Owing to low heat conduction efficiency and poor thermal conductivity effect of coal, when the temperature of surrounding coal body rises because of the oxidation reaction at position A, the change needs a long time to be transmitted to the optical fiber. In this case, the oxidation heating degree of coal in the coal gangue mountain cannot be reflected in time, so as to miss optimal fire extinguishing time. To solve this problem, a novel laying method is proposed by referring to the laying method of other fields such as power fire monitoring, dam seepage monitoring, oil and gas pipeline leakage monitoring, and combining the characteristics of coal mines, as shown in Fig. 6.71.

The advantages of the above-mentioned laying method presented in Fig. 6.71 are as follows. On the one hand, the optical fiber temperature measurement system

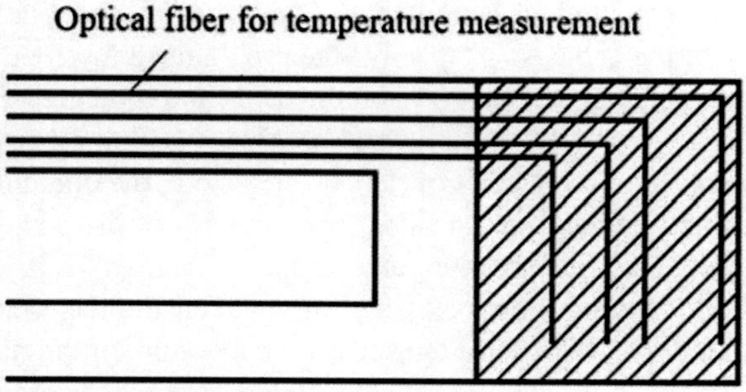

Fig. 6.70 Layout method of traditional optical fiber

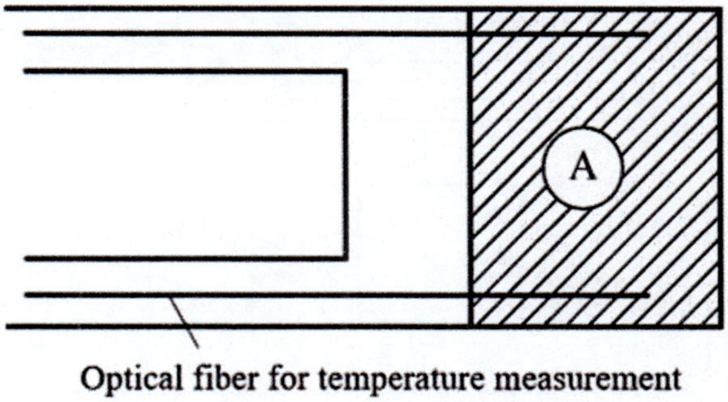

Fig. 6.71 Laying method of optical fiber in project

requires a small number of fiber channels—that is, only a few optical fibers need to be laid to complete the comprehensive monitoring of the temperature of entire goaf "three zones". On the other hand, the laying is convenient.

According to the actual project, the width of each zone of the "three zones" is as follows. The bandwidth of the ventilation and heat dissipation zone is about 16 m, the bandwidth of the oxidation zone is about 30 m, between 16 and 45 m, and the suffocation zone is after 45 m [54, 55], as shown in Fig. 6.72.

According to the division of the "three-zones" of the coal gangue mountain in the working face, four optical fibers are laid based on the designed method, as depicted in Fig. 6.73.

1# optical fiber (L-shaped, with a total length of about 400 m, connected to channel 1). Start laying when the working face is advanced to about 53 m from the mining stop line, the length of the optical fiber in the upper crossheading is about 240 m, and the length of the part perpendicular to the advancing direction at the back of the support is 50 m.

2# optical fiber (total length is about 427 m, connected to channel 2). Lay together with optical fiber 1#. One end is connected to the temperature measuring machine,

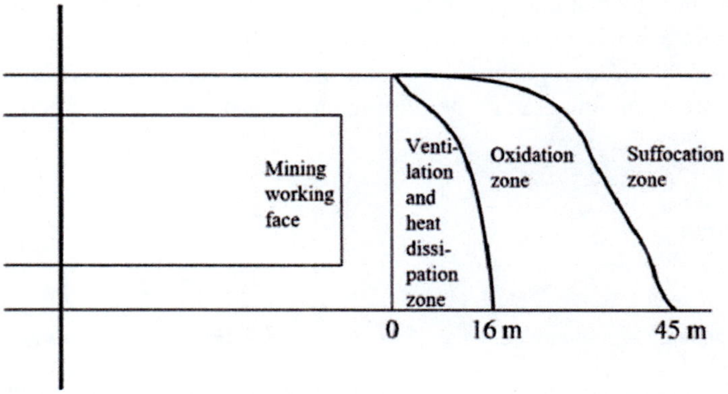

Fig. 6.72 Distribution of "three zones" of coal gangue mountains

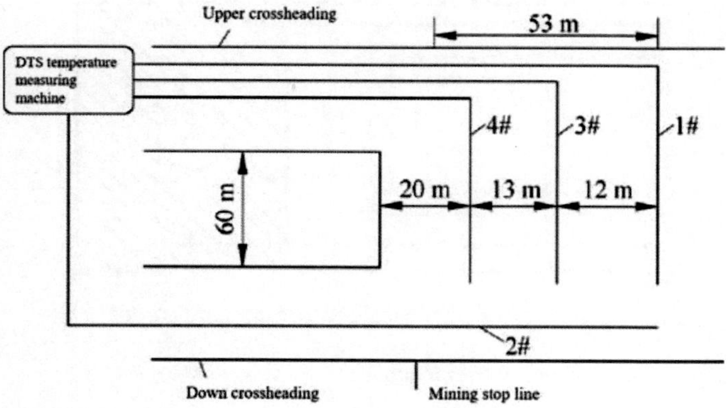

Fig. 6.73 Optical fiber layout in goaf

and the other end enters the down crossheading through the connecting lane. Then it is laid along the down crossheading to the low corner of the working face.

3# optical fiber (L-shaped, with a total length of about 367 m, connected to channel 3). Start laying when the working face is advanced to about 33 m from the mining stop line, the length of the optical fiber in the upper crossheading is about 220 m, and the length of the part perpendicular to the advancing direction at the back of the support is 60 m.

4# optical fiber (L-shaped, with a total length of about 347 m, connected to channel 4). Start laying when the working face is advanced to about 20 m from the mining stop line, the length of the optical fiber in the upper crossheading is about 200 m, and the length of the part perpendicular to the advancing direction at the back of the support is 60 m [56].

When the optical fiber layout is completed, start the optical fiber temperature measurement machine, and use the system software to set the measurement partitions of channel 1, channel 2, channel 3 and channel 4 to ensure that the system can continuously measure the temperature of goaf. The results of practical engineering application indicate that the temperature variation rule measured by the optical fiber temperature measurement machine is consistent with the temperature change of the "three zones" of the coal gangue mountain, which shows that the system can meet the requirements of temperature monitoring of the coal gangue mountain.

References

1. Namih S, Emori Y. Ultrabroad-band Raman amplifiers pumped and gain-equalized by wavelength-division-multiplexed high-power laser diodes. IEEE J Sel Top Quantum Electron. 2001;7(1):3–16.
2. Saissy A. Spontaneous Raman scattering and polarization mode coupling in polarization—maintaining optical fibers. J Lightwave Technol. 1987;5(8):1045–9.

3. Nick VDG, Susan CSD. Double-ended calibration of fiber-optic Raman spectra distributed temperature sensing data. Sensors. 2012;12(12):5471–85.
4. Farahani M, Gogolla T. Spontaneous Raman scattering in optical fibers with modulated probe light for distributed temperature Raman remote sensing. J Lightwave Technol. 1999;17(8):1379–91.
5. Li YT. The research of noise suppression and the design of human-computer interaction interface on Raman distributed temperature instrument. Taiyuan: Taiyuan University of Technology; 2017.
6. Wang ZL. Signal processing and performance improvement of distributed fiber Raman temperature sensing system. Jinan: Shandong University; 2015.
7. Cao LJ. Research on distributed optical fiber temperature measurement and data processing technology. Hefei: Hefei University of Technology; 2006.
8. Sun M. Performance improvement for distributed fiber temperature sensor system and research on fire source localization. Hefei: University of Science and Technology of China; 2017.
9. Zhang MJ, Li J, Liu Y, et al. A new temperature demodulation method for distributed fiber Raman temperature measurement. Chin J Lasers. 2017;03:219–26.
10. Li J, Xu Y, Zhang MJ, et al. Performance improvement in double-ended RDTS by suppressing the local external physics perturbation and intermodal dispersion. Chin Opt Lett. 2019;17(7):070602.
11. Li J. The research of novel temperature demodulation method and advance prewarning model for fiber Raman sensing system. Taiyuan: Taiyuan University of Technology; 2018.
12. Chen FC, Dai J, Yu CQ. Distributed Raman fiber sensing system with fiber-ring calibration and double ended Prob. Opto-Electron Eng. 2016;43(8):33–8.
13. Chai J. Brillouin optical correlation-domain reflectometry based on non-periodical signal. Taiyuan: Taiyuan University of Technology; 2015.
14. Tang WQ. Research on Brillouin frequency shift extraction technology in BOTDA optical fiber temperature sensing system. Beijing: Beijing University of Posts and Telecommunications; 2019.
15. Xu ZW. Test study on the abutment pressure of working face based on distributed optical fiber sensor]. Xian: Xi'an University of Science and Technology; 2018.
16. Chen J, Zhang Y, Mao JH, et al. Progress of distributed fiber optic sensing technology in concrete infrastructure health monitoring. Road Traffic Technol. 2018;14(05):202–5.
17. Wang W, Lv LD, Ge SW, et al. Optical fiber sensing technology and its application in intelligent cable tunnel. Distrib Utilization. 2018;35(03):25–31.
18. Zheng YT. Analysis of the application of distributed fiber optic sensing technology in intelligent dam safety monitoring. Intell City. 2018;4(02):164–5.
19. Bai Q, Wang YC, Liu X, et al. Monitoring technology and implementation of temperature, vibration, sound and strain based on distributed optical fiber. Laser Mag. 2018;39(3):47–54.
20. Zou J, Wang S, Yang ZK. A new distributed temperature measurement method based on Raman scattering. Optoelectron Lasers. 2000;05:507–10.
21. Chen RL, Wan SP, Jia P, et al. Distributed optical fiber Raman temperature measurement system based on cumulative average. Appl Opt. 2018;39(4):590–4.
22. Ma CY, Liu TG, Liu K, et al. A continuous wavelet transform based time delay estimation method for long range fiber interferometric vibration sensor. J Lightwave Technol. 2016;34(16):3785–9.
23. Hu Y, Mo WQ, Dong KF, et al. Using maximum spectrum of continuous wavelet transform for demodulation of an overlapped spectrum in a fiber Bragg grating sensor network. Appl Opt. 2016;55(17):4670–5.
24. Wang ZL, Chang J, Zhang SS, et al. An improved denoising method in RDTS based on wavelet transform modulus maxima. Sens J. 2015;15(2):1061–7.
25. Li J, Li YT, Zhang MJ, et al. Performance improvement of Raman distributed temperature system by using noise suppression. Photonic Sens. 2017;8(2):103–13.
26. Zhang YN, Zhang MJ, Zhang JZ, et al. Performance improvement of power-over-fiber system using noise-modulated laser diode. Appl Opt. 2016;55(7):1625–9.

27. Li J, Yan BQ, Zhang MJ, et al. Long-range Raman distributed fiber temperature sensor with early warning model for fire detection and prevention. IEEE Sens J. 2019;19(10):3711–7.
28. Li J, Zhang Q, Yu T, et al. R-DTS with heat transfer functional model for perceiving the surrounding temperature. Sens J. 2020;20(2):816–22.
29. Yu XD, Zhang ZX, Zhu HZ, et al. A Temperature compensating circuit for distributed optical fiber Raman temperature sensor. Chin J Lasers. 2010;37(06):1440–4.
30. Jin ZX, Cui HJ, Ning F, et al. Distributed optical fiber Raman temperature measurement system based on the dynamic multi-section temperature calibration. Acta Photonica Sinica. 2011;40(09):1297–302.
31. Zhang XL, Zhang Q, Nie FL, et al. Design of multi-parameter monitoring system for gearbox test rig based on dual ARM. Instrum Techn Sens. 2016;03:44–7.
32. Zhao QM, Xu FL, Xiao L. Design of portable instrument for impedance measurement based on STM32. Instrum Tech Sens. 2016;01:33–6.
33. Xu GP, Feng GX, Geng L. Temperature control of high density TEC based on MCU operation. Laser Infrared. 2009;39(03):254–6.
34. Huang Q. Design of temperature and bias voltage control system for distributed fiber Raman temperature sensor. Taiyuan: Taiyuan University of Technology; 2017.
35. Huang Q, Han GY, Wu RD, et al. Design of high-precious constant temperature control system based on STM32. Instrum Techn Sens. 2017;05:71–4.
36. Fang JY, Li Q, Ma X. 3D visualization data acquisition system based on LabVIEW. Comput Measu Control. 2016;24(04):26–8.
37. Yan BQ, Li J, Zhang MJ, et al. Raman distributed temperature sensors with optical dynamic difference compensation for tunnel fire detection. Sensors. 2019;19(10):2320.
38. Kishida K, Yamauchi Y, Nishiguchi K, et al. Monitoring of tunnel shape using distributed optical fiber sensing techniques. In: Proceedings of the fourth conference on smart monitoring: assessment and rehabilitation of civil structures (SMAR); 2017
39. Soga K, Schooling J. Infrastructure sensing. Interface Focus. 2016;16(23):1–17.
40. Li J, Yan BQ, Soga K, et al. Auto-correction method for improving temperature stability in a long-range Raman fiber temperature sensor. Appl Opt. 2019;58(1):37–42.
41. Zhao HZ, Li NJ, Zhao DZ. Analysis of the optimum gain of APD in distributed fiber temperature sensors. Acta Photonica Sin. 1996;11:1028–31.
42. Li SB, Wang Y, Zhao JK, et al. Research on stability of the gain of APD in fiber Raman temperature sensor. Optoelectron Technol. 2009;29(03):168–70.
43. Song JH. Digital control bias circuit of APD with temperature compensation. Opt Optoelectron Technol. 2013;11(02):12–5.
44. Xu WP, Han GY, Yang SQ, et al. The controlling of InGaAs avalanche photodiode with Geiger mode in Raman distributed temperature sensing system. China Sciencepaper. 2015;10(08):995–8.
45. Mo QY, Zhao YL. Frequency responses of communication avalanche photodiodes. Acta Phys Sin. 2011;60(07):202–11.
46. Zhang ZK. Research on on-line monitoring of natural gas storage tank leakage based on optical fiber sensing technology. Nanchang: Jiangxi University of Science and Technology; 2017.
47. Chen FQ, Li X. Fiber grating distributed sensing system and its gas pipeline leakage detection. Autom Instrum. 2009;03:114–5.
48. Wang ZQ, Sun QC. The discussion of application of distributed optical fiber temperature detection technology in gas pipeline leakage detection. Petrol Eng Constr. 2009;35(S1):77–80.
49. Zhang WP, Feng J, Hu WG, et al. Application of distributed optical fiber temperature measurement technology in high voltage cable. Opt Commun Technol. 2014;7:29–31.
50. Lei XY. Application of distributed optical fiber temperature measurement technology in power cable fault detection. China High New Technol. 2017;1(12):40–2.
51. Chen J. Research and application of optical fiber temperature measurement system in power cable [D]. Guangzhou: South China University of Technology; 2016.
52. Wang S. The discussion of application of distributed optical fiber temperature measurement technology in power cable running state monitoring. Telecom World. 2016;22:197–9.

53. Cheng GY, Tang JJ, Cao J, et al. Research on "'three zone"' of goaf based on optical fiber temperature measurement system. China Coal. 2016;42(12):107–10.
54. Xu C. The application of distributed temperature sensing technology in coal mine. Jinan: Shandong University; 2016.
55. Zhang XJ. Research on coal temperature monitoring in the goaf based on optical fiber pyrometer. Huainan: Anhui University of Science and Technology; 2015.
56. Wang JF. Research on fire monitoring system in mine Goaf. Colliery Mech Electr Technol. 2014;06:16–25.

Chapter 7
Narrow Linewidth Fiber Laser for Fiber Sensing

7.1 Research Status of Fiber Laser

As a representative of modern laser technology, fiber laser has become a research hotspot in recent years and has a wide market prospect. Single-longitudinal-mode narrow linewidth fiber laser is an important branch of fiber laser technology with many advantages, such as narrow line width, long coherence length, easy to realize high power, strong system stability and miniaturization. Therefore, it is widely used in dense wavelength division multiplexing system, optical fiber sensing system, remote measurement and control radar, remote communication system and other application fields, and plays a very important role. Especially in the field of optical fiber sensing, the coherence length of the light source directly affects the final measurement accuracy of the system, and the stability of the light source will also affect the accuracy of the measurement results. So a stable single-longitudinal-mode laser light source with long coherence length, i.e. extremely narrow linewidth, is particularly important. In addition, similar to sensing systems, the linewidth of light source directly affects the overall performance of remote monitoring and remote sensing systems in many defense and military applications, therefore requiring very high performance demanding for light sources. The performance of traditional semiconductor lasers is often difficult to meet the demand for light sources in the above application fields. Therefore, fiber lasers with good light source characteristics have increasingly become the hot button of research and application. At present, the research of fiber laser mainly focuses on narrowing the linewidth, improving the output power and realizing tunable multi-wavelength output.

M. Zhang et al., *Novel Optical Fiber Sensing Technology and Systems*, Progress in Optical Science and Photonics 28, https://doi.org/10.1007/978-981-99-7149-7_7

7.1.1 Development Background of Fiber Laser

The development of fiber laser, which uses glass fiber doped with rare earth element as gain medium, can be traced back to the 1960s. In 1958, an article entitled "Infrared and Optical Masers" was published in *Physical Review*, marking that laser has stepped onto the historical stage as a new thing. Then, following the emergence of ruby laser and He–Ne laser, E. Snitze and others obtained laser with wavelength of 1.06 μm by using rod-shaped neodymium doped (Nd^{3+}) glass waveguide in 1961 [1], the first fiber laser was born. Fiber lasers have the same structure as other types of lasers such as gas lasers and semiconductor lasers. They are composed of pump source, gain medium and resonant cavity. Among them, the fiber laser takes the optical fiber as the gain medium, and uses it as a low loss waveguide to restrict the pump light and signal light in the fiber to form a dielectric closed cavity structure [2]. Due to the limitations of fiber preparation technology and pumping technology at that time, the fiber laser was developing slowly at that time.

In the 1970s–1980s, with the emergence of low loss neodymium doped and erbium-doped fiber preparation technology, especially the lasing wavelength of erbium-doped fiber is just located in the 1550 nm low loss window of communication fiber, which makes fiber lasers, especially erbium-doped fiber lasers, show a very attractive application prospect [3]. At the same time, semiconductor laser, as the commonly used pump source of fiber laser, has also been commercialized and later wavelength division multiplexing technology has matured, which makes fiber laser attract more and more researchers' attention.

Since the 1990s, with the wide application and development of optical fiber communication system, fiber laser has developed rapidly and its performance has been continuously improved. Especially with the development and maturity of key technologies such as fiber preparation technology and pump technology, and the continuous introduction of new precision filter devices, fiber lasers are gradually replacing traditional semiconductor lasers as the dominance in the field of laser technology. Research and development on fiber laser mainly focus on the following aspects.

In terms of high power, American IPG company has developed a series of high-power fiber lasers with continuous output of 1 kW, 2 kW, 10 kW and even 20 kW, and has launched fiber lasers with higher output power. J. Nilsson and others realized stable continuous laser output with center wavelength of 1100 nm and power of 1.36 kW by using ytterbium doped fiber [4], and further commercialized kW high-power fiber laser [5]. In terms of pulsed fiber lasers, ultrashort pulse fiber lasers with pulses as narrow as 3 fs have been operated [6]. In terms of continuous tuning, researchers use high-precision adjustable filter combined with improved Mach Zehnder interferometer to obtain continuous adjustable laser output with tuning range of tens of nanometers, which has good stability and linewidth characteristics [7, 8]. In terms of multi wavelength, the successful development and application of filter devices such as birefringent fiber [9], sampled fiber Bragg grating [10], chirped grating [11], arrayed waveguide coupler [12] and Mach Zehnder interferometer [13] have promoted the development of multi wavelength laser. At the same time, in recent years, in order to obtain the narrow linewidth laser output with stable wavelength, researchers at home and abroad have put forward a variety of schemes. The research in this regard mainly focuses on optimizing the cavity structure [14–17], adopting different gain media [18–21], etc., and proposed the use of saturable absorber [22], negative feedback [23] and other methods to narrow the linewidth and suppress mode hopping. IPG has launched a narrow linewidth fiber laser based on new material fiber, whose linewidth reaches MHz. In 2008, the key programs for science and technology development titled "development of single longitudinal mode fiber laser" host by South China University of Technology made an important breakthrough, realizing a stable narrow linewidth laser output with a linewidth of 2 kHz [24]. In 2012, Professor Yan Fengping et al. from Beijing Jiaotong University used fiber Bragg grating Fabry Perot etalon and coupler as two ends of the resonator of linear cavity fiber laser respectively to obtain narrow linewidth laser output with linewidth of 10.34 MHz [25]. In 2013, Wang Xiang et al. from Taiwan University of Science and Technology used the structure of adding multiple sub-ring cavities into the linear cavity to obtain single frequency laser output with linewidth of less than 1 MHz [26]. In 2014, State Key Laboratory of Luminescent Materials and Devices in South China University of Technology proposed a new linear cavity with virtual folded cavity. A quarter wave plate is introduced into the linear cavity to cause polarization delay, effectively reducing the traveling wave polarization and spatial hole burning effect, which makes it have the advantages of linear cavity and ring cavity at the same time, and realize the ultra narrow linewidth output of 820 Hz [27]. In 2016, University of Electronic Science and technology of China successfully realized the stable single longitudinal mode laser output with linewidth of less than 1.2 kHz by using the cascaded combined ring fiber laser of phase-shifted fiber Bragg grating and saturable absorber [28]. However, the research of single frequency fiber laser with single longitudinal mode and narrow linewidth output characteristics still has a lot of development space.

7.1.2 Classification and Advantages of Fiber Lasers

Due to the rapid development of fiber lasers in recent years, there are more and more types of fiber lasers. The common types of fiber lasers include rare earth doped fiber lasers, Raman fiber lasers, fiber lasers using photonic crystal fibers as gain media, high-power fiber lasers, as well as new femtosecond fiber lasers and short-size single frequency fiber lasers, various fiber lasers have their own advantages and disadvantages. However, the different names of the above fiber lasers are often due to the focus from different perspectives.

The classification of fiber lasers varies from different perspectives. This chapter briefly summarizes and generalizes the classification of fiber lasers in the following ways. See Table 7.1 for specific classification methods.

Table 7.1 Classification of fiber lasers

Classification by resonator structure	Linear cavity fiber laser, ring cavity fiber laser, composite cavity fiber laser, "8" shaped cavity, Fabry Perot (FP) cavity, distributed Bragg reflection (DBR) fiber laser, distributed feedback (DFB) fiber laser, etc.
Classification by fiber structure	Single clad fiber laser, double clad fiber laser
Classification by gain medium	Rare earth doped fiber laser, nonlinear effect fiber laser, single crystal fiber laser, plastic fiber laser
Classification by working mechanism	Up conversion fiber laser, down conversion fiber laser
Classification by doped elements	15 kinds of erbium (Er^{3+}) doped, neodymium (Nd^{3+}) doped, praseodymium (Pr^{3+}) doped, thulium (Tm^{3+}) doped, ytterbium (Yb^{3+}) doped and holmium (Ho^{3+}) doped
Classification by output wavelength	S-band (1280–1350 nm), C-band (1528–1565 nm), L-band (1561–1620 nm)
Classification by output laser	Pulsed fiber laser, continuous wave fiber laser

It is the most intuitive and easy way to classify fiber lasers according to the resonant cavity structure. Among them, the linear cavity structure is an early fiber laser resonant cavity structure, which is mainly formed by adding mirrors at both ends of the gain medium. In contrast, the ring cavity fiber laser has attracted extensive attention of researchers because it has a larger cavity length and can obtain narrower linewidth and higher output power. This chapter will introduce the working principle of the ring cavity fiber laser in detail.

Fiber laser has become the focus of research and application in recent years for that it has very significant characteristics and advantages, mainly manifested in the following aspects:

(1) The output beam quality of fiber laser is excellent. It not only has good monochromaticity and directivity, but also has high output power stability. Moreover, the linewidth of fiber laser is usually in the order of kHz, much better than the traditional semiconductor lasers, which makes up for the shortcomings of semiconductor lasers in the fields of communication, sensing and so on.
(2) Fiber laser has the advantages of high conversion efficiency, low noise and low threshold due to its fiber structure and low loss in cavity. Therefore, it can achieve high power output. In recent years, with the development of key technologies such as pumping technology and fiber preparation technology, the output power of fiber laser is increasing. Single-mode fiber lasers with output powers on the order of several kilowatts have been commercialized. The advantages of high

power make fiber lasers widely used in laser cutting, laser marking and other fields.

(3) Fiber laser can convert 70–80% of pump light into output light, which not only means that the utilization rate of energy has been improved, but more importantly, there is only a small amount of heat generated in this process. In addition, the fiber itself has good heat dissipation. Therefore, even ultra-high-power fiber laser does not need to use a huge water cooling system, instead a simple air cooling system can ensure the heat dissipation and good operation of the laser. This also provides conditions for the integration and miniaturization of high-power fiber lasers.

(4) The fiber laser uses soft optical fiber as the main component, which avoids the defects that the traditional solid-state laser is affected by factors such as collision and vibration during the necessary handling process and affects the output characteristics. It has high stability and reliability, and is suitable for harsh environment and field construction. In addition, the corrosion resistance of the optical fiber itself makes it have the advantages of high service life and convenient maintenance.

(5) The working wavelength of fiber laser depends on the type of doped ions it adopts. By changing the doped ions such as erbium (Er^{3+}), neodymium (Nd^{3+}), praseodymium (Pr^{3+}), thulium (Tm^{3+}), ytterbium (Yb^{3+}) and other rare earth elements, its output wavelength can cover a very wide range of bands from ultraviolet to infrared, which has become the basis of all fiber communication.

7.1.3 Main Application Fields of Fiber Laser

With the continuous improvement and maturity of fiber laser technology, fiber laser has not only been commercialized, but also has been widely used in fields such as communication, energy, industrial processing, medical treatment, aerospace, military and so on. Its main applications can be summarized as follows.

(1) Military field

Fiber laser has the advantages of high brightness, high power, small irradiation area and high beam quality, making it the research focus of defense and offensive weapons in the military field, and having attracted much attention in the applications of remote positioning, remote sensing, simulated shooting, tracking guidance and so on. In addition, fiber lasers with high output power have great potential in high-energy weapons and other fields, and have attracted more and more attention. Governments of various countries are also investing more and more in the field of fiber lasers. In recent years, the U.S. government has invested tens of billions of dollars in fiber laser research in the U.S. National Fund and the National Institute of Standards and Technology. Similarly, China and European countries have invested more and more in the scientific research of fiber lasers, and the status of fiber lasers in the military field will also be higher and higher.

(2) Industrial processing

In industrial processing, fiber laser has huge potential market and economic benefits. The laser marking system has the advantages of simple operation, small volume, and its high-quality output light characteristics, which makes it show an attractive prospect in high-precision marking. At the same time, in the field of micromechanics, fiber lasers can be used for bending, decompression, collimation, welding and so on. They not only have high precision but also greatly shorten the process completion time. Besides, in the processing of microelectronic devices, the precision of micro processing using fiber laser is greatly improved compared with other traditional methods, which has become an ideal method for microelectronic manufacturing.

(3) Medical field

Fiber lasers can produce a wide range of laser wavelengths, which makes it possible for the demand for specific wavelength light sources in the medical field. Fiber laser with wavelength of 2 μm can achieve good results in the treatment of skin cancer; Thulium doped fiber laser has played a very important role in the popular ophthalmic surgery for myopia in recent years. In addition, the use of fiber lasers instead of traditional CO_2 lasers or YAG lasers as the light source of medical surgery has been also recognized by the majority of medical workers.

(4) Optical fiber communication system

Due to the advantages of single frequency, single-mode and low noise, fiber laser is also an indispensable part in the field of communication. Especially with the rapid development of wavelength division multiplexing system and dense wavelength division multiplexing system, multi-channel signal transmission can be realized in one optical fiber. The high-power fiber laser can greatly reduce the loss and interference caused by photoelectric conversion in the transmission process, and greatly promote the development of information transmission system.

(5) Optical sensing field

Application of narrow linewidth fiber laser in distributed optical fiber sensing system can realize long-distance target detection, location and classification. The system can provide low-cost and fully distributed sensing safety monitoring for nuclear power plants, oil, construction engineering, natural gas pipelines, military bases and national defense borders [29, 30]. At present, in the distributed optical fiber sensing system based on Brillouin scattering, a light source with a spectral width far less than the Brillouin gain spectrum must be used, otherwise the Brillouin gain peak will be reduced. The Brillouin gain spectral width is generally tens of MHz, a laser light source with a line width less than 1 MHz needs to be used in the system. So a stable single frequency narrow linewidth laser light source is particularly important [31, 32]. At the same time, narrow linewidth fiber lasers have long coherence lengths, which combined with coherence detection can achieve ultra-high spatial resolution and large measurement ranges.

7.2 Ring Cavity Erbium-Doped Fiber Laser

As an important branch in the field of fiber laser research, narrow line width fiber laser has many advantages, such as single frequency, narrow linewidth, long coherence length and easy to form all-fiber structure. As a high coherence light source, narrow line width fiber laser has a wide application prospect in the fields of remote communication, laser radar, optical fiber sensing and long-distance coherent detection with high precision and resolution, etc.

The difference of the ions doped in the gain medium of the fiber laser directly determines the difference of the output wavelength of the laser. The commonly used rare earth ions are erbium (Er^{3+}), neodymium (Nd^{3+}), praseodymium (Pr^{3+}), thulium (Tm^{3+}), ytterbium (Yb^{3+}), holmium (Ho^{3+}), etc. Among them, the fluorescence spectrum of erbium ion just contains the low loss window 1550 nm of the communication system, and has a gain spectrum width of 30–50 nm and a large wavelength tuning range. At the same time, it also has the advantages of higher gain and lower pump power required. So, the erbium-doped fiber used as gain medium of fiber lasers has widely attracted attention of researchers. The research on fiber lasers in this section mainly focuses on erbium-doped fiber lasers. The erbium-doped fiber is used as a gain medium to absorb the energy of the pump light, and the population inversion occurs. The light of a certain wavelength that starts to oscillate accumulates energy in the laser cavity, and finally realizes the laser output. Compared with other kinds of lasers, erbium-doped fiber lasers have excellent characteristics such as light weight, small size, easy maintenance, good beam quality, long stabilization time, wide gain range, high gain value, and slope efficiency. In this section, the basic principle of erbium-doped fiber lasers will be introduced, and then the detailed experimental research of typical erbium-doped fiber lasers will be carried out.

7.2.1 Basic Principle of Erbium-Doped Fiber Laser

The stimulated radiation proposed by Albert Einstein in 1917 laid the physical foundation for the generation of laser. When the atoms at high energy levels absorb the external energy and transit to low energy levels, it will release a photon with energy $h\upsilon$, as shown in Fig. 7.1. And the generated photons have the same propagation direction, frequency, phase and polarization direction as the external incident light, and with the increasing number of particles generated in this process, the light amplification is realized. The essence of laser generation by fiber laser is a process of optical amplification, which relies on pumping by pump light for population inversion in the gain medium, and continuously increases population inversion to make the laser oscillation reach the threshold condition. At the same time, it also needs to form a feedback cavity to reduce the loss and provide positive feedback. The feedback cavity is the resonant cavity. That is, the three basic components of laser: pump light, gain medium and resonator.

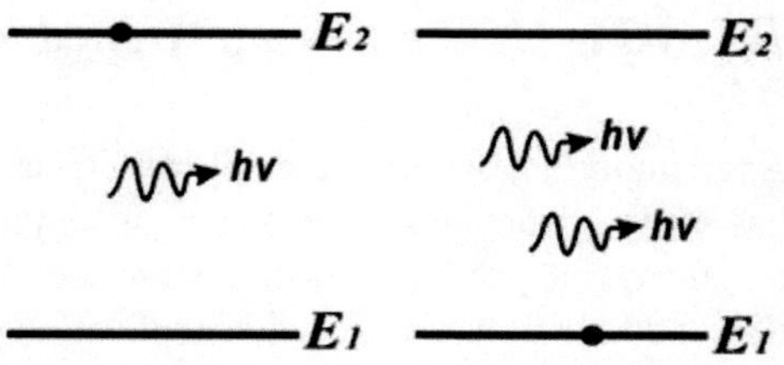

Fig. 7.1 Schematic diagram of stimulated radiation process of light

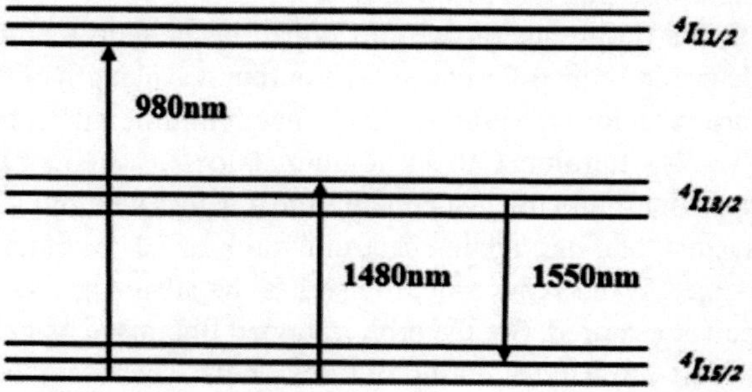

Fig. 7.2 Schematic diagram of energy level structure of erbium ion

The working wavelength of the fiber laser is determined by the type of ions doped. The gain medium used in this section is the erbium-doped fiber (EDF). The energy level structure of erbium ions is analyzed below, and then the absorption and fluorescence characteristics of erbium ions are analyzed. Figure 7.2 shows the energy level structure of erbium ions. The transition process of the lower energy level $^4I_{15/2}$ electrons to the upper energy level $^4I_{11/2}$ after absorbing the pump light is an absorption process, and the selection of the pump light wavelength of the erbium-doped fiber laser is also determined by this process. The electrons that jump to the upper energy level are unstable and will spontaneously transition to the ground state, which is a fluorescence process. The ion lifetime of $^4I_{11/2}$ is very short, so the fluorescence process is mainly the transition from $^4I_{13/2}$ to the ground state $^4I_{15/2}$. Therefore, the lasing wavelength of erbium ions is 1550 nm.

Under the action of 980 nm pump light, the above energy levels $^4I_{11/2}$, $^4I_{13/2}$, $^4I_{15/2}$ constitute the three-level structure of erbium ions, as shown in Fig. 7.3. The particle absorbs energy and jumps from the $^4I_{15/2}$ energy level to the $^4I_{11/2}$ energy level, followed by the non-radiative relaxation to the $^4I_{13/2}$ energy level, and finally forms population inversion between the $^4I_{13/2}$ and $^4I_{15/2}$ energy levels. The above process is repeated in the resonant cavity to accumulate energy and finally generate laser.

In the above case, the rate equation of the erbium-doped fiber laser can be expressed as

$$\frac{dN_3}{dt} = (N_1 - N_3)W_{13} - N_3 S_{32} \tag{7.1}$$

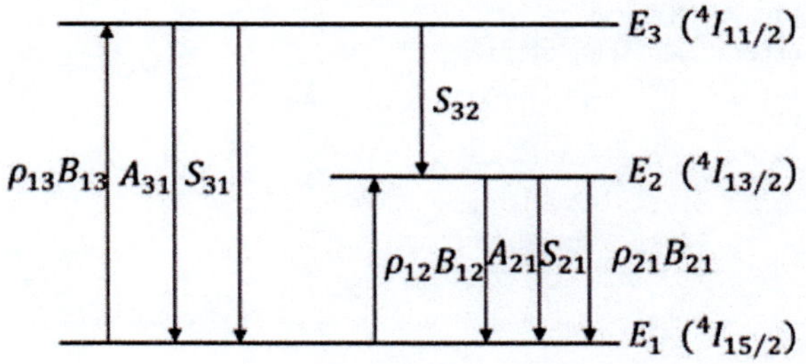

Fig. 7.3 Schematic diagram of the three-level structure of erbium ion pumped at 980 nm

$$\frac{dN_2}{dt} = N_1W_{12} - N_2W_{21} - N_2(A_{21} + S_{21}) + N_3S_{32} \tag{7.2}$$

$$N_1 + N_2 + N_3 = N \tag{7.3}$$

As mentioned above, the lifetime of particles at the E_3 level is only on the order of milliseconds (ms), so in the above formula $N_3 \approx 0$, $dN_3/dt \approx 0$, so the rate equation can be simplified as

$$\frac{dN_2}{dt} = N_1W_{12} - N_2W_{21} - N_2A_{21} + N_1W_P \tag{7.4}$$

$$N_1 + N_2 \approx N \tag{7.5}$$

If g is the energy level degeneracy, then

$$\Delta N = N_2 - \frac{g_2}{g_1}N_1 \tag{7.6}$$

Then the rate equation can be further simplified as

$$\frac{d\Delta N}{dt} = (N - \Delta N)W_p - \left(1 + \frac{g_2}{g_1}\right)W_{12}\Delta N - \left(\frac{g_2}{g_1}N + \Delta N\right)A_{21} \tag{7.7}$$

In summary, the rate equation of erbium-doped fiber laser can be expressed by formula (7.7).

7.2.2 *Typical Ring Cavity Erbium-Doped Fiber Laser*

As described in Sect. 7.2.1, the basic structure of the fiber laser has a linear cavity and a ring cavity. Since the linear cavity usually uses two fiber Bragg gratings to form a resonant cavity and to achieve wavelength selection at the same time, which makes the cavity length need to be reduced to the order of centimeters to achieve single

Fig. 7.4 Typical experimental setup of ring cavity erbium-doped fiber laser

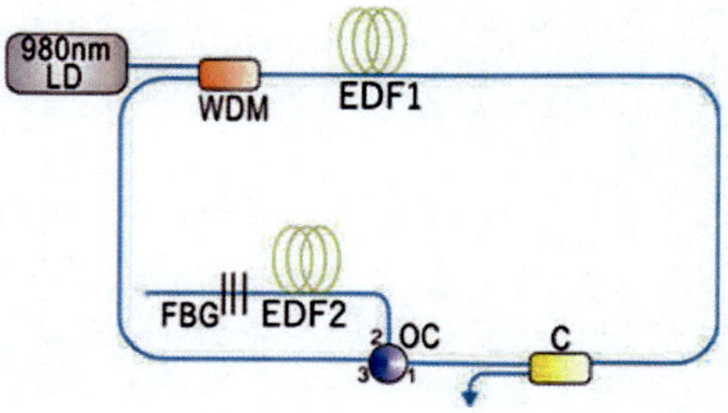

longitudinal mode output, and this condition greatly limits the output power. The ring cavity is traveling wave cavity, which is beneficial to obtain narrow linewidth and high output power. Therefore, this section mainly studies the erbium-doped fiber laser with the ring cavity structure. The typical ring cavity erbium-doped fiber laser structure is shown in Fig. 7.4.

The pump light is coupled into the loop through a 980 nm/1550 nm wavelength division multiplexer (WDM), and population inversion occurs in the erbium-doped fiber (EDF1) and spontaneous emission occurs to obtain the gain, and then travels along the optical path to reach the circulator. Filtering, mode selection and frequency stabilization are achieved by fiber Bragg grating (FBG) and a saturable absorber composed of an unpumped EDF2 in the circulator (OC) port 2. The light output from the port 3 of the circulator is again amplified by the stimulated emission through EDF1, and finally a laser output of 1550 nm is obtained. The optical circulator in the loop can also play a role in controlling the direction of optical wave transmission in the laser cavity, so that the light in the entire loop is transmitted in the clockwise direction to prevent the interference of the reverse transmission light to the loop. When the gain of the whole loop is greater than the loss, the narrow linewidth laser output is obtained at the 15% port of the coupler with the coupling ratio of 85/15.

On the basis of theoretical research, the above-mentioned typical ring cavity fiber laser experimental device was initially built. The obtained output spectrum is shown in Fig. 7.5. The center wavelength of the output laser is 1549.57 nm, which is determined by the center wavelength of the fiber Bragg grating used in the experiment. The spectral linewidth is less than 0.02 nm.

However, due to the instability of the FBG itself, that is, it is extremely susceptible to the influence of temperature, vibration and other factors in the environment, the stability of the obtained spectrum is poor, and there will be phenomenons such as mode hopping and frequency jitter in the experiment. The power stability of the measured output laser within 30 min are shown in Fig. 7.6. The maximum fluctuation is up to 0.8 dB, and the stability of the output laser is poor. It is necessary to propose an experimental scheme to further improve the stability and ensure the single longitudinal mode of the laser.

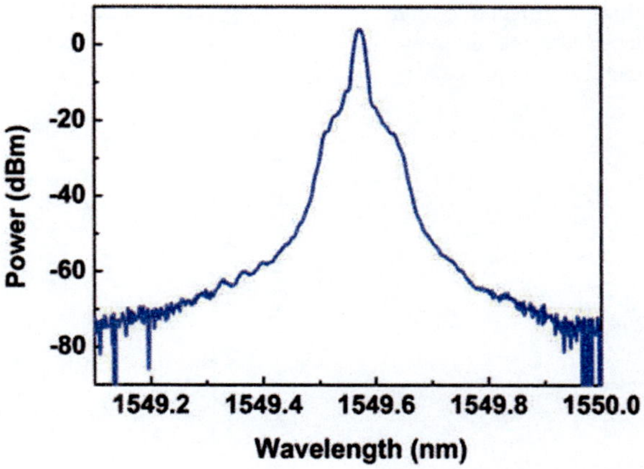

Fig. 7.5 The output spectrum of the ring cavity fiber laser under the typical structure

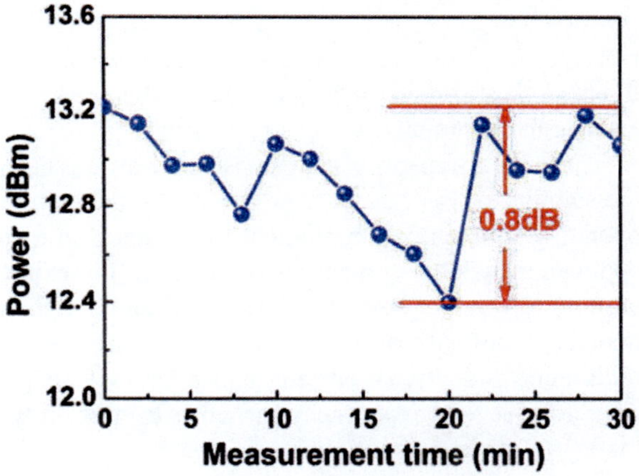

Fig. 7.6 Stability test results of fiber laser output power within 30 min

7.2.3 Selection of Optimal Length of Gain Fiber

The length of the gain fiber will affect the performance of the output laser. As stated by M. Mignon in the literature entitled "An analytical model for the determination of optimal output reflectivity and fiber length in erbium-doped fiber lasers" on PTL published in 1992, there is an optimal length for the erbium-doped fiber used in the erbium-doped fiber laser [33]. Therefore, based on experimental research and analysis, the optimal length of the erbium-doped fiber is obtained in this section. Figure 7.7 shows the change of output power with different lengths of erbium-doped fiber at the same experimental conditions. It can be seen from the figure that when the length of the erbium-doped fiber is small, the output power increases with the increase of the length of the erbium-doped fiber, and when the length of the EDF reaches a certain value, the output power basically tends to remain unchanged, and this value is the optimal length of the EDF. As shown in Fig. 7.7, the optimal length

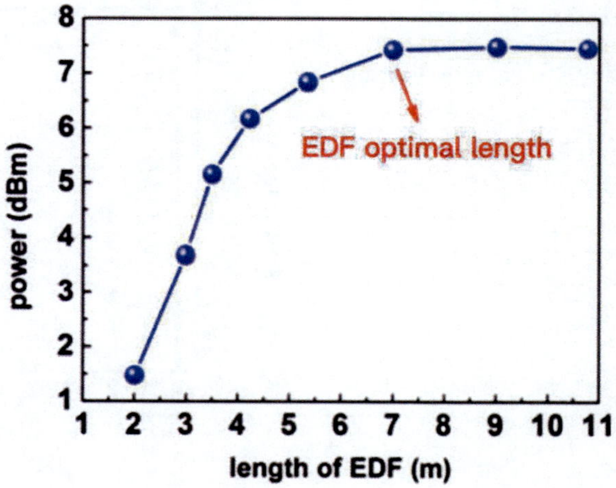

Fig. 7.7 Effect of erbium doped fibers of different lengths on output power

measured by the experiment is 7 m, that is, when the EDF length is 7 m, the pump light can be absorbed by the gain fiber to the greatest extent and the loss in the laser cavity can be minimized.

The mode competition in fiber lasers has always been an important factor affecting the stability of lasers, and fiber lasers can maintain good monochromaticity when operating in a single longitudinal mode state, so as to meet the needs of various application fields. As mentioned above, in the experimental study of the typical erbium-doped fiber laser structure, it is found that the output stability of the laser is poor, an important reason is that there is a multimode oscillation in the laser, and the multimode oscillation is unstable and there is severe mode competition phenomenon. Therefore, several experimental schemes to effectively guarantee single longitudinal mode in fiber laser are studied.

The most common and most effective methods to ensure the operation of the single longitudinal mode of the laser mainly include the following three methods.

(1) Short cavity method

Since the longitudinal mode interval in fiber lasers, that is, the free spectral range (FSR), is inversely proportional to the cavity length of the laser, shortening the cavity length as much as possible is the most direct and effective method to realize single longitudinal mode [34, 35]. The method of shortening the cavity length is widely accepted and adopted by researchers. However, it is almost contradictory to the output power of the laser. The short cavity length will inevitably lead to a great limitation of the output power of the laser. Disadvantages appear under the conditions of gradually requiring high power output.

(2) Multi-sub-ring method

As mentioned above, the most effective way to suppress multi-mode oscillation in fiber lasers is to increase the longitudinal mode interval. The FSR determined by the main ring cavity of the laser is often small. At this time, if one or more subrings are added, the effective FSR of the entire laser becomes the common multiple of the FSR determined by each of the above mentioned sub-rings. Compared with a single cavity length, this method can effectively suppress mode competition and ensure the single longitudinal mode of the output laser [36, 37].

(3) Saturable absorber method

Inserting a section of unpumped erbium-doped fiber into the laser cavity as a saturable absorber is also a commonly used method to suppress mode hopping in fiber laser [38, 39]. In the fiber laser, the saturable absorber is combined with the fiber grating or the input and output ends of the coupler are connected to the two ends of the saturable absorber, so that the light passing through the saturable absorber forms two oppositely transmitted light waves, and interference occurs in them. Under the action of saturable absorption, the light of a certain mode is least absorbed, that is, the oscillation is the strongest, which suppresses the oscillation of other modes, and finally, the single longitudinal mode oscillation of the laser is achieved.

Based on studying the basic working principle of the fiber laser, the basic structure and principle of the erbium-doped fiber laser are studied deeply, and the typical ring cavity erbium-doped fiber laser structure is used to conduct experimental research. The output spectral characteristics are analyzed, and the optimal length of erbium-doped fiber in this experiment is found out. Finally, aiming at the problem of poor stability of output laser caused by multi-mode oscillation encountered in the experiment, several effective methods to ensure the operation of the single longitudinal mode of the laser are studied and summarized, which lays a foundation for further proposing effective experimental scheme of single longitudinal mode erbium-doped fiber laser.

7.3 Single-Longitudinal-Mode Narrow Linewidth Fiber Laser

In recent years, the research on narrow linewidth fiber laser mainly focuses on the improvement of linewidth, optical signal-to-noise ratio, threshold, output power and other performance, and a variety of schemes are proposed. The research in this field mainly focuses on optimizing the cavity structures and using different gain media, and the methods of using saturable absorber and negative feedback are proposed to achieve narrow linewidth and suppress mode hopping.

Because the ring cavity fiber laser is a traveling wave cavity, it can effectively avoid the problems of small output power in the linear cavity structure and the fiber length limited by gain. However, the mode hopping phenomenon in ring cavity is the difficulty and focus of research. Researchers also put forward a variety of schemes to solve this problem. A large number of experiments and reports have confirmed the feasibility of using an unpumped erbium-doped fiber as a saturable absorber to suppress mode hopping [40], however, relying entirely on this method cannot completely suppress the phenomenon of mode hopping. Later, some researchers reported the methods of adding semiconductor optical amplifier [41] and Fabry Perot etalon [42] to the laser structure, but large insertion loss was also introduced. The sub ring cavity method is also used to ensure single longitudinal mode operation of the laser. Li Jianzhong et al. reported that three sub ring cavities composed of couplers are used to increase longitudinal mode interval [43], the single longitudinal mode output is obtained successfully. In 2013, Feng Sujuan et al. also realized the single longitudinal mode operation of the laser by adding cascaded sub-ring cavities into the ring cavity [44]. In addition, many research groups have proposed using Mach Zehnder interferometer or its improved structures to control model hopping, and achieved good results [45–47]. However, these methods usually need to accurately calculate the fiber length as small as centimeters, and sometimes need to carry out complex and difficult fiber handling processes such as fiber optic tapering. On the one hand, the requirements for experimental conditions are quite high, On the other hand, it is easy to introduce errors in the preparation of the above optical fibers. Therefore, the research on single frequency fiber laser with single longitudinal mode and narrow linewidth output characteristic still has a lot of development space.

In this section, the original ring cavity structure is further optimized, and a new erbium-doped fiber laser structure is proposed: On the basis of the ring cavity structure, two parallel fiber ring resonators embedded with common saturable absorbers and fiber gratings are used to construct Mach Zehnder interferometer, which is not only the main filter device of the device, but also the sub ring cavity in the ring cavity structure increasing the effective cavity length of the laser, thus the laser longitudinal mode interval was increased and the mode hopping phenomenon in the ring cavity structure was suppressed effectively. In addition, in cavity structure, a saturable absorber composed of an unpumped erbium-doped fiber and a fiber Bragg grating is used as the ultra-narrow band filter, the single longitudinal mode operation of laser is realized. And the laser output characteristics are studied deeply and the linewidth of the laser is analyzed and verified by the delayed self-heterodyne method.

7.3.1 Narrow Linewidth Erbium-Doped Fiber Laser Based on Parallel Double Sub-rings MZI

The experimental device diagram of narrow linewidth erbium-doped fiber laser based on parallel ring MZI and saturable absorber is shown in Fig. 7.8. A 980 nm semiconductor laser is used as the pump source, coupled into the erbium-doped fiber EDF1 through a 980 nm/1550 nm wavelength division multiplexer (WDM), and optical amplification is realized in the erbium-doped fiber, where the length of the erbium-doped fiber is 7 m. The light of excited wavelength is divided into two paths through the 50:50 coupler, respectively transmitted along the two arms of the Mach Zehnder interferometer composed of two subring cavities in the figure, and then merged into one path in the coupler (C2) for further transmission. The light excited in the laser is output 15% port of the 85:15 coupler (C6), the other 85% of the light continues to travel in the cavity. The optical isolator is introduced in the device to ensure the one-way oscillation of the laser in the ring cavity and to avoid the influence of the light transmitted in the reverse direction on the output result. The polarization of the whole laser cavity is controlled by a polarization controller, which can also adjust the loss in the cavity.

The experimental scheme is based on the existing structure of ring cavity fiber laser, the Mach Zehnder interferometer structure is introduced, which consists of two parallel fiber ring resonators. The two ring resonators are embedded with common saturable absorber and fiber grating through two circulators respectively (the insertion loss of ports 1–2 and 2–3 is 0.72 dB and 0.78 dB, respectively). The length of the two sub-ring cavities are 5 m and 5.5 m respectively, and an unpumped erbium-doped fiber with length of 2 m is used as the saturable absorber. The reflectivity of the fiber Bragg grating used is 99.73%, and its reflection spectrum is shown in Fig. 7.9. The central wavelength is 1549.57 nm, and the − 3 dB bandwidth is 0.28 nm.

The above parallel sub-ring cavity can not only increase the cavity length, but more importantly, can suppress the output optical mode hopping by using Wiener effect.

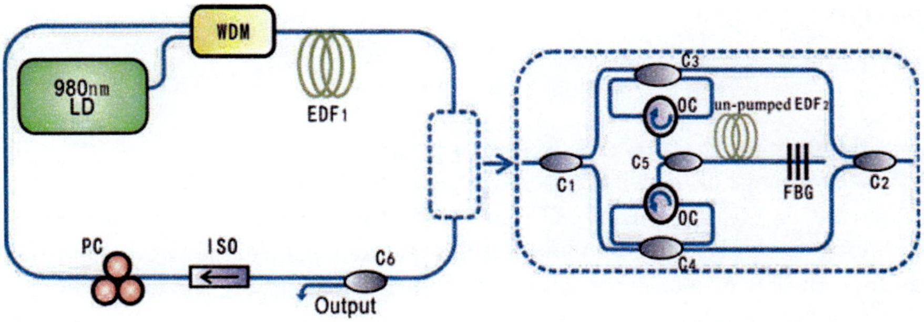

Fig. 7.8 Experimental device of single-longitudinal-mode narrow linewidth fiber laser based on double-ring MZI and saturable absorber

Fig. 7.9 Reflection spectrum of fiber Bragg gratings

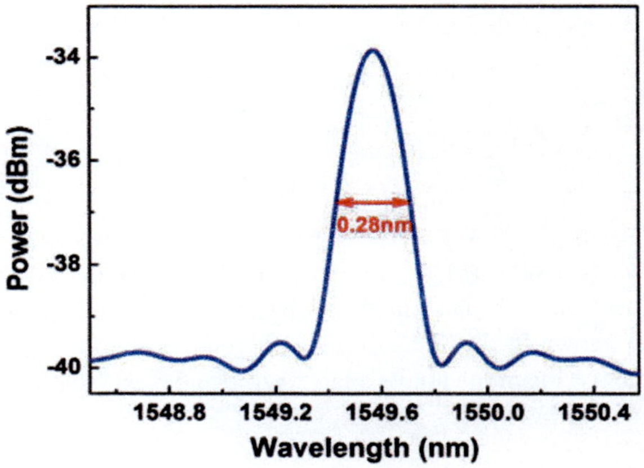

The experimental results also show that the paralleled sub-ring cavities can suppress the mode hopping and improve the stability of the output laser. The following details how it works.

The cavity length of the main ring cavity of the laser is 11.2 m, and the corresponding free spectral range is about 18 MHz. The specific calculation formula is as follows [48]:

$$FSR = \frac{c}{n_e L} \tag{7.8}$$

where c is the speed of light; n_e is the refractive index of optical fiber, about 1.468; L is cavity length.

In FBG reflection spectrum with reflection bandwidth of tens of gigahertz, FSR of 18 MHz will obviously produce multi-mode oscillation. When the parallel double-ring MZI structure proposed in this section is introduced, the FSR of its two sub-ring cavities are also calculated to be about 38 and 42 MHz, and the FSR of MZI is 205 MHz. Therefore, according to the Wiener effect, the effective FSR of the whole laser should be the least common multiple of FSR determined by the above conditions, namely:

$$FSR_e = m_1 FSR_1 = m_2 FSR_2 = m_3 FSR_3 = m_4 FSR_4 \tag{7.9}$$

In the formula, m$_i$ ($i = 1, 2, 3, 4$) is a positive integer, FSR_e is the effective free spectral range of the whole laser, and FSR_i ($i = 1, 2, 3, 4$) represents the FSR of the main cavity, the FSR of MZI and the FSR of the two subrings constituting MZI respectively. Therefore, although each of the above-mentioned FSR is only on the order of tens of megahertz, according to the Wiener effect, the effective FSR of the laser can reach tens of gigahertz. In summary, as shown in Fig. 7.10, adding MZI structure based on double rings in the laser can effectively increase longitudinal mode interval.

In addition, the phase conditions of the main ring cavity and two sub-ring cavities can be expressed as:

$$\beta L_m = 2k\pi \tag{7.10}$$

$$\beta L_i = 2l\pi \tag{7.11}$$

where L_m is the cavity length of the main ring, L_i ($i = 1, 2$) is the cavity length of the two subrings, β is the propagation constant, k and l are positive integers. The laser can only start up when all the phase conditions above are met at the same time, as shown in Fig. 7.10. Finally, there is only one mode oscillation in the whole laser cavity. From the above analysis, it can be seen that the laser can ensure single-mode operation.

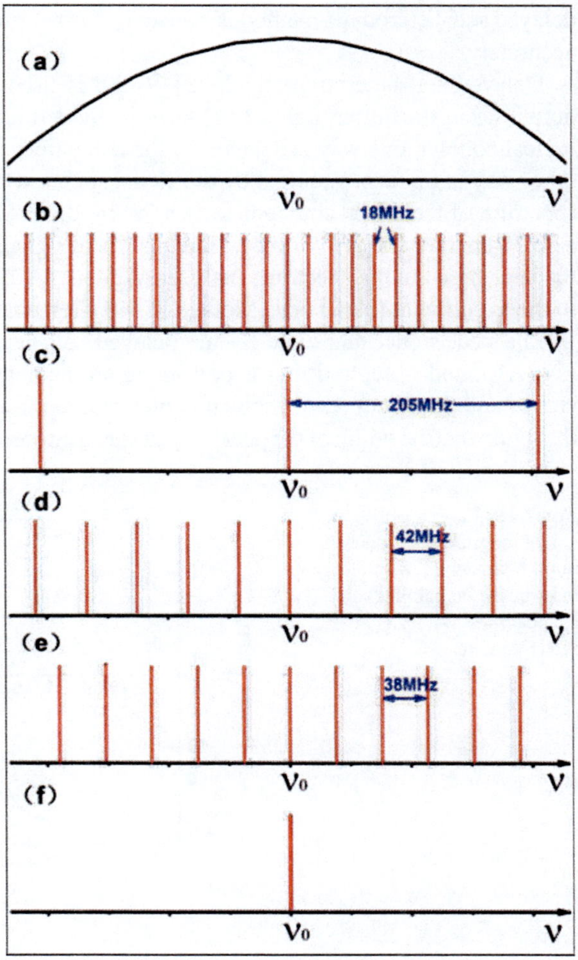

Fig. 7.10 Schematic diagram of single longitudinal mode operation of an erbium-doped fiber laser. **a** Reflection spectrum of fiber Bragg grating; **b** the laser oscillation mode and the free spectral range determined only by the length of the main ring cavity when the proposed MZ1 structure is removed from the experimental scheme; **c** the laser oscillation mode and free spectral range determined by the MZ1 structure without subring; **d** the laser oscillation mode and free spectral range determined by the cavity length of a subring cavity of MZ1; **e** the laser oscillation mode and free spectral range determined by the cavity length of another subring cavity of MZI; **f** the laser keeps stable single longitudinal mode output under the combined action of the above conditions

7.3.2 Analysis of Output Characteristics of Narrow Line-Width Erbium-Doped Fiber Laser

The output laser spectrum of the multi-ring cavity narrow-linewidth fiber laser combined with the Mach-Zehnder interferometer is shown in Fig. 7.11. The pumping power is 135 mW, and Yokogawa AQ6370C spectrometer was adopted for spectral measurement, with a maximum resolution of 0.02 nm. The central wavelength of the output laser spectrum is stable at 1549.47 nm, the peak power is 7.471 dBm, and the signal-to-noise ratio is as high as 68 dB. The -3 dB linewidth of the output laser is 0.02 nm, which is determined by the resolution bandwidth of the spectrometer. The line width of the laser detected by the spectrometer is limited by the resolution, so it is impossible to accurately measure the line width less than 0.02 nm, the delayed self-heterodyne method is need to be further used to measure laser line width accurately.

Delayed self-heterodyne method [49, 50] is used to measure the linewidth of output laser, the output light of the laser is divided into two paths through a 50:50 optical coupler, one way is delayed by the delay fiber with length of 45 km, and the other way is frequency shifted by the signal source with frequency of 50 MHz. The spectrum of the signal source is shown in Fig. 7.12. The two optical signals beat in a 50:50 coupler and are converted into electrical signals by a photodetector to detect the beat signal in the spectrum analyzer (Agilent N9020A). The photodetector used in the experiment (NewFocus Model 1554-B) has a bandwidth of 12 GHz.

The beat signal measured by the delayed self-heterodyne method is shown in Fig. 7.13, and Lorentz fitting is performed on the obtained data, and the measured curves and fitting curves are shown in blue and red respectively. It can be seen from the figure that the width of the peak of beat spectrum decreased by 20 dB is 21.68 kHz,

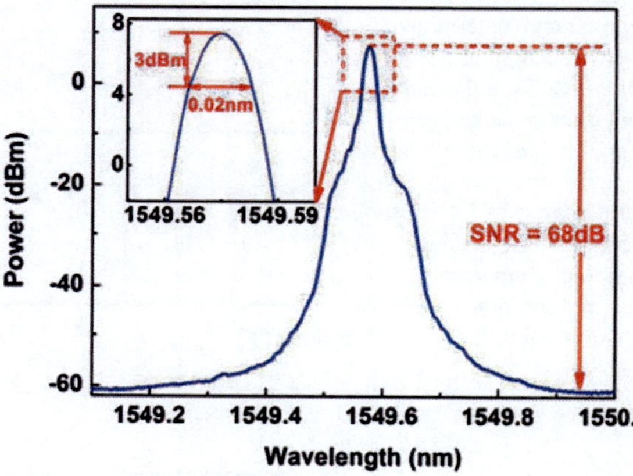

Fig. 7.11 Laser spectrum of single longitudinal mode output narrow linewidth fiber laser based on parallel sub ring cavities Mach Zehnder interferometer

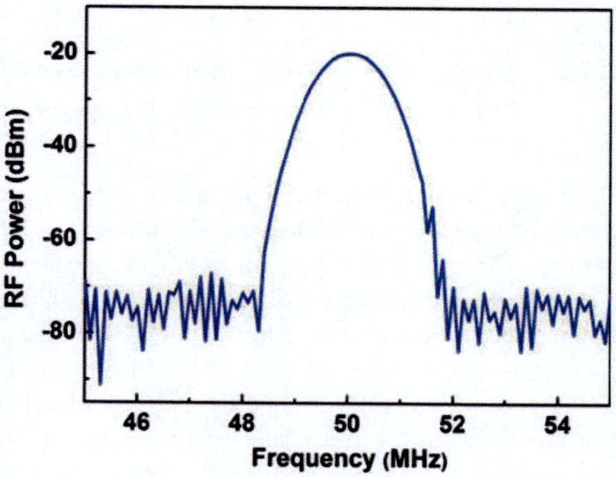

Fig. 7.12 Spectrum of a 50 MHz signal source

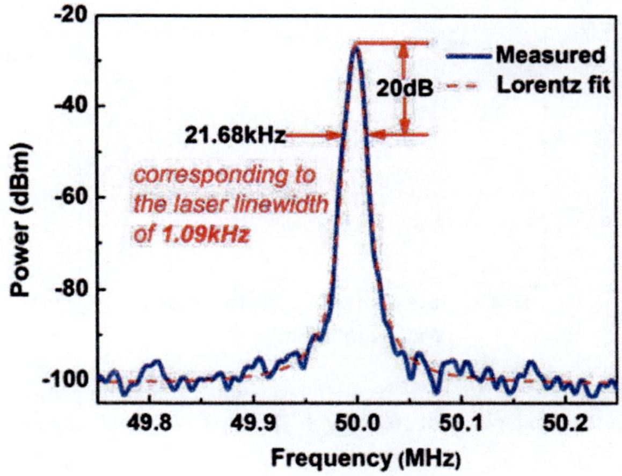

Fig. 7.13 Beat signal measured by the delayed self-heterodyne method

corresponding to the line width of the output laser obtained is 1.09 kHz. It needs to be explained that the reason why the laser linewidth is not measured by the width of the 3 dB drop at the peak is firstly because the spectral line width is narrow, measuring the beat spectrum width at − 3 dB is prone to large calculation errors. In addition, the line broadening caused by noise usually occurs near the peak value, so it is more accurate to use the width at − 20 dB to calculate the laser line width.

Furthermore, the relationship between the output power of the laser and the pump power is studied experimentally. The measured threshold power for laser is 10.7 mW, and the output spectrum of the laser is shown in Fig. 7.14a. When the pump power is greater than 10.7 mW, continue to increase pump power, and the measured threshold curve is shown in Fig. 7.14b. The output power changed linearly with the pump power. After linear fitting of the threshold curve, we got that the opto-optical conversion efficiency was 6%. When the pump power reaches the maximum output power 220 mW of the pump source used in the experiment, the output power of the laser is 12.78 mW. The maximum output power is mainly limited by the pump power, if the pump source with higher output power is used, the maximum output power of the laser can be further increased. At the same time, the optical devices in the experimental setup are mainly connected by the flange heads, and the optical conversion efficiency of the laser can be improved by using the way of fiber fusion.

In addition, the stability of the proposed erbium-doped single-longitudinal mode fiber laser with parallel sub-rings is investigated experimentally, stability measurement results for the output spectrum and power of the laser as shown in Fig. 7.15, the figure shows that in the 1 h measurement time, the spectrum no obvious fluctuation, the output power is very stable, the largest fluctuation is less than 0.2 dB; And this measurement is carried out under laboratory conditions, the measurement results are unavoidably affected by the temperature fluctuations in the experimental environment and other factors.

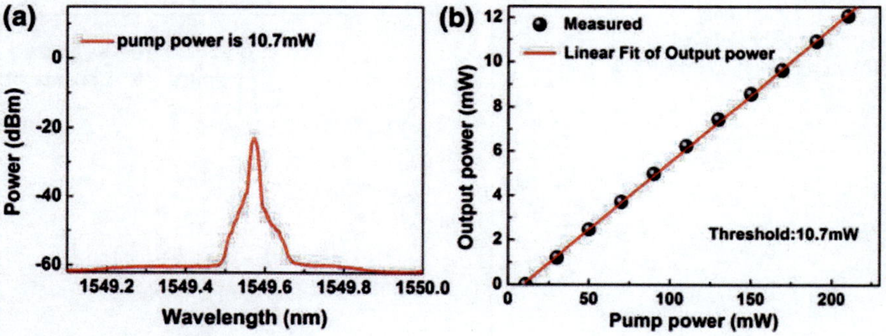

Fig. 7.14 Relationship between laser output and pump power. **a** The output spectrum of the laser at the pump power of 10.7 mW. **b** The curve of the output power changing with the pump power

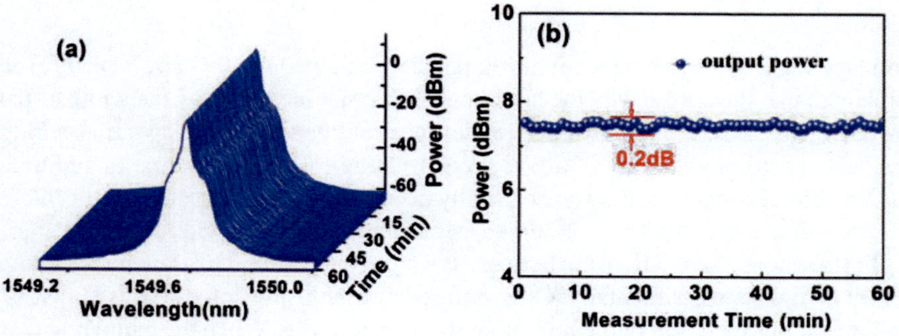

Fig. 7.15 Stability test results of the output laser, measurement time 1 h, measurement interval 1 min, pump power 135 mW. **a** Spectral stability test results of the output laser; **b** output power stability test results

7.3.3 The Performance Improvement of Output Light of Narrow Line-Width Erbium-Doped Fiber Laser

The above experimental results show that the output laser of the proposed fiber laser in this section has good performance, high stability, this is because the Mach-Zehnder structure, combined with the double sub-rings, added in the laser can suppress mode hopping very well, improve the stability of the laser, and the experimental result is consistent with the theoretical analysis mentioned above. Figure 7.16a shows the spectrum of the output laser using the classical ring cavity fiber laser structure described in Sect. 7.2.2, and Fig. 7.16b shows the spectrum of the output laser using the fiber laser structure proposed in this section. According to the spectrum shown in Fig. 7.16a, beat signal with a period of 18 MHz can be obviously observed, which is consistent with the free spectral range corresponding to the cavity length of 11.2 m. However, no obvious beat signal is observed in Fig. 7.16b, it can be seen that the fiber laser structure proposed in this section can well suppress mode hopping and achieve stable single-mode operation of the laser.

In order to evaluate the effect of adding a segment of unpumped erbium-doped fiber in a stable single-longitudinal-mode narrow linewidth fiber laser proposed in this section on the experimental results, the performance of the output laser is compared and analyzed.

Figure 7.17 shows the line width contrast result measured by the delayed self-heterodyne method. The green curve is the line width measurement result of the output laser obtained by removing the unpumped erbium-doped fiber in the experimental setup, and the blue curve is the line width measurement result of the output laser by use of the original setup. It can be clearly seen from the figure that adding a section of unpumped erbium-doped fiber can achieve the narrowing effect on the output laser line width.

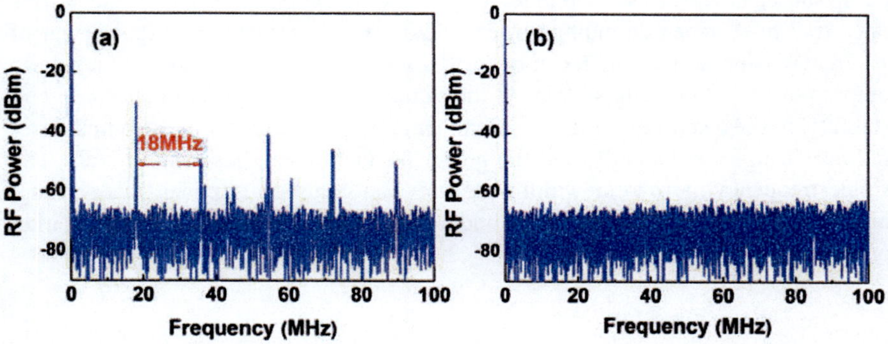

Fig. 7.16 Spectrum of the output laser. **a** Spectrum diagram without parallel double sub-rings structure; **b** spectrum with parallel double sub-ring structure

Fig. 7.17 Effect of a saturable absorber on the linewidth of the output laser

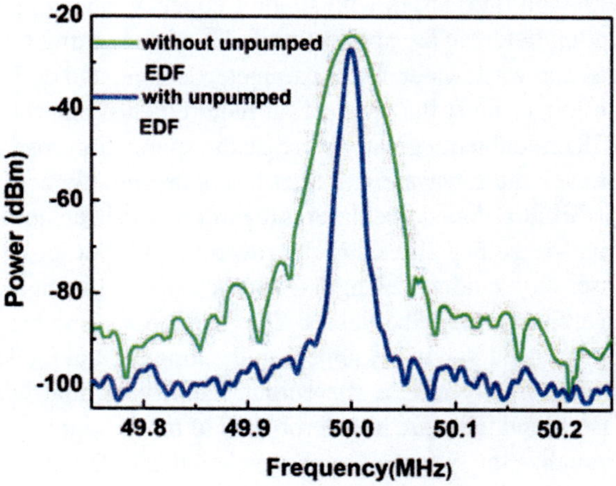

In this section, a stable single-longitudinal-mode narrow linewidth fiber laser structure is proposed. A Mach Zehnder interferometer based on parallel double sub-rings is used as the main filter of the device. This structure can not only increase the cavity length, more importantly, increase the effective free spectral range of the laser by using wiener effect, so as to suppress the mode hopping and ensure the single longitudinal mode operation. In addition, a saturable absorber composed of a section of unpumped erbium-doped fiber and a fiber grating are connected to the above double sub-rings through a coupler and a circular, thus enabling wavelength selection and line width compression of the laser. Finally, the output laser of the laser has good single frequency characteristics and stability. On the basis of a large number of experimental studies, we measure and analyze the performance indexes of the output light. The experimental results show that the single-frequency narrow linewidth laser output with a linewidth of only 1.09 kHz at 1549.57 nm is obtained by using this structure, and the optical signal-to-noise ratio is as high as 68 dB. The pump source is 980 nm semiconductor laser, and the output threshold of the laser is only 10.7 mW. When the pump power is 220 mW, the maximum output power of 12.78 mW is obtained, which can be further increased by increasing the maximum output power of the pump source. At the same time, the output laser showed good stability. In the test time of 1 h, the central wavelength and power of the output laser did not change significantly, and the power fluctuation was less than 0.2 dB. The single-frequency narrow-linewidth fiber laser can be used in optical fiber sensing, high-precision spectral measurement and radar applications, and has a broad market prospect.

7.4 Dual Wavelength Erbium-Doped Fiber Laser

With the rapid development of modern network technology and sensing technology, although fiber lasers with single frequency have high beam quality and wide application fields, in the application fields of generating high-frequency signal, detecting gas and wastewater, multi parameter sensing and detection, a single light source can no longer meet the application requirements, and the use of multiple laser sources will undoubtedly greatly increase the application cost, which makes dual wavelength or even multi wavelength fiber lasers favored by more and more researchers.

Erbium-doped fiber laser is regarded as an irreplaceable light source for generating high-frequency signals in microwave band, but it also requires erbium-doped fiber laser with dual wavelength output to operate in single longitudinal mode. However, in erbium-doped fiber lasers, a large number of mode competition and mode hopping phenomena inevitably appear in the longitudinal mode near the central wavelength, which greatly affects the output stability of dual wavelength erbium-doped fiber lasers, and there are huge problems of mode hopping and gain competition. In order to realize the output of dual wavelength single longitudinal mode laser, researchers have done a lot of research. Chen et al. have proposed a dual wavelength erbium-doped fiber laser structure based on SOA [51], but due to the problem that SOA is easily damaged by strong oscillation in the cavity, the output power of this kind of fiber laser is small. Many other methods, such as adding an adjustable intracavity calibrator into the cavity and separating the polarization state of intracavity laser light, have been reported [52, 53], but most methods require extremely short cavity length or complex intracavity polarization state control elements, which is quite inconvenient in practical application and production. Therefore, further improvement schemes need to be proposed to achieve stable dual wavelength single longitudinal mode laser output.

Based on the previous research of single longitudinal mode fiber laser, we further carried out experimental research and analysis on the proposed Mach Zehnder interferometer structure with parallel double rings, and proposed a dual wavelength single longitudinal mode erbium-doped fiber laser structure combined with the filter of Mach Zehnder structure with double rings, the single longitudinal mode operation of the output laser is verified by experiments and has good stability.

7.4.1 *Dual Wavelength Erbium-Doped Fiber Laser Based on Double Ring MZI*

The experimental setup of dual wavelength fiber laser is shown in Fig. 7.18. The pump source still adopts 980 nm semiconductor laser, after passing through the wavelength division multiplexer (WDM), the population inversion occurs in the erbium-doped fiber to realize optical amplification. Then the light successively passes through the optical isolator, Mach Zehnder interferometer based on double ring, optical coupler C5 with 85:15 split ratio and port 1 of the circulator OC. Passing through the saturable absorber and tunable fiber Bragg grating filter connected to the 2 port of the circulator, the light returned to the main ring cavity of the laser from the 3 port of the circulator to continue transmission. The output light of the laser is output from the 15% port of the coupler C5 and detected and analyzed by an optical spectrum analyzer. The polarization controller (PC) in the ring cavity is to control the polarization state of light in the whole loop. The optical isolator (ISO) after erbium-doped fiber is to prevent the light transmitted back from the coupler C1 to affect the laser and ensure the unidirectional transmission of the laser in the laser cavity. At the same time, an unpumped erbium-doped fiber inserted into the structure is used as a saturable absorber.

As described in Sect. 7.3, a Mach Zehnder interferometer with parallel twin rings added in the fiber laser structure can effectively suppress mode hopping and realize single-mode operation. This structure used in this section for experimental researches to propose a single longitudinal mode dual wavelength erbium-doped fiber laser structure based on the Mach Zehnder interferometer with twin rings. In the structure of ring cavity erbium-doped fiber laser, a Mach Zehnder interferometer composed of two parallel fiber ring resonators is added, and its arm length difference is 8 mm. The optical fiber ring resonator is mainly composed of two four port couplers, and

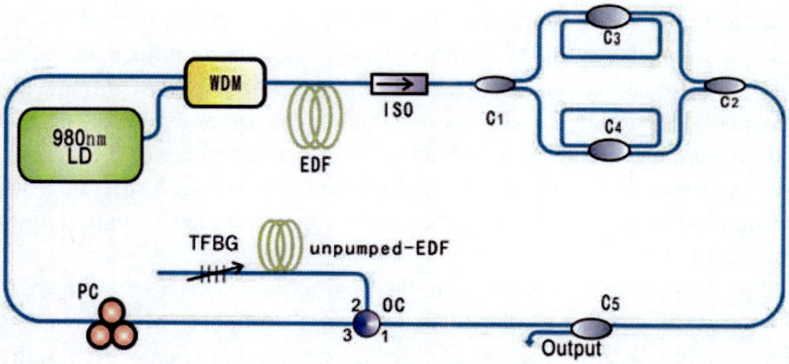

Fig. 7.18 Experimental device diagram of dual wavelength fiber laser based on twin ring Mach Zehnder interferometer; 980 nm LD: 980 nm semiconductor laser, WDM: 980/1550 nm wavelength division multiplexer, EDF: erbium-doped fiber, ISO: optical isolator, C_1–C_4: 50/50 optocoupler, C_5: 85/15 optocoupler, OC: optical circulator, unpumped-EDF: unpumped erbium-doped fiber, TFBG: tunable fiber Bragg grating filter, PC: polarization controller

a sub ring is formed by connecting two ports of the coupler. The sub ring cavity can effectively suppress the mode jump in the laser cavity and ensure the single-mode operation of the laser. The interference spectrum of MZI is observed with a spectrometer. Its adjacent wavelength interval is 0.1 nm, while the bandwidth of the tunable fiber Bragg grating filter used in the experiment is 0.2 nm. Combing the wavelength selection function of the tunable filter, the central wavelength of the filter is adjusted so that its bandwidth exactly contains two adjacent wavelength channels of the comb spectrum, thus realizing dual wavelength output.

7.4.2 *Output Characteristic Analysis of Dual Wavelength EDFL Laser*

In the experiment, the pump light power is 170 mW, and the output light of the dual-wavelength fiber laser proposed in this section is detected and analyzed by a optical spectrum analyzer. Figure 7.19 is the output spectrum of the dual-wavelength laser, the center wavelengths are 1558.67 nm and 1558.77 nm, the peak powers are − 5.508 dBm and − 5.993 dBm, respectively, and the wavelength interval is 0.1 nm, which is consistent with the wavelength interval of the Mach Zehnder interferometer described in Sect. 7.4.1.

In order to further study whether the output light of the proposed dual-wavelength fiber laser operates in a single longitudinal mode, and to verify the mode hopping suppression effect of the double sub-rings, a spectrum analyzer was used to detect the spectrum of the output light in the experiment.

Figure 7.20b is the obtained frequency spectrum of the output light. It can be seen that there is no obvious beat signal, and the laser operates in a single longitudinal

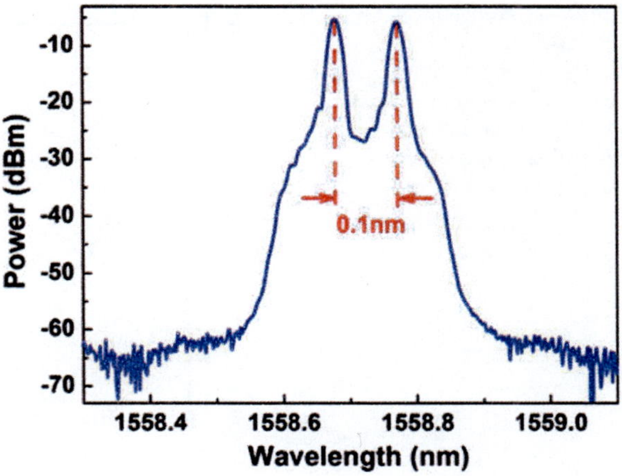

Fig. 7.19 Output spectrum of dual wavelength fiber laser with pump power of 170 mW

mode. When the double sub-rings of MZI is replaced with a fiber jumper of the same length, the obtained spectrum is shown in Fig. 7.20a, and the laser is no longer single-mode operation. It can be seen from the above analysis that the introduction of the double sub-rings can effectively suppress the mode hopping phenomenon of the laser.

We observed the output stability of the dual-wavelength laser. When the pump power was stable at 160 mW, the changes of the peak power of the two wavelengths were recorded on the spectrometer. The measurement time was 60 min, and the measurement interval was 4 min. The obtained measurement results are shown in Fig. 7.21. The red and blue curves in the figure represent the changes of the peak power of the two wavelengths respectively, and the maximum fluctuation of the power is less than 0.5 dB. In conclusion, the dual-wavelength fiber laser proposed in this section has good output stability.

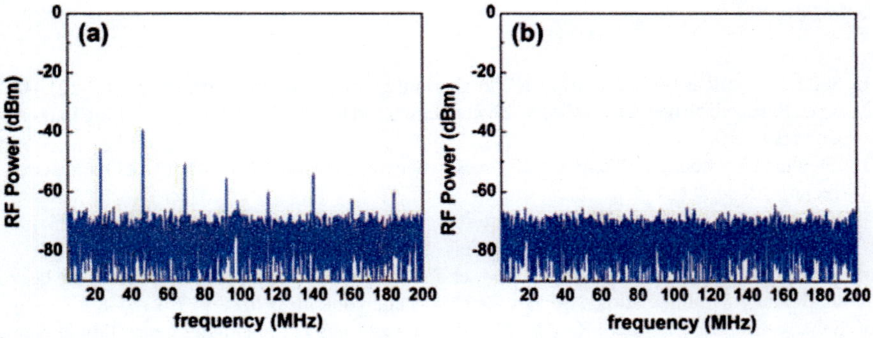

Fig. 7.20 Frequency spectrum of the output laser. **a** The frequency spectrum when the sub-ring cavity in the device is replaced by a jumper wire of equal length; **b** the frequency spectrum when the parallel double sub-rings structure is used

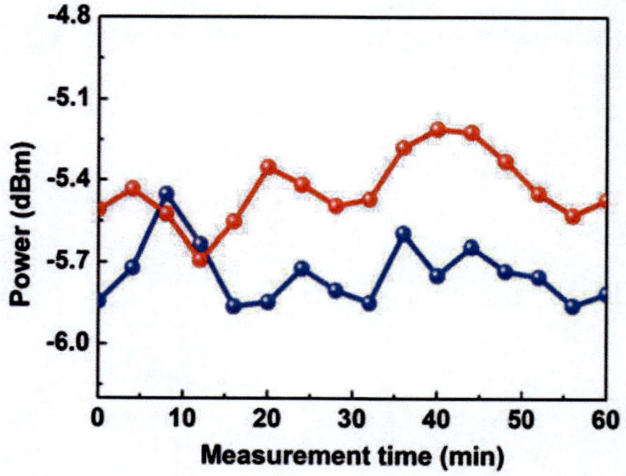

Fig. 7.21 The stability test results of the dual-wavelength output laser, the measurement time is 1 h, the measurement interval is 4 min, and the pump power is 170 mW

In this section, a dual-wavelength single-longitudinal-mode erbium-doped fiber laser setup based on a double-sub-ring Mach Zehnder interferometer is proposed. The interference spectrum of the Mach Zehnder interferometer is used as the comb filter, and the wavelength selection function of the tunable filter is combined, so that the bandwidth of the filter contains exactly two adjacent wavelengths of the comb spectrum, thereby realizing dual-wavelength output. At the same time, the two optical fiber ring resonators constituting the Mach Zehnder interferometer in the experimental setup can effectively suppress the multi-mode oscillation and achieve stable single-longitudinal-mode dual-wavelength output. The dual-wavelength output with the center wavelengths of 1558.67 and 1558.77 nm was obtained by using this setup, and the output laser showed good stability during the experiment, which has a wide range of applications and development space.

References

1. Snitzer E. Optical maser action of Nd^{3+} in a barium crown glass. Phys Rev Lett. 1961;7(12):444.
2. Ya C. Research on new narrow linewidth fiber lasers and few-mode fiber lasers. Beijing Jiaotong University; 2018.
3. Zhenhuan Y, Qihong L, Dong X. Advances in narrow linewidth fiber lasers. Opt Optoelectron Technol. 2004;2(1):1–4.
4. Jeong Y, Sahu J, Payne D, et al. Ytterbium-doped large-core fiber laser with 1.36 kW continuous-wave output power. Opt Express. 2004;12(25):6088–92.
5. Likhachev ME, Kotov LV, Bubnov MM, et al. High-power double-clad Er-doped fiber laser. In: SPIE LASE. International Society for Optics and Photonics; 2011. p. 791424–8.
6. Saha K, Okawachi Y, Shim B, et al. Modelocking and femtosecond pulse generation in chip-based frequency combs. Opt Express. 2013;21(1):1335–43.
7. Ball GA, Morey WW. Continuously tunable single-mode erbium fiber laser. Opt Lett. 1992;17(6):420–2.
8. Zou H, Lou S, Su W, et al. A wavelength-tunable fiber laser using a novel filter based on a compound interference effect. Laser Phys. 2015;25(1): 015103.

9. Takahashi H, Toba H, Inoue Y. Multiwavelength ring laser composed of EDFAs and an arrayed-waveguide wavelength multiplexer. Electron Lett. 1994;30(1):44–5.
10. Chow J, Town G, Eggleton B, et al. Multiwavelength generation in an erbium-doped fiber laser using in-fiber comb filters. IEEE Photonics Technol Lett. 1996;8(1):60–2.
11. Moon DS, Chung Y. Multi-wavelength fiber ring laser based on a sampled chirped fiber Bragg grating with a hybrid gain medium. Opt Commun. 2012;285(7):1908–10.
12. Okamura H, Iwatsuki K. Simultaneous oscillation of wavelength-tunable, singlemode lasers using an Er-doped fibre amplifier. Electron Lett. 1992;28(5):461–3.
13. An HL, Lin XZ, Pun EYB, et al. Multi-wavelength operation of an erbium-doped fiber ring laser using a dual-pass Mach-Zehnder comb filter. Opt Commun. 1999;169(1):159–65.
14. Sejka M, Varming P, Hübner J, et al. Distributed feedback Er^{3+}-doped fiber laser. Electron Lett. 1995;31(17):1445–6.
15. Chen D, Fu H, Liu W. Single-longitudinal-mode erbium-doped fiber laser based on a fiber Bragg grating Fabry-Perot filter. Laser Phys. 2007;17(10):1246–8.
16. Suzuki A, Takahashi Y, Yoshida M, et al. An ultralow noise and narrow linewidth $\lambda/4$-shifted DFB Er-doped fiber laser with a ring cavity configuration. IEEE Photonics Technol Lett. 2007;19(19):1463–5.
17. Fu ZH, Wang YX, Yang DZ, et al. Single-frequency linear cavity erbium-doped fiber laser for fiber-optic sensing applications. Laser Phys Lett. 2009;6(8):594.
18. Shirazi MR, Shahabuddin NS, Aziz SN, et al. A linear cavity Brillouin fiber laser with multiple wavelengths output. Laser Phys Lett. 2008;5(5):361–3.
19. Auerbach M, Wandt D, Fallnich C, et al. High-power tunable narrow line width ytterbium-doped double-clad fiber laser. Opt Commun. 2001;195(5):437–41.
20. Yang DZ, Liu W, Chen T, et al. Linearly-polarized Tm-doped double-clad fiber laser. Laser Phys. 2010;20(8):1752–5.
21. Hanna DC, Percival RM, Smart RG, et al. Efficient and tunable operation of a Tm-doped fiber laser. Opt Commun. 1990;75(3):283–6.
22. Yeh CH, Chow CW, Wu YF, et al. Using optimal cavity loss and saturable-absorber passive filter for stable and tunable dual-wavelength erbium fiber laser in single-longitudinal-mode operation. Laser Phys Lett. 2011;8(9):672.
23. Stepanov DY, Canning J, Bassett IM, et al. Distributed-feedback ring all-fiber laser. In: Advanced solid state lasers. Optical Society of America; 1996. IL2.
24. Li C, Xu S, Mo S, et al. A linearly frequency modulated narrow linewidth single-frequency fiber laser. Laser Phys Lett. 2013;10(7): 075106.
25. Feng T, Yan F, Li Q, et al. Stable high SNR narrow linewidth linear cavity single-wavelength Er-doped fiber laser. In: 2012 Symposium on IEEE photonics and optoelectronics (SOPO); 2012. p. 1–4.
26. Wang H, Liaw SK, Hsu HI, et al. Single-longitudinal-mode linear-cavity fiber laser using subring cavities and broadband fiber mirror. In: 2013 IEEE international symposium on IEEE next-generation electronics (ISNE); 2013. p. 79–82.
27. Mo S, Li Z, Huang X, et al. 820 Hz linewidth short-linear-cavity single-frequency fiber laser at 1.5 μm. Laser Phys Lett. 2014;11(3):035101.
28. Fan Z, Zeng X, Cao C, et al. Novel structure of an ultra-narrow-bandwidth fibre laser based on cascade filters: PGFBG and SA. Opt Commun. 2016;368:150–4.
29. Li R, Li Y. Application of distributed optical sensor in submarine cable detection. Telecommun Electr Power Syst. 2010;2:014.
30. Geng J, Spiegelberg C, Jiang S. Narrow linewidth fiber laser for 100-km optical frequency domain reflectometry. IEEE Photonics Technol Lett. 2005;17(9):1827–9.
31. Iida D, Ito F. Detection sensitivity of Brillouin scattering near Fresnel reflection in BOTDR measurement. J Lightwave Technol. 2008;26(4):417–24.
32. Xuping Z, Feng W, Yuangang L. Continuous distributed fiber optic sensing technology based on Brilouin effect. Adv Lasers Optoelectron. 2009;11:14–20.
33. Mignon M, Desurvire E. An analytical model for the determination of optimal output reflectivity and fiber length in erbium-doped fiber lasers. IEEE Photonics Technol Lett. 1992;4(8):850–2.

34. Xu O, Lu S, Feng S, et al. Single-longitudinal-mode erbium-doped fiber laser with the fiber-Bragg-grating-based asymmetric two-cavity structure. Opt Commun. 2009;282(5):962–5.
35. Ball GA, Glenn WH. Design of a single-mode linear-cavity erbium fiber laser utilizing Bragg reflectors. J Lightwave Technol. 1992;10(10):1338–43.
36. Liegeois F, Hernandez Y, Peigne G, et al. High-efficiency single-longitudinal-mode ring fiber laser. Electron Lett. 2005;41(13):729–30.
37. Feng T, Yan F, Liu S, et al. Switchable and tunable dual-wavelength single-longitudinal-mode erbium-doped fiber laser with special subring-cavity and superimposed fiber Bragg gratings. Laser Phys Lett. 2014;11(12): 125106.
38. Meng Z, Stewart G, Whitenett G. Stable single-mode operation of a narrow-linewidth, linearly polarized, erbium-fiber ring laser using a saturable absorber. J Lightwave Technol. 2006;24(5):2179.
39. Cheng Y, Kringlebotn JT, Loh WH, et al. Stable single-frequency traveling-wave fiber loop laser with integral saturable-absorber-based tracking narrow-band filter. Opt Lett. 1995;20(8):875–7.
40. Sun J, Yuan X, Zhang X, et al. Single-longitudinal-mode fiber ring laser using fiber grating-based Fabry-Perot filters and variable saturable absorbers. Opt Commun. 2006;267(1):177–81.
41. Xu L, Glesk I, Rand D, et al. Suppression of beating noise of narrow-linewidth erbium-doped fiber ring lasers by use of a semiconductor optical amplifier. Opt Lett. 2003;28(10):780–2.
42. Cheng XP, Shum P, Tse CH, et al. Single-longitudinal-mode erbium-doped fiber ring laser based on high finesse fiber Bragg grating Fabry-Pérot etalon. IEEE Photonics Technol Lett. 2008;20(12):976–8.
43. Lee CC, Chen YK, Liaw SK. Single-longitudinal-mode fiber laser with a passive multiple-ring cavity and its application for video transmission. Opt Lett. 1998;23(5):358–60.
44. Feng S, Mao Q, Tian Y, et al. Widely tunable single longitudinal mode fiber laser with cascaded fiber-ring secondary cavity. IEEE Photonics Technol Lett. 2013;25:323–6.
45. Qian J, Su J, Hong L. A widely tunable dual-wavelength erbium-doped fiber ring laser operating in single longitudinal mode. Opt Commun. 2008;281(17):4432–4.
46. Ahmad H, Dernaika M, Kharraz OM, et al. A tuneable, power efficient and narrow single longitudinal mode fibre ring laser using an inline dual-taper fiber Mach-Zehnder filter. Laser Phys. 2014;24(8): 085111.
47. Tan S, Yan F, Li Q, et al. A stable single-longitudinal-mode dual-wavelength erbium-doped fiber ring laser with superimposed FBG and an in-line two-taper MZI filter. Laser Phys. 2013;23(7): 075112.
48. Yang XX, Zhan L, Shen QS, et al. High-power single-longitudinal-mode fiber laser with a ring Fabry-Perot resonator and a saturable absorber. IEEE Photonics Technol Lett. 2008;20(11):879–81.
49. Ludvigsen H, Tossavainen M, Kaivola M. Laser linewidth measurements using self-homodyne detection with short delay. Opt Commun. 1998;155(1):180–6.
50. Richter L, Mandelberg HI, Kruger M, et al. Linewidth determination from self-heterodyne measurements with subcoherence delay times. IEEE J Quantum Electron. 1986;22(11):2070–4.
51. Chen X, Yao J, Zeng F, et al. Single-longitudinal-mode fiber ring laser employing an equivalent phase-shifted fiber Bragg grating. IEEE Photonics Technol Lett. 2005;17(7):1390–2.
52. Sun J, Yuan X, Zhang X, et al. Single-longitudinal-mode dual-wavelength fiber ring laser by incorporating variable saturable absorbers and feedback fiber loops. Opt Commun. 2007;273(1):231–7.
53. Sun Q, Wang J, Tong W, et al. Channel-switchable single-/dual-wavelength single-longitudinal-mode laser and THz beat frequency generation up to 3.6 THz. Appl Phys B. 2012;106(2):373–7.